HIGH-PERFORMANCE GRADIENT ELUTION

THE WILEY BICENTENNIAL–KNOWLEDGE FOR GENERATIONS

Each generation has its unique needs and aspirations. When Charles Wiley first opened his small printing shop in lower Manhattan in 1807, it was a generation of boundless potential searching for an identity. And we were there, helping to define a new American literary tradition. Over half a century later, in the midst of the Second Industrial Revolution, it was a generation focused on building the future. Once again, we were there, supplying the critical scientific, technical, and engineering knowledge that helped frame the world. Throughout the 20th Century, and into the new millennium, nations began to reach out beyond their own borders and a new international community was born. Wiley was there, expanding its operations around the world to enable a global exchange of ideas, opinions, and know-how.

For 200 years, Wiley has been an integral part of each generation's journey, enabling the flow of information and understanding necessary to meet their needs and fulfill their aspirations. Today, bold new technologies are changing the way we live and learn. Wiley will be there, providing you the must-have knowledge you need to imagine new worlds, new possibilities, and new opportunities.

Generations come and go, but you can always count on Wiley to provide you the knowledge you need, when and where you need it!

WILLIAM J. PESCE
PRESIDENT AND CHIEF EXECUTIVE OFFICER

PETER BOOTH WILEY
CHAIRMAN OF THE BOARD

HIGH-PERFORMANCE GRADIENT ELUTION
The Practical Application of the Linear-Solvent-Strength Model

LLOYD R. SNYDER
LC Resources, Inc., Orinda, California

JOHN W. DOLAN
LC Resources, Inc., Amity, Oregon

WILEY-INTERSCIENCE
A JOHN WILEY & SONS, INC., PUBLICATION

Copyright © 2007 by John Wiley & Sons, Inc. All rights reserved.

Published by John Wiley & Sons, Inc., Hoboken, New Jersey.
Published simultaneously in Canada.

No part of this publication may be reproduced, stored in a retrieval system, or transmitted in any form or by any means, electronic, mechanical, photocopying, recording, scanning, or otherwise, except as permitted under Section 107 or 108 of the 1976 United States Copyright Act, without either the prior written permission of the Publisher, or authorization through payment of the appropriate per-copy fee to the Copyright Clearance Center, Inc., 222 Rosewood Drive, Danvers, MA 01923, (978) 750-8400, fax (978) 750-4470, or on the web at www.copyright.com. Requests to the Publisher for permission should be addressed to the Permissions Department, John Wiley & Sons, Inc., 111 River Street, Hoboken, NJ 07030, (201) 748-6011, fax (201) 748-6008, or online at http://www.wiley.com/go/permission.

Limit of Liability/Disclaimer of Warranty: While the publisher and author have used their best efforts in preparing this book, they make no representations or warranties with respect to the accuracy or completeness of the contents of this book and specifically disclaim any implied warranties of merchantability or fitness for a particular purpose. No warranty may be created or extended by sales representatives or written sales materials. The advice and strategies contained herein may not be suitable for your situation. You should consult with a professional where appropriate. Neither the publisher nor author shall be liable for any loss of profit or any other commercial damages, including but not limited to special, incidental, consequential, or other damages.

For general information on our other products and services or for technical support, please contact our Customer Care Department within the United States at (800) 762-2974, outside the United States at (317) 572-3993 or fax (317) 572-4002.

Wiley also publishes its books in a variety of electronic formats. Some content that appears in print may not be available in electronic formats. For more information about Wiley products, visit our web site at www.wiley.com.

Library of Congress Cataloging-in-Publication Data is available.

ISBN-10 0-471-70646-9
ISBN-13 978-0-471-70646-5

Printed in the United States of America

10 9 8 7 6 5 4 3 2 1

*... every natural science involves three things: the sequence of
phenomena on which the science is based*
[experimental observation];
*the abstract concepts which call these phenomena
to mind* [a model]; *and the words in which the
concepts are expressed* [the present book].

Antoine Laurent Lavoisier [with parenthetical additions by the authors], *Traité Elémentaire de Chemie* (1789)

CONTENTS

PREFACE		xv
GLOSSARY OF SYMBOLS AND TERMS		xxi

1 INTRODUCTION 1

1.1	The "General Elution Problem" and the Need for Gradient Elution	1
1.2	Other Reasons for the Use of Gradient Elution	4
1.3	Gradient Shape	7
1.4	Similarity of Isocratic and Gradient Elution	10
	1.4.1 Gradient and Isocratic Elution Compared	10
	1.4.2 The Linear-Solvent-Strength Model	13
1.5	Computer Simulation	18
1.6	Sample Classification	19
	1.6.1 Sample Compounds of Related Structure ("Regular Samples")	19
	1.6.2 Sample Compounds of Unrelated Structure ("Irregular" Samples)	19

2 GRADIENT ELUTION FUNDAMENTALS 23

2.1	Isocratic Separation	23
	2.1.1 Retention	23
	2.1.2 Peak Width and Plate Number	24
	2.1.3 Resolution	25
	2.1.4 Role of Separation Conditions	27
	2.1.4.1 Optimizing Retention [Term a of Equation (2.7)]	27
	2.1.4.2 Optimizing Selectivity α [Term b of Equation (2.7)]	28
	2.1.4.3 Optimizing the Column Plate Number N [Term c of Equation (2.7)]	28
2.2	Gradient Separation	31
	2.2.1 Retention	32
	2.2.1.1 Gradient and Isocratic Separation Compared for "Corresponding" Conditions	34
	2.2.2 Peak Width	38
	2.2.3 Resolution	39
	2.2.3.1 Resolution as a Function of Values of S for Two Adjacent Peaks ("Irregular" Samples)	42
	2.2.3.2 Using Gradient Elution to Predict Isocratic Separation	45
	2.2.4 Sample Complexity and Peak Capacity	47
2.3	Effect of Gradient Conditions on Separation	49
	2.3.1 Gradient Steepness b: Change in Gradient Time	50
	2.3.2 Gradient Steepness b: Change in Column Length or Diameter	51

	2.3.3	Gradient Steepness b: Change in Flow Rate	55
	2.3.4	Gradient Range $\Delta\phi$: Change in Initial Percentage B (ϕ_0)	58
	2.3.5	Gradient Range $\Delta\phi$: Change in Final Percentage B (ϕ_f)	60
	2.3.6	Effect of a Gradient Delay	63
		2.3.6.1 Equipment Dwell Volume	66
	2.3.7	Effect of Gradient Shape (Nonlinear Gradients)	67
	2.3.8	Overview of the Effect of Gradient Conditions on the Chromatogram	71
2.4	Related Topics		72
	2.4.1	Nonideal Retention in Gradient Elution	72
	2.4.2	Gradient Elution Misconceptions	72

3 METHOD DEVELOPMENT 74

3.1	A Systematic Approach to Method Development		74
	3.1.1	Separation Goals (Step 1 of Fig. 3.1)	75
	3.1.2	Nature of the Sample (Step 2 of Fig. 3.1)	78
	3.1.3	Initial Experimental Conditions	79
	3.1.4	Repeatable Results	79
	3.1.5	Computer Simulation: Yes or No?	80
	3.1.6	Sample Preparation (Pretreatment)	81
3.2	Initial Experiments		81
	3.2.1	Interpreting the Initial Chromatogram (Step 3 of Fig. 3.1)	85
		3.2.1.1 "Trimming" a Gradient Chromatogram	87
		3.2.1.2 Possible Problems	88
3.3	Developing a Gradient Separation: Resolution versus Conditions		90
	3.3.1	Optimizing Gradient Retention k^* (Step 4 of Fig. 3.1)	92
	3.3.2	Optimizing Gradient Selectivity α^* (Step 5 of Fig. 3.1)	92
	3.3.3	Optimizing the Gradient Range (Step 6 of Fig. 3.1)	95
		3.3.3.1 Changes in Selectivity as a Result of Change in k^*	96
	3.3.4	Segmented (Nonlinear) Gradients (Step 6 of Fig. 3.1 Continued)	100
	3.3.5	Optimizing the Column Plate Number N^* (Step 7 of Fig. 3.1)	102
	3.3.6	Column Equilibration Between Successive Sample Injections	106
	3.3.7	Fast Separations	106
3.4	Computer Simulation		108
	3.4.1	Quantitative Predictions and Resolution Maps	109
	3.4.2	Gradient Optimization	111
	3.4.3	Changes in Column Conditions	112
	3.4.4	Separation of "Regular" Samples	114
	3.4.5	Other Features	115
		3.4.5.1 Isocratic Prediction (5 in Table 3.5)	115
		3.4.5.2 Designated Peak Selection (6 in Table 3.5)	117
		3.4.5.3 Change in Other Conditions (7 in Table 3.5)	117
		3.4.5.4 Computer-Selection of the Best Multisegment Gradient (8 in Table 3.5)	117
		3.4.5.5 "Two-Run" Procedures for the Improvement of Sample Resolution	119
	3.4.6	Accuracy of Computer Simulation	119
	3.4.7	Peak Tracking	119

				CONTENTS	ix

3.5	Method Reproducibility and Related Topics	120
	3.5.1 Method Development	121
	3.5.2 Routine Analysis	122
	3.5.3 Change in Column Volume	123
3.6	Additional Means for an Increase in Separation Selectivity	124
3.7	Orthogonal Separations	127
	3.7.1 Two-Dimensional Separations	128

4 GRADIENT EQUIPMENT   133

4.1	Gradient System Design	133
	4.1.1 High-Pressure vs Low-Pressure Mixing	133
	4.1.2 Tradeoffs	135
	4.1.2.1 Dwell Volume	135
	4.1.2.2 Degassing	136
	4.1.2.3 Accuracy	137
	4.1.2.4 Solvent Volume Changes and Compressibility	137
	4.1.2.5 Flexibility	139
	4.1.2.6 Independent Module Use	140
	4.1.3 Other System Components	140
	4.1.3.1 Autosampler	140
	4.1.3.2 Column	140
	4.1.3.3 Detector	141
	4.1.3.4 Data System	141
	4.1.3.5 Extra-Column Volume	142
4.2	General Considerations in System Selection	142
	4.2.1 Which Vendor?	143
	4.2.2 High-Pressure or Low-Pressure Mixing?	144
	4.2.3 Who Will Fix It?	144
	4.2.4 Special Applications	144
4.3	Measuring Gradient System Performance	145
	4.3.1 Gradient Performance Test	146
	4.3.1.1 Gradient Linearity	146
	4.3.1.2 Dwell Volume Determination	147
	4.3.1.3 Gradient Step-Test	147
	4.3.1.4 Gradient Proportioning Valve Test	148
	4.3.2 Additional System Checks	149
	4.3.2.1 Flow Rate Check	149
	4.3.2.2 Pressure Bleed-Down	150
	4.3.2.3 Retention Reproducibility	150
	4.3.2.4 Peak Area Reproducibility	151
4.4	Dwell Volume Considerations	151

5 SEPARATION ARTIFACTS AND TROUBLESHOOTING   153

5.1	Avoiding Problems	154
	5.1.1 Equipment Checkout	157
	5.1.1.1 Installation Qualification, Operational Qualification, and Performance Qualification	157
	5.1.2 Dwell Volume	158
	5.1.3 Blank Gradient	158
	5.1.4 Suggestions for Routine Applications	158
	5.1.4.1 Reagent Quality	159

		5.1.4.2 System Cleanliness	159
		5.1.4.3 Degassing	159
		5.1.4.4 Dedicated Columns	159
		5.1.4.5 Equilibration	159
		5.1.4.6 Priming Injections	159
		5.1.4.7 Ignore the First Injection	160
		5.1.4.8 System Suitability	160
		5.1.4.9 Standards and Calibrators	160
	5.1.5	Method Development	160
		5.1.5.1 Use a Clean and Stable Column	160
		5.1.5.2 Use Reasonable Mobile Phase Conditions	161
		5.1.5.3 Clean Samples	162
		5.1.5.4 Reproducible Runs	162
		5.1.5.5 Sufficient Equilibration	162
		5.1.5.6 Reference Conditions	162
		5.1.5.7 Additional Tests	162
5.2	Method Transfer		163
	5.2.1	Compensating for Dwell Volume Differences	163
		5.2.1.1 Injection Delay	163
		5.2.1.2 Adjustment of the Initial Isocratic Hold	164
		5.2.1.3 Use of Maximum-Dwell-Volume Methods	165
		5.2.1.4 Adjustment of Initial Percentage B	165
	5.2.2	Other Sources of Method Transfer Problems	168
		5.2.2.1 Gradient Shape	169
		5.2.2.2 Gradient Rounding	169
		5.2.2.3 Inter-Run Equilibration	169
		5.2.2.4 Column Size	169
		5.2.2.5 Column Temperature	169
		5.2.2.6 Interpretation of Method Instructions	170
5.3	Column Equilibration		170
	5.3.1	Primary Effects	171
	5.3.2	Slow Equilibration of Column and Mobile Phase	173
	5.3.3	Practical Considerations and Recommendations	174
5.4	Separation Artifacts		175
	5.4.1	Baseline Drift	176
	5.4.2	Baseline Noise	179
		5.4.2.1 Baseline Noise: A Case Study	180
	5.4.3	Peaks in a Blank Gradient	182
		5.4.3.1 Mobile Phase Water or Organic Solvent Impurities	182
		5.4.3.2 Other Sources of Background Peaks	185
	5.4.4	Extra Peaks for Injected Samples	185
		5.4.4.1 t_0 Peaks	185
		5.4.4.2 Air Peaks	186
		5.4.4.3 Late Peaks	187
	5.4.5	Peak Shape Problems	188
		5.4.5.1 Tailing and Fronting	188
		5.4.5.2 Excess Peak Broadening	188
		5.4.5.3 Split Peaks	190
		5.4.5.4 Injection Conditions	191
		5.4.5.5 Sample Decomposition	193
5.5	Troubleshooting		195

	5.5.1	Problem Isolation	196
	5.5.2	Troubleshooting and Maintenance Suggestions	197
		5.5.2.1 Removing Air from the Pump	197
		5.5.2.2 Solvent Siphon Test	197
		5.5.2.3 Premixing to Improve Retention Reproducibility in Shallow Gradients	198
		5.5.2.4 Cleaning and Handling Check-Valves	199
		5.5.2.5 Replacing Pump Seals and Pistons	200
		5.5.2.6 Leak Detection	200
		5.5.2.7 Repairing Fitting Leaks	200
		5.5.2.8 Cleaning Glassware	201
		5.5.2.9 For Best Results with TFA	201
		5.5.2.10 Improved Water Purity	201
		5.5.2.11 Isolating Carryover Problems	203
		5.5.2.12 Troubleshooting Rules of Thumb	204
	5.5.3	Gradient Performance Test Failures	206
		5.5.3.1 Linearity (4.3.1.1)	206
		5.5.3.2 Step Test (4.3.1.3)	206
		5.5.3.3 Gradient-Proportioning-Valve Test (4.3.1.4)	209
		5.5.3.4 Flow Rate (4.3.2.1)	211
		5.5.3.5 Pressure Bleed-Down (4.3.2.2)	212
		5.5.3.6 Retention Reproducibility (4.3.2.3)	212
		5.5.3.7 Peak Area Reproducibility (4.3.2.4)	213
	5.5.4	Troubleshooting Case Studies	213
		5.5.4.1 Retention Variation – Case Study 1	213
		5.5.4.2 Retention Variation – Case Study 2	218
		5.5.4.3 Contaminated Reagents – Case Study 3	220
		5.5.4.4 Baseline and Retention Problems – Case Study 4	224
6	***SEPARATION OF LARGE MOLECULES***		**228**
6.1	General Considerations		228
	6.1.1	Values of S for Large Molecules	229
	6.1.2	Values of N^* for Large Molecules	235
	6.1.3	Conformational State	236
	6.1.4	Homo-Oligomeric Samples	238
		6.1.4.1 Separation of Large Homopolymers	241
	6.1.5	Proposed Models for the Gradient Separation of Large Molecules	242
		6.1.5.2 "Critical Elution Behavior": Biopolymers	246
		6.1.5.3 Measurement of LSS Parameters for Large Molecules	247
6.2	Biomolecules		248
	6.2.1	Peptides and Proteins	248
		6.2.1.1 Sample Characteristics	249
		6.2.1.2 Conditions for an Initial Gradient Run	249
		6.2.1.3 Method Development	253
		6.2.1.4 Segmented Gradients	259
	6.2.2	Other Separation Modes and Samples	261
		6.2.2.1 Hydrophobic Interaction Chromatography	262
		6.2.2.2 Ion Exchange Chromatography	264
		6.2.2.3 Hydrophilic Interaction Chromatography	266
		6.2.2.4 Separation of Viruses	267

	6.2.3 Separation Problems	271
	6.2.4 Fast Separations of Peptides and Proteins	274
	6.2.5 Two-Dimensional Separations of Peptides and Proteins	274
6.3	Synthetic Polymers	275
	6.3.1 Determination of Molecular Weight Distribution	277
	6.3.2 Determination of Chemical Composition	278

7 PREPARATIVE SEPARATIONS 283

7.1	Introduction	283
	7.1.1 Equipment for Preparative Separation	285
7.2	Isocratic Separation	286
	7.2.1 Touching-Peak Separation	287
	7.2.1.1 Theory	287
	7.2.1.2 Column Saturation Capacity	289
	7.2.1.3 Sample-Volume Overload	292
	7.2.2 Method Development for Isocratic Touching-Peak Separation	292
	7.2.2.1 Optimizing Separation Conditions	295
	7.2.2.2 Selecting a Sample Weight for Touching-Peak Separation	297
	7.2.2.3 Scale-Up	298
	7.2.2.4 Sample Solubility	300
	7.2.3 Beyond Touching-Peak Separation	301
7.3	Gradient Separation	302
	7.3.1 Touching-Peak Separation	306
	7.3.2 Method Development for Gradient Touching-Peak Separation	306
	7.3.2.1 Step Gradients	311
	7.3.3 Sample-Volume Overload	312
	7.3.4 Possible Complications of Simple Touching-Peak Theory and Their Practical Impact	312
	7.3.4.1 Crossing Isotherms	313
	7.3.4.2 Unequal Values of S	314
7.4	Severely Overloaded Separation	315
	7.4.1 Is Gradient Elution Necessary?	316
	7.4.2 Displacement Effects	317
	7.4.3 Method Development	317
	7.4.4 Separations of Peptides and Small Proteins	318
	7.4.5 Column Efficiency	320
	7.4.6 Production-Scale Separation	320

8 OTHER APPLICATIONS OF GRADIENT ELUTION 323

8.1	Gradient Elution for LC-MS	324
	8.1.1 Application Areas	325
	8.1.2 Requirements for LC-MS	325
	8.1.3 Basic LC-MS Concepts	326
	8.1.3.1 The Interface	326
	8.1.3.2 Column Configurations	328
	8.1.3.3 Quadrupoles and Ion Traps	328
	8.1.4 LC-UV vs LC-MS Gradient Conditions	330

	8.1.5	Method Development for LC-MS	332
		8.1.5.1 Define Separation Goals (Step 1, Table 8.2)	332
		8.1.5.2 Collect Information on Sample (Step 2, Table 8.2)	334
		8.1.5.3 Carry Out Initial Separation (Run 1, Step 3, Table 8.2)	339
		8.1.5.4 Optimize Gradient Retention k^* (Step 4, Table 8.2)	339
		8.1.5.5 Optimize Selectivity α^* (Step 5, Table 8.2)	339
		8.1.5.6 Adjust Gradient Range and Shape (Step 6, Table 8.2)	340
		8.1.5.7 Vary Column Conditions (Step 7, Table 8.2)	341
		8.1.5.8 Determine Inter-Run Column Equilibration (Step 8, Table 8.2)	341
	8.1.6	Special Challenges for LC-MS	341
		8.1.6.1 Dwell Volume	342
		8.1.6.2 Gradient Distortion	342
		8.1.6.3 Ion Suppression	343
		8.1.6.4 Co-Eluting Compounds	345
		8.1.6.5 Resolution Requirements	346
		8.1.6.6 Use of Computer Simulation Software	347
		8.1.6.7 Isocratic Methods	347
		8.1.6.8 Throughput Enhancement	347
8.2	Ion-Exchange Chromatography		349
	8.2.1	Theory	349
	8.2.2	Dependence of Separation on Gradient Conditions	356
	8.2.3	Method Development for Gradient IEC	356
		8.2.3.1 Choice of Initial Conditions	356
		8.2.3.2 Improving the Separation	357
8.3	Normal-Phase Chromatography		359
	8.3.1	Theory	359
	8.3.2	Method Development for Gradient NPC	360
	8.3.3	Hydrophilic Interaction Chromatography	361
		8.3.3.1 Method Development for Gradient HILIC	361
8.4	Ternary- or Quaternary-Solvent Gradients		365

9 THEORY AND DERIVATIONS — *370*

9.1	The Linear Solvent Strength Model		370
	9.1.1	Retention	372
		9.1.1.1 Gradient and Isocratic Retention Compared	374
		9.1.1.2 Small Values of k_0	376
	9.1.2	Peak Width	378
		9.1.2.1 Gradient Compression	380
	9.1.3	Selectivity and Resolution	383
	9.1.4	Advantages of LSS Behavior	385
9.2	Second-Order Effects		386
	9.2.1	Assumptions About ϕ and k	386
		9.2.1.1 Incomplete Column Equilibration	386
		9.2.1.2 Solvent Demixing	391
		9.2.1.3 Nonlinear Plots of $\log k$ vs ϕ	393
		9.2.1.4 Dependence of V_m on ϕ	393
	9.2.2	Nonideal Equipment	393
9.3.	Accuracy of Gradient Elution Predictions		397
	9.3.1	Gradient Retention Time	397
		9.3.1.1 Confirmation of Equation (9.2)	397

		9.3.1.2 Computer Simulation	399
	9.3.2	Peak Width Predictions	399
	9.3.3	Measurement of Values of S and $\log k_0$	400
9.4	Values of S		401
	9.4.1	Estimating Values of S from Solute Properties and Experimental Conditions	402
9.5	Values of N in Gradient Elution		404

Appendix I	THE CONSTANT-S APPROXIMATION IN GRADIENT ELUTION	414
Appendix II	ESTIMATION OF CONDITIONS FOR ISOCRATIC ELUTION, BASED ON AN INITIAL GRADIENT RUN	416
Appendix III	CHARACTERIZATION OF REVERSED-PHASE COLUMNS FOR SELECTIVITY AND PEAK TAILING	418
Appendix IV	SOLVENT PROPERTIES RELEVANT TO THE USE OF GRADIENT ELUTION	434
Appendix V	THEORY OF PREPARATIVE SEPARATION	436
Appendix VI	FURTHER INFORMATION ON VIRUS CHROMATOGRAPHY	445
Index		450

PREFACE

High-performance liquid chromatography (HPLC) is today widely used for separation and analysis [1, 2]. Many samples cannot be successfully separated by the use of fixed (isocratic) conditions, but instead require *gradient elution* (also called *solvent programming*): a change in mobile phase composition during the separation, so as to progressively reduce sample retention. To take full advantage of such gradient-HPLC separations, the user needs an understanding of gradient elution comparable to that required for isocratic separation. Our reference in the present book to *high-performance gradient elution* implies such an understanding, accompanied by the use of state-of-the art equipment, columns and experimental technique. Because of the major importance of separations by reversed-phase liquid chromatography (RP-LC), this separation mode will be assumed unless otherwise stated (Sections 6.2.2, 8.2, and 8.3 discuss gradient elution with ion-exchange and normal-phase chromatography).

Several previous reviews or books ([3–8] and Chapter 8 of [2]) discuss the principles and practice of gradient elution, as these were understood at the time these accounts were written. However, these past reviews now appear dated, incomplete, and/or unnecessarily complicated for practical application. Hence the present book has been written with three different goals in mind: (a) a practical summary of what the reader needs to know in order to carry out any gradient separation; (b) a conceptual understanding of how gradient elution works; and (c) a detailed examination of the underlying theoretical framework of gradient elution, for application to special situations and to satisfy any lingering doubts of the reader. Because many readers will be interested in simply using gradient elution or developing a gradient procedure, this application is emphasized in the present book.

Of the various ways in which chromatography is applied today, few have been as misunderstood as the technique of gradient elution, which for some continues as "a riddle wrapped in a mystery inside an enigma" [9]. "Simple" isocratic separation can itself be a challenge, while gradient elution involves added complexity in terms of equipment, procedures, the interpretation of results, and a preferred method development strategy. Compared with isocratic separation, gradient elution is also regarded as (a) subject to more experimental problems and (b) inherently slower and less robust, as well as (c) presenting special difficulty for method transfer from one laboratory to another. Because of these *potentially* unfavorable characteristics of gradient elution, many workers in the past have avoided its use where possible. It is a premise of the present book that gradient elution can be much less hard to understand and much more easy to use than has been assumed previously.

Gradient elution sometimes appears to contradict our prior experience based on isocratic separation. In isocratic elution, for example, a reduction in flow rate by a factor of 2, or a 2-fold increase in column length, leads to a doubling of retention times and a 1.5- to 2-fold increase in peak widths. Similar changes in flow rate or column length when using gradient elution usually result in much smaller variations in peak retention or width. In isocratic elution, a change in flow rate or column length also has no effect on the relative spacing of peaks within the chromatogram. However, this is often not the case for gradient elution; indeed, such "surprises" are inherent in its nature. Changes in retention times and sample resolution, when flow rate, column length, or gradient time is varied in gradient elution, also depend on the nature of the sample being separated. In the latter connection, it is important to recognize four different sample groupings or classifications: "regular"/low-molecular-weight, "regular"/high-molecular-weight, "irregular"/low-molecular-weight, and "irregular"/high-molecular-weight samples. The significance for gradient elution of each of these four sample types is examined in this book. Except in Chapter 6, however, we will assume "low-molecular-weight" samples with molecular weights <1000 Da.

The essential similarity of isocratic and gradient elution is often overlooked, but once recognized it allows a much easier understanding of gradient separation, as well as an "intuitive" feeling for what will happen when some change in gradient conditions is made. In this book, we will use the *linear-solvent-strength* (LSS) model of gradient elution [3, 5, 7] as a bridge between separations by isocratic and gradient elution. This model also leads to near-exact equations for retention time, peak width, and resolution as a function of gradient conditions, as well as the widespread implementation of *computer simulation* as an aid to HPLC method development. For any sample, data from two or more experimental gradient runs can be used by the computer to predict either isocratic or gradient separation as a function of conditions, thereby facilitating the systematic improvement of the separation. Computer simulation is especially useful for developing gradient methods, and it has been used extensively in the present book as a means of more effectively illustrating the effects of different experimental conditions on gradient separation. It is also our hope that this book can prove useful "in reverse," whereby a better understanding of gradient elution may even improve our application of isocratic separation.

The beginning of the book (Chapter 1, Section 2.1, and Chapter 3) describes the application of isocratic and gradient elution for typical samples (those with molecular weights <1000 Da), with minimal digression into the derivations of important equations and little attention to less important aspects of gradient elution. Sections 2.2–2.4 provide a conceptual basis for the better interpretation and use of gradient elution, which some (but probably not all) readers will want to read prior to Chapter 3. In Chapter 4, the equipment required for gradient elution is discussed. Chapter 5 deals with experimental problems that can be encountered in gradient elution as well as related troubleshooting information. Chapter 6 recognizes important differences in gradient elution when this technique is used for macromolecular samples, for example, large peptides, proteins, nucleic acids, viruses, and other natural or synthetic polymers. Chapter 7 expands the discussion of earlier chapters to the use of gradient elution for preparative separations, that is, the injection of larger samples for recovery of purified material. Chapter 8 examines (a) separations which feature the combination of gradient elution with mass spectrometric detection

(LC-MS), (b) the application of gradient elution to normal-phase and ion-exchange separations, and (c) the use of complex gradients formed from three or more solvents. Chapter 9 concludes with a more detailed treatment of the fundamental equations of gradient elution, including attention to so-called "nonideal" contributions to gradient separation.

The present book assumes some familiarity with the principles and practice of HPLC [2]. For a quick and practical summary of the essentials of gradient elution separation, it is suggested that the reader read Chapter 1, Section 2.1, Chapter 3, and Chapter 4, in this order, then consult Chapter 5 (Troubleshooting) as needed. If greater insight into how gradient elution works is desired, Sections 2.2–2.4 provide additional background, with further detail available in Chapter 9. Biochemists may want to start with Chapters 1 and 3, plus Section 6.2, while workers engaged in the isolation of purified sample components will benefit especially from Chapter 7 (Preparative Separations). A "reading plan" for the book is suggested by Figure P.1, with the bold topics comprising a minimal introduction to gradient elution.

> No profit grows, where is no pleasure taken; In brief, sir, study what you most affect.
>
> —William Shakespeare, *The Rape of Lucrece*

The present book is heavily cross-referenced to other sections of the book, so as to allow the reader to follow up on topics of special interest, or to clarify questions that may arise during reading. Because extensive cross-referencing represents a potential distraction, *in most cases it is recommended that the reader simply ignore these invitations to jump to other parts of the book*. Some chapters include parts that are of greater academic than practical interest; these sections are in each case clearly identified (introduced with an advisory in *italics*), so that they can be bypassed at the option of the reader. We have also taken pains to provide definitions for all symbols used in this book (Glossary section), as well as a comprehensive and detailed index.

For the past 30 years, gradient elution has been a major research focus for us. During this time, we have worked together to better understand and apply this powerful experimental procedure, and we have also created commercial software (DryLab®) for the more efficient use of gradient elution by numerous workers throughout the world ("computer simulation"). For one of us (LRS), an interest in this topic extends back another 15 years into the early 1960s. The present book therefore represents the culmination of an interest of long standing, as well as an attempt at a complete and detailed account of the subject. We hope that the book will find use by practical workers throughout the world. During the past 35 years, another scientist, Pavel Jandera from the University of Pardubice, has similarly devoted much of his career to the study and elucidation of the principles and practice of gradient elution. The present book owes much to his many contributions in this area, which did not stop with the publication of *his* book on gradient elution in 1985 [6] or his recent review of the subject [8].

We very much appreciate the assistance of four co-authors, who were responsible for the preparation of Sections 6.2.2.4 [Carl Scandella (Carl Scandella

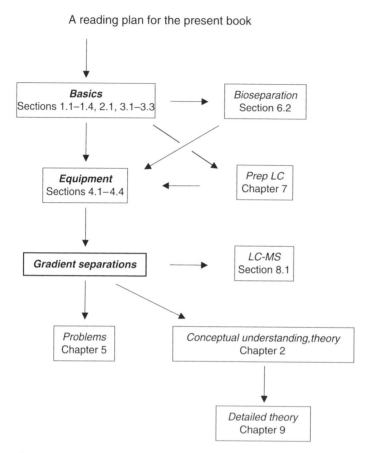

Figure P.1 How to use this book.

Consulting, Bellevue WA), Paul Shabram (Ventana Biosciences, San Diego, California), and Gary Vellekamp (Schering Plough Research Institute, Union, New Jersey)] and 7.4 [Geoff Cox (Chiral Technologies, Inc., West Chester, Pennsylvania)]. We are likewise grateful to a number of past collaborators who have greatly assisted our own research on gradient elution: Geoff Cox, Pete Carr, Julie Eble, Russel Gant, Barbara Ghrist, Jack Kirkland, Tom Jupille, Dana Lommen, Dan Marchand, Imre Molnar, Thomas Mourey, Hans Poppe, Mary Ann Quarry, Bill Raddatz, Dennis Saunders, Marilyn Stadalius, Laurie Van Heukelem, Tom Waeghe, and Peng-Ling Zhu. Finally, we very much appreciate the dedicated efforts of several reviewers of this book prior to its publication: Geoff Cox, John Ford, Pavel Jandera, Tom Jupille, John Kern, James Little, Dan Marchand, Jim Merdink, Tom Mourey, Uwe Neue, Carl Scandella, Peter Schoenmakers, Mark Stone, Tim Wehr, Loren Wrisley, Patrick Lukulay, and Jianhong (Jane) Zhao. Several of the latter reviewers have provided further assistance by supplying preprints or reprints of their own work.

REFERENCES

1. L. R. Snyder, HPLC: past and present, *Anal. Chem.* 72 (2000) 412A.
2. L. R. Snyder, J. J. Kirkland, and J. L. Glajch, *Practical HPLC Method Development*, 2nd edn, Wiley-Interscience, New York, 1997.
3. L. R. Snyder, Principles of gradient elution, *Chromatogr. Rev.* 7 (1965) 1.
4. C. Liteanu and S. Gocan, *Gradient Liquid Chromatography*, Halsted Press, New York, 1974.
5. L. R. Snyder, Gradient elution, *High-performance Liquid Chromatography. Advances and Perspectives*, Vol. 1, Cs. Horváth, ed., Academic Press, New York, 1980, Chap. 4.
6. P. Jandera and J. Churáček, *Gradient Elution in Column Liquid Chromatography*, Elsevier, Amsterdam, 1985.
7. L. R. Snyder and J. W. Dolan, The Linear-solvent-strength model of gradient elution, *Adv. Chromatogr.* 38 (1998) 115.
8. P. Jandera, Gradient elution in liquid column chromatography – prediction of retention and optimization of separation, *Adv. Chromatogr.* 43 (2004) 1.
9. Winston Churchill, *Radio Broadcast*, 1 October 1939 (originally said about Russia).

LLOYD R. SNYDER
JOHN W. DOLAN

Orinda, California
Amity, Oregon
September 2006

GLOSSARY OF SYMBOLS AND TERMS

This section is divided into "Major symbols" and "Minor symbols." "Minor symbols" refer to symbols that are used only once or twice. Most symbols of interest will be included in "Major symbols." Equations which define a particular symbol are listed with that symbol; for example, "Equation (2.18)" refers to Equation (2.18) in Chapter 2. The units for all symbols used in this book are indicated. Where IUPAC definitions or symbols differ from those used in this book, we have indicated the corresponding IUPAC term (from ASDLID 009921), for example, t_M instead of t_0.

MAJOR SYMBOLS AND ABBREVIATIONS

A	column hydrogen-bond acidity; Appendix III
A solvent	mobile phase at the start of the gradient
ACN	acetonitrile
b	intrinsic gradient steepness; Equation (2.11) (see discussion in Section 1.3.3)
B, B solvent	mobile phase at the end of the gradient; percentage B refers to the volume-percent of B in the mobile phase
B	column hydrogen bond basicity; Appendix III
C	column cation exchange capacity; Appendix III
C	concentration of the salt counter-ion in IEC (assuming a univalent counter-ion)
C^*	value of C in gradient elution (for band at column midpoint)
$(C)_f, (C)_0$	values of C at beginning (o) and end (f) of gradient
d_p	particle size (μm)
F	flow rate (mL/min)
G	gradient compression factor; Equation (9.36)
G_{12}	ratio of peak widths before and after passage of a step-gradient through a band within the column; $= W_2/W_1$ in Figure 9.4
GLP	good laboratory practice
H	plate height (mm); Equation (9.58)

xxi

H	column hydrophobicity; Appendix III
HIC	hydrophobic interaction chromatography
HILIC	hydrophilic interaction chromatography
HPLC	if you need to look up the meaning of HPLC, this is the wrong book for you
i.d.	column internal diameter (mm)
IEC	ion-exchange chromatography
IQ	installation qualification; Section 5.1.1.1
k	isocratic retention factor; Equation (2.4) (formerly called capacity factor, k')
k^*	gradient retention factor; equal to value of k for a band when it reaches the column mid-point; Equation (2.13), Figure 1.7 (previously, a different symbol was used, $\bar{k}$)
$k^*(a)$, $k^*(b)$, etc.	value of k^* for peak a, b, and so on
k_e	value of k at elution; Figure 1.7
k_i, k_j, etc.	value of k for peaks i, j, etc. Also, k_i is the instantaneous value of k for a band at any time during its migration through the column; Equation (9.1)
k_0	the value of k in gradient elution at the start of the gradient [Equation (2.10)]; *also* (Chapter 7 and Appendix V only), the value of k in isocratic elution for a small weight of injected sample (in distinction to the value of k for a large sample)
k_w	value of k for water or 0 percent B as mobile phase (ϕ_0) (extrapolated value)
k_1, k_2, etc.	value of k for solute 1, 2, and so on; also, value of k for two different values of ϕ (ϕ_1 and ϕ_2)
L	column length (mm)
LC	liquid chromatography
LCCC	liquid chromatography under critical conditions
LC-MS	liquid chromatography–mass spectrometry (Section 8.1)
LC-MS/MS	LC-MS with triple quadrupole mass spectrometer (Section 8.1)
LSS	linear-solvent-strength (model) (Sections 1.4.2, 9.1)
m	stoichiometry factor in NPC [Equation (8.8)]; also, $|z|$ in IEC
M	solute molecular weight; also counter-ion molarity in IEC
MeOH	methanol
MS	mass spectrometric
n	number of peaks in a chromatogram or sample; also the designation of the nth oligomer in an oligomeric sample
N	column plate number (isocratic); Equation (2.5); also native protein in Figure 6.4

N_0	column plate number for a small weight of sample; Equation (7.3)
N^*	column plate number (gradient); Equation (2.20)
NPC	normal-phase chromatography
ODS	octdecylsilyl; C_{18}
OQ	operational qualification; Section 5.1.1.1
p	quantity used to calculate gradient compression factor G; Equation (9.35)
P	column pressure-drop (psi); MegaPascals (MPa = 145 psi) is also commonly used, but not in the present book (the IUPAC symbol is Δp)
PC	peak capacity; the number of peaks with $R_s = 1$ that can be fit into a given chromatogram; see Figure 2.11(a) and related text.
PC_{req}	required peak capacity for the separation of a sample containing n components; see Figure 2.11(c) and related text (previously defined as "PC^*")
prep-LC	preparative liquid chromatography; Chapter 7
psi	pounds per square inch; see P
PQ	performance qualification; Section 5.1.1.1
QC	quality control
r	fractional migration of a band through the column during gradient elution; Equation (9.12)
R	equal to $1/(1+k)$ (the IUPAC symbol is κ)
R_1, R_2	equal to R for peaks 1 and 2
RP-LC	reversed-phase liquid chromatography
R_F	fractional migration of a peak through the column after the passage of one column-volume V_m of mobile phase through the column; $R_F = 1/(1+k)$
R_s	resolution of two adjacent peaks; Equation (2.6), Figure 2.1; also see Equations (2.8) (isocratic) and (2.21) (gradient); "critical" resolution refers to the value of R_s for the least well separated pair of peaks in a chromatogram
S	constant in Equation (1.2) for a given solute and experimental conditions; equal to $d(\log k)/d\phi$
S	column steric resistance to penetration; Appendix III
SA	surface area (m^2); Equation (7.5)
t	time after the beginning of a gradient run (min); Equation (9.2); also, time after the end of a gradient run (Fig. 9.5a)
T-P	"touching-peak"; preparative separation in which a large enough sample is injected to allow the desired product peak to touch an adjacent peak in the chromatogram (Section 7.1)

TFA	trifluoroacetic acid
THF	tetrahydrofuran
t_D	system dwell time (min); equal to V_D/F
t_{delay}	gradient delay time (min), corresponding to initial isocratic elution before the start of the gradient
t_{eq}	equilibration time for inter-run column equilibration in gradient elution (min); equal to V_{eq}/F
t_G	gradient time (min)
t_0	column dead time (min); retention time of an unretained peak such as thiourea (the IUPAC symbol is t_M)
t_R	retention time (min); see Figure 2.1 and related text
$t_{R,a}$, $t_{R,b}$, etc.	values of t_R for peaks a, b, etc.
$(t_R)_{avg}$	average value of t_R; Figure 3.2
t'_R	corrected retention time, equal to $t_R - t_0$
ULOQ	upper limit of quantification
USP	*United States Pharmacopeia*
UV	ultraviolet
V	volume of mobile phase that has entered the column by a given time (mL); Equation (9.1)
V_D	equipment dwell volume (mL); volume of system flowpath between inlet to gradient mixer and column inlet
V_{eq}	equilibration volume (mL) of A solvent used for inter-run column equilibration in gradient elution
V_m	column dead volume (mL); $V_m = t_0 F$; unless noted otherwise, a column internal diameter of $d_c = 4.6$ mm is assumed, in which case $V_m \approx 0.01L$, where L is column length in mm. Otherwise, $V_m \approx 0.0005$(column i.d.)^2 L, where column i.d. and L are in mm (the IUPAC symbol is V_M)
V_M	the "mixing volume" of the gradient system (mL); Table 9.2
V_R	retention volume (mL); $V_R = t_R F = V_m(1+k)$
V'_R	corrected retention volume (mL), equal to $V_R - V_m$
V_s	sample volume (mL)
W	baseline peak width (min); Figure 2.1 (IUPAC symbol is W_b)
W_i, W_j, etc.	value of W for peaks i, j, etc.
W_0	value of W for a small sample; Equation (7.2)
w_s	column saturation capacity (mg)
W_{th}	contribution to W from a sufficiently large sample weight (min); Equation (7.2)

w_x	injected weight of compound x (mg)
$W_{1/2}$	peak width at half height; Figure 2.1 (the IUPAC symbol is W_h)
x	fractional migration of a solute band through the column (Figure 1.7); also, band width in Figure 9.3
x_i, x_j	values of x for solutes i and j
z	effective charge on a sample compound in IEC
α	*selectivity* factor (isocratic); Equation (2.8)
α^*	selectivity factor (gradient) when the band-pair is at the column midpoint
α_0	the value of isocratic α or gradient α^* for a small sample
β	equal t_{G1}/t_{G2}; Equation (9.48)
δt_R	a change in retention time t_R due either to incomplete column equilibration or solvent demixing; also, an error in a calculated value of t_R; Equation (9.43)
$\delta\delta t_R$	difference in δt_R for two adjacent peaks
Δt_R	difference in retention times for two peaks (min), for example, Equation (2.24a), Figure 3.2
$\delta\phi$	error in calculated value of ϕ at elution; Equation (9.43)
$\delta\phi_m$	distortion of the gradient as a result of gradient rounding; Figure 9.7(a)
$\Delta\phi$	gradient range, equal to the final value of ϕ in the gradient (ϕ_f) minus the initial value (ϕ_0)
ϕ	volume fraction of B solvent in the mobile phase; equal to 0.01 times percentage B
ϕ_c	value of ϕ for "critical elution behavior"
ϕ_e	value of ϕ for mobile phase at the time a band elutes from the column
ϕ_f	value of ϕ for mobile phase at end of gradient; for example, for 10–80 percent B gradient, $\phi_f = 0.80$
ϕ_0	value of ϕ for mobile phase at start of gradient; for example, for 10–80 percent B gradient, $\phi_0 = 0.10$
ϕ^*	value of ϕ for mobile phase when a band is at the column mid-point
η	solvent viscosity (cPoise); Table IV.1 of Appendix IV
2-D	two-dimensional

The Jandera and Schoenmakers groups (and some other workers) have used different symbols than those employed in this book and by the authors in previous publications. Equivalent terms for these different groups are as follows.

Jandera, Shoenmakers	Present book
a	$\log k_w$
A	ϕ_0
B	$\Delta\phi/(F\, t_G)$
B'	$\Delta\phi/t_G$
k_a	k_0
m	S

MINOR SYMBOLS

ACTH	adrenocorticotropic hormone
A_{HIC}	$d(\log k)/d(C_{AS})$; Equation (6.7)
API	atmospheric pressure ionization (includes APCI and ESI)
amu	atomic mass unit; equal to 1 Da
APCI	atmospheric pressure chemical ionization interface
ASF	peak asymmetry factor
AU	absorbance units
b_A, b_Z	value of b for first peak A and last peak Z in the chromatogram [Equations (2.23) and (2.23a)]
BA	benzyl alcohol; Figure 7.13
b^*	designation of a compound in Figure 7.12
C	p-cresol; Figure 7.13
C_{AS}	concentration of ammonium sulfate in HIC; Equation (6.7)
D	fully denatured protein native protein; Figure 6.4
Da	Dalton; equal to 1 amu
d_c	column internal diameter (mm)
ESI	electrospray ionization interface (for MS)
$E_T(30)$	measure of mobile phase polarity derived from spectroscopic measurements; Equation (9.51)
F_s	column-matching function; Equation III.1 of Appendix III
h	peak height (relative units); Figure 2.1; also, reduced plate height; Equation (9.56)
$h_{1/2}$	one half of peak height; Figure 2.1
H_0	value of H for a small sample; Equation (V.4) of Appendix V
H_{th}	contribution to H of a large sample; Equation (V.2) of Appendix V
K	equilibrium constant for solute retention
k_{ACN}	value of k for pure ACN as mobile phase; Equation (6.17)

k_{H2O}	value of k for water as mobile phase in HILIC; Equation (6.14)
k_i, k_j	value of k for peaks i and j, respectively
$k_{o,A}$, $k_{o,Z}$	value of k_o for first peak A and last peak Z in the chromatogram [Equations (2.23) and (2.23a)]
k_{wi}, k_{wj}	value of k_w for peaks i and j
k_0	value of k for $C_{AS} = 0$ in HIC [Equation (6.7)]
$k_{2.5}$	the value of k for 2.5 M ammonium sulfate in HIC; Equation (6.8)
LLE	liquid–liquid extraction
m_{HILIC}	$d(\log k)/d(\log \phi_{H2O})$ in HILIC; Equation (6.14)
MRM	multiple reaction monitoring (MS/MS; Section 8.1)
MSD	mass selective detector; single-quadrupole mass spectrometer
MTBE	methyl-t-butylether
m/z	mass-to-charge ratio
P	phenol; Figure 7.13
PD	partially denatured protein; Figure 6.4
PE	2-phenylethanol; Figure 7.13
PEEK	poly-ether-ether-ketone; plastic tubing used for HPLC connections
p, q	constants in Equation (6.19)
rhGH	recombinant human growth hormone
SC	standard calibrator
S_{HIC}	equal to $-2.5\, A_{HIC}$ in HIC; Equation (6.8)
S_i, S_j	value of S for peaks i and j
SIM	selective ion monitoring; also single ion monitoring (MS)
SPE	solid-phase extraction
t_{G1}, t_{G2}, etc.	values of t_G for runs 1, 2, and so on
$t_R(1)$, $t_R(2)$	retention times of peaks 1 and 2, respectively (min)
$t_{R,A}$, $t_{R,Z}$	values of t_R for first peak A and last peak Z in the chromatogram (min)
W_b	value of W for peak b
W_i, W_j	baseline peak widths of peaks i and j, respectively (min)
w_{xn}	"loading function" in prep-LC; Equation (V.3)
δk	error in calculated value of k at elution; Equation (9.46)
Δx	fraction of a column length; Equation (9.19), Figure 9.2
ϕ_A, ϕ_B, ϕ_{AB}	values of ϕ for the mobile phase in reservoir A, B and a mixture of A and B where the volume fraction of A is ϕ_{AB} (Section 1.3)
ϕ_{HIC}	defined as $-(C_{AS} - 2.5)/2.5$; Equation (6.8)

$\phi_{H2O,f}$, $\phi_{H2O,o}$ value of ϕ_{H2O} at beginning (o) and end (f) of a HILIC gradient

σ_g surface area per unit weight of column packing (m^2/g); Equation (7.5)

υ reduced velocity; Equation (9.57)

ψ phase ratio (the IUPAC symbol is β)

CHAPTER 1

INTRODUCTION

> Begin at the beginning ... and go on till you come to the end: then stop.
> —Lewis Carroll, *Alice's Adventures in Wonderland*

1.1 THE "GENERAL ELUTION PROBLEM" AND THE NEED FOR GRADIENT ELUTION

Prior to the introduction of gradient elution, liquid chromatographic separation was carried out with mobile phases of fixed composition or eluent strength, that is, *isocratic* elution. Isocratic separation works well for many samples, and it represents the simplest and most convenient form of liquid chromatography. For some samples, however, no single mobile phase composition can provide a generally satisfactory separation, as illustrated by the reversed-phase liquid chromatography (RP-LC) examples of Figure 1.1(a, b) for the separation of a nine-component herbicide sample. We can use a weaker mobile phase such as 50 percent acetonitrile–water (50 percent B) or a stronger mobile phase such as 70 percent acetonitrile–water (70 percent B). With 50 percent acetonitrile (Fig. 1.1a), later peaks are very wide and have inconveniently long retention times. As a result, run time is excessive (140 min) and later peaks are less easily detected (in this example, peak 9 is only 3 percent as high as peak 1). The use of 70 percent acetonitrile (Fig. 1.1b) partly addresses the latter two difficulties, but at the same time it introduces another problem: the poor separation of peaks 1–3. This example illustrates the *general elution problem*: the inability of a single isocratic separation to provide adequate separation within reasonable time for samples with a wide range in retention (peaks with very different retention factors k).

Very early in the development of chromatography, Tswett introduced a practical solution to the general elution problem (cited in [1]; see also [2]). If separation is begun with a weaker mobile phase (e.g., 50 percent acetonitrile–water), a better separation of early peaks is possible within a reasonable time, following which the mobile phase can be changed (e.g., to 70 percent acetonitrile–water) for the faster elution of the remainder of the sample. This *stepwise* (or "step-gradient") elution of the sample is illustrated in Figure 1.1(c) for the same sample, with other conditions

---

High-Performance Gradient Elution. By Lloyd R. Snyder and John W. Dolan
Copyright © 2007 John Wiley & Sons, Inc.

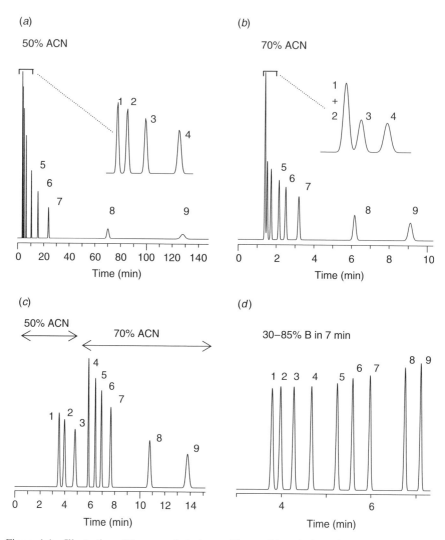

Figure 1.1 Illustration of the general elution problem and its solution. The sample is a mixture of herbicides described in Table 1.3 (equal areas for all peaks). (*a*) Isocratic elution using 50 percent acetonitrile (ACN)–water as mobile phase; 150 × 4.6 mm C_{18} column (5 μm particles), 2.0 mL/min, ambient temperature; (*b*) same as (*a*), except 70 percent ACN–water; (*c*) same as (*a*), except stepwise elution with 50 percent ACN for 5 min, followed by 70 percent ACN for 10 min; (*d*) same as (*a*), except gradient elution: 30–85 percent ACN in 7 min. Computer simulations based on the experimental data of Table 1.3.

held constant. Now, all nine peaks are separated to baseline in a total run time of only 15 min.

For the sample of Figure 1.1, stepwise elution (*c*) is an obvious improvement over the isocratic separations of Figure. 1.1(*a, b*), but it is not a perfect answer to the general elution problem. Significant differences in peak width and ease of detection still persist in Figure 1.1(c), accompanied by sizable variations in peak spacing (representing wasted space within the chromatogram). For some samples, a two-step gradient as in Figure 1.1(c) would still suffer from the problems illustrated in Figure 1.1(a, b). Furthermore, step gradients are (a) more difficult to reproduce experimentally, and (b) a potential source of "peak splitting": the appearance of two peaks for a single compound. *Gradient elution* refers to a *continuous* change in the mobile phase during separation, such that the retention of later peaks is continually reduced; that is, the mobile phase becomes steadily stronger as the separation proceeds. An illustration of the power of gradient elution is shown in Figure 1.1(d), where, all peaks are separated to baseline in a total run time of just 7 min, with approximately constant peak widths and comparable detection sensitivity for each peak.

In many cases, the advantage of gradient elution vs isocratic or stepwise elution can be even more pronounced than in the example of Figure 1.1. For several years after the introduction of gradient elution in the early 1950s, the relative merits of continuous vs stepwise elution were widely argued, with many workers expressing a preference for stepwise elution (p. 39 of [3]). For the above (and other) reasons, however, stepwise elution is much less used today, except for special applications, for example, the preparative isolation of a single compound, as described in Section 7.3.2.1 and illustrated in [4]. Software for the development of optimized multistep gradients has been described [5], although the applicability of such gradients appears somewhat limited.

The initial idea of gradient elution has been attributed to Arne Tiselius in the 1940s (cited in [1]), followed by its experimental implementation in 1950 by A.J.P. Martin [6]. Several independently conceived applications of gradient elution were reported in the early 1950s by different workers, as summarized in [3, 7]. Major credit for its subsequent rapid exploitation has been ascribed by Elberton [8] to R.J.P. Williams of the Tiselius group. Soon after the introduction of high-performance liquid chromatography (HPLC) in the late 1960s, commercial equipment became available for routine gradient elution. For further details on the early history of gradient elution, see [3, 7] and references therein. A conceptual understanding of how gradient elution works (as detailed in Chapters 2 and 9) has developed more slowly.

Temperature programming in gas chromatography (GC), which serves a similar purpose to gradient elution in liquid chromatography (LC), evolved about the same time (1952–1958) [9]. A theoretical description of these two separation procedures is remarkably similar, as can be seen from a comparison of [10] with the present book; the rate at which either the mobile phase composition (LC) or temperature (GC) is changed leads to fully analogous changes in the final separation [11, 12].

Apart from stepwise elution (Figure 1.1c), several other experimental procedures have been suggested as alternatives to gradient elution as a means of solving the general elution problem: flow programming [13–15], temperature programming [16, 17], and column switching [18]. However, for reasons summarized in Table 1.1, none of these alternative LC techniques is able to fully duplicate the advantages of gradient elution for the separation of wide-range samples. For a further discussion and comparison of these different programming techniques, see [18, 19].

TABLE 1.1 Alternatives to Gradient Elution

Procedure	Basis	Comment
Flow programming [13–15]	Increase in flow rate during separation	Very limited ability to deal with wide-range samples
		Much reduced peak heights and areas for later peaks
		For most detectors, peak area varies with small changes in flow rate[a]
		Ability to use this approach is limited by the pressure tolerance of the system
Temperature programming [16, 17]	Increase in temperature during separation	Limited ability to deal with wide-range samples, because temperature has less effect on retention than a change in %B
		Possible sample reaction during separation of later peaks, due to their elution at higher temperature
		Many columns will not tolerate large changes in temperature
Column switching [18]	Transfer of sample fraction from a first column to a second column	Similar disadvantages as for stepwise elution
		More complicated method development and equipment
		Less reproducible method transfer

[a]At constant flow, the analyte mass under the peak is proportional to the peak area multiplied by the flow rate. When flow rate is programmed, the flow rate during the time each peak is eluted becomes less controllable, as does peak area.

1.2 OTHER REASONS FOR THE USE OF GRADIENT ELUTION

Apart from the need for gradient elution in the case of wide-range samples like that of Figure 1.1, a number of other applications of this technique exist (Table 1.2). Large molecules, such as proteins or synthetic polymers, cannot be conveniently separated by isocratic elution, because their retention can be extremely sensitive to small changes in mobile phase composition (%B). For example, the retention factor k of a 200 kDa polystyrene can change by 25 percent as a result of a change in the mobile phase of only 0.1 percent B [21]. This behavior can

TABLE 1.2 Reasons for the Use of Gradient Elution

Problem	Application
General elution problem	Samples with a wide retention range
Compounds whose retention changes markedly for small changes in mobile phase %B	Large biomolecules and synthetic polymers
Generic separation	A large number of samples of variable and/or unknown composition; the development of separate procedures for each sample would be economically prohibitive
Efficient method development	All samples; the final method can be either isocratic or gradient
Sample preparation needed	Samples that contain extraneous material that might interfere with HPLC separation
Tailing peaks	Especially for samples that are prone to exhibit tailing peaks, such as protonated bases

make it extremely difficult to obtain reproducible separations of macromolecules from one laboratory to another, or even within the same laboratory. Furthermore, the isocratic separation of a mixture of macromolecules usually results in the immediate elution of some sample components (with no separation), and such slow elution of other components that it appears that they never leave the column. With gradient elution, on the other hand, there is a much smaller problem with irreproducible retention times for large molecules, and their resulting separation can be fast, effective, and convenient (Chapter 6).

In some applications of RP-LC, a single generic separation procedure is needed that can be used for samples composed of different components, for example, compounds *A*, *B*, and *C* in sample 1, compounds *D*, *E*, and *F* in sample 2. Typically, each sample will be separated just once within a fixed run time, with no further method development for each new sample. In this way, hundreds or thousands of unique samples can be processed in minimum time and with minimum cost. Generic separations by RP-LC (with fixed run times, for automated analysis) are only practical by means of gradient elution and are commonly used to assay combinatorial libraries [22] and other samples [23]. Generic separation is also often combined with mass spectrometric detection [24], which allows both the separation and identification of the components of samples of previously unknown composition (Section 8.1).

Efficient HPLC method development is best begun with one or more gradient experiments (Section 3.2). A single gradient run at the start of method development can replace several trial-and-error isocratic runs as a means for establishing the best solvent strength (value of %B) for isocratic separation. An initial gradient run can also establish whether isocratic or gradient elution is the best choice for a given sample.

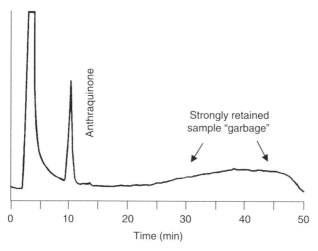

Figure 1.2 Illustration of a gradient separation that eliminates the need for sample pretreatment. The sample is a wood-pulp extract that contains anthraquinone. Conditions: 250 × 4.6 mm C_{18} column (10 μm particles); A, solvent, water; B, solvent, methanol; 20–20–100–100 percent B in 0–15–20–25 min. Adapted from [26].

Many samples are unsuitable for direct injection followed by isocratic elution. Typically, some kind of sample preparation (pretreatment) is needed [25], in order to remove interfering peaks and prevent the buildup of strongly retained components on the column. In some cases, however, gradient elution can minimize (or even eliminate) the need for sample preparation. As an example, consider the HPLC analysis of wood-pulp extracts for anthraquinone with UV detection [26]. These samples can be separated isocratically with 20 percent by volume methanol–water as mobile phase. A sharp anthraquinone peak results, which is well separated from adjacent peaks in the chromatogram. However, the continued isocratic analysis of these samples results in a gradual deterioration of separation, due to a buildup on the column of strongly retained sample components that are of no interest to the analyst. A separate sample pretreatment *could* be used to remove these strongly retained sample constituents prior to analysis by RP-LC, and this is often the preferred option. However, when gradient elution is used for these samples (Fig. 1.2), any strongly retained material is washed from the column *during* each separation, so that column performance does not degrade rapidly over time. In this example, the use of gradient elution eliminates the need for sample pretreatment, while minimizing column deterioration.

An early goal of gradient elution was the reduction of peak tailing during isocratic separation [27]. Because of the increase in mobile phase strength during the time a peak is eluted in gradient elution, the tail of the peak moves faster than the peak front, with a resulting reduction in peak tailing and peak width. This peak compression effect is illustrated in Figure 1.3 for (*a*) isocratic and (*b*) gradient separation of the same sample by means of anion-exchange chromatography. Note the pronounced tailing in the isocratic run (*a*) of peaks 12 and 13 (asymmetry factor, ASF = 2–4), but their more symmetrical shape (ASF = 1.2) in the gradient run (*b*).

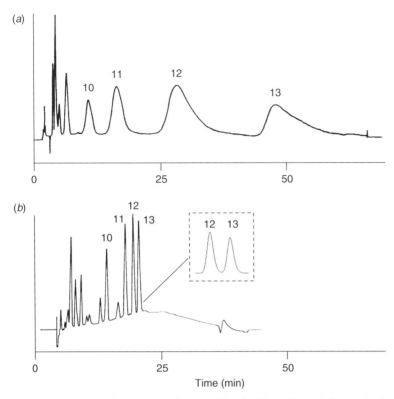

Figure 1.3 Illustration of reduced peak tailing in (*b*) gradient elution vs (*a*) isocratic elution. Separations of a mixture of aromatic carboxylic acids by anion-exchange chromatography. Conditions: (*a*) 0.055 M NaNO$_3$ in water; (*b*) gradient from 0.01 to 0.10 M NaNO$_3$ in 20 min. Adapted from [28].

1.3 GRADIENT SHAPE

By "gradient shape," we mean the way in which mobile phase composition (%B ≡ percentage by volume organic in RP-LC) changes with time during a gradient run. Gradient elution can be carried out with different gradient shapes, as illustrated in Figure 1.4(*a*–*f*). Most gradient separations use linear gradients (*a*), which are strongly recommended during the initial stages of method development. Curved gradients (*b*, *c*) have been used in the past for certain kinds of samples, but today such gradients have been largely replaced by segmented gradients (*d*). Segmented gradients can provide all the advantages of curved gradients, and also furnish a greater control over separation (as well as freedom from the need for specialized gradient formers). The use of segmented gradients as a means of enhancing separation is examined in Section 3.3.4. Gradient delay, or "isocratic hold" (*e*), and a step gradient (*f*) are also illustrated in Figure 1.4.

A linear gradient can be described (Fig. 1.4*g*) by the initial and final mobile phase compositions, and gradient time (the time during which the mobile phase is changing). We can define the initial and final mobile phase compositions in terms

8 CHAPTER 1 INTRODUCTION

of %B, or we can use the volume-fraction ϕ of solvent B in the mobile phase (equal to 0.01 percent B), that is, values ϕ_0 and ϕ_f, respectively, for the beginning and end of the gradient. The change in %B or ϕ during the gradient is defined as the *gradient range* and designated by $\Delta\phi = \phi_f - \phi_0$. In the present book, values of %B and ϕ

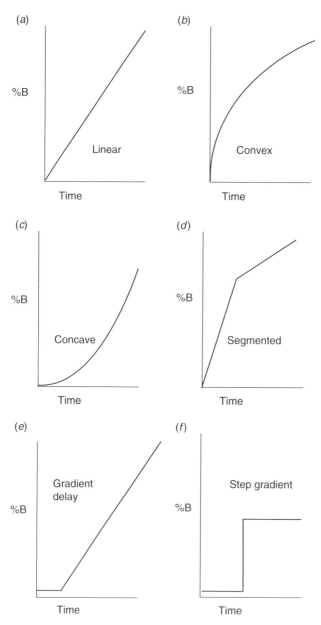

Figure 1.4 Illustration of different gradient shapes (plots of %B at the column inlet vs time). See text for details.

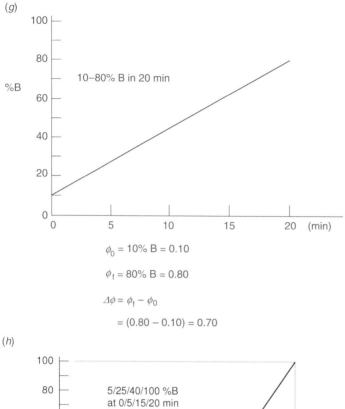

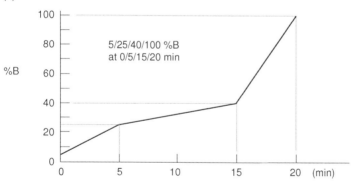

Figure 1.4 (*Continued*)

will be used interchangeably; that is, ϕ always equals 0.01 percent B, and 100 percent B means pure organic solvent ($\phi = \phi_f = 1.00$). For reasons discussed in Chapter 5, the A- and/or B-reservoirs may contain *mixtures* of the A- and B-solvents, rather than pure water and organic, respectively, for example, 5 percent acetonitrile–water in the A-reservoir and 95 percent acetonitrile–water in the B-reservoir. For the latter example, a 0–100 percent B gradient would correspond to 5–95 percent acetonitrile.

By a *gradient program*, we refer to the description of how mobile phase composition changes with time. Linear gradients represent the simplest example, for example, a gradient from 10 to 80 percent B in 20 min (Fig. 1.4g), which can also be described as 10–80 percent B in 0–20 min (10 percent B at time 0 to

10 CHAPTER 1 INTRODUCTION

80 percent B at 20 min). Segmented programs are usually represented by values of %B and time for each linear segment in the gradient, for example, 5–25–40–100 percent B at 0–5–15–20 min (Fig. 1.4h).

1.4 SIMILARITY OF ISOCRATIC AND GRADIENT ELUTION

A major premise of the present book is that isocratic and gradient separations are fundamentally similar, so that well-established concepts for developing isocratic methods can be used in virtually the same way to develop gradient methods [25]. This similarity of isocratic and gradient elution is hinted at in the examples of Figure 1.1. Thus, the stepwise gradient in Figure 1.1(c) is seen to represent a combination of the two isocratic separations of Figure 1.1(a, b). As the number of isocratic steps is increased from two (as in Fig. 1.1c) to a larger number, the separation eventually approaches that of a continuous gradient (as in Fig. 1.1d).

1.4.1 Gradient and Isocratic Elution Compared

The movement of a band through the column as a function of time proceeds in similar fashion for both isocratic and gradient elution (Figs 1.5 and 1.6, respectively). First consider Figure 1.5 for an isocratic separation, where the position of the

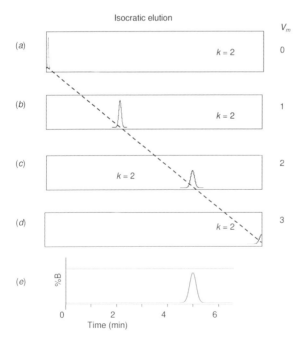

Figure 1.5 Illustration of band migration within the column during isocratic elution. See text for details.

band within the column is noted during its migration from column inlet to outlet. In (*a*), the solute band is shown at the column inlet just after sample injection. In (*b*), one column-volume (V_m) of mobile phase has moved through the column, and the band has broadened while migrating one-third of the way through the column [note that the fractional migration R_F is equal to $1/(1+k)$; in Figure 1.5, $k=2$ is assumed]. Here k refers to the retention factor of the solute (Section 2.1.1). In (*c*), a second column volume has entered the column, and the band has now migrated two-thirds of the way through the column with further broadening. After the passage of a third column volume in (*d*), the band has arrived at the column outlet and is ready to leave the column and appear as a peak in the final chromatogram (*e*); %B [dotted line in (*e*)] does not change with time (isocratic elution). Note that the band moves at constant speed through the column in isocratic elution, as indicated by the dashed, straight line of Figure 1.5 through the band centers at each stage of peak migration.

Figure 1.6 shows the similar separation of a band during gradient elution. Most sample-compounds in a gradient separation are initially "frozen" at the column inlet, because of their strong retention in the starting (relatively weak) mobile phase. However, as the separation proceeds, the mobile phase becomes progressively stronger, and the value of k for the band continually decreases. The example of Figure 1.6 begins (*a*) after five column volumes have passed through the column; because of the strong initial retention of the band, only a limited migration has occurred at this point in the separation. When the sixth column

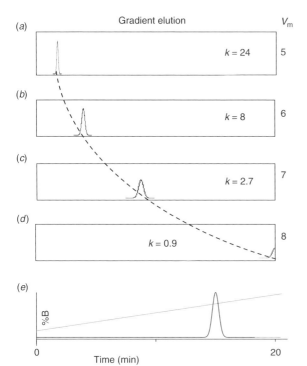

Figure 1.6 Illustration of band migration within the column during gradient elution. See text for details.

volume passes through the column (b), with an average value of $k = 8$, the band now moves appreciably further through the column (with additional broadening). For the next column volume (c), with an average $k = 2.7$, the extent of band migration and broadening is considerably greater, because of the stronger mobile phase. Finally, after the eighth column volume (d), the band reaches the end of the column and appears as a peak in the chromatogram of (e); note that %B (dotted line) increases with time (gradient elution). The dashed curve in Figure 1.6 marks the continued acceleration of the peak as it moves through the column in gradient elution, in contrast to its constant migration rate in isocratic elution (Fig. 1.5).

Figure 1.7 extends the example of Figure 1.6 for two different compounds (i and j) that are separated during gradient elution. Consider first the results for the initially eluted compound i in (a). The solid curve ["$x(i)$"] marks the fractional migration x of the band through the column as a function of time (note that $x = 1$ on the y-axis corresponds to elution of the band from the column; $x = 0$ corresponds to the band position at the start of the separation). This behavior is similar to that shown by the dashed curve in Figure 1.6, that is, accelerating migration with time, or an upward-curved plot of x vs t. Also plotted in Figure 1.7(a) is the instantaneous value of k for band i [dashed curve, "$k(i)$"] as it migrates through the column. The quantity k_i is the value of k if the compound were run under the isocratic

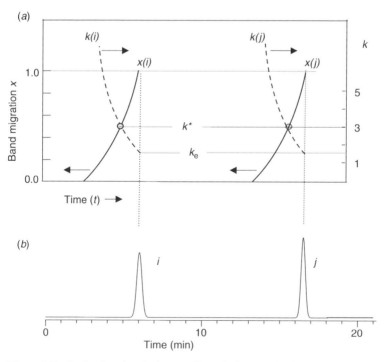

Figure 1.7 Peak migration during gradient elution; (a) band-migration plots showing average (k^*) and final values of k at elution (k_e); (b) resulting chromatogram. See text for details.

conditions present at that instant in time, that is, using a mobile phase whose composition (%B) is the same as the mobile phase in contact with the band at a given time during band migration. As we will see in Chapter 2, peak retention and resolution in gradient elution depend on the median value of k (i.e., the instantaneous value of k when the band has migrated halfway through the column, defined as the *gradient retention factor* k^*), while peak width is determined by the value of k when the peak is eluted or leaves the column (defined as k_e).

A comparison in Figure 1.7(*a*) of the two compounds *i* and *j* shows a generally similar behavior for each band as it migrates through the column, apart from a greater delay in the migration of band *j*. Specifically, values of k^* and k_e for both early and late peaks in the chromatogram are about the same for *i* and *j*, suggesting that resolution and peak spacing will not decrease for earlier peaks; compare the gradient separation of Figure 1.1(*d*) with the isocratic separation of Figure 1.1(*b*). Small values of k (or k^*) generally result in poorer separation (e.g., the early portion of Fig. 1.1*b*), whereas larger values give better separation (e.g., later peaks in Fig. 1.1*b*). Constant (larger) values of k^* in gradient separation should improve separation throughout the chromatogram. Values of k_e are also usually similar for early and late peaks in gradient elution, meaning that peak width will be similar for both early and late peaks in the chromatogram (as also observed in the gradient separation of Fig. 1.1*d*). *The relative constancy of values of k^* and k_e for a given gradient separation contributes to the pronounced advantage of gradient over isocratic elution for the separation of many samples.*

1.4.2 The Linear-Solvent-Strength Model

The linear-solvent-strength (LSS) model for gradient elution is based on an approximation for isocratic retention in RP-LC as a function of *solvent strength*. In terms of the retention factor k and the percentage-volume of organic solvent in the water–organic mobile phase (%B),

$$\log k = a - b(\%\text{B}) \qquad (1.1)$$

Here, a and b are usually positive constants for a given compound, with only %B varying. Equation (1.1) is an empirical relationship that was cited in almost a dozen separate reports in the mid-1970s [25], not to mention its earlier recognition in analogous thin-layer chromatography separations [29].

Equation (1.1) is illustrated in Figure 1.8(*a*) for nine different solutes (1–9); for examples of plots with individual data points, see Figure 6.1. Equation (1.1) is more often represented by

$$\log k = \log k_w - S\phi \qquad (1.2)$$

where ϕ is the volume-fraction of organic solvent in an RP-LC mobile phase (or %B expressed in decimal form; $\phi = 0.01$ percent B), S is a constant for a given compound and fixed experimental conditions (other than ϕ), and k_w is the (extrapolated) value of k for $\phi = 0$ (i.e., water as mobile phase). Values of log k_w and S for the compounds of Figure 1.8 are listed in Table 1.3 ("regular" sample).

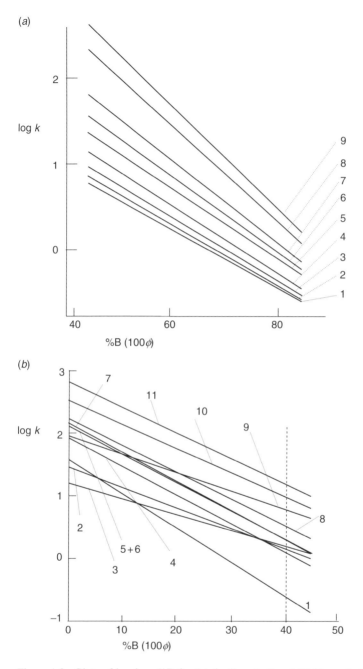

Figure 1.8 Plots of log k vs %B for (a) the "regular" and (b) "irregular" samples of Table 1.3. Separation conditions defined in the footnotes of Table 1.3.

TABLE 1.3 Values of log k_w and S for the Representative "Regular" and "Irregular" Samples Used in this Book

"Regular" sample[a]			"Irregular" sample[b]		
Compound	log k_w	S	Compound	log k_w	S
1. Simazine	2.267	3.41	1. Phthalic acid	1.58	5.46
2. Monolinuron	2.453	3.65	2. 2-Nitrobenzoic acid	1.47	3.34
3. Metobromuron	2.603	3.746	3. 4-Chloroaniline	1.23	2.50
4. Diuron	2.816	3.891	4. 2-Fluorobenzoic acid	1.90	4.46
5. Propazine	3.211	4.222	5. 3-Nitrobenzoic acid	2.12	4.55
6. Chloroxuron	3.602	4.636	6. 2-Chlorobenzic acid	2.12	4.55
7. Neburon	3.920	4.882	7. 3-Fluorobenzic acid	2.17	4.61
8. Prometryn	4.731	5.546	8. 2,6-Dimethylbenzoic acid	2.22	4.29
9. Terbutryn	5.178	5.914	9. 2-Chloroaniline	1.95	2.90
			10. 3,4-Dichloroaniline	2.52	3.80
			11. 3,5-Dichloroaniline	2.81	4.02

[a]Data of [20] for methanol–water mixtures as mobile phase and a 5 μm C_{18} column; ambient temperature.
[b]Data of [38] for acetonitrile–buffer mixtures as mobile phase and a 5 μm C_{18} column; the buffer was 25 mM citrate (pH 2.6); 32.1°C.

If a linear gradient is used (Fig. 1.4a or g), %B is related to time t as

$$\%B = c + dt \tag{1.3}$$

where c and d are also constants. The combination of Equations (1.1) and (1.3) then gives

$$\log k = a - bc - bdt \equiv (\text{constant}) - (\text{constant})t \tag{1.4}$$

Here, k refers to the value the solute would have at the column inlet, for an isocratic mobile phase having a composition ϕ at time t (this ignores the migration of the solute during the gradient). A gradient in which retention is described by Equation (1.4) is referred to as a *linear-solvent-strength* gradient. Equation (1.4) predicts a linear decrease in log k during the gradient with either time or the volume of mobile phase that has left (or entered) the column. Equations (1.2) and (1.4) are never exact relationships, especially for large changes in k or ϕ, but Equation (1.4) nevertheless allows accurate predictions of separation in gradient elution for "practical" experimental conditions. The advantages of a linear gradient and Equation (1.4) include (a) the easy determination of constants a–d, followed by quantitative predictions of separation, and (b) the ability of Equation (1.4) to relate gradient separations to corresponding isocratic separations. As a result, chromatographers who have experience in the development and use of isocratic RP-LC methods can apply this knowledge directly to the development and troubleshooting of corresponding gradient methods. The extension of the LSS model to non-reversed-phase HPLC methods is possible, with a slight decrease in reliability or some increase in mathematical complexity (Sections 8.2 and 8.3).

16 CHAPTER 1 INTRODUCTION

When Equation (1.4) is combined with a fundamental equation of gradient elution [Equation (9.2) in Chapter 9], it becomes possible to accurately predict retention time, peak width, and resolution as functions of experimental conditions, and to express these results in terms that are equivalent to those used in isocratic elution. We will next present a qualitative picture of an important relationship: the dependence of values of the gradient retention factor k^* (Fig. 1.7) on gradient conditions. In Chapter 2, we will see that Equation (1.5) below can be used to conveniently compare gradient and isocratic separation. *The discussion of the following three paragraphs [ending with the paragraph that contains Equation (1.7) below] can be skipped if the reader is primarily interested in the practical application of gradient elution, rather than insight into its fundamental basis.*

In Figure 1.6, the value of k when the band reaches the column mid-point (k^*) is determined by how fast k changes during the migration of the band through the column. This can be seen by comparing Figure 1.6 with Figure 1.9. In Figure 1.6, values of k are reduced 3-fold after the passage of each successive column volume. As a result, the

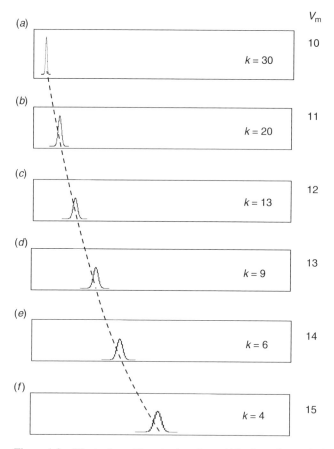

Figure 1.9 Illustration of band migration within the column during gradient elution (shallower gradient and slower peak migration than in Fig. 1.6). See text for details.

peak reaches the column mid-point in Figure 1.6 with a value of $k^* \approx 2$. In Figure 1.9, the change in k is slower (i.e., a flatter gradient), with only a 1.5-fold decrease in k^* for each successive column volume; the resulting value of $k^* \approx 4$. That is, for a slower change in k during gradient elution, the value of k when the peak reaches the column mid-point (k^*) will be larger than for a steeper gradient. The reason for a larger value of k^*, when k changes more slowly, is that (for a given change in %B or ϕ) a larger volume of mobile phase passes through the column, carrying the band further along the column. The faster migration of the band (in terms of the change in ϕ) therefore results in an earlier arrival of the band at the column midpoint, with a smaller value of ϕ for the adjacent mobile phase, and a larger value of $k = k^*$ [i.e., Equation (1.2)]. In the limit, for the flattest possible gradient or no decrease in k during band migration (isocratic elution), the value of k^* is equal to k at the start of elution (the largest possible value of k^* in gradient elution).

Conversely, for a faster change of k during gradient elution, the value of k^* will be smaller. Thus, the value of k^* is determined by the rate of change in k during gradient elution, or by *gradient steepness*. The steeper the gradient, the smaller is k^*. This picture of gradient elution suggests a fundamental definition of gradient steepness, namely the rate of change in log k during the gradient per volume of mobile phase passing through the column. We will define the latter quantity as the *intrinsic gradient steepness b*. The quantity b is therefore determined by the total change in log k during the gradient divided by the number of column-volumes of mobile phase that have passed through the column during the gradient (t_G/t_0). The change in ϕ during the gradient (final value of ϕ minus the initial value) will be defined as $\Delta\phi$, so that the change in log k during the gradient is $\Delta\phi S$ [difference in values of log k for initial vs final values of ϕ in the gradient; Equation (1.2)]. Therefore, b is given by $(\Delta\phi S)/(t_G/t_0)$, or

$$b = t_0 \Delta\phi S / t_G \qquad (1.5)$$

Because column dead-time $t_0 = V_m/F$, where F is flow rate, Equation (1.5) can also be written as

$$b = V_m \Delta\phi S / t_G F \qquad (1.6)$$

For all but early-eluting peaks in gradient elution, we can show (Section 9.1.1) that $k^* = 1/1.15b$, and values of k^* determine peak width and resolution in gradient elution.

Note that b in Equation (1.6) is determined by gradient conditions ($\Delta\phi$, t_G), column dead-volume V_m, flow rate F, and the parameter S from Equation (1.2). Values of S for a given sample and separation conditions can be obtained either from two isocratic measurements with ϕ varied [Equation (1.2)] or from two gradient experiments where gradient time is varied (Section 9.3.3). This in turn allows the calculation of values of b and k^* for each solute in each of the two gradient runs, which can be used to predict gradient separation as a function of experimental conditions. Later chapters will show that the intrinsic gradient steepness b is of fundamental importance in understanding gradient elution.

Values of S in RP-LC can be approximated (Section 6.1.1) by

$$S \approx 0.25(\text{molecular weight})^{1/2} \tag{1.7}$$

so that typical small molecules with molecular weights around 200 have $S \approx 4$, while larger molecules have larger values of S; for example, $S \approx 11$ for a solute with a molecular weight of 2000. Macromolecules such as proteins, DNA, or synthetic polymers (with molecular weights $> 10^4$) can have very large values of S, with important consequences for their gradient separation (Chapter 6).

Some potential advantages of LSS gradients were first recognized in 1964 [30], and the LSS model presented here was then developed over the next 35 years [31–34]. For further information and a detailed, current review of this concept and its application to gradient elution, see Chapters 2 and 9.

1.5 COMPUTER SIMULATION

The LSS model allows the reliable calculation of separation in gradient elution as a function of experimental conditions (Chapters 2 and 9). Thus, there is a predictable effect on separation of (a) gradient steepness (%B/min), (b) initial and final values of %B in the gradient, (c) gradient shape, (d) flow rate, and (e) column dimensions; other conditions such as temperature, mobile phase pH, etc. are assumed to be held constant while the latter conditions are changed. Calculations of gradient separation in this manner require values of k_w and S for each compound in the sample [Equation (1.2)], which can be obtained from two gradient runs for a given sample, where only gradient time t_G is varied (Section 9.3.3). Computer simulation makes use of the foregoing calculations as a result of two or more initial gradient separations (experimental "calibration runs"), in order to then predict isocratic or gradient separation as a function of different experimental conditions. Computer simulation begins with the entry of (a) experimental data from the calibration runs, plus (b) separation conditions. The computer program then determines values of k_w and S for each compound in the sample, following which separation can be predicted as a function of the above (and other) experimental conditions. In this way, the process of developing a gradient RP-LC method can be made more efficient, with resulting methods that are better, as well as less costly to develop; see Section 3.4 for further details, as well as some examples of computer simulation.

A good way to demonstrate the various principles of gradient elution is also provided by computer simulation, which we will use extensively in the present book (DryLab® software, Rheodyne LLC, Rohnert Park, CA, USA [35–37]). By selecting representative samples, it is possible to generate simulated chromatograms which show what happens when gradient conditions are changed. Because predicted separations from computer simulation are usually quite reliable (Section 9.3), they can be considered equivalent to corresponding experimental runs. Computer simulation can be assumed for chromatograms shown in the present book, unless a literature reference (for a "real" separation) is given.

1.6 SAMPLE CLASSIFICATION

As we will see, samples can differ in two important respects: sample type ("regular" vs "irregular"), and sample molecular weight (greater or less than 1000 Da). The nature of the sample (defined in this way) plays an important role in determining gradient separation as a function of experimental conditions. The present book deals mainly with samples of "low" molecular weight (<1000 Da), while Chapter 6 can be consulted for the separation of high-molecular-weight samples. A definition of what we mean by sample type, and an examination of some of its consequences for gradient elution, will be considered next.

1.6.1 Sample Compounds of Related Structure ("Regular Samples")

For compounds of highly related structure (e.g., homologs), plots of log k vs %B vary in a regular fashion, as illustrated by the sample of Figure 1.8(a).We will refer to samples of this kind as "regular," in contrast to "irregular" samples as in Figure 1.8(b) (note that this definition of "regular" samples differs from that used in [25]). The relative retention of a "regular" sample does not change when %B is varied in isocratic elution; thus, regardless of %B, the order of peak elution for the sample of Figure 1.8(a) is always $1 < 2 < 3 < \cdots < 9$. As we will see in Chapter 2, there is also no change in relative retention when gradient conditions are changed for the separation of "regular" samples (e.g., Fig. 2.5). In this book, we have chosen a particular sample that is representative of "regular" samples: a mixture of nine herbicides (phenylureas and triazines) summarized in Table 1.3, whose retention as a function of %B is illustrated in Figure 1.8(a) (this "regular" sample was also used for the examples of Fig. 1.1). Separations of this sample as a function of both isocratic and gradient conditions will be illustrated in later chapters.

1.6.2 Sample Compounds of Unrelated Structure ("Irregular" Samples)

Samples which contain compounds which are structurally diverse often exhibit "irregular" retention vs %B, in contrast to the "regular" sample of Figure 1.8(a). This is illustrated in Figure 1.8(b) for the "irregular" sample of Table 1.3, where it is seen that relative retention changes as %B is varied. This differing behavior of the "regular" and "irregular" sample is better illustrated in Figure 1.10, for the separation of compounds 1–7 of the "regular" sample in (a) and (b), and the separation of compounds 1–5 of the "irregular" sample in (c) and (d). For a change in mobile phase from 40 to 60 percent B for the "regular" sample, peaks are more bunched together in (b), but there is no change in retention sequence: peak $1 < 2 < 3 < 4 < 5 < 6 < 7$. For the separation of the "irregular" sample, there is a considerable change in retention order for a change of mobile phase from 20 to 40 percent B: (20 percent B) peak $1 < 3 < 2 < 4 < 5$; (40 percent B) peak $1 < 2 = 4 < 3 < 5$. Thus, by changing %B it is possible to change the order of peak elution for an "irregular" sample (but not a "regular" sample), and (more important) improve resolution (cf. Fig. 1.10c vs d). As will be seen in Chapter 2, a change in

20 CHAPTER 1 INTRODUCTION

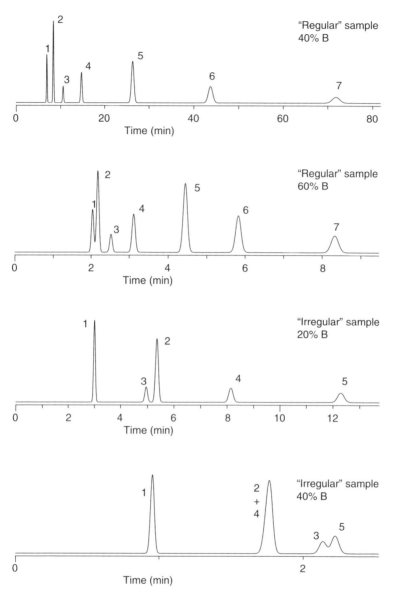

Figure 1.10 Separation of "regular" and "irregular" samples of Table 1.3 as a function of %B. Conditions: 150 × 4.6 mm C_{18} columns; 2.0 mL/min; other conditions in Table 1.3 or in figure.

gradient conditions can also result in dramatic changes in separation order for "irregular" samples (Figs 2.10 and 2.16), similar to the effects of a change in isocratic %B as in Figure 1.10(*c*, *d*). In later chapters, we will contrast the gradient separation of "regular" vs "irregular" samples when changing different experimental conditions (e.g., column length, flow rate, gradient time and range). *It is important to note that*

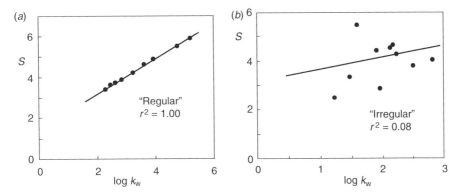

Figure 1.11 Plots of S vs log k_w for (a) "regular" sample and (b) "irregular" sample. Data of Table 1.3.

the "structural similarity" of the components of a sample, as it relates to sample "irregularity," is usually not obvious. As illustrated in the following chapters, sample "regularity" or "irregularity" is best identified on the basis of observed chromatographic behavior, rather than molecular structure; if there are noticeable changes in relative retention when only gradient time is changed, then the sample should be regarded as "irregular."

Plots of log k vs %B for individual compounds in a "regular" sample (as in Fig. 1.8a) exhibit slopes [values of $S = d(\log k)/d\phi$] that increase regularly for more retained compounds (larger values of k_w), while this is not the case for the "irregular" sample of Figure 1.8(b). This is illustrated in Figure 1.11 for the "regular" (a) and "irregular" (b) samples of Table 1.3. Values of S vs log k_w for the "regular" sample are highly correlated ($r^2 = 1.00$), while a similar plot for the "irregular" sample exhibits considerable scatter ($r^2 = 0.08$).

The specific "regular" and "irregular" samples of Table 1.3 will be used extensively in following chapters as representative examples of each sample type (all separations of these samples shown in this book are created by computer simulation). Most samples share the characteristics of both "regular" and "irregular" samples, resulting in an intermediate behavior as %B is changed in isocratic elution, or gradient conditions are changed for gradient separation. As a result, changes in isocratic %B or gradient time t_G often result in potentially useful changes in compound separation order or selectivity.

REFERENCES

1. R. L. M. Synge, *Discussions Faraday Soc.* 7 (1949) 164.
2. L. S. Ettre, *LCGC* 21 (2003) 458.
3. L. R. Snyder, *Chromatogr. Rev.* 7 (1965) 1.
4. J. W. Dolan and L. R. Snyder, *LCGC* 17 (1999) S17 (April supplement).
5. H. Xiao, X. Liang, and P. Lu, *J. Sep. Sci.* 24 (2001) 186.
6. G. A. Howard and A. J. P. Martin, *Biochem. J.* 46 (1950) 532.
7. C. Liteanu and S. Gocan, *Gradient Liquid Chromatography*, Wiley, New York, 1974.

8. P. Lebreton, *Bull. Soc. Chim. France* (1960) 2188.
9. L. E. Ettre, *LCGC* 20 (2002) 128.
10. W. E. Harris and H. W. Habgood, *Programmed Temperature Gas Chromatography*, Wiley, New York, 1967.
11. D. E. Bautz, J. W. Dolan, W. D. Raddatz, and L. R. Snyder, *Anal. Chem.* 62 (1990) 1560.
12. J. W. Dolan, L. R. Snyder, and D. E. Bautz, *J. Chromatogr.* 541 (1991) 21.
13. R. P. W. Scott and J. G. Lawrence, *J. Chromatogr. Sci.* 7 (1969) 65.
14. L. R. Snyder, *J. Chromatogr. Sci.* 8 (1970) 692.
15. K. Aitzetmüller, *J. High Resolut. Chromatogr.* 13 (1990) 375.
16. G. Hesse and H. Engelhardt, *J. Chromatogr.* 21 (1966) 228.
17. T. Greibokk and T. Andersen, *J. Chromatogr. A* 1000 (2003) 743.
18. P. Schoenmakers, in *Handbook of HPLC*, E. Katz, R. Eksteen, P. Schoemnakers, and N. Miller, eds, Marcel Dekker, New York, 1998, p. 193.
19. P. Jandera, *Adv. Chromatogr.* 43 (2004) 1.
20. T. Braumann, G. Weber, and L. H. Grimme, *J. Chromatogr.* 261 (1983) 329.
21. M. A. Quarry, R. L. Grob, and L. R. Snyder, *Anal. Chem.* 58 (1986) 907.
22. F. Leroy, B. Presle, F. Verillon, and E. Verette, *J. Chromatogr. Sci.* 39 (2001) 487.
23. S. R. Needham and T. Wehr, *LCGC Europe* 14 (2001) 244.
24. J. Ayrton, G. J. Dear, W. J. Leavens, D. N. Mallet, and R. S. Plumb, *J. Chromatogr. B* 709 (1998) 243.
25. L. R. Snyder, J. J. Kirkland, and J. L. Glajch, *Practical HPLC Method Development*, 2nd edn, Wiley-Interscience, New York, 1997.
26. K. H. Nelson and D. Schram, *J. Chromatogr. Sci.* 21 (1982) 218.
27. L. Hagdahl, R. J. P. Williams, and A. Tiselius, *Arkiv. Kemi* 4 (1952) 193.
28. J. Aurenge, *J. Chromatogr.* 84 (1973) 285.
29. E. Soczewiński and C. A. Wachtmeister, *J. Chromatogr.* 7 (1962) 311.
30. L. R. Snyder, *J. Chromatogr.* 13 (1964) 415, 15 (1964) 344.
31. L. R. Snyder and D. L. Saunders, *J. Chromatogr. Sci.* 7 (1969) 195.
32. L. R. Snyder, J. W. Dolan, and J. R. Gant, *J. Chromatogr.* 165 (1979) 3.
33. J. W. Dolan, J. R. Gant, and L. R. Snyder, *J. Chromatogr.* 165 (1979) 31.
34. L. R. Snyder and J. W. Dolan, *Adv. Chromatogr.* 38 (1998) p. 115.
35. P. Haber, T. Baczek, R. Kaliszan, L. R. Snyder, J. W. Dolan, and C. T. Wehr, *J. Chromatogr. Sci.* 38 (2000) 386.
36. T. H. Jupille, J. W. Dolan, L. R. Snyder, W. D. Raddatz, and I. Molnar, *J. Chromatogr. A* 948 (2002) 35.
37. Molnar, *J. Chromatogr. A* 965 (2002) 175.
38. P. L. Zhu, L. R. Snyder, J. W. Dolan, N. M. Djordjevic, D. W. Hill, L. C. Sander, and T. J. Waeghe, *J. Chromatogr. A* 756 (1996) 21.

CHAPTER 2

GRADIENT ELUTION FUNDAMENTALS

> There is something fascinating about science. One gets such wholesale returns of conjecture out of such a trifling investment of fact.
>
> —Mark Twain, *Life on the Mississippi*

In this chapter we will make use of the LSS model (Section 1.4.2) in order to (a) lay a foundation for understanding gradient elution, and (b) compare isocratic and gradient separation. Isocratic elution is simpler and more easily understood [1], so we will examine it first. An understanding of isocratic separation will also serve as a starting point for our following discussion of gradient elution. Finally, we will see that gradient and isocratic separations can be interpreted and developed in much the same way, once we see how to interpret gradient elution in terms of what we know about isocratic elution. Further details concerning the theory of gradient elution are provided in Chapter 9. *For an immediate practical application of gradient elution theory, see Chapter 3 after reading Section 2.1 on isocratic separation.*

2.1 ISOCRATIC SEPARATION

2.1.1 Retention

Band migration during isocratic elution was illustrated in Figure 1.5. The fractional migration of a band through the column (per column-volume V_m of mobile phase) is

$$R_F = 1/(1+k) \tag{2.1}$$

where k refers to the band (or peak) retention factor. The volume of mobile phase required to elute the peak from the column (the retention volume V_R) is then

$$V_R = V_m/R_F = V_m(1+k) \tag{2.2}$$

where V_m is the column dead-volume. Similarly, if we divide both sides of Equation (2.2) by the flow rate F, and define the retention time of a nonretained compound

---

High-Performance Gradient Elution. By Lloyd R. Snyder and John W. Dolan
Copyright © 2007 John Wiley & Sons, Inc.

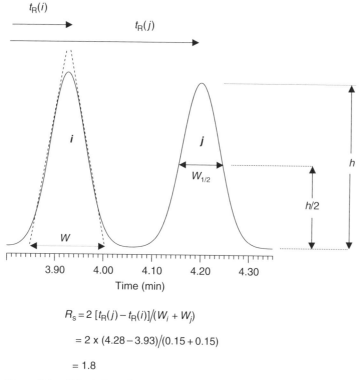

$$R_s = 2\,[t_R(j) - t_R(i)]/(W_i + W_j)$$
$$= 2 \times (4.28 - 3.93)/(0.15 + 0.15)$$
$$= 1.8$$

Figure 2.1 Calculation of retention time, peak width, and resolution in isocratic elution.

(the column dead time) as $t_0 = V_m/F$, the retention time $t_R = V_R/F$ is

$$t_R = t_0(1 + k) \tag{2.3}$$

From Equation (2.3), k can be calculated from an experimental value of t_R as

$$k = (t_R - t_0)/t_0 \tag{2.4}$$

Thus, for the two peaks shown in Figure 2.1, the retention times are $t_R(i) = 3.93$ min (peak i) and $t_R(j) = 4.20$ min (peak j), while the column dead time $t_0 = 1.54$ min (not shown). From Equation (2.4) we then obtain the retention factors for peaks i and j, respectively: $k_i = 1.55$ and $k_j = 1.73$. The column dead time t_0 is equal to the retention time of a nonretained compound, for example, thiourea (preferred), or uracil.

2.1.2 Peak Width and Plate Number

Narrow (taller) peaks favor improved separation and increased detection sensitivity. So-called *baseline peak width*, W, can be measured (Fig. 2.1) from tangents drawn to each side of the peak. Other measurements of peak width can be used, of which the *half-height* peak width, $W_{1/2}$, is more common and more convenient to measure. $W_{1/2}$ is measured half-way between the baseline and the top of the peak ($W = 1.70\ W_{1/2}$).

Retention time t_R and peak width W can be combined to define the plate number N, which is the usual measure of column *efficiency*, or the general ability of a column to separate different samples. The column plate number N is given as

$$N = 16(t_R/W)^2 \tag{2.5}$$

or

$$N = 5.54(t_R/W_{1/2})^2 \tag{2.5a}$$

The plate number for peak i in Figure 2.1 is $N = 16(3.93/0.15)^2 = 11,000$ plates [Equation (2.5)]. Values of N are roughly constant for the different peaks in an isocratic chromatogram, but vary with column dimensions and experimental conditions; see Section 9.5 for details. Symmetrical (Gaussian) peaks are assumed in the present book, unless noted otherwise; see Section 5.4.5 and Appendix III for a brief discussion of distorted peaks.

2.1.3 Resolution

The separation of two peaks i and j as in Figure 2.1 is usually described in terms of their *resolution* R_s

$$R_s = \text{(difference in retention times)/(average peak width)}$$
$$= 2[t_R(j) - t_R(i)]/(W_i + W_j) \tag{2.6}$$

W_i and W_j are the baseline widths W for peaks i and j, respectively. Better separation (increased resolution) results from a larger difference in peak retention times and/or narrower peaks. Accurate quantitative analysis based on a separation as in Figure 2.1 is favored by *baseline resolution*, where the valley between the two peaks returns to the baseline. For two peaks of comparable size, baseline resolution corresponds to $R_s > 1.5$. For preparative separations (Chapter 7), baseline resolution also allows a near-complete recovery of each peak in ~100 percent purity. A common goal of HPLC method development is the separation of every peak of interest from other peaks with $R_s \geq 2$, corresponding to a 1 : 3 safety factor. The goal of $R_s \geq 2$ takes into account minor peak tailing, peaks of dissimilar size, and the usual slow deterioration of the column over time (with decrease in N) – all of which decrease R_s. The resolution of two peaks can be varied by changes in either (*a*) column efficiency (the value of N) or (*b*) peak relative retention, that is, the difference in retention times $[t_R(j) - t_R(i)]$.

When more than two peaks are to be separated, the goal is usually a resolution $R_s \geq 2$ for the least-well-separated peak-pair. This peak-pair is referred to as the *critical* peak-pair, and its resolution is referred to as the *critical resolution* of the separation, for example, overlapping peak-pair $1 + 2$ in Figure 2.2(*a*), for which $R_s < 0.5$ (meaning that two peaks look like a single peak). Method development usually strives for an acceptable resolution of the critical peak-pair, which then means an adequate separation of other peaks as well. In this book, *when we refer*

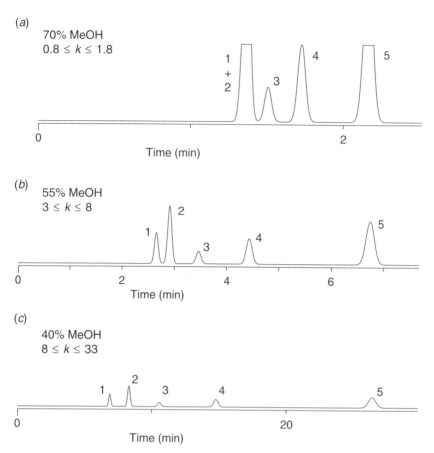

Figure 2.2 Isocratic separation of a "regular" sample as a function of mobile phase %B. Sample: compounds 1–5 of "regular" sample of Table 1.3. Conditions: methanol (B)–water (A) mobile phase; 150 × 4.6 mm C_{18} column (5 μm particles); 2.0 mL/min; 30°C. Note that peak heights areas are as observed (*not* normalized to 100 percent for tallest peak).

to the resolution of the total sample (for either isocratic or gradient elution), we mean its critical resolution R_s.

For method development purposes, it is convenient to derive an alternative, approximate expression for resolution from Equations (2.3), (2.5), and (2.6) (assuming equal widths for the two peaks)

$$R_s = \underset{(a)}{(1/4),} \underset{(b)}{[k/(1+k)],} \underset{(c)}{(\alpha - 1) \text{ and } N^{1/2}} \quad (2.7)$$

Here, resolution is expressed as a function of the retention factor k for the first peak i (term *a*), the separation factor α (term *b*), and column efficiency N (term *c*). The separation factor α (a measure of so-called separation *selectivity* or relative retention) is defined as

$$\alpha = k_j/k_i \qquad (2.8)$$

The quantities k_i and k_j are the values of k for adjacent peaks i and j, as in Figure 2.1. The value of α for the separation of Figure 2.1 is therefore equal to $(1.73/1.55) = 1.116$. For the separation of Figure 2.1, Equation (2.7) gives $R_s = (1/4)(1.55/2.55)(1.116 - 1)(11{,}000^{1/2}) = 1.8$, which can be compared to the value calculated from Equation (2.6) in Figure 2.1 ($R_s = 1.8$). Note that Equation (2.7) is an approximate relationship, which becomes less accurate for values of $\alpha \geq 1.2$; it will be used mainly for a qualitative understanding of how resolution depends on various experimental conditions. When values of R_s are reported in this book, they are calculated from Equation (2.6).

2.1.4 Role of Separation Conditions

The development of an isocratic HPLC method proceeds by systematically adjusting experimental conditions until adequate separation is achieved (preferably, with a critical resolution $R_s \geq 2$). Equation (2.7) provides a useful guide for isocratic method development, because each of terms a–c can be controlled by varying certain separation conditions. Usually an appropriate value of k [term a of Equation (2.7)] is selected first, followed by optimizing selectivity [α, or term b of Equation (2.7)]. Finally, the column plate number N can be adjusted [term c of Equation (2.7)] for a best compromise between increased resolution or decreased run time.

2.1.4.1 Optimizing Retention [Term a of Equation (2.7)] The first step in isocratic method development is to achieve values of k for the sample that are neither too small nor too large. Sample retention k in isocratic elution is usually controlled by varying mobile phase composition (%B). The usual goal is $k < 10$ for all peaks, because this corresponds to narrow, taller peaks (for better detection), and short run times (so that more samples can be analyzed each day). Values of $k \ll 1$ result in small values of term a of Equation (2.7) and generally poor resolution, as well as the possible overlap of analytes with matrix interferences that typically accumulate near t_0. Therefore, $1 \leq k \leq 10$ is usually the goal for all peaks. However, at the option of the chromatographer, it is possible to expand this preferred retention range somewhat, for example, $0.5 \leq k \leq 20$. Alternatively, regulatory agencies may recommend $k \geq 2$ for all peaks of interest in the chromatogram [2], in order to minimize possible interference from sample excipients or other non-assayed peaks that elute near t_0.

The effect of a change in %B is illustrated in Figure 2.2 for the isocratic separation of compounds 1–5 of the "regular" sample of Table 1.3. With 70 percent B as mobile phase (*a*), values of k are small ($0.8 \leq k \leq 1.8$), and as a result early peaks 1 and 2 are unresolved. With 40 percent B [(*c*); $8 \leq k \leq 33$], the sample is well resolved, but run time is excessive and later peaks are wide, with reduced detection sensitivity. An intermediate mobile phase [55 percent B (*b*)] provides an acceptable range in k ($3 \leq k \leq 8$) with a reasonable compromise among resolution, detection sensitivity and run time.

Changes in sample retention with change in %B (as in Fig. 2.2) are governed by the empirical relationship

$$\log k = \log k_w - S\phi \tag{2.9}$$

where ϕ is the volume-fraction of the B-solvent (equal to $0.01 \times \%B$), k_w is the extrapolated value of k for compound X with water as the mobile phase (i.e., $\phi = 0$), and S is a constant for compound X when only ϕ is varied. For "small" molecules with molecular weights of 100–500, $S \approx 4$ [3]. An increase in ϕ by 0.1 unit (e.g., a change in the mobile phase by 10 percent B) will therefore result in an average increase in k for all peaks in the sample by a factor of $10^{(0.1 \times 4)}$, or about 2.5-fold. The constant $S = d(\log k)/d\phi$ can differ somewhat for two adjacent peaks in an "irregular" sample (Section 1.6.2), meaning that peak spacing (selectivity) may also vary with changes in %B (some examples are given in the following sections). Section 9.4 provides a further discussion of Equation (2.9).

2.1.4.2 Optimizing Selectivity α [Term b of Equation (2.7)] For a further improvement of the separation, peak spacing (selectivity or α) is next adjusted by varying conditions such as mobile phase composition or temperature (see Table 3.4 in Chapter 3 and the related text for additional options). Figure 2.3 provides an illustration for the separation of six compounds (1–6) from the "irregular" sample of Table 1.3. For a mobile phase of 29 percent B and a temperature of 36°C [Fig. 2.3(a)], peak-pairs 3–4 and 5–6 are poorly resolved ($R_s = 0.6$–0.7). An *increase* in %B (to 32 percent B, Fig. 2.3b) results in a better separation of peaks 3 and 4 (with a reversal of peak position), but the poor resolution of peaks 5 and 6 is unchanged. The better separation of peaks 3 and 4 in (*b*) vs (*a*) may appear counter-intuitive, since resolution for a "regular" sample (as in Fig. 2.2) *decreases* for an increase in %B. However, a change in %B for an "irregular" sample can result in changes in α [term *b* of Equation (2.7)] which may more than compensate for the usual decrease in R_s with increase in %B due to term *a* of Equation (2.7); see Section 2.2.3.1 for a detailed analysis.

If temperature is increased from 36 to 40°C (Fig. 2.3c, same %B as Fig. 2.3a), an improved separation of both peak-pairs (3–4 and 5–6) results ($R_s = 1.1$ for critical peak-pair 5–6). Simultaneous adjustment of both %B and temperature yields the final separation of Figure 2.3(*d*) ($R_s = 2.6$), in this case with better resolution and little increase in run time compared with the starting conditions (Fig. 2.3a). The goal of improving selectivity (as in the examples of Fig. 2.3) can be either an increase in resolution, a decrease in run time, or (usually) both. The selection of conditions for acceptable separation will usually emphasize changes in selectivity, because of its major effect on resolution.

2.1.4.3 Optimizing the Column Plate Number N [Term c of Equation (2.7)] When selectivity has been adjusted for optimum peak spacing and maximum sample resolution, an adequate separation will often result. However, a further improvement in separation is possible by a change in the column plate number N.

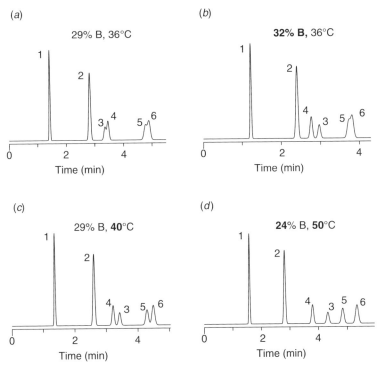

Figure 2.3 Effect of mobile phase %B and/or temperature on the isocratic separation of an "irregular" sample. Sample: compounds 1–6 of "irregular" sample of Table 1.3. Conditions: mobile phase containing acetonitrile (B)–pH 2.6 buffer (A) (see Table 1.3 for the buffer); 150 × 4.6 mm C_{18} column (5 μm particles); 2.0 mL/min; see figure-text for values of %B and temperature (changed conditions in bold in b–d). Note that peak heights are normalized to 100 percent for tallest peak in each chromatogram.

N can be varied by changes in *column conditions*: column length, flow rate, and/or particle size. Note that relative retention and peak spacing (values of k and α) should remain the same when only column conditions are changed in isocratic separation (Fig. 2.4), meaning that the optimized peak spacing achieved previously by varying α [term b of Equation (2.7)] will not be compromised by changes in column conditions. The chromatogram of Figure 2.4(a) is taken from Figure 2.3(c), based on a 150 mm column of 5 μm particles and a flow rate of 2.0 mL/min. When the initial critical resolution is $R_s < 2$ (as in Fig. 2.4a), resolution can be increased moderately by an increase in column length, decrease in flow rate, or decrease in particle size. Figure 2.4(b–d) shows the resulting changes in separation when column length (b), flow rate (c), or particle size (d) are varied, so as to increase N and resolution. These examples are intended as illustrations of the effects of different column conditions on resolution and run time, but it can be seen that a desired resolution of $R_s \geq 2$ is not achieved for this sample. This in turn suggests further changes in conditions that affect α (as in Fig. 2.3d), rather than trying to achieve a large relative increase in resolution by column conditions alone.

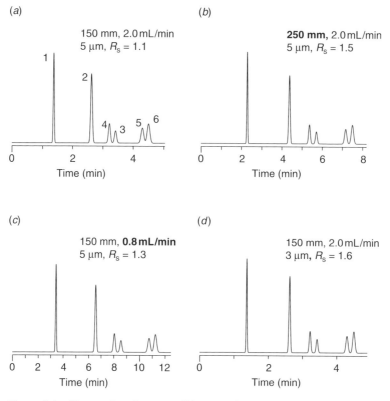

Figure 2.4 Changes in column conditions can increase resolution. "Irregular" sample and conditions as in Figure 2.3(c) (24 percent B, 50°C), except where noted otherwise (changed condition in bold in b–d). See text for details. Note that peak heights are normalized to 100 percent for tallest peak.

When $R_s \gg 2$, run time can be reduced substantially by sacrificing excess resolution, while maintaining $R_s \geq 2$; this is most easily achieved by reducing column length and/or increasing flow rate. This is illustrated in Figure 2.5, starting with the separation of Figure 2.3(d), which is repeated as Figure 2.5(a). The excess starting resolution ($R_s = 2.6$) in Figure 2.5(a) can be traded for a shorter run time by decreasing column length from 150 to 100 mm (Fig. 2.5b). Run time is shortened by a third, from 5.4 to 3.6 min, while resolution is decreased ($R_s = 2.1$), but is still acceptable. A further shortening of run time is possible by increasing flow rate from 2.0 to 4.0 mL/min (Fig. 2.5c). Run time is reduced by another 50 percent, to 1.8 min. A slight further reduction in resolution occurs ($R_s = 1.9$), but this should prove acceptable when a very short run time is important. Note that the column pressure changes with change in column length or flow rate: (a) 900 psi, (b) 600 psi, and (c) 1200 psi. In this case, the higher pressure in (c) is of little concern.

A change in column size, and especially particle size, may occasionally result in unintended changes in values of α, because of batch-to-batch differences in the stationary phase and associated changes in column selectivity. Random changes

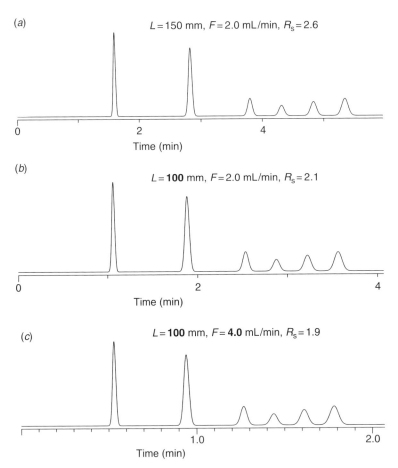

Figure 2.5 Changes in column conditions can be used to decrease run time. "Irregular" sample and conditions as in Figure 2.3(*d*) (29 percent B, 40°C), except where noted otherwise (changed condition in bold in *b* and *c*). See text for details. Note that peak heights are normalized to 100 percent for tallest peak.

in selectivity, after selectivity has been optimized, are generally undesirable, because they can more than cancel the intended benefit of a change in column. However, many manufacturers correctly claim that column selectivity should not vary significantly from batch to batch of columns which vary in column size or particle diameter (e.g., [4, 5] and Fig. 5.21 of [1]).

2.2 GRADIENT SEPARATION

For many chromatographers who have experience with isocratic elution, separations by gradient elution may appear initially puzzling. When compared with previous isocratic separations, a change in gradient conditions often results in unexpected changes in the chromatogram. Fortunately, the use of the LSS model can help us to more reliably anticipate the consequences of a change in gradient conditions, by demonstrating the essential similarity of isocratic and gradient separation. Predictions for gradient elution also become more reliable when we distinguish "regular" from "irregular" samples (Section 1.6). "Regular" samples are more likely to consist of compounds of similar molecular structure (e.g., homologs), whereas "irregular" samples are often more diverse in terms of their molecular composition. However, it is usually not possible to recognize a "regular" or "irregular" sample from its molecular composition; most samples exhibit an intermediate separation behavior (i.e., plots of S vs log k_w that are *slightly* scattered, vs the two extreme examples of Fig. 1.10). The following examples of the effects on separation of changes in gradient conditions will be based on both the "regular" and "irregular" samples of Table 1.3.

Reliable, quantitative relationships based on the LSS model and Equation (2.9) can be derived for gradient elution separation (Chapter 9). In the following discussion, several of these equations will be presented for the interpretation and further improvement of an initial gradient separation. Our "regular" sample can serve as an example for a *general* discussion of gradient elution as a function of experimental conditions; separations of the "regular" sample will therefore be discussed first. As we will see, the separations of "irregular" samples as a function of conditions may at first glance appear similar, yet differ in apparently minor (but potentially important) ways when compared with the behavior of "regular" samples.

Finally, an understanding of gradient separation as a function of different experimental conditions is most easily acquired by beginning with approximate, general relationships which hold for all samples and gradients (best illustrated with the "regular" sample). Deviations from these general relationships can then be recognized for certain samples ("irregular" vs "regular"), and for special cases: (a) sample bands that are *not* strongly retained at the start of the gradient, as opposed to the more common case where bands are initially strongly retained, and (b) separations where there is an initial isocratic hold [either because of an intentional gradient delay (Fig. 1.4d) or an appreciable equipment hold-up or "dwell volume" (see below)]. We believe that gradient separation can be interpreted most easily by successive refinements of an initial (simplified) picture which assumes a linear-gradient separation of a "regular" sample, strongly retained bands at the start of the gradient, and zero equipment dwell volume. This approach is followed in the treatment below.

2.2.1 Retention

Retention time t_R for a given peak in liner gradient elution can be derived for LSS gradients (Section 9.1.1):

$$t_R = (t_0/b) \log(2.3k_0 b + 1) + t_0 \qquad (2.10)$$

Here, k_0 is the value of k at the start of the gradient (equal to k_w for a gradient that starts at 0 percent B), and b is the *intrinsic gradient steepness* (Section 1.4.2):

$$b = V_m \Delta \phi S/(t_G F) \qquad (2.11)$$
$$= t_0 \Delta \phi S/t_G \qquad (2.11a)$$

V_m is the column dead volume (mL, proportional to column length and internal diameter squared), and $\Delta\phi$ is the change in ϕ during the gradient, that is, $\Delta\phi = 1$ for a 0–100 percent B gradient. S is defined by Equation (2.9), t_G is gradient time (min), F is the mobile phase flow rate (mL/min), and the column dead time $t_0 = V_m/F$ (min). The value of k for a given compound at the start of the gradient can be obtained from Equation (2.9):

$$\log k_0 = \log k_w - S\phi_0 \qquad (2.11b)$$

where ϕ_0 is the value of ϕ at the start of the gradient (see Fig. 1.4g for an illustration).

Equation (2.10) ignores the equipment hold-up volume ("dwell" volume, V_D). V_D is the volume of the gradient mixer plus the additional equipment volume in the flow path between the mixer and the column inlet; see Figures 4.1 and 4.2, and Sections 2.3.6.1, 4.3.1.2, and 9.1.1.2 for details. Values of k_0 are usually >10, which leads to a simplified form of Equation (2.10) that also takes dwell volume into account:

$$t_R \approx (t_0/b) \log (2.3 k_0 b) + t_0 + t_D \qquad (2.12)$$

The "dwell time" $t_D = V_D/F$. Equation (2.12) is adequate for most practical applications; for a more general equation that is accurate for both large and small values of k_0, see Section 9.1.1.2.

Retention in gradient elution varies with gradient time t_G in similar fashion to isocratic retention varying with %B; this is seen in Figure 2.6 for the separation of the "regular" sample, which can be compared with Figure 2.2 for isocratic separation. Thus, larger values of t_G (corresponding to smaller values of %B in isocratic elution, for an increase in k or k^*) lead to longer run times, better overall resolution, and wider, shorter peaks. This apparent similarity of isocratic and gradient elution can be better seen by comparing k in isocratic elution with the *median* or *effective* value of k in gradient elution (defined as k^*, the *gradient retention factor*). k^* is the value of k when the band is at the column midpoint (Fig. 1.7); note the values of k^* shown in Figure 2.6.

Values of k^* provide a convenient basis for understanding and controlling resolution in gradient elution, because the separation of two adjacent peaks will be similar in both isocratic and gradient elution, when their average values of k and k^* are equal. Likewise, similar changes in either k (isocratic) or k^* (gradient) as a result of a change in conditions will result in similar changes in peak height and resolution. Thus, just as in isocratic elution it is advantageous to maintain k within certain limits (e.g., $1 \leq k \leq 10$), for similar reasons $1 \leq k^* \leq 10$ is recommended in gradient elution (compare Figs 2.2 and 2.6).

Values of k^* are given [Equations (9.14) and (9.16)] by

$$k^* = 1/1.15b$$
$$= t_G F/(1.15 V_m \Delta\phi S) \qquad (2.13)$$

A qualitative justification of Equation (2.13) was provided in Section 1.4.2; for a quantitative derivation, see Section 9.1. For many samples, values of S increase somewhat with increasing retention or values of k_w [6]; see also Section 9.4.1.

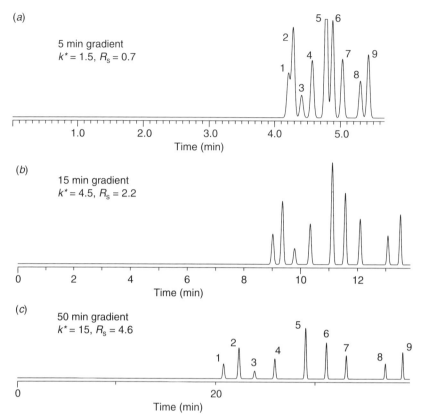

Figure 2.6 Separations of the "regular" sample for different gradient times t_G. Sample numbering as in Table 1.3 (note that separation order does not change as t_G is varied). Conditions: 150 × 4.6 mm C_{18} column (5 μm particles), 2.0 mL/min, 30°C, 0–100 percent B in (a) 5 min, (b) 15 min, (c) 50 min. Note that peak heights are as observed (*not* normalized to 100 percent for tallest peak).

This is especially the case for homologous mixtures and other "regular" samples (e.g., Fig. 1.8a). However, values of S for other samples often do not vary much from the first to last peaks in the chromatogram (e.g., the "irregular" sample of Fig. 1.8b). A useful first approximation is to assume that values of S can be regarded as similar (but not identical) for each compound in a typical sample. *This means that values of b and k^* will be approximately constant for different compounds in a linear-gradient separation* (i.e., for specific values of gradient time t_G, flow rate F, column size V_m, and gradient range $\Delta\phi$). However, k^* can vary in different runs for changes in t_G, F, V_m, or $\Delta\phi$ [Equation (2.13)].

2.2.1.1 Gradient and Isocratic Separation Compared for "Corresponding" Conditions

We will next show that isocratic and gradient elution can provide similar resolution for a given sample, when the separation conditions are "comparable." By comparable or "corresponding" conditions, we mean that the column, temperature, and mobile phase are the same – except that %B varies in gradient elution and is constant in isocratic separation. A further requirement of "corresponding" separations by gradient and isocratic elution is that the average value of k for the isocratic separation is approximately equal to k^* for the gradient run. More specifically, average values of k and k^* should be equal for the critical peak pair, if critical resolution is to be similar for the two separations.

It will prove instructive to compare plots of $\log k$ vs %B (or ϕ) with resulting isocratic chromatograms for different values of %B; see the hypothetical separation of a "regular" sample in Figure 2.7. For a given value of %B, corresponding values of k for solutes 1–3 are located on a vertical line in Figure 2.7(a) (e.g., the two dashed lines for 30 and 50 percent B). The separations of this sample for 30 and 50 percent B, respectively, are shown in Figure 2.7(b and c). A larger value of %B in (c) vs (b) results in shorter retention times, narrower (taller) peaks, and a decrease in resolution (as in Fig. 2.2).

We can restate Equation (2.9) (isocratic separation) for gradient elution with "corresponding" conditions as

$$\log k^* = \log k_w - S\phi^* \qquad (2.14)$$

Here, ϕ^* is the median value of ϕ for mobile phase in contact with the band when it reaches the column midpoint. Values of k^* and ϕ^* in gradient elution have the same significance as values of k and ϕ in isocratic elution. For the same sample and "corresponding" conditions, the quantities $\log k_w$ and S in Equations (2.9) and (2.14) each have the same value for a given sample compound (Section 9.1.1.1). Thus, the same $\log k$ vs ϕ plots apply (for a given sample and same conditions other than %B) for both isocratic and gradient elution, when only ϕ is varied; compare the identical plots of $\log k$ vs ϕ and $\log k^*$ vs ϕ^* in Figures 2.7(a) and 2.8(a), respectively.

A major difference between isocratic and gradient elution is that isocratic values of k vary for different peaks in the chromatogram, while values of k^* in gradient elution remain approximately the same [Equation (2.13); assumes $S \approx$ constant for all peaks, the usual case]. Thus, isocratic separation corresponds to constant ϕ for all peaks, whereas gradient separation corresponds approximately to constant k^* (but note the minor qualification of Appendix I). This is illustrated for gradient elution (Fig. 2.8a) by the horizontal dashed lines labeled "$k^* = 3$" and "$k^* = 10$," corresponding to $t_G = 10.5$ and 35 min, respectively (see resulting chromatograms in Fig. 2.8b and c). In both Figures 2.7 (isocratic) and 2.8 (gradient), an increase in k or k^* results in longer run times, wider and shorter peaks, and (for "regular" samples as in Figs 2.7 and 2.8) improved resolution. Note again in the comparisons of Figures 2.7 and 2.8 that isocratic separation corresponds to constant %B (vertical dashed line in Fig. 2.7a), while gradient separation corresponds to constant k^* (horizontal dashed line in Fig. 2.8a).

The retention of a peak can also be expressed as the value of ϕ at elution (ϕ_e):

$$\phi_e = \phi_0 + (\Delta\phi/t_G)(t_R - t_0 - t_D) \qquad (2.15)$$

A value of ϕ_e can in turn be used to calculate a value of ϕ^* in Equation (2.22a) below.

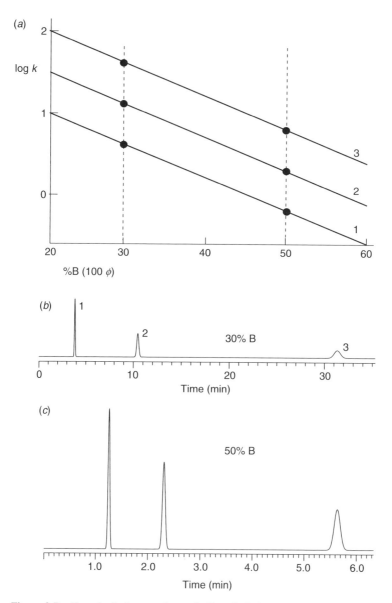

Figure 2.7 Hypothetical separations of a "regular" three-component mixture by isocratic elution as a function of %B; assumes 150 × 4.6 mm C_{18} column (5 μm particles) with a flow rate of 2.0 mL/min. Sample description: log k_w = 1.8, 2.3, and 2.8 for compounds 1, 2, and 3, respectively; S = 4.0 for each compound; (a) plots of log k vs %B for compounds 1–3; (b) separation for 30 percent B; (c) separation for 50 percent B. See text for details. Note that actual peak heights are shown (*not* normalized to 100 percent for tallest peak).

2.2 GRADIENT SEPARATION

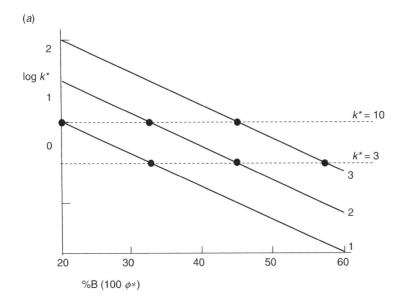

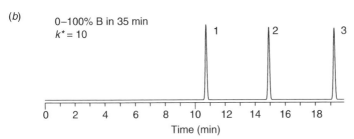

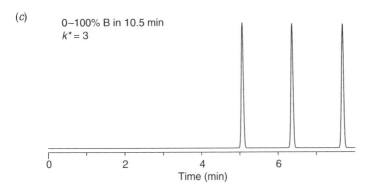

Figure 2.8 Hypothetical separation of the sample of Figure 2.7 as a function of gradient time; other conditions constant and equipment hold-up volume equal zero. (*a*) Plots of log *k* vs %B for compounds 1–3; (*b*) separation for 35 min gradient; (*c*) separation for 10.5 min gradient. See text for details. Note that peak heights are as observed (*not* normalized to 100 percent for tallest peak).

2.2.2 Peak Width

Peak width in isocratic elution can be obtained from Equations (2.3) and (2.5):

$$\text{(isocratic)} \quad W = 4N^{-1/2}t_0(1+k) \tag{2.16}$$

A similar expression applies for gradient elution, where k in Equation (2.16) is replaced by the value of k at the time that the band elutes from the column (k_e; see Section 9.1.2):

$$\text{(gradient)} \quad W \approx 4N^{-1/2}t_0(1+k_e) \tag{2.17}$$

It can be shown that $k_e = k^*/2$ (Section 9.1.1), so that

$$\text{(gradient)} \quad W \approx 4N^{-1/2}t_0(1+0.5k^*) \tag{2.18}$$

or

$$\text{(gradient)} \quad W \approx 4N^{-1/2}t_0(1+[1/2.3b]) \tag{2.19}$$

Equation (2.19) overlooks "gradient compression" (Section 9.1.2.1), which predicts a further decrease in values of W for gradient elution (especially for $b > 0.5$ or $k^* < 2$). However, experimental values of W in gradient elution usually fall within ± 10–20 percent of values predicted by Equation (2.19) [7, 8], so that Equation (2.19) is adequately reliable for practical application.

It is interesting to compare Equation (2.18) (gradient) with Equation (2.16) (isocratic). If isocratic and gradient retention are made equal for a single peak (i.e., $k = k^*$ in "corresponding" separations), peaks in gradient elution are predicted to be narrower than in isocratic elution, by the factor $(1 + 0.5k^*)/(1 + k^*)$. Thus, for large values of k^*, peaks in gradient elution will be about half as wide as peaks in isocratic elution. This suggests that narrower peaks and improved detection sensitivity are possible in gradient vs isocratic elution (other factors equal), especially for peaks with larger k and k^*. However, because baseline noise in gradient elution is often greater than in isocratic elution, detection sensitivity in gradient elution is usually no better (and often worse) than in isocratic elution. On the other hand, gradient elution provides similar values of peak width for *all* peaks in the chromatogram, which normally means increased detection sensitivity for later peaks (where typically $k \gg k^*$).

Note also that an experimental value of N in gradient elution cannot be calculated by the usual expression for isocratic separation [Equation (2.5)] [9]. If Equation (2.5) is used for gradient elution, resulting values of N are usually much too large (e.g., $N > 100{,}000$) and can vary by 10-fold or more for different peaks in the chromatogram. Values of N in gradient elution should instead be determined from Equation (2.18), which rearranges to

$$N \approx 16[t_0(1+0.5k^*)/W]^2 \tag{2.20}$$

Equation (2.20) requires a value of k^* for the estimation of N in gradient elution [Equation (2.13)]. Since peak widths and values of k^* are approximately constant for different peaks in a gradient run, values of N calculated from

Equation (2.20) will also tend to be similar. In practice, it is more convenient to monitor peak width (rather than N) over time (as a measure of column aging), since N is proportional to $1/W$.

2.2.3 Resolution

> And thus the native hue of resolution is sicklied o'er with the pale cast of thought.
>
> —William Shakespeare, *Hamlet*

Resolution in gradient elution can be measured in the same way as for isocratic separation [Equation (2.6) and Fig. 2.1]. An equation for gradient resolution that corresponds to Equation (2.7) for isocratic elution can also be derived (Section 9.1.3):

$$R_s = (1/4)\,\underbrace{[k^*/(1+k^*)]}_{(a)}\,\underbrace{(\alpha^*-1)}_{(b)}\,\underbrace{N^{*1/2}}_{(c)} \qquad (2.21)$$

Here, α^* refers to the median value of α (for when a critical band-pair reaches the midpoint of the column); for bands i and j, $\alpha^* = k_j^*/k_i^*$. The median plate number N^* in gradient elution will be about the same as N for isocratic elution, when $k = k^*$ (Section 9.5). Resolution in gradient and isocratic elution can be compared in terms of Equations (2.21) and (2.7) for "corresponding" separations, that is, where the sample and all conditions except %B are kept the same for both isocratic and gradient elution, and $k = k^*$. For "corresponding" separations, the resolution of any two (e.g., "critical") adjacent peaks should be approximately the same for both isocratic and gradient elution, when the average values of k (isocratic) and k^* (gradient) are equal for the peak pair (Section 9.1.3).

"Corresponding" separations by isocratic and gradient elution can also be expected to be similar for the entire sample, when the range in isocratic k is sufficiently narrow, for example, $1 < k < 10$. This similarity of corresponding separations by isocratic and gradient elution is suggested by the examples of Figures 2.7 and 2.8 (for a "regular" sample), which further implies that *method development in gradient elution can be carried out in essentially the same way as for isocratic elution* [1]. Thus, resolution in either isocratic or gradient elution should be affected in the same way by changes in the column, mobile phase, or temperature, when values of k and k^* are approximately the same for both isocratic and gradient separation. Consequently, whatever changes in conditions prove successful in isocratic method development should also be applicable to gradient method development, for example, changes in the mobile phase, temperature, flow rate, or column. Finally, the same approach used to develop an isocratic separation (Section 2.1.4) can be used for the development of a gradient method (Section 3.3): First optimize k^*, then α^*, and finally N^*.

For the case of "irregular" samples, there is also a similarity of isocratic and gradient elution for "corresponding" conditions – but with some minor differences, compared with the separation of "regular" samples as in Figures 2.7 and 2.8. The separations of Figures 2.9 and 2.10 provide respective examples of the isocratic and gradient separation of a narrow-range "irregular" sample ("irregular"

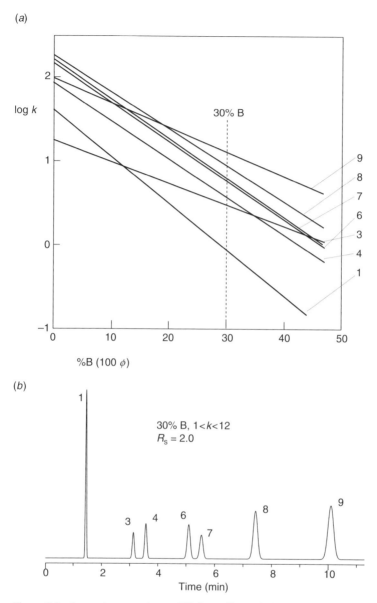

Figure 2.9 Isocratic separation vs %B for an "irregular" sample. Sample: compounds 1, 3, 4, 6–9 of "irregular" sample of Table 1.3. Conditions: 150 × 4.6 mm C_{18} column (5 μm particles), 30°C, 2.0 mL/min flow rate, acetonitrile–buffer mobile phases; (b) isocratic separation of sample for 30 percent B.

compounds 1, 3, 4, 6–9 of Table 1.3) and "corresponding" conditions. For the isocratic separations of Figure 2.9, near-optimal peak spacing, maximum resolution, and acceptable retention (approximately $1 \leq k \leq 10$) is suggested for an intermediate mobile phase composition: 30 percent B (dashed vertical line; $R_s = 2.0$). The separation of the same sample by gradient elution is illustrated in Figure 2.10,

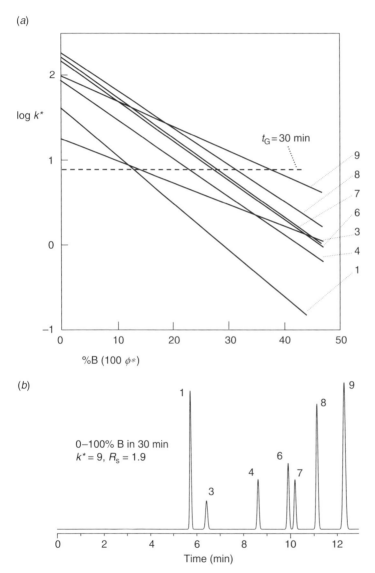

Figure 2.10 Gradient separation as a function of gradient time t_G and k^* for an "irregular" sample. Sample and conditions as in Figure 2.9. (*a*) Plots of log k vs %B for different sample compounds; (*b*) gradient separation for 0–100 percent B in 30 min.

where the best separation (30 min gradient, $k^* = 9$) provides a similar resolution: $R_s = 1.9$. The gradient separation of Figure 2.10 is seen to be comparable to the isocratic separation of Figure 2.9, although there are some differences in peak spacing for "noncritical" peak pairs (note the relative position of peak 3 vs peaks 1 and 4 in the two chromatograms).

On the basis of unreported studies [10] that comprised a large number of narrow-range samples, we conclude that the maximum resolution that is achievable for isocratic vs gradient separations (same sample, "corresponding" conditions, $1 \leq k \leq 10$ and $1 \leq k^* \leq 10$) is usually similar, sometimes favoring isocratic elution, sometimes favoring gradient elution. However, isocratic separation tends to give better resolution for a narrower range in k (smaller ratio k_z/k_a, where k_a is the value of k for the first peak in the chromatogram, and k_z is the value for the last peak). The latter observation supports the general rule that isocratic elution is preferred for narrow-range samples, while gradient elution is usually necessary for wide-range samples.

2.2.3.1 Resolution as a Function of Values of S for Two Adjacent Peaks ("Irregular" Samples)
Changes in relative retention and resolution for "irregular" samples, as a result of deliberate changes in isocratic %B or gradient time t_G, represent an important (and sometimes unappreciated) feature of both isocratic and gradient elution. Similar changes of relative retention can occur in gradient elution when flow rate or column size are varied without concomitant changes in gradient time, so that k^* varies. For this reason, the present section will examine this phenomenon in detail. For a practical summary of such changes in relative retention with gradient conditions, see Figure 3.9 and the accompanying discussion.

In Figures 2.7 and 2.8 for the separation of a "regular" sample (in this case with equal values of S for all peaks), a change in %B (k) or gradient time (k^*) does not change the relative retention of the sample, that is, the order of retention is always $1 < 2 < 3$, and the relative spacing of the peaks stays the same. In Figures 2.9 and 2.10 for the "irregular" sample (where the slopes S of plots of log k vs ϕ vary), relative retention and elution order change markedly for different values of ϕ or t_G. This can be seen for isocratic separation in Figure 2.9 by shifting the 30 percent B (dashed) line right or left, and noting the resulting changes in elution order (relative values of k for each compound) as %B is varied. Similarly, in Figure 2.10 the (dashed) line for $t_G = 30$ min can be shifted up or down, corresponding to changes in gradient time – again, with changes in retention order. The variation of retention order for this ("irregular") sample with gradient time is further illustrated in Figure 2.11. As gradient time t_G increases from (a) to (c), peaks 3 and 9 (with flatter slopes, or smaller values of S, in Fig. 2.10) are seen to move toward the front of the chromatogram, relative to other peaks.

A closer look at changes in relative retention for a change in %B or gradient time is provided by Figure 2.12. In Figure 2.12(a), isocratic plots of log k vs ϕ are shown for two hypothetical peaks, i and j (identical gradient plots result for log k^* vs ϕ^* for these two "corresponding" separations). The slope for peak i is steeper, so $S_i > S_j$. For mobile phase compositions of 30 and 50 percent

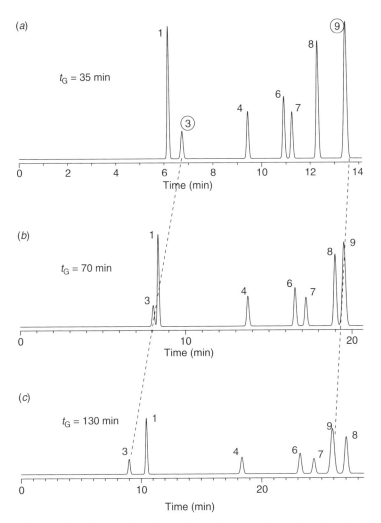

Figure 2.11 Changes in peak spacing with change in gradient time for "irregular" sample of Figure 2.9. Same conditions as in Figure 2.10. Note that peak heights are as observed (*not* normalized to 100 percent for tallest peak).

B, corresponding to a decrease in k, the relative retention of the two compounds is observed to reverse (Fig. 2.12b). Similarly, for gradient times of 25 and 5 min, a similar decrease in k^* and reversal in peak position occurs (Fig. 2.12c). The relative positions of the two peaks in either isocratic or gradient elution will be similar (as will their resolution), when the average value of k^* for the two peaks equals the average value of k. The average value of k^* and k for the 30 percent B and $t_G = 25$ min runs is indicated by a solid circle in Figure 2.12(a), as is the average value of k for the 50 percent and $t_G = 5$ min runs. Since the average values of k and k^* are approximately equal in each case, the resulting resolution for these two pairs

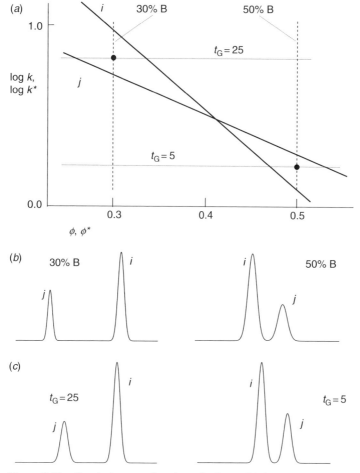

Figure 2.12 Resolution as a function of values of S for two adjacent peaks i and j: comparison of isocratic and gradient elution for this "irregular" sample and "corresponding" conditions. Hypothetical separations: (*a*) log k–ϕ (or log k^*–ϕ^*) plots for "corresponding" isocratic and gradient separations; (*b*) isocratic separations for 30 and 50 percent B mobile phases; (*c*) gradient separations for 25 and 5 min gradients. See text for details. Note that peak heights are normalized to 100 percent for tallest peak.

of "corresponding" runs (30 percent B and $t_G = 25$; 50 percent B and $t_G = 5$) are each quite similar.

For the isocratic elution of two compounds i and j (with values of k_w and S indicated by subscripts i and j), the separation factor α as a function of ϕ is given as

$$\log \alpha = \log (k_j/k_i) = (\log k_j - \log k_i)$$

which with Equation (2.9) for peaks i and j gives

$$\log \alpha = (\log k_{wj} - \log k_{wi}) - (S_j - S_i)\phi \tag{2.22}$$

where k_{wi} and k_{wj} refer to values of k_w for peaks i and j, respectively; S_i and S_j refer to values of S for peaks i and j. If $S_i \neq S_j$, the value of α will change with change in ϕ [or %B; Equation (2.22)]. That is, selectivity will vary with %B, and this variation will be greater with greater difference in values of S for the two peaks. In similar fashion, the spacing of two bands in gradient elution can vary with gradient time or values of k^* (as in Fig. 2.12).

$$\log \alpha^* = (\log k_{wj} - \log k_{wi}) - (S_j - S_i)\phi^* \tag{2.22a}$$

The value of ϕ^* decreases for an increase in gradient time (corresponding to larger k^*), as can be seen from Equations (2.13) plus (2.14), or Figure 2.12.

For gradient separations where only gradient time is changed (as in Fig. 2.11), if the relative positions of two peaks vary with t_G (e.g., peaks 1 and 3), their value of α^* will be determined by ϕ^* [Equation (2.22a)], and therefore also by k^*. For the specific example of Figure 2.11 and peaks 1 and 3, an increase in k^* is seen to lead to a decrease in $\alpha^* = k^*(3)/k^*(1)$, that is, as k^* increases, peak 3 becomes less retained compared with peak 1. Similar changes in the relative retention of these two peaks can now be predicted for many other changes in separation conditions. For example, an increase in flow rate or decrease in column length (other conditions held the same) each correspond to an increase in k^* [Equation (2.13)], resulting therefore in increased retention of peak 1 relative to peak 3. Other examples similar to this will be noted in Sections 2.3.1–2.3.6. See Figure 3.9 for the convenient prediction of changes in peak spacing for "irregular" samples and various changes in separation conditions (after a change in peak spacing has been observed for any one of these changes in conditions).

It is sometimes assumed for "regular" samples that a change in gradient time will have no effect on elution order and therefore no effect on α^* for two adjacent peaks. This is only approximately correct, since values of S for "regular" samples tend to increase for more retained peaks. For example, the values of S for compounds 1 and 2 of the "regular" sample are 3.41 and 3.65, respectively (Table 1.3). This means that a change in isocratic %B that is large enough to increase k for compound 1 from 1 to 10 (change in ϕ by -0.29 units) will increase $\log \alpha$ for these two compounds by 0.070 units [Equation (2.22a)], meaning a 17% increase in $(\alpha - 1)$ and resolution. Thus, a change in gradient time with resulting change in k^* for a "regular" sample will usually have a slightly larger effect on resolution than is predicted by term a of Equation (2.21) alone [since term b of Equation (2.22) also tends to increase as k^* increases].

2.2.3.2 Using Gradient Elution to Predict Isocratic Separation

In Chapter 3, we discuss the development of a gradient method or procedure. The first step in RP-LC method development should be to carry out a gradient run which can provide a reasonable initial separation, for example, with $k^* \approx 5$ [based on Equation (2.13) for k^* as a function of conditions]. Then, before continuing with further method development experiments, it is important to consider the possibility of isocratic separation. An isocratic separation, if practical, can have several advantages: (a) the use of simpler, more commonly accessible equipment; (b) fewer experimental problems or demands on the operator; (c) a possibly shorter run time; and (d) a greater likelihood of successful method transfer. It is possible to assess the feasibility of isocratic separation for given sample from an initial gradient run.

In order to choose between isocratic and gradient elution, there are two questions we first need to answer: (a) Can isocratic separation provide acceptable retention for all peaks of interest, for example, $1 \leq k \leq 10$ (preferred)? (b) If isocratic separation is possible with an acceptable range in values of k, what %B will provide this acceptable retention range? Answers to these two questions can be arrived at through the use of Equation (2.12) for retention in gradient elution as a function of gradient conditions and the chromatographic properties of the sample [its values of k_w and S from Equation (2.9) or (2.14)].

Consider first whether isocratic separation is possible with an acceptable range in k. For $k_0 > 10$ (our initial assumption), gradient retention time can be approximated by Equation (2.12). This equation applies to the first peak "A" in the initial gradient chromatogram, and the last peak "Z":

$$t_{R,A} = (t_0/b_A) \log(2.3 k_{0,A} b_A) + t_0 + t_D \quad (2.23)$$

$$t_{R,Z} = (t_0/b_Z) \log(2.3 k_{0,Z} b_Z) + t_0 + t_D \quad (2.23a)$$

Here, $t_{R,A}$ and $t_{R,Z}$ represent values of t_R for peaks A and Z, $k_{0,A}$ and $k_{0,Z}$ are values of k_0 for compounds A and Z, and b_A and b_Z are corresponding values of b for each peak [Equation (2.11)]. We will assume as a second approximation that values of S for each peak are approximately equal, which then means that $b_A = b_Z \equiv b$. Also, an average value of $S \approx 4$ can be assumed for samples with molecular weights between 100 and 500 [3], which in turn allows gradient conditions to be specified, such that $k^* \approx 5$ (as in the recommended conditions of Table 3.2).

For a maximum range in isocratic retention of $1 \leq k \leq 10$, and equal values of S for peaks A and Z, it is required that $k_Z/k_A \leq 10$, or $\log(k_Z/k_A) \leq 1$. If $b_A = b_Z$, then $k_Z/k_A = k_{0,Z}/k_{0,A}$. With the substitution of b for b_A and b_Z in Equations (2.23) and (2.23a), we can obtain

$$\begin{aligned}(t_{R,Z} - t_{R,A}) &= (t_0/b) \log(k_{0,Z}/k_{0,A}) \\ &= (t_G \Delta\phi/S) \log(k_Z/k_A)\end{aligned} \quad (2.24)$$

or, if we define $\Delta t_R \equiv (t_{R,Z} - t_{R,A})$,

$$\Delta t_R/t_G = (\Delta\phi/S) \log(k_Z/k_A) \quad (2.24a)$$

For a full-range gradient ($\Delta\phi = 1$), and $S \approx 4$, the requirement for $1 \leq k \leq 10$ [or $\log(k_Z/k_A) \leq 1$] in an isocratic separation is then $\Delta t_R/t_G \leq 0.25$ (assuming that all experimental conditions except ϕ are the same for isocratic and gradient elution). Thus, the region of the chromatogram occupied by peaks must not comprise more than one-quarter of the total gradient chromatogram. If the requirement for isocratic elution is widened to $0.5 \leq k \leq 20$, the corresponding value of $\log(k_Z/k_A) \leq 1.6$, and then the maximum allowable value of $\Delta t_R/t_G$ is $\Delta t_R/t_G \leq 0.40$. Corresponding maximum values of $\Delta t_R/t_G$ (for isocratic separation) for other values of $\Delta\phi$ can be determined from Equation (2.24a).

Once it has been established that isocratic separation is possible, we need an estimate of the value of %B for an isocratic separation that will exhibit acceptable retention (either $1 \leq k \leq 10$ or $0.5 \leq k \leq 20$). The use of an initial gradient run (as above) for this purpose is described in Section 3.2.1; the corresponding theoretical background is detailed in Appendix II.

2.2.4 Sample Complexity and Peak Capacity

By "sample complexity," we mean the relative number n of compounds in the sample. As n increases (a more "complex" sample), the baseline separation of all sample components becomes progressively more difficult. Similar to the separation of a given sample as in Figure 2.11, gradient time and other conditions can be varied to obtain the maximum resolution. The probability that a successful separation can be achieved in this way is related to (a) the number of compounds in the sample and (b) the *peak capacity PC* of the separation [11]. The peak capacity of a separation is defined as the maximum number of adjacent peaks that can be separated with $R_s = 1$, as illustrated in Figure 2.13(a) for a gradient separation with $t_G = 10$ min. In this case, 50 peaks can be accommodated within the duration of the gradient, so $PC = 50$ for this example.

The peak capacity of a gradient separation can be expressed as

$$PC = (t_G/W) + 1 \quad (2.25)$$

with W given by Equation (2.18). Thus, peak capacity increases with separation time (t_G) and is approximately proportional to $N^{*1/2}$. The plate number N^* in turn depends on column dimensions, particle size, and flow rate. For a judicious selection of both gradient and column conditions ("best" choices of k^*, column length, particle size, and flow rate; see Section 9.5), values of peak capacity as a function of run time can be estimated (Fig. 2.13b). In Figure 2.13b, the solid curves are for various values of N (5000, 10,000, and 20,000, respectively). The dashed curve in Figure 2.13b represents the maximum peak capacity as a function of gradient time, for a maximum column pressure of 1500 psi (for a given gradient time, maximum PC corresponds to a specific value of N^*). Values of $PC = 200-250$ are seen to be possible within run times of 30–60 min. Larger PC values require inconveniently long run times, although the combined use of ultra-high pressures with very small particles can provide somewhat larger values of PC for a given gradient time [12]; the maximum value of N^* increases with pressure P as $P^{1/2}$ [13], so maximum PC increases as $P^{1/4}$; see Section 9.5 for further details. Experimental values of PC as large as 400 have been reported for the separation of peptide mixtures at pressures <3000 psi [14, 15], which may be attributable to larger values of S for these higher-molecular-weight compounds (and narrower resulting bands; Equation (2.19), as well as the approximate nature of Figure 2.13(b). The combination of a very high pressure (20,000 psi) with a long run time (15 h) has resulted in $PC = 1200$ for the peptides in a tryptic digest [16]. For a further discussion of peak capacity in gradient elution, see [17, 18].

A sample containing n components might appear to be separable with $R_s = 1$ when $PC = n$, but this is almost never the case. Such a result would require an

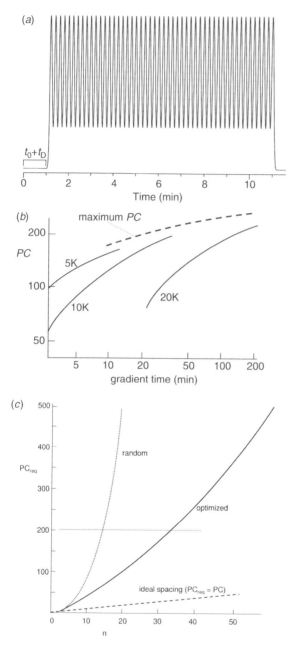

Figure 2.13 Peak capacity in gradient elution: (a) hypothetical example of gradient separation with peak capacity $PC = 50$; (b) dependence of PC on gradient time and $N^* = 5000$ ("5K"), 10,000 ("10K"), or 20,000 ("20K"), $k^* = 1$, $t_G/t_0 = 4$; the dashed curve is the maximum possible PC for a column pressure of 1500 psi; adapted from [11] for gradient time $= 4t_0$; (c) required peak capacity for different cases: "random," typical of a "first" separation before attempting to optimize peak spacing and resolution; "optimized," after optimizing gradient time and temperature to improve peak spacing and resolution; "ideal," best possible peak capacity for evenly space peaks in the chromatogram. See text for details.

exactly even spacing of every peak within the chromatogram (as in Fig. 2.13a), which is highly improbable [19]. We can define the *required peak capacity* PC_{req} for the separation of a sample containing n components (Fig. 2.13c) [11]. Thus, if $PC \geq PC_{req}$ for a given separation, a critical resolution of $R_s \geq 1$ is likely. Values of PC_{req} for a given value of n can vary considerably, depending on whether or not separation conditions are first varied ("optimized") in order to improve peak spacing (values of α) and critical resolution. For an initial separation, prior to the adjustment of conditions for improved peak spacing, we expect a more or less "random" arrangement of peaks within the chromatogram. Assuming a maximum value of $PC = 200$ (dotted horizontal line in Fig. 2.13c) and random peak spacing (· · · ·) in Fig. 2.13c) that fills the chromatogram for a 0–100 percent B gradient, the maximum number of peaks that is likely to be separable with $R_s = 1$ in a first gradient run is $n \approx 13$. The number of separated peaks is usually much smaller, however, because only a fraction of the total peak capacity is normally utilized, namely the peak capacity measured between the first and last peaks in the chromatogram [the *sample peak capacity* PC^{**}, equal to PC × $[t_{RZ} - t_{RA}]/t_G$, where t_{RZ} and t_{RA} refer to retention times for the first peak A and last peak Z in the chromatogram]; see [11] for details.

However, if peak spacing (selectivity) is first improved by varying gradient time and temperature ("optimized" curve in Fig. 2.13c), the allowed value of n for $R_s = 1$ increases from 13 to about 30. An "optimized" separation (with a similar maximum value of n) can also be achieved by holding temperature and gradient time constant, while varying the proportions of both acetonitrile and methanol in mixtures of these two solvents [11, 20]. Further optimization by simultaneous changes in several separation conditions (see Tables 3.4 and 3.6) can increase the allowable value of n still more. However, because of the large number of experimental runs required in order to optimize three or more separation conditions simultaneously, and because further increases in resolution are likely to be modest at best, few attempts of this kind have been reported (see isocratic examples of [21, 22]). When a critical resolution $R_s \geq 2$ is required, the maximum possible value of n is reduced 2-fold, to a value of $n \approx 15$ for "optimized" conditions where peaks are distributed across the entire chromatogram ($PC \approx PC^{**}$). This implies that samples with $n > 15$ will usually be difficult to separate with $R_s \geq 2$, which agrees with the experience of most chromatographers. Peak capacity can be greatly increased by the use of two-dimensional (2-D) separation (Section 3.7.1).

2.3 EFFECT OF GRADIENT CONDITIONS ON SEPARATION

"Gradient conditions" include the initial and final %B in the gradient, gradient shape, and those conditions which determine the intrinsic gradient steepness b [Equation 2.11)] – namely gradient time t_G, flow rate F, gradient range $\Delta\phi$, and column dead volume V_m (as determined by column length and internal diameter). The effect of gradient conditions on separation will also be influenced by the nature of the sample, for example, "regular" vs "irregular" samples. During method development, column length and flow rate should be varied *after* other conditions have been selected for a final gradient procedure or method (just as for isocratic method development, Section 2.14), as a means of either increasing resolution (by an increase in N^*) or reducing run time (with a decrease in N^*). When column length L and flow rate F are varied, gradient time should be changed at the same time so as to keep k^* for the separation constant [Equation (2.13)], in order not to change previously optimized separation selectivity (values of α^*). This will also simplify the interpretation of experiments where L and F are varied, as will be seen in Sections 2.3.2 and 2.3.3 (where L and F are varied *without* changing gradient time).

The remainder of this section is fairly long and detailed, in order to fully expose the reader to the consequences of different changes in conditions, for both "regular" and "irregular" samples. On the other hand, once a few general principles have become apparent, many of the examples discussed below should be fairly obvious. For this reason, *some readers may wish to skip or skim parts of this section*. A final section (2.3.8) provides a summary and overview of how separation varies with different gradient conditions.

2.3.1 Gradient Steepness *b*: Change in Gradient Time

The effect of gradient steepness on separation is illustrated in Figure 2.6 for the "regular" sample and in Figure 2.14 for the "irregular" sample. When gradient time is varied, elution order remains the same for "regular" samples, while it can vary for "irregular" samples (note the changing relative retention of peaks 3 and 9 in Fig. 2.14 as t_G increases). Critical resolution always increases with gradient time for a "regular" sample (as in Fig. 2.6), whereas maximum resolution for "irregular" samples often occurs for an intermediate gradient time – as in the example of Figure 2.14 where the 15 min gradient yields the largest resolution. Maximum resolution with $1 \leq k^* \leq 10$ is usually the initial goal when adjusting gradient time during method development.

In both Figures 2.6 and 2.14, an increase in gradient time results in better *average* resolution (increased peak capacity, Section 2.2.4), a decrease in peak heights with reduced ease of detection, and longer run times. Note also that increasing gradient time 10-fold (Fig. 2.14c vs a) leads to an increase in retention for the last peak which is less than 10-fold (21 min in c vs 4 min in a, or only 5-fold). This is typical of gradient elution with "small-molecule" samples, and is equivalent to elution of each peak at lower values of ϕ as gradient time increases. This decrease in ϕ at elution (ϕ_e) necessarily follows from the decrease in b with increase in gradient time [Equation (2.11)] and the resulting increase in k at elution [k_e, Equation (9.14a)]. Since k_e increases with t_G, ϕ_e must decrease [Equation (2.14)], where k_e and ϕ_e replace k^* and ϕ^*, respectively]. A value of ϕ_e can also be obtained from Equations (2.12) and (2.15):

$$\phi_e = \phi_0 + (1/S)[\log 2.3 k_0 + \log(t_0 \Delta \phi S / t_G)] \qquad (2.26)$$

For an increase in t_G (other conditions remaining the same), ϕ_e is seen to decrease, and this change in ϕ_e is greater for smaller values of S [by the factor $(\log S)/S$]. When S is very large (for high-molecular-weight solutes; Section 6.1.1), little change in ϕ_e will occur as t_G is varied.

2.3 EFFECT OF GRADIENT CONDITIONS ON SEPARATION

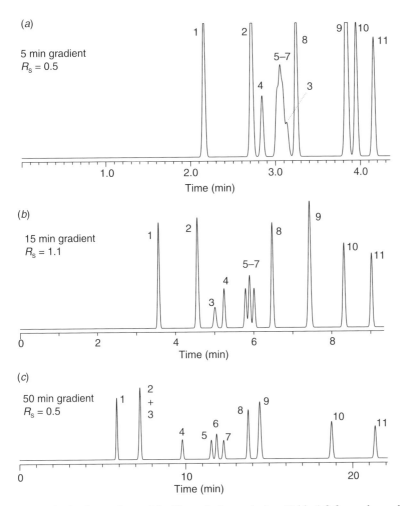

Figure 2.14 Separations of the "irregular" sample (see Table 1.3 for peak numbering) for different gradient times t_G. Conditions: 150 × 4.6 mm C_{18} column (5 μm particles), 2.0 mL/min, 42°C, 5–100 percent B in (*a*) 5 min, (*b*) 15 min, and (*c*) 50 min. Note that peak heights are as observed (*not* normalized to 100 percent for tallest peak).

2.3.2 Gradient Steepness *b*: Change in Column Length or Diameter

An increase in column length *L* alone increases V_m, t_0, and gradient steepness *b* [Equation (2.11)], that is, column volume V_m is proportional to *L*, provided that the column diameter is not changed. Because peak width increases with V_m and t_0, and decreases with gradient steepness *b* [Equation (2.19)], these two effects cancel approximately, so that peak height does not change much when column length (alone) is changed. An illustration of the consequences of a change in

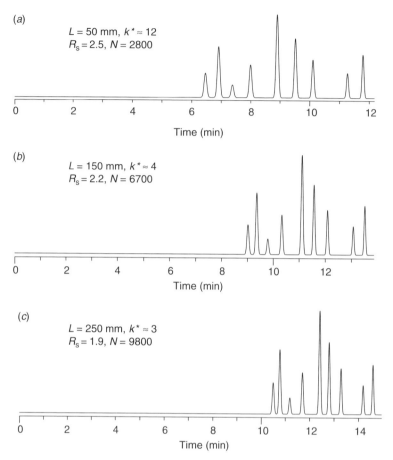

Figure 2.15 Effect of a change in column length for the "regular" sample (other conditions held constant) in gradient elution. Conditions: 0–100 percent B in 15 min; methanol (B)–water (A) mobile phase; C_{18} column (5 μm particles); 2.0 mL/min; 30°C; (a) 50 × 4.6 mm C_{18} column; (b) 150 × 4.6 mm column; (c) 250 × 4.6 mm column. Note that peak heights are as observed (*not* normalized to 100 percent for tallest peak).

column length is provided in Figure 2.15 for the "regular" sample. Here, gradient time (15 min), gradient range (0–100 percent B), and flow rate (2.0 mL/min) are held constant, while columns of the same diameter but varying length are used (50, 150, and 250 mm). The result of an increase in column length in the example of Figure 2.15 is a moderate decrease in resolution, with little change in run time or peak heights. The pressure increases in proportion to column length: 4400 psi in (c) vs 880 psi in (a). Note that relative retention and elution order do not change with column length for a "regular" sample, that is, the sequence of peaks is the same in each chromatogram of Figure 2.15, just as in the case of Figure 2.6 where gradient time was varied. That is, relative retention does not change for a "regular" sample when column length (or other conditions) are changed so as to result in change in k^* [Equation (2.13)].

The moderate *decrease* in resolution observed in Figure 2.15 for a longer column might appear strange, based on our experience with isocratic elution where an increase in column length always increases resolution (Figs 2.4*b* vs *a*). The decrease in resolution of Figure 2.15(*c*) vs (*a*) is the result of a smaller value of k^* (larger V_m, Equation (2.13), which decreases term *a* of Equation (2.21)] – and this is only partly compensated by the increase in N^* for longer columns [which increases term *c* of Equation (2.21)]. When column length is varied for different values of gradient time and/or flow rate vs the values of t_G and F used in Figure 2.15 (15 min, 2.0 mL/min), resolution can change slightly in either direction with increase in column length. The slight increase in retention time for the longer column is the result of a lower value of k^*, which also means a higher value of ϕ^* and of ϕ at elution (ϕ_e); retention time increases for larger values of ϕ_e [Equation (2.15)].

To conclude, for "regular" samples and gradient elution, a change in column length has only a minor effect on retention time, sample resolution or peak heights, *when all other conditions are held constant*. Several experimental examples as in Figure 2.15 have been reported [23–28], often with comments about these "unusual" results when compared with isocratic separation. In view of the foregoing comparisons, one might ask why not *generally* use very short columns? The answer is that shorter columns *per se* do not result in shorter run times or any other advantage – except a reduction in the amount of column packing that is required. Very short columns are also limited in their ability to provide improved resolution for longer gradient times (the time allowed for a gradient separation determines the "best" column length, as discussed in Sections 3.3.7 and 9.5).

If the column diameter d_c is varied in either isocratic or gradient elution, the column dead volume changes:

$$V_m \approx 5 \times 10^{-4} L d_c^2 \qquad (2.27)$$

where L and d_c are in mm. When the column diameter is changed for either "regular" or "irregular" samples, it is often desired to maintain the same relative retention, resolution, run time, and column pressure. In this case, the flow rate should also be changed, in proportion to V_m or the value of d_c^2 for the new vs old columns. For simultaneous changes in column diameter and flow rate in this way, the same separation (same retention times, peak heights, and resolution) will result, if extra-column band broadening can be neglected (Section 9.2.2). Also, if the same weight of sample is injected, peak heights should vary inversely with d_c^2; thus, narrow-diameter columns are a means of increasing peak heights and improving detection sensitivity, when the amount of sample is limited (assumes injection of a large part of the available sample). Alternatively, when sample weight is changed in proportion to d_c^2, peak heights will remain constant. These observations and recommendations for gradient elution apply equally for isocratic separation.

When column length alone is changed for the gradient separation of an "irregular" sample, the above generalizations still apply: little change in *average* resolution, run time, or peak widths. However, because an increase in column length results in a decrease in k^*, relative retention and critical resolution can vary. This is illustrated in Figure 2.16 for the "irregular" sample, where it is seen that peaks 3 and 9 change their relative positions within the chromatogram as column length is varied. As a result, it is possible to see a better resolution of some peak pairs (e.g., 3 and 4) when shorter columns are used. These changes in relative retention are equivalent to those seen when gradient time is varied for the "irregular"

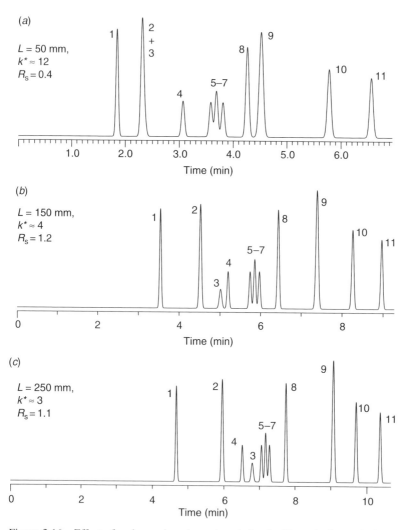

Figure 2.16 Effect of a change in column length for the "irregular" sample in gradient elution (other conditions held constant). Conditions: 5–100 percent B in 15 min (acetonitrile–buffer mobile phase); 43°C; 2.0 mL/min; (a) 50 × 4.6 mm C_{18} column (5 μm particles); (b) 150 × 4.6 mm column; (c) 250 × 4.6 mm column. Note that peak heights are as observed (*not* normalized to 100 percent for tallest peak).

sample (Fig. 2.14), and for the same reason [unequal S-values of adjacent peaks accompanied by changes in k^*; or ϕ^*, Equation (2.22a)]. Because an increase in column length means a decrease in k^* [Equation (2.13)], we see from the separations of Figure 2.16 for this sample and peaks 3 and 4 that $\alpha^* = k^*(4)/k^*(3)$ increases for larger k^*, that is, the relative retention of peak 4 vs peak 3 decreases for a longer column or smaller k^*. Because k^* *increases* with larger t_G, the relative positions of peaks 3 and 4 change in an opposite direction in Figure 2.14 as gradient time increases. The latter observation can be generalized for "irregular" samples, so as to allow the prediction of changes in relative retention for various changes in gradient conditions (see Fig. 3.12 and the related text).

As discussed in Section 3.3.5 for method development, changes in column length are sometimes worthwhile in gradient elution, as a means of achieving an advantageous change in either resolution or run time. Such improvements in separation should be carried out *after* adjusting conditions for optimum peak spacing and maximum resolution (i.e., optimum k^* and b). *Consequently, it is then important to maintain gradient steepness b (and k^*) constant during subsequent changes in column dimensions, by varying gradient time in proportion to V_m* [Equation (2.11); e.g., if column length is increased 2-fold, increase gradient time 2-fold].

2.3.3 Gradient Steepness *b*: Change in Flow Rate

The effect of a change in flow rate in gradient elution (holding other conditions constant) is illustrated in Figure 2.17 for the "regular" sample. As flow rate is decreased from (*a*) to (*c*), resolution decreases, peak heights increase, retention time increases slightly, and column pressure drops by a factor of 4. The contrast between the effects of flow rate in gradient vs isocratic elution (cf. Fig. 2.4*a* and *c* with Fig. 2.17) arises from simultaneous changes in k^*, t_0, and N^* when flow rate is varied in gradient elution (other conditions remaining constant), similar to the case of varying column length in Figure 2.15. Thus, as flow rate decreases, there is a proportional decrease in k^* [Equation (2.13)] and increase in t_0, combined with a more modest increase in N^*; these changes partly cancel so far as peak width W is concerned [Equation (2.18)], but peak *volume* (for a given value of W) is proportional to flow rate. A reduced peak volume means a proportionately higher solute concentration (other considerations equal), and a proportionately higher peak. Compared with a change in column length (Section 2.3.2), resolution (for a change in flow rate) is more affected by the decrease in k^* (and a related decrease in α^*; Section 2.2.3.1) than by the relatively small increase in N^*. The moderate increase in retention time as flow rate is decreased is due to the corresponding decrease in k^*, similar to the effect of column length on sample retention discussed above (Section 2.3.2). A 4-fold decrease in flow rate as in Figure 2.17 decreases pressure 4-fold for both gradient and isocratic elution.

In the case of "irregular" samples, the above *general* effects also apply (decrease in resolution and increase in peak height for lower flow rates), *accompanied* by changes in relative peak spacing because of the associated changes in k^* (as in Fig. 2.16 for changing column length for the separation of the "irregular" sample). For example, from the above observation (based on Fig. 2.16) that the

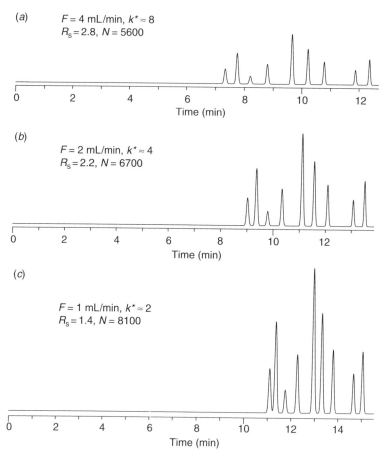

Figure 2.17 Effect of a change in flow rate for the "regular" sample in gradient elution (other conditions held constant). Conditions: 50 × 4.6 mm C_{18} column (5 μm particles); 0–100 percent B in 15 min [methanol (B)–water (A)]; 30°C; (a) 4.0 mL/min; (b) 2.0 mL/min; (c) 1.0 mL/min. Note that peak heights are as observed (*not* normalized to 100 percent for tallest peak).

retention of peak 3 decreases relative to peak 4 as k^* increases, we can predict a similar decrease in the relative retention of peak 3 vs 4 as flow rate (and k^*) are increased for the same sample. A remarkable example of changes in peak spacing with flow rate is shown in Figure 2.18, for the separation of a peptide digest by gradient elution. In Figure 2.18(a), the entire chromatograms are shown for flow rates of 0.5 and 1.5 mL/min (other conditions unchanged). In Figure 2.18(b), corresponding portions of the chromatograms of (a) (marked by arrow) are shown in expanded form. For this change in flow rate, initially well-resolved peaks 5 and 5a (0.5 mL/min, Fig. 2.18b) become completely overlapped at 1.5 mL/min; initially overlapped peaks 6 and 6a become partly resolved at 1.5 mL/min, and initially well-resolved peaks 6b and 7 change positions when the flow rate is changed from 0.5 to 1.5 mL/min.

2.3 EFFECT OF GRADIENT CONDITIONS ON SEPARATION 57

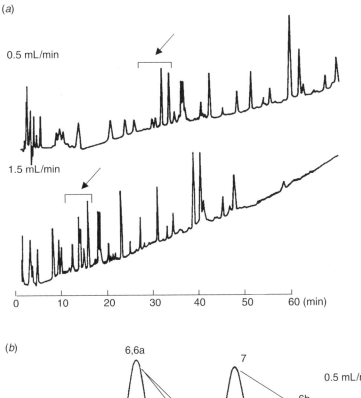

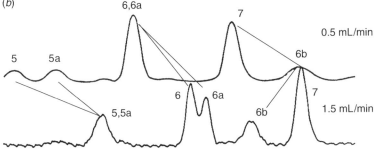

Figure 2.18 Effect of flow rate on the gradient separation of peptides from the tryptic digest of myoglobin. Conditions: 80 × 6.2 mm C_8 column (5 μm particles); 10–70 percent ACN–buffer gradient in 60 min; (*a*) complete chromatograms; (*b*) expanded portions of chromatograms of (*a*). Reprinted with permission from [30]. Peak heights are normalized to 100 percent for tallest peak.

The examples of Figures 2.16 and 2.18 might suggest the initial variation of column length and/or flow rate for the purpose of optimizing peak spacing and resolution. However, this is *not* a good practice, as other means of controlling peak spacing are more useful, systematic, and convenient (Sections 3.3.2 and 3.6). As developed more fully in Section 3.3.5, the best choice of column length and flow rate should be determined in the same way for both gradient and isocratic separation, that is, after optimizing values of k^* and α^*. *When either column length or flow rate are varied for gradient elution, while maintaining k^* constant by simultaneously* *adjusting gradient time [Equation (2.13)], the effects on resolution, run time, or column pressure will be the same in both isocratic and gradient elution*). That is, the plate number N or N^*, resolution, run time, and pressure will all change in the same proportion for either isocratic or gradient elution, as long as k^* does not change.

According to Equation (2.13), when only flow rate is varied, values of k^* are proportional to the *gradient volume* $t_G F$. If the gradient volume is kept constant during changes in F, then the relative retention of peaks in the chromatogram will also be maintained constant [29].

2.3.4 Gradient Range $\Delta\phi$: Change in Initial Percentage B (ϕ_0)

The effect of a change in gradient range $\Delta\phi$, without change in gradient time or other conditions, is to change gradient steepness b proportionately and k^* inversely [Equations (2.11) and (2.13)]. In the present and following sections, we will examine the effects of a change in initial or final %B in the gradient, while holding b and k^* constant by varying gradient time t_G in proportion to $\Delta\phi$, that is, holding $(\Delta\phi/t_G)$ constant. *Keep in mind that if only gradient range is changed, while holding other variables constant, resulting changes in separation will represent the combined effect of change in gradient steepness b and the value of initial %B.* It is much easier to interpret and optimize separation as a function of $\Delta\phi$, if b is held constant while ϕ is varied.

Figure 2.19 illustrates the effects of a change in gradient range for the separation of the "regular" sample. The value of %B at the start of the gradient is varied, while simultaneously adjusting gradient time t_G so as to keep $\Delta\phi/t_G$ and gradient steepness b constant [Equation (2.11)]. For an increase in initial %B from 0 to 20 percent (*b*), with a proportionate shortening of gradient time by 10 min, the separation remains essentially the same as in (*a*), except that all peaks leave the column 10 min earlier (and the run time is 10 min shorter). When the initial %B is increased further to 40 percent B (*c*), a slight change in the separation of peaks 1 and 2 is observed: The height of these peaks has increased slightly, and their resolution has decreased to $R_s = 2.7$ from $R_s = 4.0$ in (*b*). Finally, in (*d*), the initial %B is increased further to 60 percent, with a considerable increase in the heights of early peaks 1–5, as well as markedly decreased resolution for peaks 1 and 2 ($R_s = 0.9$).

Because the first peak (1) elutes fairly late in the initial 0–100 percent B gradient (Fig. 2.19*a*), this means that it and subsequent peaks were initially strongly retained at the column inlet (k_0 for peak 1 = 180). As a result, values of k^* for peaks 1–4 are given by Equation (2.13) (average $k^* \approx 3.7$). When the initial %B of the gradient is increased to 20 percent B (*b*), the initial peak is still well retained ($k_0 = 33$), as are later peaks. Because of this reduced value of k_0, the value of k^* for the initial four peaks is slightly decreased [$k^* \approx 3.2$; Equation (9.18)]; however, there is little change in the resolution or peak heights of early peaks in the chromatogram. The 10 min required in the gradient of (*a*) to go from 0 to 20 percent B is eliminated in (*b*), with the result that the chromatogram of (*a*) has simply been shifted 10 min to the left in (*b*).

2.3 EFFECT OF GRADIENT CONDITIONS ON SEPARATION 59

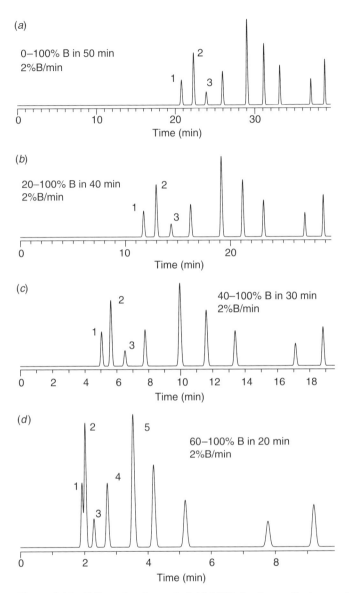

Figure 2.19 Effect of a change in initial %B for the gradient separation of the "regular" sample. Conditions: 150 × 4.6 mm C_{18} column (5 μm particles); 30°C; 2.0 mL/min; methanol–water mobile phase; gradient time adjusted to maintain $k^* = 4$. Other conditions indicated in the figure; see text for details. Peak heights *not* normalized to 100 percent.

For an initial %B of 40 percent (Fig. 2.19c), peaks 1–4 are even less strongly retained at the beginning of the gradient ($k_0 = 8$–18). As a result, k^* for these peaks has decreased to $k^* \approx 2.8$ [Equation (9.18)], with a small, but noticeable, decrease in their resolution and increase in their peak heights. Later peaks show a further decrease in retention time of 10 min [total of 20 min compared with (a)], but no other change – because their values of k_0 are still relatively large, and their values of k^* have not changed much.

For an initial %B of 60 percent B (Fig. 2.19d), k_0 values for peaks 1–4 are now all relatively small ($k_0 \leq 1.7$), and values of k^* are even more significantly reduced ($k^* \approx 1.4$). As a result, the resolution of these peaks has decreased further, accompanied by a continuing increase in their peak heights. Thus, the effect of an increase in the initial %B of the gradient, while holding gradient steepness b constant, is a reduction in the retention times of all peaks, accompanied by a lowering of values of k^* for early peaks. The effect on early peaks is then similar to an increase in %B for isocratic elution (with a similar lowering of values of k), that is, increased peak heights and decreased resolution.

Figure 2.20 shows the effect of increasing the initial %B for the "irregular" sample, from 5 to 20 to 30 percent B, while reducing gradient time so as to maintain $\Delta\phi/t_G$ [and values of k^*; Equation (2.13)] constant. In going from the separation of (a) to (b) to (c), there is a gradual increase in the crowding of early peaks 1–4, with a modest decrease in resolution and increase in peak height (as in Fig. 2.19, but less pronounced, due to smaller changes in initial %B). More importantly, significant changes in relative retention can be seen in the early part of the chromatogram, especially for peaks 3 vs 4. This change in peak spacing can be attributed to a decrease in k^* for peaks 3 and 4 as the initial %B increases [from an average $k^* = 4.6$ in (a) to $k^* = 2.2$ in (c); Equation (9.18)]. From our earlier discussion of Figure 2.16, where the retention of peak 3 increases relative to that of peak 4 as column length increases (and k^* decreases), we can similarly predict that an increase in initial %B (decrease in k^*) will have the same effect on the relative retention of these two peaks – an increase in the relative retention of peak 3 vs 4, as is indeed observed in Figure 2.20. For a lower initial %B in (a), and relatively larger values of k^*, peak 3 elutes prior to peak 4. For higher initial %B values in (b) and (c), and lower values of k^*, peak 3 elutes after peak 4.

From the foregoing discussion, it can be appreciated that in some cases an increase in %B at the start of the gradient can result in an *increased* resolution of two or more early-eluting peaks. An example is shown in Figure 2.21 for the separation of a herbicide mixture [31]. In the separation of (a), using a 15–100 percent B gradient in 45 min, peaks 1 and 2 are poorly resolved. For an increase in initial %B (b), the spacing of peaks 1–3 is altered, so that all three peaks are now much better separated. Note, however, that there are *two* changes to the gradient in (b) vs (a): an increase in initial %B (which tends to decrease k^*), and a decrease in $\Delta\phi$ [which tends to increase k^*; Equation (2.13)]. It cannot be determined which of these two (opposing) effects is more important for improving the separation of Figure 2.21(b).

2.3.5 Gradient Range $\Delta\phi$: Change in Final %B (ϕ_f)

Figure 2.22 illustrates the effect of changing the final %B in the gradient for the "regular" sample. The separation in (a) is for a gradient of 0–100 percent B. Subsequent changes in the final %B value are accompanied by changes in

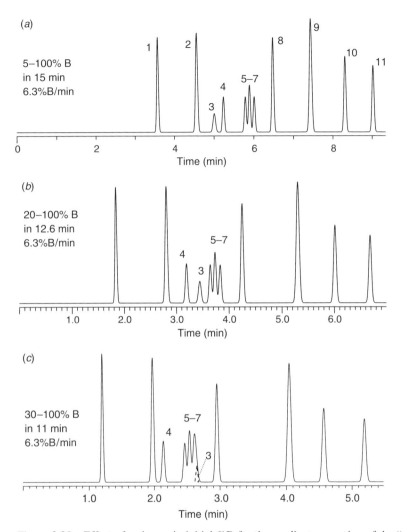

Figure 2.20 Effect of a change in initial %B for the gradient separation of the "irregular" sample. Conditions: 150 × 4.6 mm C_{18} column (5 μm particles), 42°C, 2.0 mL/min flow rate, acetonitrile–buffer mobile phase; gradient time adjusted to maintain $k^* = 4$ or $b = 0.048$ S (see figure for details). Peak heights as observed (*not* normalized to 100 percent). Other conditions indicated in the figure; see text for details.

gradient time so as to keep $(\Delta\phi/t_G)$ and k^* constant (as in Fig. 2.19 for changes in initial %B). For a gradient of 0–80 percent B in (*b*), there is no change in the separation, because the last peak in the sample leaves the column before the gradient has ended (see arrow). A further shortening of the gradient to 0–60 percent B (*c*), however, results in elution of peaks 7–9 *after* the end of the gradient, so that these peaks leave the column under isocratic conditions. As a result, peak

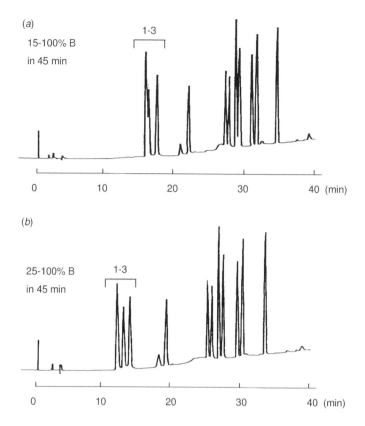

Figure 2.21 Effect of a change in initial %B for a phenylurea herbicide sample. Samples: 1, hydroxymetoxuron; 2, desfenuron; 3, fenruon; 4, metoxuron; 5, fluometuron; 6, chlorotoluron; 7, isoproturon; 8, diuron; 9, linuron; 10, chlorbromuron; 11, neburon. Conditions: 300 × 4.2 mm C_{18} column (5 μm particles); methanol–water gradients; 1.0 mL/min; ambient temperature. Adapted from [31]. Note that peak heights are normalized to 100 percent for tallest peak.

width and resolution increase for peaks 7–9 [because values of k for these peaks are greater than their values of k^* in the separations of (a) and (b)], as does run time. Figure 2.22, where the final %B is reduced, can be contrasted with Figure 2.19 where the initial %B is increased. An increase in initial %B or a decrease in final %B results, respectively, in decreased k^* values for later peaks (but only if the gradient ends prior to the elution of the last peak).

As long as the last peak leaves the column before the end of the gradient, there is no effect of a change in the final %B on separation, other than to decrease run time for smaller values of final %B. In most cases, it will be advisable to end the gradient as soon as the last peak leaves the column, but not before. Elution of peaks after the gradient wastes run time and leads to undesirable peak broadening. The effect of the final %B on separation is similar for

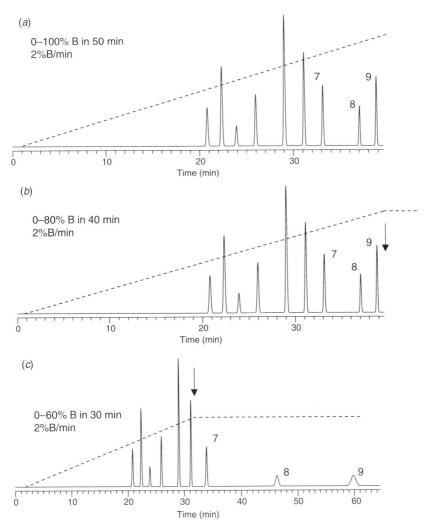

Figure 2.22 Effect of a change in final %B for the gradient separation of the "regular" sample. Conditions: 150 × 4.6 mm C_{18} column (5 μm particles); 30°C; 2.0 mL/min; methanol–water mobile phase; gradient time adjusted to maintain $k^* = 4$ (see figure for details). Peak heights *not* normalized to 100 percent; arrows mark the end of the gradient at the end of the column. Gradient indicated by dashed line. Other conditions indicated in the figure; see text for details.

both "regular" and "irregular" samples (no change in relative retention or elution order), as long as any late elution of peaks is avoided.

2.3.6 Effect of a Gradient Delay

Gradient delay refers to initial isocratic elution for some period of time, prior to the start of the gradient. The effect of a gradient delay is illustrated in Figure 2.23 for

64 CHAPTER 2 GRADIENT ELUTION FUNDAMENTALS

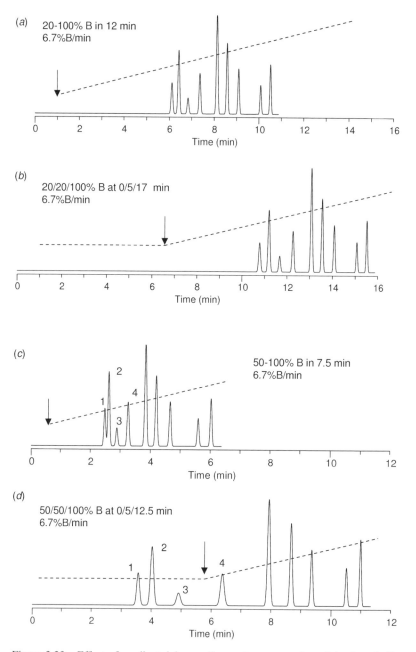

Figure 2.23 Effect of gradient delay on the gradient separation of the "regular" sample. Conditions: 150 × 4.6 mm C_{18} column (5 μm particle); 30°C; 2.0 mL/min (methanol–water mobile phase); arrows mark beginning of gradient (as it leaves the column). Peak heights *not* normalized to 100 percent. Other conditions indicated in the figure; see text for details.

the "regular" sample. Figure 2.23(a) shows a chromatogram for a 20–100 percent B gradient without a gradient delay, where the first peak in the chromatogram does not leave the column until well after the start of the gradient at the *outlet* of the column (indicated by the arrow). When a 5 min gradient delay is added (Fig. 2.23b), the effect is to increase the retention times of all peaks by 5 min, but the two chromatograms of (a) and (b) are otherwise virtually indistinguishable. In this example, the isocratic retention of the initial peak in a mobile phase of 20 percent B is $k \equiv k_0 = 38$, so there is little migration of this peak during the initial isocratic elution (gradient delay) in (b), and even less migration of later peaks. Under these conditions, the entire sample effectively remains at the head of the column (inlet) during the gradient delay, after which the same gradient separation results as in the absence of a gradient delay. Run time is increased by the delay time.

When initial peaks are *not* strongly retained at the start of the gradient, a gradient delay will have a more noticeable effect on the separation. This is illustrated in the similar examples of Figure 2.23(c) and (d) (same "regular" sample, same gradient steepness $\Delta\phi/t_G$, but higher initial %B). In the gradient separation of (c) with no gradient delay, the retention of the initial peak in 50 percent B is $k \equiv k_0 = 3.6$, and peaks 2–4 have $k < 10$. Therefore, significant elution of these four peaks will occur during the 5 min gradient delay of (d); in fact, peaks 1–3 leave the column during the gradient delay. As can be seen in these two examples, the initial two peaks are poorly separated in (c) with $R_s = 1.1$, whereas in (d) the separation of these two peaks has increased to $R_s = 2.3$. The better resolution of early peaks in (d) is due to larger values of k^* for these peaks compared with the separation of (c), because the gradient delay has interrupted (and therefore reduced) the normal decrease in sample retention k after the start of the gradient (i.e., during the gradient).

A gradient delay is used sometimes to increase the resolution of early peaks in the chromatogram, by increasing k^*. For cases such as that of Figure 2.23(c), however, resolution can best be improved by simply reducing the initial value of %B in the gradient [compare separations (a) and (c)]. However, when the initial %B of the gradient is close to zero (and a significant reduction in initial %B is therefore not feasible), a gradient delay may be the only alternative – apart from a change in the column or the use of ion-pairing [1]. This need for a gradient delay often arises in the separation of small, polar molecules which have small values of k, even with water as the mobile phase.

The effect of a gradient delay on the separation of "regular" samples is to increase the resolution of early peaks (as in Fig. 2.23), but relative retention is unchanged (no change in elution order). As illustrated in Figure 2.24, this is not necessarily the case for early peaks in "irregular" samples when the gradient is delayed. The separation in Figure 2.24(a) has no gradient delay, while there is a gradient delay of 5 min in (b); the gradient itself is unchanged (5–100 percent B in 15 min). Whereas in the similar separation of the "regular" sample in Figure 2.23 there was no change in relative retention but an increase in resolution, a similar gradient delay for the "irregular" sample of Figure 2.24 leads to changes in relative retention and (for this sample) a *decrease* in resolution; note the faster elution of peak 3 vs 4 in (b) compared with (a). The reason for this behavior is

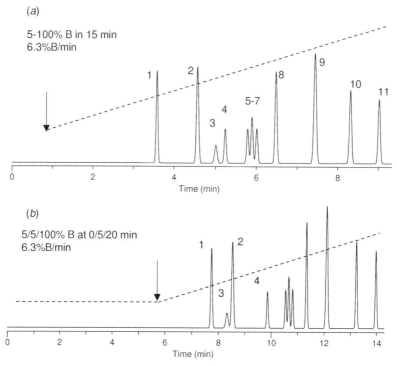

Figure 2.24 Effect of gradient delay on the gradient separation of the "irregular" sample. Conditions: 150 × 4.6 mm C_{18} column (5 μm particles); 42°C; 2.0 mL/min (acetonitrile–buffer mobile phase); arrows mark start of gradient (as it leaves the column). Peak heights *not* normalized to 100 percent. Other conditions indicated in the figure; see text for details.

that the gradient delay increases values of k^*, and we have seen that an increase in k^* can lead to faster relative migration of peak 3 vs 4 in other separations (e.g., Fig. 2.16). For later-eluted peaks (peaks 5–11), the only change observed between (*a*) and (*b*) is an increase in retention equal to the delay, because k^* is unchanged for these peaks. *When a gradient delay is used with an "irregular" sample, changes in relative retention plus either an increase or decrease in resolution can result for early peaks in the chromatogram.* For some "irregular" samples, a gradient delay of the right length can provide a major increase in the resolution of early peaks in the chromatogram, by an advantageous change in selectivity (values of α^*).

2.3.6.1 Equipment Dwell Volume

Every instrument used for gradient elution will have a certain hold-up volume (so-called *dwell volume* V_D), equal to the volume of the gradient mixer plus that of the mobile phase flow-path between the mixer and the column inlet (Figs 4.1 and 4.2; Sections 4.4 and 9.2.2). Values of V_D can vary for different gradient equipment, from a fraction of a milliliter for modern equipment to several milliliters for older equipment. The existence of a dwell volume is equivalent to the intentional use of a gradient delay, and the effects of varying dwell time on separation can therefore be inferred from Figures 2.23 and 2.34.

The equipment dwell time t_D is equal to V_D/F, where F is flow rate. If k_0 for the first-eluted peak is reasonably large for separations with a nonzero dwell time, retention times are then given by Equation (2.12). When k_0 for a peak is <10, however, Equation (2.12) should be replaced by Equation (9.24), corresponding to an increase in k^* for early peaks in the chromatogram. The gradient programmed into the instrument controller corresponds to the gradient that enters the gradient mixer (Figs 4.1 and 4.2). The gradient entering the column is delayed by a time equal to the dwell time t_D, while the gradient leaving the end of the column is delayed by a further time equal to the column dead time t_0 (Fig. 2.25). Values of V_D for a given gradient equipment can be determined in various ways (Sections 4.3.1.2 and 9.2.2); problems relating to equipment dwell volume are discussed in Section 5.2.1.

2.3.7 Effect of Gradient Shape (Nonlinear Gradients)

Apart from the use of a gradient delay, nonlinear gradients have been used in the past for the separation of oligomeric samples such as synthetic polymers (Section 6.1.4). These "regular" samples are characterized by a continuous, significant increase in values of S for more retained compounds, as in Figures 1.8(a) and 1.11(a). When an oligomeric sample is separated by means of a linear gradient, resolution is observed to decrease regularly for later peaks in the chromatogram, sometimes leading to complete peak overlap. An example is shown in Figure 2.26(a) for the separation of a 2000 Da polystyrene sample which contains significant amounts of each oligomer between $n = 2$ and $n = 44$. In the separation

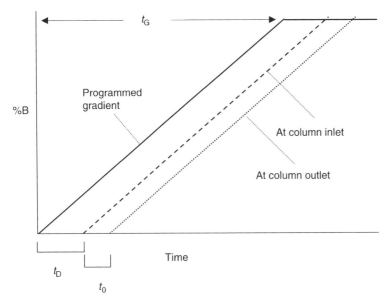

Figure 2.25 The effect of dwell volume on the gradient leaving the column. See text for details.

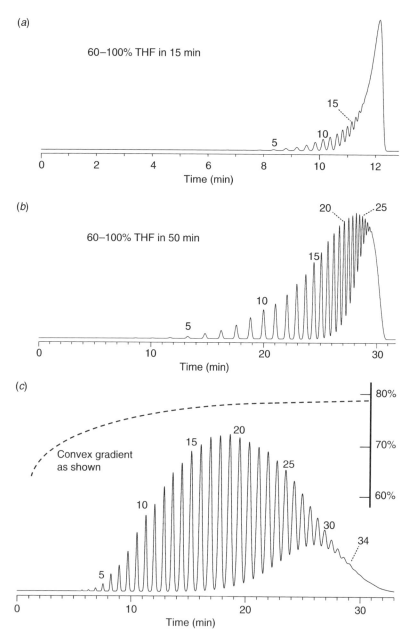

Figure 2.26 Use of convex gradients for the separation of a 2300 Da polystyrene sample. Conditions: 250 × 4.6 mm C_{18} column (5 μm particles); A, solvent, water; B, solvent tetrahydrofuran; 35°C; 1.0 mL/min. (a) 60–100 percent B in 15 min; (b) 60–100 percent B in 50 min; (c) convex gradient (62–79 percent B) as indicated in figure. Simulated chromatograms based on data of [42]. Peak heights are normalized to 100 percent for tallest peak.

2.3 EFFECT OF GRADIENT CONDITIONS ON SEPARATION

of Figure 2.26(a), individual peaks as large as $n = 17$ can be distinguished, while higher oligomers are merged into a single broad peak. For an increase in gradient time from 15 to 50 min (b), the resolution of later peaks is much improved, and it is possible to recognize separated oligomers as large as $n = 28$. At the same time, the resolution of early peaks is $R_s \gg 2$, that is, there is excess (wasted) resolution at the beginning of the chromatogram, and insufficient resolution at the end. For the separation of oligomeric samples such as that of Figure 2.26, previous workers have recognized that early peaks can be resolved adequately with a fairly steep gradient, while later peaks required a much flatter gradient. This suggests the use of a convex gradient, which becomes progressively flatter as it proceeds. As shown in Figure 2.26(c), a convex gradient for this sample results in better resolution for later peaks and recognizable separation up to $n = 34$, as well as more than adequate resolution of early peaks, with no increase in run time compared with the linear separation of Figure 2.26(b). For past comparisons of the relative advantage of convex vs linear gradients for oligomeric samples, see [32–34] and p. 373 of [1].

For various reasons, curved gradients as in Figure 2.26(c) are used less often today, compared with before 1990. A better alternative to curved gradients is the use of *segmented* gradients that are composed of two or more linear sections. For example, it is possible to duplicate many of the advantages of convex gradients for oligomeric samples by means of segmented gradients [35], with less trial-and-error experimentation. Segmented gradients are commonly used for several other purposes:

- to clean the column between sample injections;
- to shorten run time;
- to increase resolution by adjusting selectivity for different parts of the chromatogram (only for "irregular" samples).

Cleaning the column and shortening run time (by means of segmented gradients) are discussed in Section 3.3.4. The principle of segmented gradients for increasing resolution in different parts of the chromatogram is illustrated in Figure 2.27, for a hypothetical sample composed of compounds 1–4. As seen in the log k^*–ϕ^* plots of Figure 2.27(a), peaks 1 and 2 are best separated with a flat gradient ($k^* = 10$), shown in Figure 2.27(d). Similarly, peaks 3 and 4 are best separated with a steep gradient ($k^* = 1$; Fig. 2.27b). Peak pairs 1–2 and 3–4 are each potentially critical, so that the best critical resolution occurs for an intermediate gradient steepness (indicated by the dashed horizontal line in Fig. 2.27a for log $k^* \approx 0.5$, or $k^* = 3$), and shown in Figure 2.27(c). However, the resulting sample resolution for the latter separation is rather low; $R_s = 1.3$. If compounds 1 and 2 could be separated with $k^* = 10$, while compounds 3 and 4 were separated with $k^* = 1$, a much better critical resolution would result for the entire sample. The latter possibility can be approximated by use of a segmented gradient where the initial segment is flat, followed by changing to a steeper segment after peaks 1 and 2 leave the column; this is shown in Figure 2.27(e), with $R_s = 1.7$. Further (real) examples of the use of segmented gradients for this purpose are described in Sections 3.3.4.1 and 6.2.1.4.

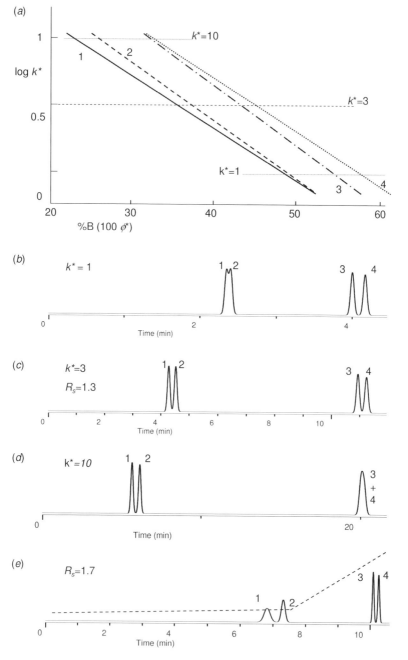

Figure 2.27 Use of segmented gradients to improve the separation of certain samples. Log k vs %B for a hypothetical sample. (a) log $k^*-\phi^*$ plot for sample compounds: solid line, peak 1; dashed line, peak 2; dashed–dotted line, peak 3; dotted line, peak 4. (b–e), Separations for different gradient conditions. See text for details. Peak heights are normalized to 100 percent for tallest peak.

Despite the potential advantage of segmented gradients for increasing the critical resolution of some samples (Fig. 2.27), segmented gradients are not often used for this purpose. An increase in critical resolution from the use of segmented gradients requires two critical pairs that elute, respectively, early and late in the chromatogram. Otherwise, the partial migration of the second peak pair (by the time the first peak pair leaves the column) under the influence of the initial gradient segment can result in little or no overall advantage from the use of the second gradient segment. However, this limitation of segmented gradients for an increase in sample resolution becomes less important for higher-molecular-weight samples, as discussed in Section 6.2.1.4. The use of segmented gradients for the purpose of increasing critical resolution is therefore more practical for the case of large-molecule samples.

A more detailed examination of the use of segmented gradients in this way is offered in [36]. Computer programs have also been reported for the automated development of optimized segmented gradients [37–39]. Stepwise elution can be regarded as a simple (if less generally effective) kind of segmented gradient, and a theory of such separations has also been reported [40].

2.3.8 Overview of the Effect of Gradient Conditions on the Chromatogram

By "gradient conditions," we refer to any separation condition that affects gradient retention k^*, for some or all peaks in the chromatogram. These conditions can be summarized as follows:

- gradient time t_G;
- column length L (or volume V_m);
- flow rate F;
- gradient range $\Delta\phi$;
- initial %B (ϕ_0);
- gradient delay or change in the dwell volume V_D.

A change in t_G, L, F, or $\Delta\phi$ results in a change in k^* that is predicted by Equation (2.13); k^* increases with increase in t_G or F, and decreases with increase in L or $\Delta\phi$. When k^* decreases, the *general* effect on separation is to reduce resolution [Equation (2.21)] and increase peak height [or reduce peak width, Equation (2.18)]. For irregular samples, however, a change in k^* also causes changes in relative retention that often result in maximum critical resolution for intermediate values of k^*.

A decrease in initial %B, the insertion of a gradient delay (isocratic hold at the start of the gradient), or an increase in system dwell volume V_D all serve to increase values of k^* for early peaks in the chromatogram. This selective increase in k^* has its usual effects: (a) an *average* increase in resolution for early peaks in the chromatogram; (b) a decrease in the heights of early peaks; and (c) for irregular samples, possible changes in relative retention that can either increase or decrease the resolution of certain early-eluting peaks.

2.4 RELATED TOPICS

2.4.1 Nonideal Retention in Gradient Elution

By "nonideal" retention behavior, we mean the apparent failure of Equation (2.12) for various reasons. Differences in experimental vs predicted values of t_R from Equation (2.12) can arise from faulty equipment, nonlinear plots of log k vs %B [deviations from Equation (2.9)], small values of k at the start of the gradient (k_0), as well as other causes. For a detailed discussion of such errors in predicted values of t_R, as well as their correction, see Sections 9.2 and 9.3. However, such failures of Equation (2.12) are seldom of practical importance, being important mainly when accurate predictions of retention and separation are required, as for computer simulation using commercial software (Sections 3.4 and 9.3).

2.4.2 Gradient Elution Misconceptions

Workers using gradient elution for the first time often encounter apparently puzzling results, many of which have been illustrated in this chapter. Usually these surprises arise from expectations based on prior experience with isocratic separation. For example, if flow rate is decreased, or column length increased, retention times, peak widths, and resolution generally increase significantly for isocratic elution (Fig. 2.4), but much less so for gradient elution (Figs 2.13 and 2.15). The apparent failure of resolution to increase with increasing column length in gradient elution (while holding other conditions fixed) has surprised many workers in the past [23–28, 41]. A further misconception in gradient elution is that column plate number can be calculated by the same equation as for isocratic separation [Equation (2.5)]; resulting (incorrect) values of N in gradient elution are still sometimes reported, based on Equation (2.5). Similar mistakes in interpretation have led others to propose that the separation of large molecules by gradient elution is based on a fundamental process that is quite different than for small molecules (Section 6.1.5). Misconceptions such as these, which have persisted well beyond the dissemination of a general theory of gradient elution in the early 1980s, should eventually be laid to rest. Hopefully the present book can bring that time a little closer.

REFERENCES

1. L. R. Snyder, J. J. Kirkland, and J. L. Glajch, *Practical HPLC Method Development*, 2nd edn, Wiley-Interscience, New York, 1997.
2. *Reviewer Guidance. Validation of Chromatographic Methods*, Center for Drug Evaluation and Research, Food and Drug Administration, November 1994.
3. L. R. Snyder and J. W. Dolan, *J. Chromatogr. A* 721 (1996) 3.
4. J. J. Kirkland, *Am. Lab.* 26 (1994) 28K.
5. U. D. Neue, D. J. Philips, T. H. Walter, M. Capparella, B. Alden, and R. P. Fisk, *LCGC* 12 (1994) 468.
6. K. Valkó, L. R. Snyder, and J. L. Glajch, *J. Chromatogr.* 656 (1993) 501.

7. P. Jandera and J. Churáček, *Gradient Elution in Column Liquid Chromatography*, Elsevier, Amsterdam, 1985, p. 101.
8. J. D. Stuart, D. D. Lisi, and L. R. Snyder, *J. Chromatogr.* 485 (1989) 657.
9. J. W. Dolan, J. R. Gant, and L. R. Snyder, *J. Chromatogr.* 165 (1979) 31.
10. L. R. Snyder, unreported observations.
11. J. W. Dolan, L. R. Snyder, N, M. Djordjevic, D. W. Hill, L. Van Heukelem, and T. J. Waeghe, *J. Chromatogr. A* 857 (1999) 1.
12. K. D. Patel, A. D. Jerkovich, J. C. Link, and J. W. Jorgenson, *Anal. Chem.*, 76 (2004) 5777.
13. L. R. Snyder and G. B. Cox, *J. Chromatogr.* 483 (1989) 85.
14. U. D. Neue, J. L. Carmody, Y. F. Cheng, Z, Lu, C. H. Phoebe, and T. E. Wheat, in *Design of Rapid Gradient Methods for the Analysis of Combinatorial Chemistry Libraries and the Preparation of Pure Compounds*, P. Brown and E. Grushka, eds, Marcel Dekker, New York, 2001, p. 93.
15. M. Gilar, A. E. Daly, M. Kele, U. D. Neue, and J. C. Gebler, *J. Chromatogr. A* 1061 (2004) 183.
16. Y. Shen, R. Zhang, R., R. J. Moore, J. Kim, T. O. Metz, K. K. Hixson, R. Zhao, E. R. Livesay, H. R. Udseth, and R. D. Smith, *Anal. Chem.* 77 (2005) 3090.
17. U. D. Neue, *J. Chromatogr. A* 1079 (2005) 1053.
18. X. Wang, D. R. Stoll, A. P. Schellinger, and P. W. Carr, *Anal. Chem.* 78 (2006) 3406.
19. J. M. Davis and J. C. Giddings, *Anal. Chem.* 53 (1983) 418.
20. D. P. Herman, H. A. H. Billiet, and L. DeGalan, *Anal. Chem.* 58 (1986) 2999.
21. J. L. Glajch, J. C. Gluckman, J. G. Charikofsky. J. M. Minor, and J. J. Kirkland, *J. Chromatogr* 318 (1985) 23.
22. J. S. Kiel, S. L. Morgan, and R. K. Abramson, *J. Chromatogr.* 485 (1989) 585.
23. F. E. Regnier, *Science* 222 (1983) 245.
24. H. Engelhardt and H. Müeller, *Chromatographia* 19 (1984) 77.
25. R. M. Moore and R. R. Walters, *J. Chromatogr.* 317 (1984) 119.
26. M. Verzele, Y.-B. Yang, C. Dewaele, and V. Berry, *Anal. Chem.* 60 (1988) 1329.
27. J. Koyama, J. Nomura, Y. Shiojima, Y. Ohtsu, and I. Horii, *J. Chromatogr.* 625 (1992) 217.
28. Y.-B. Yang, K. Harrison, D. Carr, and G. Guiochon, *J. Chromatogr.* 590 (1992) 35.
29. H. Engelhardt and H. Elgass, *J. Chromatogr.* 158 (1978) 249.
30. J. L. Glajch, M. A. Quarry, J. F. Vasta, and L. R. Snyder, *Anal. Chem.* 58 (1986) 280.
31. P. Jandera and M. Špaček, *J. Chromatogr.* 366 (1986) 107.
32. Sj. van der Wal and L. R. Snyder, *J. Chromatogr.* 255 (1983) 463.
33. G. Vivó-Truyols, J. R. Torres-Lapasií, and M. C. Garcia-Alvarez-Coque, *J. Chromatogr. A* 1018 (2003) 183.
34. Y. Baba and M. K. Ito, *J. Chromatogr.* 485 (1989) 647.
35. B. F. D. Ghrist and L. R. Snyder, *J. Chromatogr.* 459 (1989) 63.
36. B. F. D. Ghrist and L. R. Snyder, *J. Chromatogr.* 459 (1989) 25.
37. S. V. Galushko and A. A. Kamenchuk, *LCGC Int.* 8 (1995) 581.
38. H. Xiao, X. Liang, and P. Lu, *J. Sep. Sci.* 24 (2001) 186.
39. T. Jupille, L. Snyder, and I. Molnar, *LCGC Europe* 15 (2002) 596.
40. W. Markowski and W. Golkiewicz, *Chromatographia* 25 (1988) 339.
41. B. G. Belenkii, A. M. Podkladenko, O. I. Kurenbin, V. G. Mal'tsev, D. G. Nasledov, and S. A. Trushin, *J. Chromatogr. A* 645 (1993) 1.
42. J. P. Larmann, J. J. DeStefano, A. P. Goldberg, R. W. Stout, L. R. Snyder, and M. A. Stadalius, *J. Chromatogr.* 255 (1983) 163.

CHAPTER 3

METHOD DEVELOPMENT

> I pass with relief from the tossing sea of Cause and Theory to the firm ground of Result and Fact.
>
> —Winston Churchill, *The Story of the Malakind Field Force*

3.1 A SYSTEMATIC APPROACH TO METHOD DEVELOPMENT

This chapter will describe the systematic development of a gradient elution separation, primarily for sample analysis, and mainly for the separation of "small-molecule" samples (molecular weight <1000). Chapter 6 should be consulted for the separation of larger molecules, especially biomolecules and synthetic polymers with molecular weights >5000 Da. Much of the information in the present chapter will be applicable to preparative separations, where the goal is to obtain purified components from the sample. However, Chapter 7 contains a more complete account of preparative separations by means of gradient elution. No attention is given in this chapter to the use of procedures other than reversed-phase liquid chromatography, although the general principles revealed here will prove applicable for other separation modes when using gradient elution. See Sections 6.2.2, 8.2, and 8.3 for a discussion of normal-phase, ion-exchange, and other separation modes. Finally, this chapter is mainly concerned with gradient separation *per se*, rather than with such related topics as sample pretreatment or HPLC equipment (including detector options). See Chapter 4, Section 8.1, and [1, 2] for a more complete account of the latter topics. The present chapter assumes no use of mass spectrometric detection (Section 8.1), which requires special buffers.

Figure 3.1 outlines our recommended procedure for developing a gradient method that will be used for sample analysis. It may be unnecessary to fully explore each of the seven steps of Figure 3.1; rather, the intent of method development is a final separation procedure that meets the goals of the chromatographer – at which point further experiments aimed at improving separation are no longer needed. Rarely is a fully optimized separation ever achieved, or even necessary (in the present book we will use the terms "optimization" and "improvement of separation" synonymously, as is common in the literature).

High-Performance Gradient Elution. By Lloyd R. Snyder and John W. Dolan
Copyright © 2007 John Wiley & Sons, Inc.

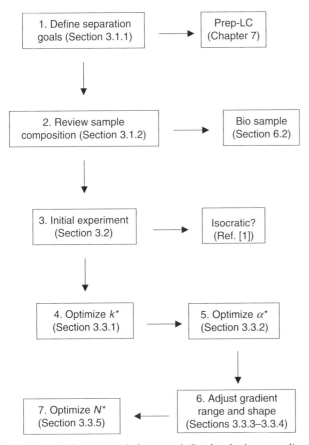

Figure 3.1 Recommended approach for developing a gradient elution separation.

At the start of method development, some thought should be given to

- separation goals;
- nature of the sample;
- experimental conditions for the first separation;
- repeatability of results;
- the availability of HPLC simulation software;
- possible need for sample preparation ("pretreatment").

3.1.1 Separation Goals (Step 1 of Fig. 3.1)

A final separation should aim for some required resolution of peaks of interest, within an acceptable run time. When a large number of samples must be assayed (at acceptable cost), short run times can be paramount. For example, there are increasing reports of so-called "ballistic gradients," which can be completed in a minute or less. In addition, consideration should be given to method ruggedness and detection requirements. Separations which are sensitive to small changes in experimental conditions are generally undesirable; some applications may place a premium on method ruggedness and robustness. Finally, the choice of detector (UV, MS, etc.) usually constrains the choice of mobile phase. Mobile phases that are transparent at low wavelengths (e.g., <210 nm) are preferred for UV detection, while for MS detection the mobile phase must be volatile and not adversely affect the sensitivity of the mass spectrometer (Section 8.1). The expected column pressure drop (referred to hereafter as "column pressure" or "pressure") should also be kept in mind. Many HPLC systems work best at pressures <3000 psi (<200 bar, <20 MPa), and allowance must be made for the usual increase in column pressure during use (although the use of replaceable column filters and guard columns can minimize pressure buildup).

Gradient elution can support any of several separation objectives:

- preliminary sample assessment;
- development of a routine assay procedure;
- development of a "generic" or "fingerprint" separation;
- development of a preparative separation.

A preliminary sample assessment can be carried out with a single, well-chosen gradient separation (Section 3.2). Such an experiment can answer such questions as:

- Is isocratic separation (often preferable) possible for this sample? If so, what mobile phase %B would be a good choice?
- Is gradient elution promising with the conditions used initially?
- Should the initial and/or final %B values of subsequent gradient experiments be changed?
- Does the resulting chromatogram support our preliminary information on the nature of the sample (Section 3.1.2)?

The initial gradient run can be used to decide whether isocratic or gradient elution will be the best choice for the final separation. This first experiment may also suggest that gradient elution with the starting conditions does not look promising, for example, bunching or late-elution of peaks (Section 3.2.1.2). An initial gradient separation can also be used to shorten run time in subsequent method development experiments, by reducing the gradient range (difference in initial and final %B $\equiv \Delta\phi$; see Fig. 1.4g). Finally, an examination of this first separation can be compared with what we know (or think we know) about the sample. If some number n of sample components is expected, the presence of a greater number of peaks in the chromatogram may suggest further questions about the sample before proceeding further with method development (where did the "extra" peaks come from?). When the number of peaks is $<n$, hidden (overlapped) peaks can be assumed, that is, peaks that will need to be separated in the final gradient method. The initial gradient separation may also indicate a sample that is too complex (too many peaks) to be baseline separated by a single gradient procedure.

The *development of a routine assay procedure* is the most likely starting goal, where the final method will be used to assay samples of interest. Here, the usual requirements are $R_s \geq 2$ for all analytes, the shortest possible run time, and (usually) an operating pressure $\leq$2000 psi for a new column. It is often possible to reduce run time by further changes in separation selectivity (combined with shortening the column and an increase in flow rate, e.g., Fig. 3.12), which means that possible savings from a shorter run time during routine analysis should be weighed against the cost of further method development. Thus, the final method depends on the relative importance of (a) resolution, (b) run time, and (c) the amount of time that can be justified for method development. In the case of liquid chromatography–mass spectrometry (LC-MS) separation (Section 8.1), resolution becomes less important, because of the selective detection provided by the mass spectrometer. The present chapter is devoted mainly to the development of routine assay procedures that will use UV-detection, as outlined in Figure 3.1 and further detailed in Table 3.1.

There is increasing interest in the development of "*generic*" *separation procedures*. By a "generic" procedure, we mean that exactly the same experimental conditions will be used for different samples, each of which may contain compounds that are unique to that sample. For example, thousands of different samples may be generated by means of combinatorial chemistry, each of which contains a different reaction product plus some related impurities. What is needed for such samples is a single separation procedure that can provide at least partial separation of product from impurities, regardless of the chemical structure of the product. Similarly, a single gradient procedure is often desirable for the partial separation of (a) peptide digests from the enzymatic hydrolysis of different proteins or (b) mixtures of different compounds present in plant or animal extracts. Reversed-phase gradient elution is generally the best option to meet this challenge, especially in combination with mass spectrometric detection (Section 8.1).

Fingerprint procedures are similar to generic separations. For example, in the production of recombinant proteins, it is important to assess both the purity and identity of the protein product. This can be accomplished by preparing an enzymatic digest of the sample, followed by a gradient separation of the peptides in the digest. By comparing the chromatograms of standard and assay samples, any small differences in the product can be visually identified and related to possible contamination or failure of the manufacturing process. Another application of fingerprinting is in the separation of various plant-derived materials for purposes of identifying the source or species [3], by a comparison of the sample chromatogram with reference chromatograms for different species. Fingerprint assays need not require baseline separation of all the peaks in the chromatogram.

The *development of a preparative separation* becomes necessary when it is required to isolate purified product from a mixture. The recovery of small quantities ($<$100 µg) of purified product can be carried out on the basis of the present chapter, usually with a goal of $R_s \geq 1.5$ for the product peak. Separations to be used for obtaining larger amounts of purified product will benefit from $R_s \gg 2$ (Chapter 7).

TABLE 3.1 Outline for the Development of a Routine Gradient Separation (Compare with Fig. 3.1)

Step (as in Fig. 3.1)	Comment
1. Define separation goals	Section 3.1.1
2. Collect information on sample	a. Molecular weight >5000 Da? (see Chapter 6) b. Mobile phase buffering required? c. Sample pretreatment required?
3. Carry out initial separation (run 1)	a. 15 min gradient (run 1); Table 3.2 b. Any problems? (Section 3.2.1.2, Fig. 3.5) c. Isocratic separation possible? (Figs 3.2 and 3.3)
4. Optimize gradient retention k^*	The conditions of Table 3.2 should yield an acceptable value of $k^* \approx 5$
5. Optimize separation selectivity α^*	Increase gradient time by 3-fold (run 2, 45 min); increase temperature by 20°C (runs 3 and 4); see Figure 3.6
5a. If best resolution from step 5 is $R_s < 2$, or if very short run times are required, vary further conditions to optimize peak spacing (for maximum R_s or minimum run time)	a. Replace acetonitrile by methanol and repeat runs 1–4 b. Replace column and repeat runs 1–4 c. Change pH and repeat runs 1–4 d. Consider use of segmented gradients (Section 3.3.4)
6. Adjust gradient range and shape	a. Select best initial and final values of %B for minimum run time with acceptable R_s (Section 3.3.3) b. Add a steep gradient segment to 100 percent B for "dirty" samples (e.g., Fig. 3.8b) c. Add a steep gradient segment to speed up separation of later, widely spaced peaks (Fig. 3.8c)
7. With best separation from step 5 or 6, choose best compromise between resolution and run time	Vary column conditions (Section 3.3.5)
8. Determine necessary column equilibration between successive sample injections	Using the procedure developed above, carry out successive, identical separations while varying the equilibration time between runs; select an equilibrium time which results in no change in separation between adjacent runs (Section 5.3)

3.1.2 Nature of the Sample (Step 2 of Fig. 3.1)

The possible need for a pretreatment of the sample (Section 3.1.6) should be considered prior to any sample injections, although in many cases gradient elution can minimize or even eliminate the need for sample pretreatment. The chemical nature and molecular structure of each compound present in the sample also require attention before proceeding further. Some sample characteristics which need to be considered include:

- molecular weight;
- presence of acids or bases (and their approximate pK_a values);
- presence of isomers or enantiomers;
- presence of very polar compounds;
- presence of very hydrophobic compounds.

Samples with molecular weights >10,000 Da usually require columns with pore diameters >10 nm, as well as the use of less-steep gradients (Chapter 6). If acids or bases are present in the sample, the mobile phase should be buffered in most cases; a buffer is not required for non-ionizable samples. The separation of isomers is often possible by means of RP-LC [1, 4], although normal-phase separation on a silica column is usually more effective. The separation of enantiomers requires special conditions; usually a so-called "chiral" column will be necessary [1]. Very polar compounds such as sugars, as well as molecules that are ionized over a wide pH-range (quaternary ammonium compounds, sulfonic acids, etc.), may not be retained sufficiently in RP-LC separation for their adequate separation; ion-pairing [1] or normal-phase chromatography (Section 8.3) is often required for such samples. Finally, very hydrophobic compounds may require the use of (a) nonaqueous reversed-phase chromatography (NARP) [1], (b) normal-phase chromatography, or (c) more polar columns such as phenyl or cyano. Polar RP-LC columns are characterized by low hydrophobicity values **H** (Appendix III).

3.1.3 Initial Experimental Conditions

The choice of column and mobile phase depends on the nature of the sample (see preceding section). In most cases, a C_8 or C_{18} type-B column is recommended (Appendix III), with particle pore diameters of 8–12 nm. In the absence of prior information on the sample to be separated, a buffered, near-full-range gradient (e.g., 5–100 percent B) should be run initially, while keeping in mind the need for column equilibration prior to each sample injection (10 column volumes between successive gradients is suggested, but see Section 5.3 for further guidance). For additional details, see the recommendations of Table 3.2 and Section 3.2.

3.1.4 Repeatable Results

Method development (whether for isocratic or gradient elution) requires repeatable experiments, that is, yielding the same retention times for all peaks when an experiment is replicated. Apart from the usual contributions to method variability (poorly controlled conditions and equipment performance), gradient elution is susceptible to two unique repeatability problems: (a) differences in the hold-up volume of different gradient systems ("dwell" volume, Sections 2.3.6.1 and 5.2.1), and (b) slow column

TABLE 3.2 Preferred Conditions for the Initial Experiment In Gradient Method Development. A Small-Molecule Sample (100–500 Da) is Assumed

Column[a]	Type	C_8 or C_{18} (type B)
	Dimensions	150 × 4.6 mm
	Particle size	5 μm
	Pore diameter	8–12 nm
Mobile phase	Sample contains no acids or bases	Acetonitrile–water
	Sample contains acids and/or bases	Acetonitrile–aqueous buffer (pH 2.5–3.0)[b]
Flow rate		2.0 mL/min
Temperature		30 or 35°C
Gradient		5–100 percent B in 15 min
Sample	Volume	≤50 μL
	Weight	≤10 μg
k^*		≈5

[a]Alternatively, use a 100 × 4.6 mm column of 3 μm particles.
[b]See Section 3.2 for details on buffer composition, for example, aqueous 0.5 g/L phosphoric acid.

equilibration between successive gradient runs (Section 5.3). Gradient elution experiments for use in method development are best carried out with a single HPLC system, with adequate equilibration of the column prior to each sample injection [e.g., flushing the column with initial-%B mobile phase ($\phi = \phi_0$) for 5–10 min between gradient runs]. Duplicate runs of each experiment should be carried out to confirm that the column is adequately equilibrated, and the initial experimental run (Section 3.2) should be repeated at the end of each day, in order to confirm that column performance has not changed. The dwell volume V_D of the equipment used to develop a gradient method should be known or measured (Section 4.3.1.2). See Section 3.5 for additional information relating to reproducible results.

3.1.5 Computer Simulation: Yes or No?

When developing a gradient method, planning for the initial experiments depends in part on whether or not simulation software is available (Section 3.4). Computer simulation allows the use of as few as two initial experiments, in order to make accurate predictions of separation for changes in one or more experimental conditions. Because of the added complexity of gradient elution (compared with isocratic separation), computer simulation can save a great deal of time in gradient method development, as well as providing shorter run times and/or better resolution for the final method. In Sections 3.2–3.3, dealing with gradient method development, we will assume that simulation software is *not* available, so that a "trial-and-error" strategy becomes necessary. In Section 3.4 the results of trial-and-error method development are compared with the use of computer simulation, and some further uses of computer simulation are explored.

3.1.6 Sample Preparation (Pretreatment)

A need for sample pretreatment prior to injection can result from

- the presence of particulates in the sample;
- the presence of sample components that are not eluted by the gradient and which can build up on the column;
- the presence of interfering compounds;
- samples that are too dilute;
- samples dissolved in a solvent that is too "strong."

A detailed discussion of sample pretreatment is provided in [1]; some of the above five sample characteristics may be less important for gradient elution. Sample particulates need to be removed before the sample enters the column; however, the use of in-line filters and guard columns can provide sufficient protection in many cases. For gradients that end with 100 percent B, the retention of strongly held sample compounds is less likely (e.g., the example of Fig. 1.2). If a gradient ending in 100 percent B does not wash the entire sample from the column, the use of a stronger B-solvent such as tetrahydrofuran can sometimes overcome this problem, or a final column wash with a stronger solvent can be added to the original gradient. The need to remove early-eluting interferences from the sample is reduced in gradient elution, because of improved resolution at the front of the chromatogram (compared to isocratic elution). For samples that are strongly retained by water or aqueous buffer as mobile phase, large volumes of dilute aqueous samples can often be injected in gradient elution with on-column concentration, in order to provide enough sample for the detection of trace components of interest (this approach may be less effective or inconvenient with isocratic elution). Finally, the sample received for analysis may have been dissolved in a solvent that interferes with separation [e.g., dimethylsulfoxide (DMSO) or other organic]. In some cases, dilution of the sample with water may allow its direct injection, by converting the injection solvent to a composition $\phi \approx \phi_0$; in other cases, replacement of the solvent with water or an aqueous buffer may be required.

3.2 INITIAL EXPERIMENTS

The equipment used in gradient elution is discussed in Chapter 4. Initial experiments will preferably involve linear gradients formed from a weaker A solvent and a stronger B solvent. The A solvent is typically water or (for samples that contain acids or bases) an aqueous buffer; the B solvent is usually unbuffered ACN or methanol. For UV detection, buffers and organic solvents are selected to be transparent at the detection wavelength. Table 3.3 summarizes the approximate absorbance of various possible mobile phase components as a function of wavelength. Because a phosphate buffer with acetonitrile as B solvent allows detection at wavelengths as low as 200 nm, this is a preferred choice for samples that can only be detected below 210 nm. A low-pH mobile phase and a type-B alkylsilica column (Appendix III)

TABLE 3.3 UV Absorbance of Reversed-Phase Mobile-Phase Components as a Function of Wavelength

	Absorbance (AU) at wavelength (nm) specified									
	200	205	210	215	220	230	240	250	260	280
SOLVENTS										
Acetonitrile	0.05	0.03	0.02	0.01	0.01	<0.01				
Methanol	2.06	1.00	0.53	0.37	0.24	0.11	0.05	0.02	<0.01	
Degassed	1.91	0.76	0.35	0.21	0.15	0.06	0.02	<0.01		
Isopropanol	1.80	0.68	0.34	0.24	0.19	0.08	0.04	0.03	0.02	0.02
Tetrahydrofuran										
Fresh	2.44	2.57	2.31	1.80	1.54	0.94	0.42	0.21	0.09	0.05
Old	>2.5	>2.5	>2.5	>2.5	>2.5	>2.5	>2.5	>2.5	2.5	1.45
ACIDS AND BASES										
Acetic acid, 1%	2.61	2.63	2.61	2.43	2.17	0.87	0.14	0.01	<0.01	
Hydrochloric acid, 6 mM (0.02%)	0.11	0.02	<0.01							
Phosphoric acid, 0.1%	<0.01									
Trifluoroacetic acid										
0.1% in water	1.20	0.78	0.54	0.34	0.20	0.06	0.02	<0.01		
0.1% in acetonitrile	0.29	0.33	0.37	0.38	0.37	0.25	0.12	0.04	0.01	<0.01
Ammonium phosphate, dibasic, 50 mM	1.85	0.67	0.15	0.02	<0.01					
Triethylamine, 1%	2.33	2.42	2.50	2.45	2.37	1.96	0.50	0.12	0.04	<0.01
BUFFERS AND SALTS										
Ammonium acetate, 10 mM	1.88	0.94	0.53	0.29	0.15	0.02	<0.01			
Ammonium bicarbonate, 10 mM	0.41	0.10	0.01	<0.01						
EDTA (ethylenediaminetetraacetic acid), disodium, 1 mM	0.11	0.07	0.06	0.04	0.03	0.03	0.02	0.02	0.02	0.02

Buffer / Detergent										
HEPES [N-(2-hydroxyethyl)piperazine-N'-2-ethanesulfonic acid], 10 mM pH 7.6	2.45	2.50	2.37	2.08	1.50	0.29	0.03	<0.01		
MES[2-(N-morpholino)ethanesulfonic acid], 10 mM, pH 6.0	2.42	2.38	1.89	0.90	0.45	0.06	<0.01			
POTASSIUM PHOSPHATE										
Monobasic, 10 mM	0.03	<0.01								
Dibasic, 10 mM	0.53	0.16								
Sodium acetate, 10 mM	1.85	0.96	0.52	0.30	0.15	0.03	<0.01			
Sodium chloride, 1 M	2.00	1.67	0.40	0.10	<0.01					
Sodium citrate, 10 mM	2.48	2.84	2.31	2.02	1.49	0.54	0.12	0.03	0.02	0.01
Sodium formate, 10 mM	1.00	0.73	0.53	0.33	0.20	0.03	<0.01	0.01		
Sodium phosphate, 100 mM, pH 6.8	1.99	0.75	0.19	0.06	0.02	0.01	0.01	<0.01		
Tris–hydrochloric acid, 20 mM										
pH 7.0	1.40	0.77	0.28	0.10	0.04	<0.01	<0.01			
pH 8.0	1.80	1.90	1.11	0.43	0.13	<0.01				
DETERGENTS										
Brij 35 (23 lauryl ether), 1%	0.06	0.03	0.02	0.02	0.01	<0.01	0.01			
CHAPS (3-[3-cholamidorophyl)dimethyl-ammonio]-1-propanesulfonate), 0.1%	2.40	2.32	1.48	0.80	0.40	0.08	0.04	0.02	0.02	0.01
SDS (sodium dodecyl sulfae), 0.1%	0.02	0.01	<0.01							
Triton X-100 (octoxynol), 0.1%	2.48	2.50	2.43	2.42	2.37	2.37	0.50	0.25	0.67	1.42
Tween 20 (polyoxyethlenesoritan monolaurate), 0.1%	0.21	0.14	0.11	0.10	0.09	0.06	0.05	0.04	0.04	0.03

Source: J.B. Li, *LCGC* 10 (1992) 856, with permission.

are recommended for initial experiments, as a means of minimizing peak tailing for ionizable sample components.

The solubility of the buffer in the B solvent is also an important consideration. Buffers are generally more soluble in methanol than in acetonitrile. If the buffer is insoluble in the B solvent, it is usually necessary to stop the gradient short of 100 percent B. However, it should be noted that the buffer is often only added to the A solvent, so its concentration in the B solvent (organic) will be zero. Potassium salts tend to be more soluble than sodium salts, and are preferred for this reason. One study [5] found that potassium phosphate as buffer begins to precipitate near 90 percent ACN, suggesting that gradients with this buffer extend no further than 90 percent ACN, depending on the concentration of the buffer in the A solvent. However, whether or not a buffer precipitates in high-%B mobile phase depends on both buffer concentration and the design of the gradient mixer. For operation at pH ≤ 2.7, $0.1-0.5$ g/L phosphoric acid can be used as buffer [6], which presents no problem with buffer precipitation for 0–100 percent B gradients. For more information on buffer solubility as a function of organic solvent, mobile phase pH and buffer type, see [5].

If a buffer is required, it is recommended that initial experiments be carried out with a mobile phase pH < 3 in order to minimize tailing of protonated basic compounds, for example, either 0.5 g/L phosphoric acid or 10 mM potassium phosphate with pH = 2.5 (but the gradient should be limited to ≤ 85 percent B potassium phosphate). See the further discussion of peak tailing in Section 5.4.5 and Appendix III. For mass-spectrometer detection (which is being used increasingly), volatile buffers and solvents are required, and analytes may need to be ionized in the mobile phase (Section 8.1).

Ideally, a full-range gradient (0–100 percent B) is preferred for the initial experiment. However, as discussed above, it may be necessary to truncate the gradient due to reduced buffer solubility in the B solvent. Similarly, initiating the gradient at 0 percent B can cause problems with some columns, due to nonwetting of the stationary phase by organic-free water [7, 8]. For this reason, it is recommended to initiate the gradient at 5 percent B or higher, unless it is known that the column can tolerate 0 percent B. Problems with stationary-phase wetting are more likely for heavily bonded C_{18} columns (larger **H**; Appendix III), and unlikely for columns that are lightly bonded, which contain embedded polar groups, or are end-capped with polar groups.

The first gradient run is important as a means of (a) assessing the likely difficulty of method development and (b) planning further experiments. Some general considerations for the design of this initial run were described in Section 3.1.3; Table 3.2 recommends specific starting conditions: 5–100 percent ACN–buffer in 15 min, 150 × 4.6 mm C_8 or C_{18} column, 30 or 35°C, and 2.0 mL/min. Other column configurations also can be used, for example, a 250 × 4.6 mm column with a flow rate of 1.0 mL/min, or smaller-diameter columns with flow rates reduced in proportion to column-diameter-squared. A representative initial gradient run is shown in Figure 3.2, based on the "regular" sample of Table 1.3 and the experimental conditions of Table 3.2.

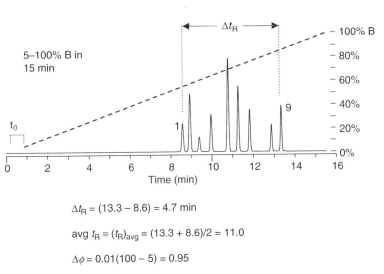

$\Delta t_R = (13.3 - 8.6) = 4.7$ min

avg $t_R \equiv (t_R)_{avg} = (13.3 + 8.6)/2 = 11.0$

$\Delta \phi = 0.01(100 - 5) = 0.95$

Figure 3.2 Use of a standard gradient run to determine whether isocratic or gradient elution is best for the sample. In this example, the "regular" sample was separated with the recommended initial conditions of Table 3.2 (5–100 percent acetonitrile in 15 min, 150 × 4.6 mm C$_{18}$ column, 5 μm particles, 2.0 mL/min, 30°C). A 250 × 4.6 mm column and 1.0 mL/min can be used instead, if desired (with little change in interpretation of the chromatogram).

3.2.1 Interpreting the Initial Chromatogram (Step 3 of Fig. 3.1)

An initial gradient run can be used to answer two questions [9–12]: (a) Will gradient elution be required for the sample? (b) If isocratic separation is possible, what isocratic mobile phase should be tried first in order to achieve $1 \leq k \leq 10$ for all peaks? The following discussion describes a simple approach that should prove adequate for separations of "small-molecule" samples with molecular weights in the range of 100–500 Da. Samples containing larger molecules usually require gradient elution.

In the initial separation of Figure 3.2, retention times for the first and last peaks (1 and 9) are equal to 8.6 and 13.3 min, respectively. The latter retention times determine whether or not isocratic separation is feasible. First, calculate the *difference* in retention times (Δt_R) for peaks 1 and 9 (13.3–8.6), or $\Delta t_R = 4.7$ min. Also, calculate the *average* retention time for the first and last peak: $(t_R)_{avg} = (8.6 + 13.3)/2 = 11.0$ min. Samples which have small values of Δt_R can be separated isocratically with acceptable values of k ($1 \leq k \leq 10$), while samples with larger values of Δt_R may require gradient elution. An approximate rule for deciding whether to use isocratic or gradient elution is as follows (Section 2.2.3.2): if $\Delta t_R/t_G \leq 0.25$, use isocratic elution. If $\Delta t_R/t_G \geq 0.40$, use gradient elution. For intermediate values of $\Delta t_R/t_G$, either isocratic or gradient elution may prove best.

A somewhat more accurate assessment of the possibility of isocratic elution is as follows, because acceptable (i.e., maximum) values of Δt_R for isocratic separation

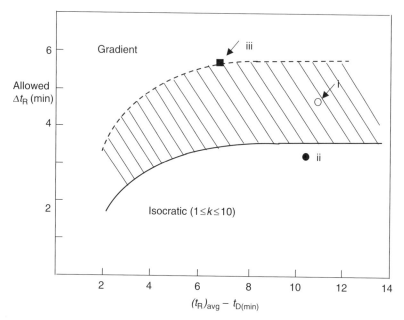

Figure 3.3 Determining whether isocratic elution is possible for a given sample, based on the measurement of retention times as in Figure 3.2. Assumes 5–100 percent B gradient in 15 min, sample molecular weight of 100–500 Da. See text for details.

also depend on the average value of t_R in the gradient separation. Figure 3.3 relates maximum allowable values of Δt_R for isocratic elution to values of $(t_R)_{\text{avg}}$, assuming a 5–100 percent B gradient in 15 min (for other gradients, see Appendix II). The region below the cross-hatch area denotes samples that can be separated isocratically with $1 \leq k \leq 10$, while the region above the cross-hatch area indicates samples that require gradient elution. For samples that fall within the cross-hatched region, either isocratic or gradient elution may be feasible, but with a range in k for isocratic separation as large as $1 \leq k \leq 20$. For the example of Figure 3.2, $\Delta t_R = 4.7$ min and $(t_R)_{\text{avg}} = 11$. This sample is designated by the open circle and arrow (labeled "*i*") within the cross-hatched region. Therefore, isocratic separation is possible, but not with ($1 \leq k \leq 10$).

When isocratic separation is feasible, the recommended %B for the isocratic mobile phase can be estimated from the value of $(t_R)_{\text{avg}}$ [Equation (II.11) of Appendix II]:

$$\text{isocratic } \%B \approx 6.3[(t_R)_{\text{avg}} - t_D] - 2 \tag{3.1}$$

Here, t_D is the hold-up or "dwell" time of the gradient system, equal to the dwell volume V_D divided by flow rate (Section 4.3.1.2). For the separation of Figure 3.2, $t_D = 0$. As an example, consider the example of Figure 3.2 with peaks 8 and 9 omitted; resulting values of Δt_R and average t_R are then 3.2 and 10.2 min, respectively. Referring to Figure 3.3 again (solid point labeled "*ii*"), we see that the latter sample (minus peaks 8 and 9) *can* be separated isocratically

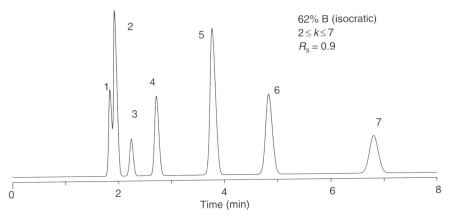

Figure 3.4 Isocratic separation of the "regular" sample (peaks 1–7 only) as predicted from the initial gradient separation in Figure 3.2. Conditions as in Figure 3.2 except 62 percent B (isocratic); see Table 3.2 for other conditions and the text for details.

with $1 \geq k \geq 10$. From Equation (3.1), the best mobile phase for an isocratic separation of this sample is 62 percent B. This separation is shown in Figure 3.4, where $2 \leq k \leq 7$ (acceptable for isocratic separation). The first two peaks have marginal resolution ($R_s = 0.9$), which might be improved by varying conditions so as to provide better peak spacing or selectivity (Section 2.1.4.2). For other gradients (values of t_G, $\Delta\phi$, and initial %B other than 5–100 percent in 15 min), see Equation (II.12) of Appendix II.

The above discussion concerns only whether isocratic elution is possible for the RP-LC separation of a given sample, not whether isocratic elution should be preferred. When isocratic elution is possible, many previous workers have avoided gradient elution for reasons cited in the Preface. Among these reasons is the assumption that gradient separation will always be slower than isocratic elution, primarily because of the need to equilibrate the column between successive runs (Section 5.3). More recently, the latter assumption has been challenged [13]; for most assays by gradient elution it now appears possible to reduce the time required for column equilibration to a small fraction of the gradient time (Section 9.2.1). Therefore, the need for column equilibration may no longer be a reason to avoid gradient elution for routine analysis. When isocratic separation is possible, we recommend that the choice of gradient vs isocratic elution should be made on the basis of several considerations: comparative run times, resolution and detection sensitivity for isocratic vs gradient elution, the availability of gradient equipment, and experience with gradient elution in laboratories in which the method will be carried out, and so on. Very likely, isocratic elution will continue to be preferred when the sample can be separated with $1 \leq k \leq 10$.

3.2.1.1 "Trimming" a Gradient Chromatogram

If we decide to separate the sample of Figure 3.2 (peaks 1–9) using gradient elution, it can be seen that there is "wasted" space in the chromatogram prior to the elution of peak 1 and following elution of peak 9. By "trimming" the chromatogram or reducing the gradient range, we minimize this wasted time. It is seen from Figure 3.2 that the first peak (1) leaves the column at $\phi_e = 54$ percent B, while the last peak (9) leaves at 85 percent B. This suggests that the gradient can be shortened to 54–85 percent B, although this is only a very rough (and not very good) approximation. For reasons to be discussed (Section 3.3.3), the gradient range needs to be somewhat wider than suggested by this example (54–85 percent B). In any case, *it is recommended to leave values of initial and final %B unchanged until certain other steps in method development (Sections 3.3.1 and 3.3.2) are complete*, because of (a) the possibility of peaks leaving the column too early or too late (see examples of Figs 2.19 and 2.20), and (b) ease of planning further experiments that might be carried out unattended. The main exceptions are separations of higher-molecular-weight samples such as peptides, proteins, and synthetic polymers (Chapter 6), which typically require longer gradients; for these separations, trimming the gradient at the start of method development can save considerable time in subsequent experiments (with little effect on separation, other than reduced run times).

3.2.1.2 Possible Problems Tailing peaks or other separation problems may be encountered in the initial gradient separation. In such cases, it is important to correct the problem (Section 5.4.5, Appendix III) before carrying out the further experiments of Section 3.3 below. If the correction of peak tailing is delayed until later in method development, resulting changes in relative retention (selectivity) may require additional experiments that could otherwise have been avoided.

Figure 3.5 illustrates three other potential problems that may be apparent from an initial gradient run. *Early elution* of peaks as in Figure 3.5(*a*) is not uncommon for small, polar molecules, especially ionized acids or bases. Some improvement in separations such as that of Figure 3.5(*a*) can be obtained by a reduction in initial %B for the gradient (if feasible), or by the use of an initial isocratic hold as in Figure 2.23(*d*). For samples containing acids or bases, a change in mobile phase pH will generally be more effective; a decrease in pH will result in the decreased ionization of acids and their increased retention, and vice versa for bases. The addition of an ion-pair reagent is an alternative way to increase the retention of early-eluting, ionized compounds [1]; however, the use of ion-pairing with gradient elution is generally not recommended because of possible slow column equilibration (Section 3.3.7). (An exception to the latter rule is the use of trifluoroacetic acid for peptide and protein separations; Section 6.2.1.2.) Neutral, very polar molecules, such as carbohydrates, are also less retained in reversed-phase chromatography; the separation of such samples with good retention may require the use of normal-phase or (preferably) hydrophilic interaction chromatography ("HILIC," Section 8.3.3 and [1]).

Late elution, as in Figure 3.5(*b*), suggests that the sample may be too nonpolar for separation by the usual reversed-phase conditions. In such cases, an acetonitrile–water gradient can be replaced by a gradient from acetonitrile to a less polar solvent, such as tetrahydrofuran or (better) methyl-*t*-butyl ether, either of which is a stronger RP-LC solvent than acetonitrile (buffer solubility should be checked for either of the latter two gradients, but a buffer is often not required for very nonpolar samples).

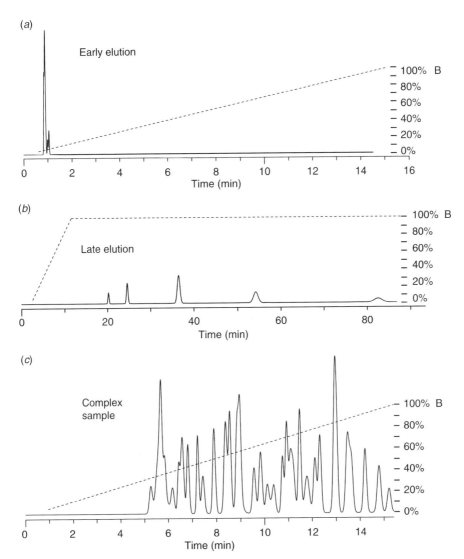

Figure 3.5 Examples of potential problems in an initial gradient run. (*a*) Sample is too nonretentive; (*b*) sample is excessively retentive; (*c*) sample contains too many components to be separated in a single run.

Alternatively, a less hydrophobic column (lower value of **H**; Appendix III) or normal-phase chromatography (Section 8.3) can be used.

Complex samples with numerous components can result in crowded chromatograms, as in Figure 3.5(*c*). For such samples, it is unlikely that a single reversed-phase separation will be able to separate all peaks to baseline (Section 2.2.4). If all sample components are of interest, it may be necessary to develop a more powerful separation scheme. Two-dimensional chromatography is the most commonly used option (Section 3.7.1), where fractions from an initial run are further resolved in a second "orthogonal" separation. The orthogonal separation should possess a markedly different selectivity compared to the initial run (Section 3.7).

3.3 DEVELOPING A GRADIENT SEPARATION: RESOLUTION VERSUS CONDITIONS

The *gradient retention factor* k^* is the key to understanding and improving a gradient separation. Values of the retention factor k in *isocratic* elution can be approximated by

$$\log k = \log k_w - S\phi \quad (3.2)$$

where ϕ is the volume fraction of the B solvent in the mobile phase ($\phi = 0.01 \times \%B$), k_w is an extrapolated value of k for water as mobile phase ($\phi = 0$), and S is a constant for a given sample compound and fixed experimental conditions (except for %B, which is allowed to vary). $S \approx 4$ for typical "small-molecule" samples with molecular weights of 100–500 Da.

Values of k^* in gradient elution depend on gradient time t_G, flow rate F, column dead volume V_m, the change in ϕ during the gradient $\Delta\phi$ (Fig. 1.4g), and the value of S for a given sample compound [from Equation (3.2)]:

$$k^* = (t_G F)/(1.15 V_m \Delta\phi S) \quad (3.3)$$

An equivalent expression for k^* is

$$k^* = t_G/(1.15 t_0 \Delta\phi S) \quad (3.3a)$$

Here, t_0 is the column dead-time, equal to V_m/F. For the separation of Figure 3.2, $t_G = 15$ min, $F = 2.0$ mL/min, $S \approx 4$, $\Delta\phi = 0.95$ (5–100 percent B gradient), and the column-volume $V_m = 1.5$ mL. For an internal column diameter of 4.6 mm, $V_m \approx 0.01$ times the column length L in mm; for other column diameters,

$$V_m \approx 0.0005 (\text{column i.d.})^2 L \quad (3.3b)$$

where column i.d. and L are each in mm. The average value of k^* in Figure 3.2 [Equation (3.3)] is therefore $(15 \times 2)/(1.15 \times 1.5 \times 0.95 \times 4) \approx 4.6$. Just as we prefer $1 \leq k \leq 10$ in an isocratic separation (for acceptable run time and resolution, and improved detection), a similar recommendation exists for gradient elution, that is, $1 \leq k^* \leq 10$, although $0.5 \leq k^* \leq 20$ may be acceptable for some applications of gradient elution. Values of k^* are most conveniently varied by varying gradient time t_G:

$$t_G = 1.15 V_m \Delta\phi S k^*/F$$
$$= 1.15 t_0 \Delta\phi S k^* \quad (3.3c)$$

Thus, for a 50 mm column (with $V_m = 0.5$ mL), a flow rate of 4 mL/min, a gradient range of 0–60 percent B ($\Delta\phi = 0.6$), $S \approx 4$, and a (minimum) value of $k^* = 1$, the recommended gradient time would be $1.15 \times 0.5 \times 0.6 \times 4 \times 1/4 = 0.3$ min. It is seen that very short gradient times can be achieved by the combination of short columns, a reduced gradient range, lower values of k^*, and high flow rates. Similar conditions might well be useful for the rapid screening of a large number of random samples, as in the analysis of samples from combinatorial chemistry.

Resolution for a gradient separation was described in Section 2.2.3 as a function of the gradient retention factor k^*, the "effective" (median) separation factor α^* of two adjacent peaks, and the "effective" (median) column plate number N^* by the relationship

$$R_s = (1/4)\underbrace{[k^*/(1+k^*)]}_{(i)}\underbrace{(\alpha^*-1)}_{(ii)}\underbrace{N^{*1/2}}_{(iii)} \qquad (3.4)$$

The use of Equation (3.4) for guiding the development of a gradient separation is almost identical to the use of an equivalent relationship for isocratic elution (Section 2.1.4 and [1]):

$$R_s = (1/4)[k/(1+k)](\alpha-1)N^{1/2} \qquad (3.4a)$$

We recommend to optimize k first, then α for the critical peak pair, and finally (if necessary) N. Different separation conditions allow the separate control of k, α, and N. Because changes in conditions in either isocratic or gradient elution will have similar effects on k and k^* (as well as on separation), method development in gradient elution can be carried out in a similar fashion to that for isocratic separation [1]. Furthermore, the resolution of two adjacent peaks in gradient and isocratic elution will be similar (Section 9.1.3), when k(isocratic) $\approx k^*$(gradient), assuming otherwise similar ("corresponding") conditions; see Section 2.2.1.1.

When developing a gradient method, first adjust retention k^* for $1 \leq k^* \leq 10$, then improve selectivity α^*, and finally adjust column plate number N^*. In addition, other changes in the gradient program may prove useful: changes in the value of %B for the beginning and/or end of the gradient, the use of segmented gradients, and so on. An example of this method-development process will be presented next, using the "irregular" sample of Table 1.3 and following the series of steps laid out in Figure 3.1. The data of Table 1.3 were used to simulate separations for this sample (computer simulation, Sections 1.5 and 3.4), in place of additional trial-and-error experiments.

Our primary objective in developing a gradient elution method is to achieve acceptable sample resolution, usually defined as $R_s > 1.5$ (baseline separation), and preferably $R_s \geq 2$. At the same time, a short run time is desirable; in some cases the shortest possible run may be paramount. If an acceptable separation results after we achieve a satisfactory range in values of k^*, we may decide to stop method development at this point. Likewise, if separation is acceptable after optimizing α^*, there may be no need for further experiments involving change in the column or flow rate. Keep in mind that "optimization" is a relative, rather than absolute, term; also, "better" is the enemy of "good enough."

3.3.1 Optimizing Gradient Retention k^* (Step 4 of Fig. 3.1)

The initial gradient conditions recommended in Table 3.2 will result in an average value of $k^* \approx 5$ for most small-molecule samples (those with molecular weight <500), that is, a value of k^* in the middle of the desirable range $1 \leq k^* \leq 10$. Thus, unlike isocratic method development, the first gradient elution experiment can be carried out so as to guarantee an optimized sample retention k^*. The initial separation of the "irregular" mixture with these conditions is shown in Figure 3.6(a). We note first that the difference in retention times between first and last peaks is $9.7 - 4.0$ min, or $\Delta t_R = 5.7$ min, and the average value of t_R is 6.8 min. From Figure 3.3 (solid square, "*iii*"), we estimate that isocratic separation is barely possible for $0.5 \leq k \leq 20$, making gradient elution a much more likely choice. The initial separation of Figure 3.6(a) is reasonably promising, with only one overlapping peak pair (5 + 6, arrow). The next step is to change separation conditions so as to improve peak spacing (selectivity) and resolution.

3.3.2 Optimizing Gradient Selectivity α^* (Step 5 of Fig. 3.1)

Changes in values of α^* can be achieved by varying any of the conditions of Table 3.4 (see the detailed discussion of Section 3.6). A growing body of evidence [4, 14–20] suggests that gradient time and temperature should be changed first, as the preferred means for adjusting values of α^* during initial method-development experiments (while maintaining k^* within a reasonable range of values). Our usual recommendation for the second method-development experiment is to increase gradient time by a factor of 2–3. For the example of Figure 3.6, a gradient time of 45 min was chosen, holding other conditions constant (Fig. 3.6b, $k^* \approx 15$). A longer gradient time means an increase in k^* and often a better overall separation (e.g., Fig. 2.6 and accompanying text). A change in gradient time can also affect selectivity (relative retention or values of α^*), which is of primary interest at this point in our discussion.

The effect of a change in gradient time on relative retention (values of α^*) can be seen in the examples of Figure 3.6(a) vs (b) for the "irregular" sample. While at first glance there may appear to be little change in relative retention (no change in retention order), a closer look is more informative. Thus, peaks 3 and 9 are seen to move relative to surrounding peaks 1/2 and 8/10, respectively. However, a change in gradient time has not noticeably improved the separation of overlapping peaks 5 and 6 in this example. If a 3-fold change in gradient time does not appear to change the resolution of the critical peak pair, further changes in gradient time are unlikely to provide much additional benefit—as long as other conditions are held constant.

The next step is a change in temperature. The third and fourth method-development runs are illustrated in Figure 3.6(c and d), where the runs of (a) and (b) are each repeated with a change in temperature from 30 to 50 °C. Note that a change in temperature has little effect on values of S or the average value of k^* [21], but usually shifts the chromatogram to slightly lower retention times (due to a decrease in values of log k_w). Because peak pair 5–6 was unresolved in the first

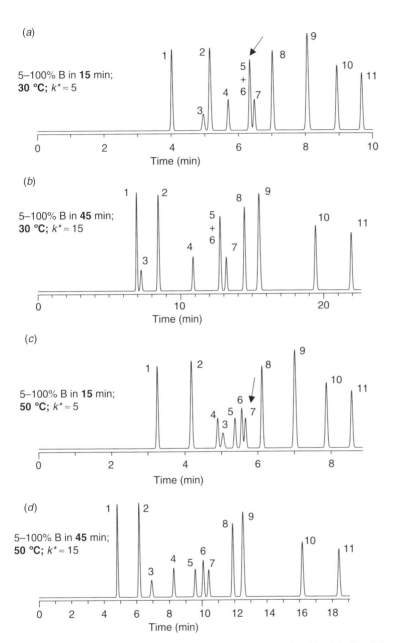

Figure 3.6 Initial separations of the "irregular" sample of Table 1.3. Conditions: 150 × 4.6 mm C_{18} column (5 μm particles); flow rate, 2.0 mL/min; 5–100 percent acetonitrile–buffer gradients (pH 2.6) in indicated times and for indicated temperatures.

TABLE 3.4 Experimental Conditions that Affect Selectivity in RP-LC Gradient Elution

Condition	Comment
1. Gradient time t_G	Usually maximum change in $t_G \leq 10$-fold
2. Temperature T	Maximum change in T limited by thermal stability of sample and column (usually a change of $\leq 30°C$)
3. *Solvent type*	
Acetonitrile	Preferred initial solvent
Methanol	Usually acceptable alternative
Tetrahydrofuran[a]	Unstable; avoided by many laboratories (although may be required for very hydrophobic samples)
4. Column type	Very wide range of columns available (Appendix III); change in selectivity is discontinuous (less useful), unlike other selectivity variables such as temperature or mixtures of two organic solvents
5. Mobile phase pH	Potentially large change in α for ionizable compounds; negligible change in α for other compounds
6. Mobile phase buffer type or concentration	Modest change in α for ionized compounds; little change for other compounds
7. Ion-pair reagents	Potentially large change in α for ionized compounds; little change for other compounds; usually not recommended for gradient elution

[a]Methyl-*t*-butylether may be a better alternative.

two runs, the primary question concerning runs (*c*) and (*d*) is whether peaks 5 and 6 can be separated at the higher temperature. A large increase in resolution for peaks 5 and 6 is seen in Figure 3.6(*c*) ($R_s = 2.0$), but peaks 6 and 7 are now critical ($R_s = 1.1$). Critical resolution increases further in Figure 3.6(*d*), for an increase in gradient time from 15 to 45 min ($R_s = 1.7$), as was the case for peak pair 6–7 at 30°C (Fig. 3.6*a* and *b*). The increased resolution of critical peaks 6 and 7 as gradient time increases (Fig. 3.6*d* vs *c*) suggests a further increase in gradient time. A few additional experiments were carried out at 50°C, as summarized in Figure 3.7(*a* and *b*). The best gradient time appears to be about 50 min (separation of Fig. 3.7*a*, with $R_s = 1.8$), due to a decrease in the resolution of peaks 8 and 9 as gradient time increases (arrows in Fig. 3.7 mark the critical peak pair). The separation of Figure 3.7(*a*) is close to acceptable in terms of resolution ($R_s = 1.8$; our original goal was $R_s \geq 2.0$), suggesting a change in column conditions to increase N^* and resolution (Section 3.3.5) – rather than further attempts at improving selectivity by changes in the other conditions of Table 3.4. Alternatively, further changes in peak spacing (values of α^*) could be explored at this time by varying other conditions in Table 3.4 (see discussion of Section 3.6), especially if a shorter run time is needed (excess resolution can be converted to a shorter run time by a change in column conditions; Section 3.3.5).

Note that the matching of various peaks in the chromatogram (peak numbering in the examples of Fig. 3.6) is *quite* important in method development, and may require additional information beyond that obtainable from the chromatogram. This additional information might be supplied by the use of a mass-spectrometer

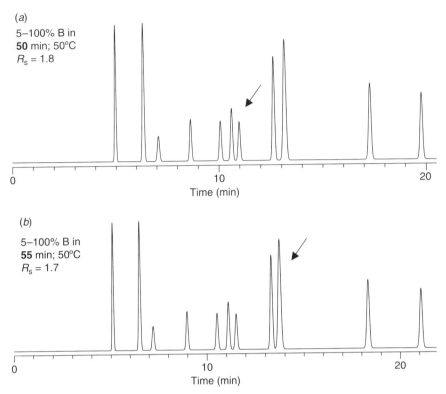

Figure 3.7 Separations of the "irregular" sample of Figure 3.6 with indicated changes in conditions: (*a*) separation as in Figure 3.6 except 5–100 percent B in 50 min at 50°C; (*b*) same as (*a*), except 55 min gradient. Arrow marks critical peak pair.

detector, diode-array detector, quantitative comparisons of peak areas, or the separate injection of individual standards. While peak *identification* is not required for further experiments in method development, it is important to be able to *match* (or "track") peaks among different method development runs (e.g., match compound "1" in run 3.6*a* with compound "1" in run 3.6*b*). Note that peak order can change as a result of these (or other) changes in conditions, for example, peak 3 in Figure 3.6(*c*). See Section 3.4.7 and [1] for further details on peak tracking.

3.3.3 Optimizing the Gradient Range (Step 6 of Fig. 3.1)

The next step in gradient method development is to consider (a) whether the gradient range $\Delta \phi$ can be shortened (with a decrease in run time), and (b) whether the use of a segmented gradient (Section 3.3.4) might lead to either a faster separation or better resolution. It can be seen in Figure 3.7(*a*) that the last peak leaves the column at about 20 min, while the gradient ends at 50 min. In such cases, it is recommended to terminate the gradient just after the elution of the last peak. In the example of Figure 3.7(*a*), the gradient can be continued for an additional minute or so after the last peak is eluted, for a total time of 21 min. If the gradient time is shortened in this way, the final %B in the gradient must also be reduced proportionately in order to maintain k^* constant (so as to preserve the optimum peak spacing of Fig. 3.7a). That is, $t_G/\Delta\phi$ in Equation (3.3) must be held constant. The initial value of $t_G/\Delta\phi$ in Figure 3.7(a) is $50/0.95 = 52.6$, which must equal the final value for $t_G = 21$ min, that is, $21/\Delta\phi = 52.6$, or $\Delta\phi = 0.40$. The final value of ϕ then equals $0.40 + 0.05 = 0.45$, corresponding to a 5–45 percent B gradient. A useful relationship in the present connection is the value of ϕ at the time a peak elutes from the column (ϕ_e):

$$\phi_e = \phi_0 + \Delta\phi(t_R - t_0 - t_D)/t_G \qquad (3.5)$$

where ϕ_0 is the value of ϕ at the start of the gradient.

The first peak in Figure 3.7(a) leaves the column in a mobile phase of 13 percent B, which means that a significant increase in initial %B (from 5 percent B) could further reduce run time (while keeping $\Delta\phi/t_G$ constant). However, an increase in initial %B is not always feasible, due to a possible loss in resolution as a result of smaller values of k^* for early peaks (Section 2.3.4) This in fact proved to be the case for the present separation, meaning that 5 percent B remains the best choice for the start of the gradient. The resulting separation with shortened gradient (5–45 percent B in 21 min) is shown in Figure 3.8(a). Note that, in some cases, an increase in initial %B can *increase* resolution, as in Figure 2.21. When the critical peak pair elutes early in the chromatogram, further experiments in which initial %B is varied can prove worthwhile.

3.3.3.1 Changes in Selectivity as a Result of Change in k*

The following section describes the visual interpretation of two chromatograms where relative retention and resolution for one or more peak pairs differ, as a result of a change in some condition that affects k^* (gradient time, flow rate, etc.). Based on this interpretation of two initial chromatograms, our goal is to achieve convenient (but reliable) predictions of how relative retention will change for a change in any other condition that affects k^*. This section comprises a less important part of gradient method development, and the discussion is somewhat more complicated. You may therefore choose to skip to Section 3.3.4, and return to this section at a later time or as needed. However, this treatment may add to the reader's intuitive understanding of gradient elution, as well as find occasional practical application.

Changes in k^* for "irregular" samples can result from a change in any of the experimental conditions summarized in Equation (3.3) (t_G, F, V_m or column length L, $\Delta\phi$), as well as a change in initial %B or the introduction of a gradient delay (while holding $\Delta\phi/t_G$ constant). Any change in k^* will result in similar changes in relative retention and resolution, regardless of how k^* is changed. This is illustrated in the remainder of this section by the examples of Figure 3.9 for an "irregular" sample and various changes in the gradient.

A starting separation of peak pairs 2–3 and 8–9 of the "irregular" sample is shown in Figure 3.9(a). These two peak pairs have been selected because their resolution responds in opposite fashion to a change in k^*. Consider first an *increase in gradient time* from 5 to 20 min (Fig. 3.9b), corresponding to an increase in k^* from 5 to 20. As a result, the retention of peak 2 increases *relative* to that of peak 3, and the

3.3 DEVELOPING A GRADIENT SEPARATION   97

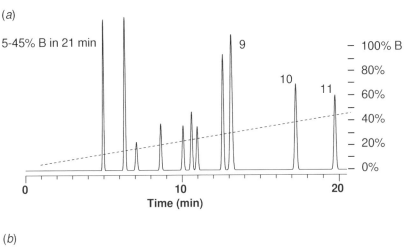

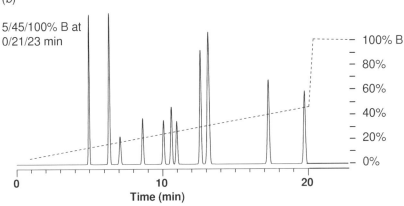

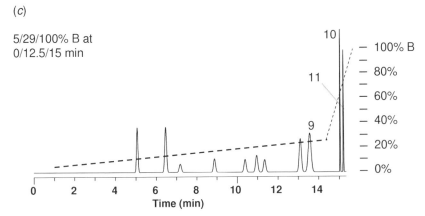

Figure 3.8 Separations of the "irregular" sample of Figure 3.6. (*a*) Separation as in Figure 3.7(*a*), except separation time reduced by narrowing the gradient range; (*b*) same as in (*a*), except steep gradient segment added in order to remove strongly retentive "junk" from the column; (*c*) same as in (*a*), except a second gradient segment is added in order to accelerate elution of the last two peaks in the chromatogram. See text for details.

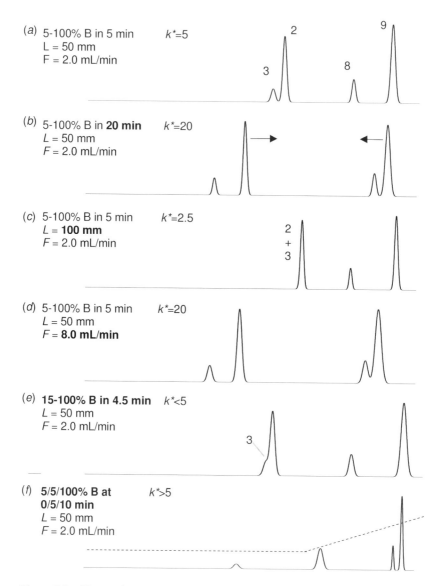

Figure 3.9 Changes in peak spacing with changes in gradient conditions. Sample consists of peaks 2, 3, 8, and 9 of "irregular" samples. Changes relative to (a) in bold. Conditions: 28°C; see figure for other conditions and see text for details.

resolution of peak-pair 2–3 increases. At the same time, the relative retention of peak 9 decreases relative to peak 8 when gradient time is increased, and the resolution of this peak-pair decreases. Similar changes in relative retention and resolution for these two peak-pairs can be expected for changes in any other condition which results in an increase in k^*. Opposite changes in retention will occur when k^* is decreased.

In Figure 3.9(c), *column length is increased* from 50 to 100 mm, while other conditions remain the same as in (a); the value of k^* decreases by a factor of 2 to $k^* = 2.5$ [Equation (3.3), since $V_m \propto L$]. As expected from this decrease in k^* [relative to the separation of (a)], the changes in relative retention seen in (b) vs (a) are reversed: peak 2 now moves *toward* peak 3 with a decrease in resolution, while peak 9 has moved *away* from peak 8, with an increase in resolution.

The effect of an *increase in flow rate* is seen in Figure 3.9(d), from 2.0 mL/min in (a) to 8.0 mL/min in (d) [other conditions the same as in (a)]. Because k^* has increased from 5 to 20 [just as in (b)], a similar change in relative retention is expected as for an increase in gradient time: again, peak 2 moves away from peak 3 with an increase in resolution, and peak 9 moves toward peak 8, with a decrease in resolution.

When *%B at the start of the gradient* (ϕ_0) is increased while holding $\Delta\phi/t_G$ constant, k^* from Equation (3.3) remains the same. However, values of k_0 are decreased for each peak:

$$\log k_0 = \log k_w - S\phi_0 \qquad (2.11b)$$

Resulting values of k_0 that are sufficiently small can affect the value of k^*, apart from the value calculated from Equation (3.3) (which assumes k_0 is large, Section 9.1.1.2). As a result, an increase in initial percent B for the gradient then results in a decrease in the values of k^* for early-eluting peaks (despite holding $t_G/\Delta\phi$ and b constant). This can in turn affect relative retention (selectivity) and resolution (see the further discussion of Section 2.3.6).

In Figure 3.9(e), the initial %B for the gradient is increased from 5 percent in (a) to 15 percent in (e), while gradient time is shortened to 4.5 min. Because $t_G/\Delta\phi$ remains constant [other conditions of (a) unchanged], k^* from Equation (3.3) does not change. However, the increase in initial %B means a smaller value of k_0, and a decrease in k^* for peaks that elute early in the gradient [Equation (3.6)]. As a result, $k^* < 5$ for early peaks in the chromatogram, for example, peaks 2 and 3. Consequently, relative retention changes in similar fashion as for an increase in column length (Fig. 3.9c): peak 2 has moved toward peak 3 with a decrease in resolution, while peak 9 has moved (slightly) away from peak 8, with a small increase in resolution (i.e., a smaller change in k^*, with less effect on relative retention, for these later peaks). Note that if the initial %B is increased *without* changing gradient time, $\Delta\phi$ decreases and k^* increases [Equation (3.3)]; at the same time, k_0 becomes smaller, which *decreases* the actual value of k^*. Because of these counteracting effects on k^*, the effect on relative retention would be difficult to predict for a change in initial %B only.

Finally, in Figure 3.9(f), a *gradient delay* of 5 min is introduced into the separation of Figure 3.9(a). As in the preceding example (e), the value of k^* from Equation (3.3) is unchanged ($k^* = 5$), but the effect of a gradient delay is to reduce the effect of the gradient on initial peaks in the chromatogram. This in turn means an *effectively* higher value of k_0, and a larger value of k^*. As a result, a similar change in relative retention and resolution results as in (b), for an increase in gradient time – but to a lesser extent for later peaks 8 and 9 (whose values of k^* are less affected by a gradient delay or change in initial %B).

From the examples of Figure 3.9, we see that predictable changes in relative retention and resolution occur as a function of k^* for the separation. It is necessary to first change some condition that affects k^*, in order to determine how the spacing and resolution of different peak pairs in the chromatogram will be affected by a change in k^*. Figure 3.9 assumes that gradient time is increased first (the usual case). However, similar conclusions can be drawn from an initial increase *or* decrease in *any* of the conditions of Figure 3.9, since two runs with different values of k^* determine the effect on peak spacing of a change in k^*.

Finally (as a footnote), the various examples of Figure 3.9 emphasize changes in relative retention and α^*, rather than related changes in resolution. If the selectivity (α^*) for two adjacent peaks does not change when k^* is changed, but the initial value of $k^* < 1$, a *further* decrease in k^* can result in decreased resolution because of the resulting small value of term *i* in Equation (3.4) [$k^*/(1 + k^*)$]. This means that the use of steep gradients with $k^* < 1$ can result in a misinterpretation of changes in peak spacing with k^*, especially if resolution is considered to the exclusion of relative retention. *The conclusions of this section cannot be extended to changes in conditions 2–7 listed in Table 3.4, since these conditions affect values of α^* and relative retention separately from any effect on k^*.*

3.3.4 Segmented (Nonlinear) Gradients (Step 6 of Fig. 3.1 Continued)

The preceding discussion of gradient elution assumes that we are dealing with linear gradients. There are various reasons for the possible use of a segmented gradient in place of a linear gradient:

- to clean the column between sample injections;
- to shorten run time;
- to improve separation by adjusting selectivity for different parts of the chromatogram.

Segmented gradients for cleaning the column are often employed for environmental or biological samples, because extraneous, strongly retained sample components may be present that can foul the column. When separating samples of this kind, *and* where the gradient required to elute all peaks of interest ends at less than 100 percent B (as in Fig. 3.8*a*), it is customary to follow the initial gradient with a steep gradient-segment that ends at or near 100 percent B. This is illustrated for the separation of Figure 3.8(*b*) (which is a modification of the gradient for Fig. 3.8*a*). Typically, the second segment is completed as a single step in 0.1 min, and the final segment may be followed by a brief gradient hold at 100 percent B. A "column flush" of this kind need not add appreciably to the overall run time.

Segmented gradients can also be used to shorten run time, when the end of the chromatogram has widely separated peaks (excess resolution, and therefore wasted space). This is true of the region between 13 and 20 min in Figure 3.8(*a*). A second, steeper gradient segment beginning just after 13 min can remove the last two peaks (10 and 11) from the column within a shorter time, while preserving a resolution of $R_s \geq 2$ for peaks 9–11. Figure 3.8(c) shows such a separation. Note that the initial gradient segment in Figure 3.8(c) retains the same slope ($\Delta\phi/t_G$) as in Figure 3.8(a) until peak 9 has eluted, in order to maintain the optimized peak spacing and resolution of peaks 1–9. Shortening run time in this way (the use of a steeper gradient for the elution of peaks 10 and 11) results in a decrease in resolution for peaks 10 and 11 from $R_s = 10$ in (a) to $R_s = 3.7$ in (c), which is still more than adequate. Run time is shortened from 21 min in Figure 3.8(a) to about 16 min in Figure 3.8(c), while also incorporating "column cleaning" as in Figure 3.8(b). The use of a segmented gradient in this way for shortening run time works best for compressing the end of the gradient chromatogram – as in the present example. Compressing the middle of the chromatogram is generally less successful and more difficult to achieve. Compressing the front end of the chromatogram is usually best achieved by the use of a higher %B at the start of the gradient, as in Figure 2.19.

For some samples, *segmented gradients can be used to improve resolution* by selecting an optimum gradient steepness for different parts of the chromatogram. An example is shown in Figure 3.10, where peak pair 3–4 is better separated with

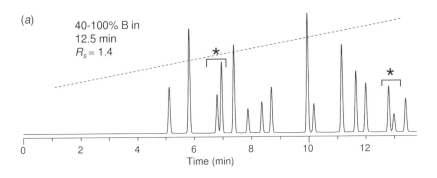

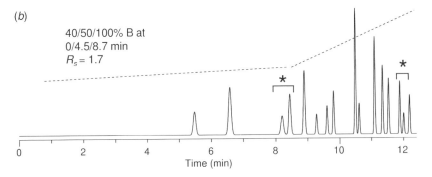

Figure 3.10 Separation of a mixture of 16 polycyclic aromatic hydrocarbons, adapted from Figure 8.13 of [1]. Conditions: 150 × 4.6 mm C_{18} column (5 μm particles); 35°C; 2.0 mL/min. (a) Separation with an optimized linear gradient; (b) separation with an optimized two-segment gradient. See [1] for further details.

a flatter gradient, while the separation of peaks 14 and 15 improves for a steeper gradient. In Figure 3.10(a), the slope of a linear gradient has been selected for maximum critical resolution of the sample (peak pairs 3–4 and 14–15 are critical, marked by asterisks). Maximum critical resolution corresponds to equal resolution for each of these two peak pairs, because a change in gradient time will increase resolution for one peak pair while decreasing resolution for the other. However, the resolution of each peak pair can be improved by the segmented gradient shown in Figure 3.10(b), which provides a flatter gradient for peaks 3 and 4, and a steeper gradient for peaks 14 and 15. For a more detailed discussion of segmented gradients for increasing resolution, see Section 2.3.7.

The selection of a segmented gradient by trial and error, for either shortening or improving separation, can be somewhat tedious and time-consuming. On the other hand, the use of computer simulation can make the design of a segmented gradient relatively easy (Section 3.4.5.4). It should be noted that segmented gradients can be more difficult to reproduce when the gradient equipment is changed (a potential problem in method transfer; Section 5.2). Furthermore, there is usually not a very large advantage in the use of segmented gradients for improving resolution (e.g., the example of Fig. 3.10, where the increase in R_s equals only 0.3 units). Consequently, segmented gradients should be used cautiously, except for purging the column as in Figure 3.8(b).

3.3.5 Optimizing the Column Plate Number N^* (Step 7 of Fig. 3.1)

The column plate number N^* is affected by column dimensions, particle size, and flow rate (so-called *column conditions*, Section 9.5), and (to a lesser extent) sample molecular weight (Section 6.1.2). For isocratic separation, an increase in column length or decrease in flow rate usually results in an increase in N, resolution, and run time (Figs 2.4 and 2.5); the price of increasing sample resolution in this way is a longer run time. After varying conditions for improved selectivity α^* (step 5 of Fig. 3.1), and adjusting gradient range and shape (step 6 of Fig. 3.1), the resulting separation may exhibit a resolution that is either (a) slightly too low ($R_s < 2$) or (b) much greater than needed ($R_s > 2$). In the first case ($R_s < 2$), an increase in column length and/or a decrease in flow rate can be used to increase resolution moderately, at the cost of a significant increase in run time (just as for isocratic separation). In the second case ($R_s > 2$), a decrease in column length with proportional increase in flow rate can be used to shorten run time substantially (with some allowable decrease in resolution, as long as $R_s \geq 2$).

In isocratic elution, changes in column length or flow rate do not affect relative retention or selectivity, because values of k are not changed when column conditions are varied. When varying column length or flow rate in gradient elution, however, a change in either of these two conditions alone will result in a change in k^* [Equation (3.3)] and selectivity. Since selectivity for a gradient method will have been optimized prior to a change in column conditions (step 5 of Fig. 3.1), it is important to maintain the same values of k^* (and α^*) when changing column conditions and

N^*. This can be achieved by maintaining $(t_G F/V_m)$ constant [Equation (3.3)]; for example, if column length (proportional to V_m) is doubled, gradient time must also be doubled so that t_G/V_m stays constant; if flow rate is decreased by half, gradient time must be doubled so as to keep $t_G F$ constant. In the following examples, we will assume that column diameter and gradient range $\Delta\phi$ remain constant when other column conditions are changed.

For an increase in resolution by changing column conditions, there are three options: (a) a decrease in flow rate, (b) an increase in column length, or (c) the use of a smaller-particle column. A decrease in flow rate is most convenient, while a decrease in particle size assumes that relative retention (column selectivity) does not change when particle size is changed for the same kind of column packing (this may not be the case!). Figure 3.11 shows the results of changes in each of these three column conditions, *while varying gradient time in order to maintain k^* constant*. The original separation of Figure 3.8(a) is repeated in Figure 3.11(a). Figure 3.11(b) shows the effect of a decrease in flow rate from 2.0 in (a) to 1.0 mL/min in (b). Resolution increases only slightly (from $R_s = 1.8$ to 1.9), the pressure drops from 910 psi in (a) to 455 psi in (b), and the run time is doubled from 21 to 42 min. This result is not atypical for a starting column with 5 μm or smaller particles, and a decrease in flow rate is often ineffective for a significant increase in resolution. Figure 3.11(c) shows the effect of an increase in column length from 150 to 250 mm (same column diameter). Resolution increases significantly (from $R_s = 1.7$ to 2.3), pressure increases from 900 to a still acceptable 1500 psi, and run time increases from 21 to 35 min. Again, this is a rather typical result: a significant increase in resolution, at the expense of some increase in both run time and pressure.

Finally, Figure 3.11(d) shows the effect of a decrease in particle size from 5.0 to 3.5 μm. Resolution increases significantly (from $R_s = 1.7$ to 2.2), with an increase in pressure to 1850 psi (still acceptable), but no increase in run time; option (d) appears preferable for this example. Because 3.0–3.5 μm columns provide a generally better trade-off between resolution, run time, and pressure, many laboratories begin method development with a 100 or 150 mm column packed with particles of this size. In this way, they avoid possible problems due to changes in selectivity for columns of different particle size (which would require a re-optimization of selectivity, often the major effort in gradient method development). When column length or flow rate is changed for gradient elution, and gradient time is adjusted to maintain k^* constant, changes in these conditions will have identical effects on run time, sample resolution, and column pressure for both isocratic and gradient elution, for example, a doubling of run time when column length is doubled or flow rate is cut in half. This is another example of the equivalence of isocratic and gradient elution, when k^* is held constant for a change in these conditions. The foregoing discussion ignores extra-column peak broadening, which becomes increasingly important as column volume and/or particle size decrease (Section 4.1.3.5).

Because resolution changes only with the square root of column length, and fairly slowly with flow rate, *the variation of these column conditions is much more effective for reducing run time* when $R_s > 2$, than for increasing resolution when $R_s < 2$. A decrease in run time in this way is illustrated in Figure 3.12 for

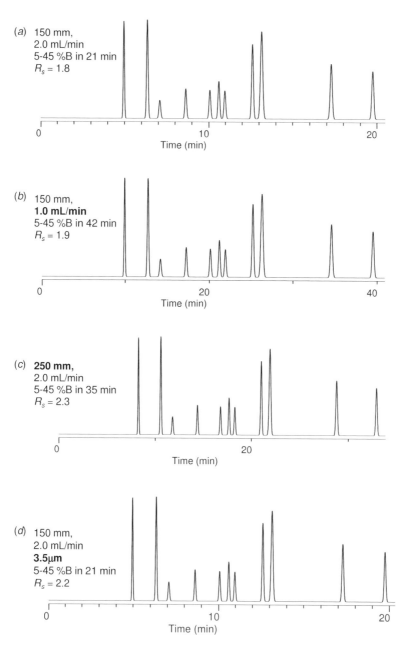

Figure 3.11 Use of a change in column conditions in order to improve separation, holding k^* and selectivity constant. Separation conditions as in Figure 3.7 except where indicated otherwise. (*a*) Same separation as that of Figure 3.8(*a*); (*b*) flow rate reduced from 2.0 to 1.0 mL/min, gradient time increased from 21 to 42 min; (*c*) column length increased from 150 to 250 mm, gradient time increased from 21 to 35 min; (*d*) particle size reduced from 5.0 to 3.5 μm (only). Changed conditions are bold in (*b*)–(*d*).

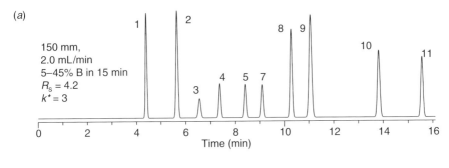
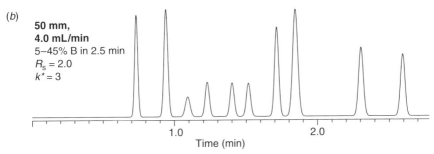

Figure 3.12 Separation of compounds 1–5 and 7–11 of the irregular sample. Conditions: 150 × 4.6 mm C_{18} column (5 μm particles); 2.0 mL/min. (*a*) Separation optimized for maximum resolution by varying gradient time and temperature: 5–45 percent B in 15 min, 51 °C; (*b*) fast separation based on a decrease in column length and gradient time, and increase in flow rate, with other conditions the same with k^* constant. Separation conditions indicated within the figure, see text for details. Changed conditions are in bold in (*b*).

the separation of the "irregular" sample of Figure 3.11 *minus* peak 6 (a much easier separation, by virtue of the missing peak). After optimizing selectivity (gradient time and temperature) for this sample, and adjusting the gradient range, the separation of Figure 3.12(*a*) results, with a resolution of $R_s = 4.2$ and a run time of 15 min. Run time can be shortened by sacrificing excess resolution, while maintaining the previously optimized selectivity (i.e., keeping k^* constant). This is achieved by maintaining $(t_G F/V_m)$ constant, when reducing column length L and/or increasing flow rate F. The much shortened separation of Figure 3.12(*b*) involves a reduction in column length from 150 to 50 mm, and an increase in flow rate from 2 to 4 mL/min. Because k^* should be held constant during changes in flow rate or column size, $t_G F/L$ must also remain constant [Equation (3.3)]. For the initial run (*a*), $t_G F/L = (15 \times 2)/15 = 2$. The gradient time for the shortened run (*b*) is then $t_G = 2L/F = (2 \times 5)/4 = 2.5$ min. The separation of Figure 3.12(*b*) with these changes in column conditions maintains an adequate resolution ($R_s = 2.0$), but run time has been reduced from 15 to only 2.5 min, with a pressure of only 600 psi [vs 900 psi in (*a*)].

Resolution can often be increased to $R_s \gg 2$ by further changes in selectivity for maximum α^* (Section 3.6). Therefore, when a short run-time is important, *additional time should be spent on step 5 of Figure 3.1* in order to maximize resolution before shortening run time as in Figure 3.12.

The column pressure drop also varies with mobile phase composition and temperature, due to changes in mobile phase viscosity (Appendix IV). Thus, a temperature increase of 20°C will result in ~30 percent decrease in both viscosity and pressure. For acetonitrile–water gradients, viscosity and pressure are highest at the beginning of a run and decrease during the gradient; with methanol or tetrahydrofuran as the B solvent, a maximum pressure is reached for about 50 percent B. When a pressure is cited in this book for a gradient elution run, the value refers to the maximum pressure during the separation. The pressure drop during gradient elution is significantly lower for acetonitrile as B solvent, compared with methanol or tetrahydrofuran. This and the lower UV absorbance of acetonitrile at low wavelengths are important reasons for preferring acetonitrile as B solvent.

For further details on selecting conditions for maximum N^* in minimum time, see Section 9.5.

3.3.6 Column Equilibration Between Successive Sample Injections

After method development is complete, the resulting separation usually will be used for routine sample analysis. During this application of the method, sufficient time must pass between the end of one gradient run and the start of the next. This column-equilibration step is intended to allow for (a) the hold-up volume of the gradient equipment (dwell time t_D) and (b) the slow equilibration of the column when switching from high %B at the end of the gradient to low %B at the beginning of the next gradient. For method development experiments, 10 column volumes or more of the starting mobile phase ($\phi = \phi_0$) should be passed through the column before starting the next gradient run. Otherwise, any change in the time between successive experiments (often the case in method development), can result in variable column equilibration and resulting differences in retention and separation. For routine gradient separations, however, the volume of solvent required for column equilibration can be greatly reduced, because partial column equilibration is usually acceptable, if the extent of equilibration is the same for each gradient run (this assumes that the first gradient run will be discarded). Whatever column equilibration time is allowed, it should be confirmed that repetitive sample injections yield the same (acceptable) chromatograms, except possibly for the first run. The nature of and requirements for column equilibration are discussed further in Sections 5.3 and 9.2.1.

3.3.7 Fast Separations

Very fast separations are needed for high-volume testing, where thousands of samples must be analyzed at acceptable cost (and therefore minimum run time). As run time is decreased below a few minutes, the performance of the equipment can become limiting. Fast separations require very small values of the dwell volume V_D, sample injections that can be performed within a few seconds, fast detector response, an ability to carry out very steep gradients (e.g., 1–2 percent B/s), and rapid or off-line data processing (Section 4.2.4). Fast separations usually involve somewhat smaller values of the column plate number N^* (Section 9.5), but moderate plate numbers may still allow adequate separation in many situations:

- samples with fewer, easily separated components;
- following exhaustive optimization of separation selectivity (Section 3.6);
- separations with a tolerance for small values of R_s, because of either selective detection (e.g., LC-MS) or an acceptance of reduced accuracy in assay results.

Apart from the equipment needed for fast separation, the choice of column dimensions, particle size, and flow rate determine the maximum value of N^* that is achievable within a given separation time ($\sim t_G$). For gradient separations which are completed in a minute or less, it becomes important to select experimental conditions which provide maximum N^* within the time allowed for separation. Table 9.3 summarizes some approximate experimental conditions for various separation times. For example, a 1 min gradient time with maximum N^* would require a 30 × 4.6 mm column of 1.5 μm particles, and a flow rate of about 1 mL/min. This assumes a full-range gradient ($\Delta \phi \approx 1$), a sample with a molecular weight <500 Da, and a maximum column pressure of 2000 psi. With the advent of HPLC pumping systems which can operate reliably at higher pressures (e.g., 15,000 psi), it is possible to reduce the time of separation further, without sacrificing column efficiency (see Fig. 9.11 and related discussion; also [22]). Run time (without loss in resolution) will be inversely proportional to pressure, provided that column length, particle size, and flow rate are optimized for each pressure. This could mean a possible reduction in run time by a factor of 10, for an increase in pressure by 10-fold, for example, from 1500 to 15,000 psi. Alternatively, for the same run time, an increase in pressure by 10-fold could mean an increase in resolution by $\sim 10^{1/4} \approx 2$.

Several reports provide both examples and further experimental details [23–28] for "fast" or "high-resolution" separation. Figure 3.13 shows three such separations of peptides and/or proteins in times of 0.5–3 min. In (a) the high-resolution separation of a mixture of four proteins is achieved in 40 s. Example (b) shows the partial separation of a mixture of peptides and proteins in 25 s. In (c) the partial separation of more than 50 peaks from a complex protein digest is obtained in <3 min. Each of these separations was carried out at a temperature of 70 °C with columns packed with small, pellicular (surface coated) particles. Higher temperatures and pellicular packings can each provide faster separations than corresponding fully porous particles at lower temperatures; see also the review in [29]. With sufficiently "easy" separations and appropriate equipment, run times of <10 s can be achieved with little difficulty ("easy" mainly refers to large values of α for all peak pairs in the chromatogram).

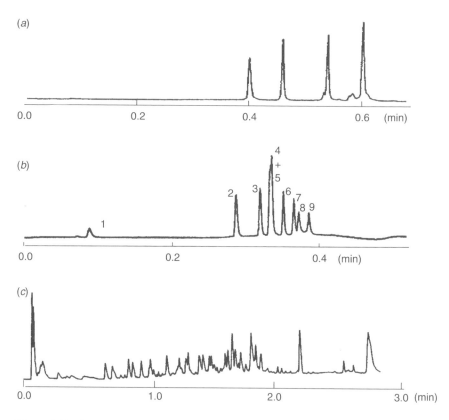

Figure 3.13 Fast gradient separations of peptide and protein mixtures using a 75 × 4.6 mm pellicular C_{18} column (5 μm particles) at 70°C. Other conditions: (a) 15–100 percent acetonitrile–buffer in 1.0 min; 2 mL/min; (b) 0–100 percent acetonitrile–buffer in 0.25 min; 4.0 mL/min; (c) 0–100 percent acetonitrile–buffer in 0.25 min; 2.0 mL/min. Pressures <2500 psi. Adapted from [25].

3.4 COMPUTER SIMULATION

Computer simulation refers to the use of a computer program that allows the prediction of either isocratic or gradient separation as a function of changes in experimental conditions. Several programs of this kind (with varying capabilities) are commercially available. The simulation of gradient separation usually begins with two initial experiments where only gradient time is varied (e.g., Fig. 3.6a and b). Retention times and (optionally) peak widths from these two "calibration" runs are entered into the computer, as are experimental conditions for each run (column dimensions, particle size, flow rate, initial and final %B, etc.). It is now possible to predict separation as a function of different gradient conditions (gradient time, gradient shape, initial and final %B for the gradient), column dimensions and flow rate, using fundamental relationships that are described in Chapter 9. The two initial experiments in which gradient time is varied can also be combined with additional experiments where other separation conditions that affect peak spacing are changed, such as temperature or mobile phase pH (e.g., Fig. 3.6c and d, where the temperature has been changed). The latter experiments then allow predictions of separation for additional changes in conditions. The computer program can display its predictions in the form of chromatograms, tabular data, resolution plots, and so on.

In the present section, we will describe some of the features of one, widely used computer-simulation program: DryLab® [30, 31] (Rheodyne LLC, Rohnert Park, CA, USA). Because several companies offer similar software, and because the capabilities of such software continue to advance, the following discussion will be limited to the more important (and presently available) capabilities of computer simulation, primarily for use in developing an acceptable gradient separation. Computer-simulation software also allows predictions for isocratic separation (Section 3.4.5.1), although usually with reduced accuracy when gradient "calibration" runs are used [32, 33]. Computer simulation software is also available from several other companies or groups, for example, ChromSword (Merck KGaA, Darmstadt, Germany), ChromSmart (Agilent, Palo Alto), ACD/LC Simulator (Advanced Chemistry Development, Toronto, Ontario, Canada), Osiris [34], and Preopt-W [35]. See also several reviews and/or comparisons of software of this kind [36–39]. For a detailed review of some technical requirements of computer simulation, see [40].

While the remainder of Section 3.4 addresses computer simulation specifically, *the various examples also offer insights that should prove useful for manual method development*, that is, without the help of computer simulation.

3.4.1 Quantitative Predictions and Resolution Maps

Consider the two experimental runs of Figure 3.6(c and d) for the "irregular" sample, where only gradient time is changed between the two runs. When data for these two runs are used for computer simulation, it becomes possible to predict separation as a function of gradient time, gradient shape, and column conditions (column length, flow rate, and particle size). This can be done by trial-and-error simulations, but a more efficient process is the initial use of *resolution maps* that are provided by the computer simulation software. Figure 3.14(a) shows a resolution map for the separation of the "irregular" sample as a function of gradient time (50°C, 150 mm column, 2 mL/min, 5–100 percent B). Maximum resolution within a range in gradient time of 5–100 min occurs for a gradient time of 53 min. This can be compared with a "best" gradient time of 50 min that was determined experimentally by trial and error (Figs 3.6c, 3.7a, and 3.7b). Note that the critical peak pair changes with gradient time. For $12 < t_G < 19$ min, peak pair 3–4 is critical; for $19 < t_G < 53$ min, 6–7 is critical; for $t_G > 53$ min, peaks 8 and 9 are critical (see numbering on plot of Fig. 3.14a, which identifies the critical peak pair for different values of t_G).

Resolution maps are also useful as indicators of non-robust separation conditions. Thus, in the example of Figure 3.14(a), an "optimum" gradient time of

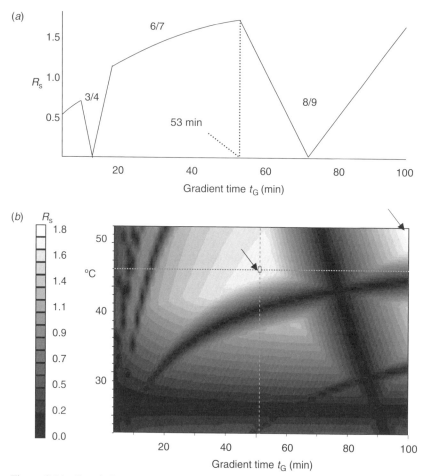

Figure 3.14 Resolution maps for separations of the "irregular" sample. Conditions as in Figure 3.6 except where noted otherwise (5–100 percent B gradients). (a) Resolution map for separation at 50°C and different gradient times; (b) resolution map for separation at different temperatures and gradient times (shading code is at left). Arrows designate optimum conditions for maximum resolution; numbers on graph indicate critical peak pairs.

53 min can be compared with sitting on the edge of a cliff. A small increase in gradient time (as a result of equipment malfunction or the use of different equipment) leads to a rapid loss in resolution, due to the convergence of peaks 8 and 9 with increasing gradient time. This suggests that a preferred gradient time will be somewhat less than 53 min, for example, the 50 min gradient selected in Figure 3.7(a) by trial-and-error, with a resolution that is virtually as good as for a 53 min gradient, but with a greater tolerance for experimental variability.

Resolution maps simplify the identification of "best" conditions for a given separation, usually leading to better separations with less effort. The value of resolution maps increases markedly when two separation conditions that independently affect peak spacing and α^* are changed simultaneously. Thus, from the four runs of Figure 3.6 (two different temperatures and two different gradient times), computer simulation provides the resolution map of Figure 3.14(b) as a function of gradient time and temperature. Resolution maxima are seen for two different conditions (marked by arrows): 5–100 percent B in 51 min at 47°C, and 5–100 percent B in 100 min at 50°C. Since $R_s = 1.9$ for both runs, the shorter gradient (51 min) is preferred. The latter separation is only slightly better than the optimized run of Figure 3.7(a) ($R_s = 1.8$), but the use of a resolution map allows a faster, more systematic approach to these final conditions, with the assurance that the best possible conditions have been selected, with a minimum experimental effort. For some samples, the simultaneous optimization of both gradient time and temperature can be fairly difficult if carried out manually, but relatively easy using computer simulation.

An additional example of the optimization of gradient time and temperature by means of computer simulation is shown in Figures 3.15 and 3.16 for a proprietary sample from a pharmaceutical laboratory. Four gradient separations from 0 to 100 percent methanol were carried out in gradient times of 20 and 60 min, and temperatures of 30 and 50°C (Fig. 3.15), and a resolution map was generated (Fig. 3.16a). The best choice of conditions ($t_G = 48$ min, $T = 37°C$) gives the separation of Figure 3.16(b) with a resolution of $R_s = 2.4$, a significant improvement over the best input run (Fig. 3.15a, $R_s = 1.6$). In this example, peak 5 exhibits major changes in relative retention as gradient time and temperature are varied, while peaks 6 and 7 experience significant, but smaller changes in relative retention.

3.4.2 Gradient Optimization

Another feature of computer simulation is its ability to *automatically* search for the best combination of (a) initial and final %B values for the gradient, (b) gradient time, and (c) temperature (based on data for four input runs, as in Fig. 3.6). A separation similar to that of Figure 3.16(b) can therefore be achieved *without* operator intervention, at the same time providing optimized values of initial and final %B for the gradient – after the results from Figure 3.15 are entered into the computer. The result of an automatic search for optimum gradient conditions for this same sample is shown in Figure 3.16(c); a 5-fold shorter run-time is achieved with slightly *greater* resolution ($R_s = 2.5$). It is unlikely that these conditions could be identified by trial-and-error experimentation in an acceptable amount of time.

A large number of segmented gradients can be explored intuitively in a very short time by means of computer simulation, allowing the easy (manual) development of a final method that meets the needs of the user. This is illustrated in Figure 3.17, which shows the DryLab computer screen used to design such gradient shapes. The example of Figure 3.17 is the final gradient for the separation of Figure 3.8(c). During gradient design, each of the three points (arrows) shown in Figure 3.17 can be dragged via a computer mouse to any desired value of time and %B (thereby changing the gradient program), while the resulting chromatogram and its resolution immediately displayed as in Figure 3.17.

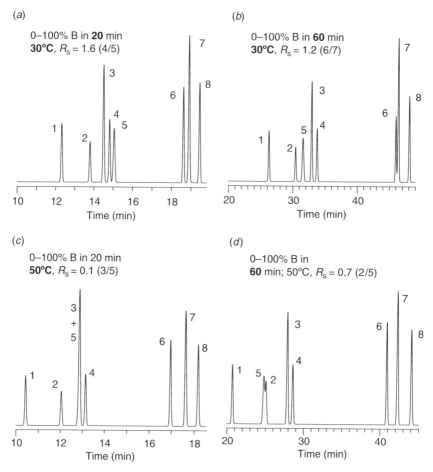

Figure 3.15 Experimental separations of a proprietary pharmaceutical sample at the start of method development. Conditions: 75 × 4.6 mm Zorbax Bonus-RP (embedded-polar-group) column; 0–100 percent MeOH–buffer gradients; 1.5 mL/min flow rate. Recreated chromatograms using computer simulation, based on data from [42].

3.4.3 Changes in Column Conditions

After separation conditions have been selected for optimum selectivity, as in Figure 3.16, changes in column conditions can be explored by entering different values of column length, particle size, or flow rate. Because relative peak spacing (selectivity) has already been optimized, it is important to maintain k^* constant while changing column length or flow rate, by making appropriate changes in gradient time t_G [Equation (3.3)]. DryLab software recognizes this need by automatically adjusting the value of t_G to maintain k^* constant, when a change in column length or flow rate is made. The result for each such computer simulation will include values of resolution, run time, and column pressure drop.

3.4 COMPUTER SIMULATION   113

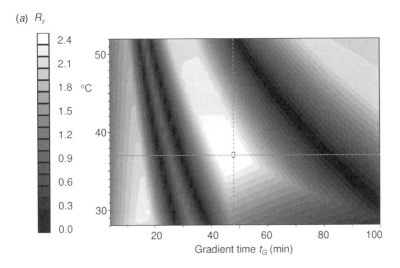

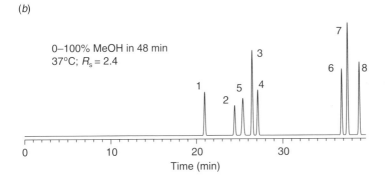

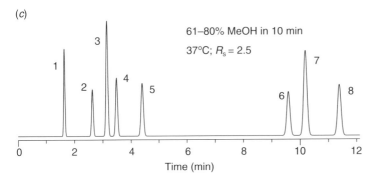

Figure 3.16   Optimized separation of proprietary pharmaceutical sample of Figure 3.15. (*a*) Resolution map for varying temperature and gradient time; (*b*) optimized separation for indicated gradient time and temperature; (*c*) further optimization of initial and final percent B. See text for details.

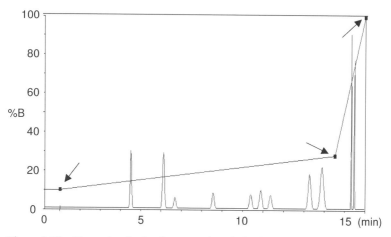

Figure 3.17 Computer display for manual optimization of a segmented gradient. Points marked by arrows can be moved using a mouse to change the gradient program. See text for details.

Columns with 3–3.5 μm particles are becoming more popular, because they are consistently better performing than 5 μm particles for run times less than about an hour. For this reason, *beginning* method development with a 100 × 4.6 mm, 3 or 3.5 μm column (instead of the 150 × 4.6 mm, 5 μm column of Table 3.2) is a reasonable alternative. The pressure drop for the latter two column configurations will be about the same, so there is no need to change flow rate when substituting one column for the other at the start of method development. Note that if a column with a different particle size is substituted for the original column *after* optimizing separation selectivity (and resolution), the selectivity of the new column may not be exactly the same as that of the original column. In this case, it may be necessary to repeat some of the earlier experiments that were used to optimize separation selectivity. Alternatively, having established how the resolution of different peak pairs changes with different changes in conditions, it may be possible to estimate what change in conditions could compensate for any adverse change in column selectivity (with no need to repeat previous experiments).

Predictions of peak width and resolution by computer simulation are most reliable when peak widths from the experimental ("calibration") runs are entered into the computer (along with retention times and separation condition), instead of using the computer program to estimate peak widths or plate numbers. In the absence of experimental peak width data, however, predictions of peak width and resolution by computer simulation are usually adequate for method development.

3.4.4 Separation of "Regular" Samples

Method development for a "regular" sample (unlike that for the "irregular" sample described above) proceeds in similar fashion, but with important differences. Figure 3.18 shows chromatograms for the separation of a mixture of 10 fatty acid methyl esters. It is seen that there is not much change in relative retention as gradient

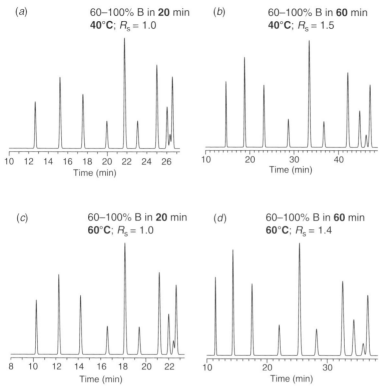

Figure 3.18 Separations of a "regular" sample (mixture of fatty acid methyl esters) Sample: compounds 1–5, 7–12 of Table 2 in [20]; conditions, 0–80 percent acetonitrile–water; 150 × 4.6 mm C_{18} column, 2.0 mL/min. Simulated separations based on data of [41].

time is changed, unlike the example of Figure 3.15 (an "irregular" sample). The resolution map for the separation of Figure 3.18 is correspondingly less "interesting" (Fig. 3.19). Although relative retention and selectivity are less dependent on gradient conditions for "regular" samples, run time and resolution still vary somewhat with gradient time and gradient shape [mainly due to changes in term i of Equation (3.4)], making computer simulation useful even for these samples.

3.4.5 Other Features

Computer-simulation software offers a number of other options to the user, as summarized in Table 3.5.

3.4.5.1 Isocratic Prediction (5 in Table 3.5)

Once computer simulation has been initiated on the basis of the four gradient runs of Figure 3.6, it is possible to predict isocratic separation as a function of %B and temperature by means of Equation (3.2). Values of log k_w and S for each sample compound are calculated by the computer at the two temperatures (30 and 50°C in Fig. 3.6; see discussion

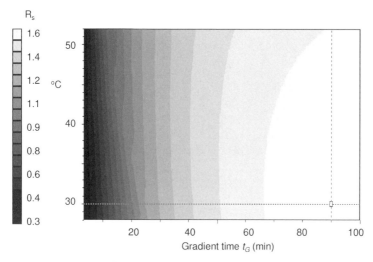

Figure 3.19 Resolution map for fatty acid methyl ester separations of Figure 3.18.

TABLE 3.5 Some Computer Simulation Options

Option	Comment
1. Operator simulation of a chromatogram for different gradient conditions and temperature	See discussion of Section 3.4.1
2. Use of resolution maps for easy selection of optimized conditions	See discussion of Section 3.4.1
3. Automatic selection of an optimum linear gradient	See discussion of Section 3.4.2
4. Operator selection of best column conditions	See discussion of Section 3.4.3
5. Predictions of isocratic separation	Computer simulation of changes in gradient time and temperature also allow predictions of isocratic elution as a function of %B and temperature (Section 3.4.5.1)
6. Selection of peaks of interest	Resolution and resolution maps calculated only for peaks of interest; if a peak is not of interest and it does not overlap a peak of interest, its resolution is ignored (Section 3.4.5.2)
7. Simulation for change in other conditions that can affect selectivity	One- or two-dimensional separation can be simulated for any of the variables of Table 3.2. A change in column requires experimental data for that column (Section 3.4.5.3)
8. Computer-controlled design of optimized two-segment gradients	Allows separate adjustment of selectivity and resolution in each gradient segment (Section 3.4.5.4)

in Section 9.3.3), and values of log k_w and S can then be estimated at other temperatures by interpolation. The ability to predict isocratic separation allows a more complete assessment of isocratic elution as a possible alternative to gradient separation. As the range in k for the sample decreases, isocratic elution usually becomes a more promising option, for example, for $2 \leq k \leq 4$. For a further discussion of the prediction of isocratic elution from gradient runs, see [9–14].

3.4.5.2 Designated Peak Selection (6 in Table 3.5) Many chromatograms contain peaks that are of no interest to the chromatographer. For example, the analysis of biological or environmental samples for specific compounds may be complicated by the presence of numerous interfering peaks. Similarly, in preparative separation (Chapter 7), we are concerned with the recovery of some peaks, but not others. When only some peaks are of interest, it is important to separate these peaks from the remaining peaks, but the separation of interfering peaks from each other is unnecessary. As an illustration of designated peak selection, consider the example of Figure 3.6(a), and assume it is required to assay only peaks 3, 8, and 10 in the presence of the remaining "interfering" peaks. A final separation is therefore required that will separate these three peaks from each other *and* from any other peaks that might overlap 3, 8, and 10 (this is much easier than the separation of all 11 peaks). Using computer simulation, we can designate peaks 3, 8, and 10 as "peaks of interest." When a resolution map for "designated peaks" only (e.g., 3, 8, and 10) is next requested, values of the critical resolution R_s will be plotted vs temperature and gradient time for *just* peaks 3, 8, and 10. This is illustrated in Figure 3.20(a), and the best separation (cross-hairs and arrow in a) is shown in Figure 3.20(b) (numbers mark peaks of interest). As anticipated, the possible critical resolution for the separation of only three of the 11 peaks in this sample is much greater ($R_s = 4.7$) than for the separation of all 11 peaks ($R_s = 1.8$ in Fig. 3.14b). This "excess" resolution can be traded for a much shorter run time, as in the example of Figure 3.12.

3.4.5.3 Change in Other Conditions (7 in Table 3.5) Other conditions in Table 3.4 that affect selectivity also can be modeled by computer simulation, varying either one or two different conditions at a time. A best choice of conditions to use for adjusting separation selectivity can be inferred from the nature of the sample, the experience of the chromatographer, and the results of previous method-development experiments with the sample. See the further discussion in Section 3.6.

3.4.5.4 Computer-Selection of the Best Multisegment Gradient (8 in Table 3.5) Other than for cleaning the column as in Figure 3.8(b), two-segment gradients are used primarily for one of three general applications: (1) the resolution of bunched peaks at the beginning of the gradient often can be improved with an isocratic hold (e.g., Fig. 2.23c and d); (2) separation of critical peak pairs at the beginning and end of a separation that respond differently to changes in the gradient may improve with segments of differing slope (e.g., Figs 2.27 and 3.9); and (3) runs with excessive resolution at the end often can be shortened by using a steeper

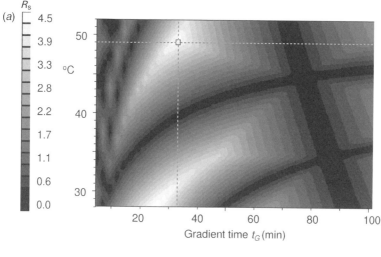

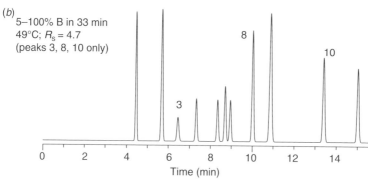

Figure 3.20 Optimizing the separation of selected peaks in the "irregular" sample; use of computer simulation to select optimum values of gradient time and temperature for the separation of peaks 3, 8, and 10 of Figure 3.6 from remaining peaks. (*a*) Resolution map; (*b*) best separation for 5–100 percent B in 33 min and 49°C.

gradient to compress the end of the gradient (e.g., Fig. 3.8*c*). Computer simulation can provide a trial-and-error examination of a large number of two-segment gradients (as illustrated in Fig. 3.17), followed by the selection of a gradient that yields the best selectivity for maximum resolution or shortest run time. However, the application of this approach to several samples [43] suggests that the advantage of two-segment gradients for further improvement in resolution (as in the example of Fig. 3.9) is often marginal – it is rare for a segmented gradient to improve resolution by as much as 0.5 units.

When compared with the simultaneous optimization of gradient time and separation temperature (a preferred approach), there appears to be little advantage in the use of two-segment gradients for increasing resolution. A minor exception is the use of segmented gradients for the separation of samples that contain large molecules such as proteins (Section 6.2.1.4). A disadvantage of segmented gradients is that they can contribute to problems in method transfer, because of gradient rounding (Sections 5.2.2.2 and 9.2.2). The computer design of segmented gradients allows the use of more than two segments [43–46], but such separations can be expected to be more susceptible to differences in gradient equipment, with little evidence that they can achieve better separations than result from linear gradients with optimized gradient time and temperature [43].

3.4.5.5 "Two-Run" Procedures for the Improvement of Sample Resolution
It can be difficult to achieve the adequate separation (e.g., $R_s \geq 2$) of some samples, especially if the sample contains more than 15 components (Section 2.2.4). An alternative (occasionally successful) approach for such samples is the use of two *different* gradient procedures (or "runs") for the same assay [47], but with no change in either the column or the A and B solvents. For example, the two runs might each use a different gradient time and a different temperature – which allows the two runs to be carried out without any additional operator intervention (all samples would be assayed with one set of conditions, followed by their reanalysis using the second set of conditions). The goal is adequate resolution of every peak of interest in one or the other of the two runs, which then allows an assay of all the peaks in the sample (based on a composite of the two runs). Using computer simulation, it is relatively easy to select best conditions for the two runs, such that overall "critical" resolution for the sample is maximized [47]. A similar approach has been described for isocratic separation, based on two so-called "complementary" runs [48].

3.4.6 Accuracy of Computer Simulation

The accuracy of computer simulation (based on the linear-solvent-strength model) has been examined systematically and found to be generally acceptable for purposes of method development [33, 37, 45–47]. Additionally, a number of examples of practical method development based on computer simulation have been reported [46–75], each of which shows close agreement between simulated and experimental gradient runs. Predictions of resolution are usually reliable within ± 10 percent, which is generally adequate for the purpose of method development. For further details, see Section 9.3 and the reviews of [30, 31] for additional applications of the DryLab software.

3.4.7 Peak Tracking

Computer simulation begins with experimental chromatograms, as in Figure 3.6 or Figure 3.15. In order to determine values of log k_w and S for each compound, so as to allow predictions of separation by the computer, it is necessary to match peaks for each compound in the different runs. Thus, if peak 1 in run 1 corresponds to compound "A" (whose chemical structure may or may not be known), it is necessary to know which peak in run 2 also corresponds to "A." This is not too difficult for the examples of Figure 3.6, where relative retention does not

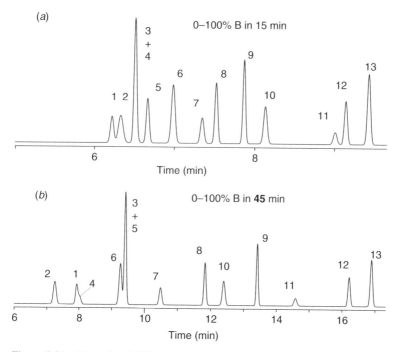

Figure 3.21 Example of difficult peak tracking. Separation of miscellaneous drug compounds described in [6]. Conditions: 250 × 4.6 mm C_{18} column, buffer–acetonitrile gradients, 2.0 mL/min, 30°C.

change much and the areas of adjacent peaks are sufficiently different to allow matching on the basis of peak areas. Manual peak tracking can take advantage of peak area, peak shape, and the observation that, when retention order changes, it usually does not change much (i.e., a peak usually appears in the same region of the chromatogram).

Peak tracking can be much more difficult in other cases, however, for example, the two separations of Figure 3.21, which show numerous changes in relative retention for the first six peaks, as well as overlapped peaks which involve different peak-pairs [peaks 3 and 4 in (a), 3 and 5 in (b)]. While several workers have suggested ways to improve peak tracking when using UV detection [76–80], none of these procedures has proven adequate for all samples. More and more method development studies now make use of mass spectrometric detection (LC-MS), which largely eliminates problems in peak tracking because of the ability of MS detection to assign a (usually unique) molecular mass to each peak [76].

3.5 METHOD REPRODUCIBILITY AND RELATED TOPICS

The ability to obtain the same separation each time a new sample is injected is important during both the development of a method and its later routine application. A number of different contributions to separation variability can be anticipated:

1. Poor control of experimental conditions from run to run.
2. Malfunctioning or poorly performing equipment.
3. Changes in column performance with time.
4. Insufficient column equilibration between gradient runs.
5. Differences in equipment dwell volume when using different HPLC systems.
6. Batch-to-batch differences in columns.

The above contributions to separation variability can impact both method development *per se*, and the subsequent routine use of a gradient method – as discussed below. Additionally, the accuracy and precision of gradient assays (and the interpretation of method development experiments) can be compromised by drifting baselines and artifactual peaks that are not associated with the sample. For this reason, it is recommended to begin every series of gradient runs with a blank gradient (Section 5.1.3).

3.5.1 Method Development

Consider first the need for repeatable data during method development, where it is advisable to repeat each experiment so as to verify that the data obtained are reproducible from run to run (this is especially important for gradient elution experiments). Retention times in duplicate, back-to-back runs should not vary by more than some set amount, for example, ± 0.02 min or $\pm 0.1\%$, whichever is larger.

Poorly controlled experimental conditions and malfunctioning equipment (items 1 and 2) fall largely under the heading of good laboratory technique. For purposes of the present discussion, we will assume that all experimental conditions are controlled within limits necessary for repeatable separation. We will also assume that the equipment is operating properly, and that column performance meets the manufacturer's specifications (Section 4.3). Apart from operator and equipment issues, however, a major objective of method development should be final assay procedures that can tolerate small changes in conditions, such as temperature and mobile phase composition (%B, pH, etc.). Method robustness should be established during method development by determining the effect on separation of likely variations in different conditions, for example, a change in pH of ± 0.05 units, ± 1.0 percent B, or $\pm 1-2°C$. A robust method should tolerate changes in conditions as large as this (with no significant loss in resolution). If a method appears not to be robust, efforts should be made to reduce the dependence of the method on conditions, by examining method robustness and resolution as a function of conditions. In order to achieve acceptable method robustness, it is sometimes necessary to sacrifice resolution (recall the above discussion of Fig. 3.14*a*).

Because of possible *changes in column performance with time* (item 3), it is strongly recommended that the development of each gradient method start with a new column, that is, one that has not been used previously. It is also recommended that the initial gradient run (as in Fig. 3.2) be repeated at the end of each day, in order to confirm that no change in column performance has occurred during the day. Some laboratories routinely start method development with nominally "good" columns of uncertain history, that is, columns that have been used previously for other samples. Although a column might not appear to have been degraded by its previous use (exhibiting a satisfactory plate number and no peak tailing for a test mixture), more subtle changes in the column may go unrecognized – changes that might result in irreproducible retention during method development experiments.

Incomplete column equilibration (item 4) is a major source of variable retention in gradient elution, so a column-equilibration step between each run or experiment is necessary. During method development, we recommend a minimum 5–10 min flush of the column by the starting mobile phase (i.e., with composition ϕ_0) between each gradient run. Retention time repeatability should be checked initially for two replicate, successive runs that use the selected minimum equilibration time (e.g., 10 min) between the two runs, but with a 2-fold longer equilibration prior to the first run; in some cases, a longer equilibration of the column may be required.

Differences in equipment dwell volume (item 5) can significantly affect experimental results (Section 2.3.6.1). For this reason it is strongly recommended to carry out all method development experiments for a given sample on the same (or equivalent) equipment, so that no difference in dwell volume exists among experiments.

Batch-to-batch column differences (item 6) are unlikely for most columns sold today [79–81], but exceptions are possible – depending on the nature of the sample, column type, and column source. Therefore, at the conclusion of method development it is customary to verify that the method performs adequately on columns of the same kind from several different batches. It also may be appropriate to identify one or more alternative column types that can provide equivalent separation. Recommendations have been presented [84] for the selection of alternative columns of similar selectivity, that is, columns from a different source which will give the same separation as the original column. Based on these guidelines, it is possible to select one or more alternative columns of equivalent selectivity, and to confirm their suitability at the conclusion of method development (see Appendix III). Alternatively, by means of computer simulation [85], it is possible during method development to select gradient conditions such that equivalent separations can be achieved on two or more columns of moderately different selectivity.

3.5.2 Routine Analysis

During method development, it is necessary to anticipate possible changes in the separation that might inadvertently occur when the method is transferred to another laboratory for routine analysis. *Variation in experimental conditions, malfunctioning equipment,* and *changes in column performance* with time (items 1–3 above) are usually handled by system suitability tests (Section 5.1.4.8). It is also customary to carry out a blank gradient run before injecting samples (Section 5.1.3). When the method no longer meets system suitability, the column should be replaced. *Variable column equilibration* (item 4) should be handled differently in routine analysis vs method development. During method development, a between-run equilibration time of at least 10 min is usually acceptable (variable equilibration times, which are common in method development, should then have little effect on separation repeatability). For routine analysis, where the equilibration time between runs is generally fixed, it is possible to shorten the equilibration time to a few minutes or less, in order to minimize the overall run time (Sections 5.3 and 9.2.1.1).

Differences in equipment dwell volume (item 5) are a common reason for the failure of a gradient method during method transfer or routine application on a different HPLC system. The dwell volume V_D can vary significantly between different gradient systems; older systems usually have larger values of V_D. A second gradient system will often be used to carry out routine assays by a method developed on the original equipment. If the second system has a different dwell volume V_D compared with the original system, unacceptable changes in separation can result (Section 5.2.1, Fig. 2.24), especially for "irregular" samples. When V_D for the second system is smaller, this difference in dwell volumes can be compensated for by adding a gradient delay time t_{delay} to the separation carried out on the second system, since this is equivalent to an increase in dwell volume (Section 2.3.6). The length of this gradient delay in minutes should be made equal to the difference in dwell times t_D (dwell volume divided by flow rate). A second gradient system with a larger dwell volume presents a more difficult problem (see discussion of Section 5.2.1), *but an effort should be made in method development to anticipate the maximum dwell volume likely to be encountered in other laboratories.* The original method can be developed with a total gradient delay ($t_D + t_{delay}$) that effectively increases dwell time to the maximum value expected in other laboratories to which the method will be transferred; the value of t_{delay} can then be varied so as to compensate for differences in dwell volume relative to the original equipment.

Batch-to-batch column differences (item 6) can be anticipated to some extent at the conclusion of method development, by testing columns from several different batches. However, a method that is used over a long period of time may still encounter a column batch that is different, so as to result in unacceptable separation. If alternative columns that can provide the same separation were not confirmed in method development, then the same approach as described above for the replacement of the original column by an "equivalent" column can be followed. Alternatively, it is possible to adjust separation conditions to compensate for changed column selectivity [86].

3.5.3 Change in Column Volume

A situation similar to method transfer (above) is presented by a change in column dimensions for various purposes, for example, (a) a reduction in column diameter for improved detection sensitivity of trace components of the sample, or compatibility with MS detection (Section 8.1), or (b) an increase in column diameter for preparative separation (Section 7.2.2.3). In either case, flow rate should be varied first, in proportion to column-diameter-squared; this will maintain an equivalent separation and pressure drop as for the original method. An additional consideration is the effect of the system dwell volume on separation, which should be maintained proportional to the column dead volume V_m when changing column size. If the system dwell volume remains the same when the column volume is changed, changes in relative retention and resolution can occur for "irregular" samples [87, 88]. Consequently, to maintain the same effect of the dwell volume on peak spacing and resolution, it is necessary that V_D/V_m be held constant. This implies a need for either a reduction in V_D when smaller columns are used, or the addition of a gradient delay t_{delay} for larger columns, so as to maintain the equivalent dwell volume ($[V_D + t_{delay}F]/V_m$) constant. For many samples, resolution and separation may still be acceptable – even when V_D/V_m is allowed to vary. This is fortunate, since many workers seem unaware of the possible need for constant V_D/V_m when changing column size.

The reason for possible changes in relative retention and resolution when V_D/V_m is allowed to vary (despite holding k^* constant) can be seen as follows. The effect of the dwell volume is to move each sample compound a certain distance x through the column, before the arrival of the gradient. The value of x is given as

$$x = V_D/V_m k \qquad (3.6)$$

where k is the isocratic retention factor for mobile phase that corresponds to the initial %B of the gradient. Unless V_D/V_m is maintained constant when column size and V_m are changed, the initial isocratic migration x of each compound will vary. This in turn can lead to changes in relative retention and resolution, similar to that observed when a gradient delay is introduced (Fig. 2.23c and d). Another way of looking at the effect of V_D/V_m on peak spacing is that a reduction in this quantity is equivalent to a reduction in V_D when V_m is held constant (Section 2.3.6.1).

Practical alternatives to the adjustment of V_D for constant V_D/V_m include (a) delay of the injection by t_D, so that the sample is injected just as the gradient reaches the head of the column (a feature available on some equipment), and (b) selection of new equipment (or replumbing of existing systems) for minimum V_D (e.g., $V_D < 0.5$ mL) in order to minimize the influence of t_D on the gradient (e.g., as described in Section 8.1.6.1).

3.6 ADDITIONAL MEANS FOR AN INCREASE IN SEPARATION SELECTIVITY

There are several ways of changing separation selectivity (Table 3.4) besides varying temperature and gradient time. We recommend that gradient time t_G be explored first as a means of adjusting peak spacing and maximizing resolution. If additional improvement in selectivity is needed, simultaneous change in both t_G and temperature T should be investigated (Section 3.3.2). Further changes in selectivity are not often required, but if needed, the column can be changed and the optimization of T and t_G repeated [18]. For a maximum change in column selectivity, Appendix III and [42] should be consulted. For example, when a C_8 or C_{18} column is used initially, it might be replaced either by a column with embedded polar groups, or a phenyl column. Alternatively, the B solvent can be changed (e.g., methanol replacing acetonitrile), or mixtures of two or more organic solvents can be used as B solvent (see the examples of Figs 8.18 and 8.19).

After changes in column type, pH, and/or the B solvent have been made, gradient time and temperature should be re-optimized for maximum resolution [89]. When replacing the starting acetonitrile–buffer gradient with a methanol–buffer gradient, keep in mind that the UV cutoff for methanol is somewhat higher than for acetonitrile. A further increase in resolution (or decrease in run time) can be pursued by the *simultaneous* optimization of three or more conditions from Table 3.4. However, this approach can require a formidably large number of experiments. A simpler alternative has been described for isocratic separation [90, 91] that is also applicable to gradient elution; this approach requires only one additional experiment for each separation condition that is to be optimized, but allows only a limited change in each condition ("restricted multi-parameter method development").

Other means for changing selectivity are listed in Table 3.4. The conditions of Table 3.4 can be varied initially, or after making changes in t_G and T. For changes in selectivity, the choice of any one of the conditions of Table 3.4 depends on (a) the relative ability of a change in the condition to change selectivity, as well as (b) possibly adverse effects on detection sensitivity, column stability, or method robustness. With the possible exception of changes in buffer concentration, all of the other conditions in Table 3.4 are usually able to provide *adequate* changes in selectivity for some samples. However, changes in separation conditions other than gradient time and temperature can have offsetting disadvantages [6].

The use of mixtures of methanol, acetonitrile, and tetrahydrofuran (THF) as B solvent in gradient elution allows large, continuous changes in selectivity, as shown by several studies [92–94]. By an appropriate choice of experiments involving these three solvents, it is possible to map resolution as a function of the B solvent and quickly determine the optimum mixture of solvents in the final gradient separation. However, certain problems associated with THF as a mobile phase component (e.g., high UV cutoff, incompatibility with PEEK tubing, formation of peroxides with aging) make its use less desirable, compared with the above options for varying selectivity. Resolution can also be mapped as a function of mobile phase pH. Six experiments with varying pH (over a range in pH of 1–1.5 units) and gradient time allow the best combination of pH and gradient time to be obtained by computer simulation [95]. The use of additional experiments at other pH values can extend the pH range to be explored.

Table 3.6 summarizes information on the relative effectiveness of different conditions as a means of varying selectivity. In the case of a change in gradient time, temperature or mobile phase pH, a maximum (somewhat arbitrary) change in each variable is specified, for example, 6-fold for a change in gradient time, corresponding to a variation of values of k^* by a maximum factor of 6. Note that the relative effectiveness of each condition depends on the kind of sample being separated: "neutral," "ionic," or "mixed." "Ionic" samples contain acids and/or bases, whereas "mixed" samples contain both ionizable and neutral compounds. The relative change in selectivity or values of α for change in certain conditions

TABLE 3.6 A Comparison of Different Separation Conditions that Affect Selectivity; See Text for Details

Condition	Maximum change	Relative change in α (for different samples)			Comment
		Neutral	Ionic	Mixed	
1. Gradient time t_G	6-fold[a]	1.0	2.3	1.6	k^* is proportional to t_G
2. Temperature T	30°C	0.4	1.7	1.0	Retention decreases at higher T
3a. MeOH replaces ACN		1.3	2.9	2.6	Small decrease in retention with ACN
3b. THF replaces ACN		1.8	1.8	2.4	Moderate decrease in retention with THF
4. Column type[b]				2.5	Decreased retention for columns other than C_8 or C_{18}
5. Mobile phase pH	1.0 unit	0.0	6.8	3.9	Decreased retention of acids and increased retention of bases at higher pH
6. Buffer concentration	2-fold	0.0	0.2	0.1	Little change in retention for change in buffer concentration
7. Ion-pair reagent					Increased retention of acids and decreased retention of bases at higher pH

[a]Assumes initial $k^* \approx 3-5$.
[b]C_8 or C_{18} is replaced by a fluoro-substituted ("fluoro") column or a column with embedded polar groups.

has been compared [96]. The data of Table 3.6 are based on a subsequent study [97] of a 67-component sample comprised of acids, bases, and neutrals of widely varying molecular structure.

For samples that contain acids or bases, we can rank the seven conditions of Table 3.6 in approximate order of their relative ability to change selectivity:

buffer concentration (least) < temperature < gradient time < change in B solvent (use of methanol or tetrahydrofuran instead of acetonitrile) ≈ change of column type < mobile phase pH (greatest)

For neutral samples, changes in buffer concentration or pH are relatively ineffective as a means of varying selectivity. In the past, changes in conditions which have the largest likely effect on selectivity have been preferred, for example, a change in column type, B solvent, or pH. For many samples, however, a large change in selectivity is not the most important consideration. Also, the optimization of selectivity by simultaneous changes in gradient time and temperature is quite convenient, free of any significant disadvantages, and adequate for most samples. A change in other separation conditions as a means of varying selectivity may have a significant downside. Thus, a change to methanol or THF as B solvent precludes detection below 210 nm, and THF is unstable (maintaining stable baselines in gradient elution with THF as B solvent can be a problem). A change in column type can provide a significant change in selectivity, but because column selectivity is not continuously variable, a change in the column alone seldom improves overall separation – unless a change in column is followed by optimizing gradient time and temperature. A change in columns also can be more expensive than a change in other variables. See Appendix III for a further discussion of column selectivity, including means for identifying columns of very different selectivity. The variation of mobile phase pH as a means of changing selectivity can complicate peak tracking (Section 3.4.7) based on relative retention, peak size, and/or UV spectra, while methods based on pH optimization tend to be less robust – because of the difficulty in controlling mobile phase pH within an often required $\pm 0.01-0.02$ units [98]. Nevertheless, very large changes in selectivity can often be achieved by a change from low pH (<3) to high pH (≥ 7) conditions. A change in buffer concentration alone usually is not very effective as a means for varying selectivity. Ion-pair reagents are seldom used in gradient elution because of slow column equilibration (Section 5.3). An exception is the use of trifluoroacetic acid (TFA) as combination buffer *and* ion-pair reagent [99], because of rapid column equilibration with this compound [100].

3.7 ORTHOGONAL SEPARATIONS

By an "orthogonal" separation, we mean a second separation procedure with quite different selectivity compared to an initial or "primary" procedure [42, 76]. Orthogonal methods can be used for various purposes. A common use of orthogonal separation is for confirming that the primary separation has resolved all the components in a sample, that is, to show that there are no peak overlaps. Thus, if a first procedure has been developed for optimized selectivity (e.g., by varying both gradient time and temperature), it is possible for two sample compounds to remain unresolved (and undetected) in all the experiments leading to the final method. This possibility suggests the development of an orthogonal procedure whose selectivity is chosen to differ substantially from that of the primary method; it is then unlikely that the same peak pair will be unresolved in *both* procedures. Orthogonal methods can also be used at a later time, during routine application of the primary procedure, in order to check that no new sample impurities or degradation products might be present *and* undetected, because of overlap with another peak. Assuming that an orthogonal method is judged necessary, *we recommend that it be developed at the same time that the primary HPLC procedure is created.*

A fairly simple procedure for the development of an orthogonal method has been described [42]. This approach begins by considering the conditions for the primary method, whose selectivity is to differ from that of the orthogonal method. A column of very different selectivity from the primary column is first selected for the orthogonal method (Appendix III), followed by the replacement of the original B solvent. If acetonitrile is used for the primary method, methanol is selected for the orthogonal method (and vice versa if methanol is selected for the primary method). The separation is next developed further by optimizing gradient time and temperature. In this way, an adequate separation of the sample by means of the orthogonal method usually can be obtained.

An example of the development and application of an orthogonal separation is shown in Figure 3.22. The primary method used an Aquasil C_{18} column and acetonitrile as the B solvent (Fig. 3.22a), while the development of the orthogonal method started with a Betamax Acid C_{18} column and methanol as B solvent. The second column was selected on the basis of its very different selectivity, determined as previously described [84] (Appendix III). Temperature and gradient time were varied next for the orthogonal method, so as to arrive at the final optimized conditions of Figure 3.22(b). Following the development of the separation of Figure 3.22(b), it was found that the sample separated in Figure 3.22(a) contains a sixth, previously unrecognized component (peak 6) which overlaps peak 3 in the primary separation. This particular orthogonal method has thus performed as intended. The evaluation of a primary and orthogonal gradient method in terms of selectivity needs to be carried out for the sample being separated. If retention times (in min) for the orthogonal method are plotted vs retention times for the primary method, a standard deviation SD can be determined. If the quantity SD(b/t_0) > 0.07 [or SD($\Delta\phi S/t_G$) > 0.07], it is likely that two peaks that overlap in the primary method will not overlap in the orthogonal method. See [42] for additional details. An example of such a plot for the separations of Figure 3.22 is shown in Figure 3.22(c), with SD(b/t_0) = 0.35 (i.e., 5-fold larger than the minimum value of 0.07 for an orthogonal separation that is likely to separate two overlapped peaks in the primary method). For a further discussion of orthogonal separations see [42, 101–105].

3.7.1 Two-Dimensional Separations

A single gradient separation is limited in its ability to resolve all the peaks in the chromatogram, when the number of sample components n is large. As discussed in Section 2.24 dealing with "peak capacity" PC, when n exceeds about 30, it is unlikely that all sample components can be resolved adequately by gradient RP-LC (i.e., with $R_s \geq 1$). Two dimensional separation refers to the successive application of two different separations to the same sample, with fractions from the first separation being further resolved in the second separation. If the two separations were completely orthogonal (they never are), and if enough fractions are collected in the first separation for further analysis in the second separation, the total number of *compounds* in the sample that could each be separated with $R_s > 1$ would increase to about $30^2 \approx 1000$, a number that has never been remotely approached in practice. Bear in mind that it is possible to separate as many as 1000 *peaks* in a 2D separation, when many of the peaks are doublets or triplets, that is, where $n > 1000$.

Several groups have reported on 2D separation in the past decade [106–109], often for the separation of complex biological samples such as peptide digests or mixtures of proteins. Usually the two separations do *not* each involve RP-LC, although 2D separations based on RP-LC alone are possible (using two orthogonal methods as in Section 3.7). An alternative (or augmentation) to 2D separation is

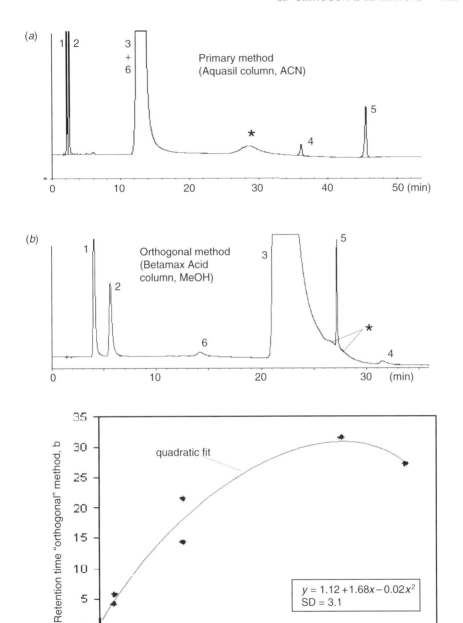

Figure 3.22 Separation of a sample by (*a*) original and (*b*) orthogonal methods. Column dimensions are 150 × 4.6 mm, flow rate is 1.0 mL/min; other conditions: (*a*) 0–0–10 percent ACN at 0–24.5–44.5 min, Aquasil C_{18} column, 30°C; (*b*) 0–0–8–35 percent methanol at 0–10–23–25 min, Betamax Acid column, 23°C; (*c*), plot of retention times from (*b*) vs retention times from (*a*). Adapted from [42]. *Gradient artifact.

provided by reversed-phase gradients with mass spectrometric detection (LC-MS), as discussed in Section 8.1.

> Now this is not the end. It is not even the beginning of the end. But it is, perhaps, the end of the beginning.
>
> —Winston Churchill, *Speech at the Lord Mayor's Day Luncheon, London*
> (*about the progress of World War II*) (*10 November, 1942*)

REFERENCES

1. L. R. Snyder, J. J. Kirkland, and J. L. Glajch, *Practical HPLC Method Development*, 2nd edn, Wiley-Interscience, New York, 1997.
2. J. W. Dolan and L. R. Snyder, *Troubleshooting HPLC Systems*, Humana Press, Clifton, NJ, 1989.
3. L. Van Heukelem and C. S. Thomas, *J. Chromatogr. A* 910 (2001) 31.
4. L. R. Snyder and J. W. Dolan, *J. Chromatogr. A* 892 (2000) 107.
5. P. Schellinger and P. W. Carr, *LCGC* 22 (2004) 544.
6. P. L. Zhu, L. R. Snyder, J. W. Dolan, N. M. Djordjevic, D. W. Hill, L. C. Sander, and T. J. Waeghe, *J. Chromatogr. A* 756 (1996) 21.
7. Z. Li, S. C. Rutan, and S. Dong, *Anal. Chem.* 68 (1996) 124.
8. R. Majors, *LCGC*, 20 (2002) 516.
9. L. R. Snyder and J. W. Dolan, *J. Chromatogr. A* 721 (1996) 3.
10. P. J. Schoenmakers, H. A. H. Billiet, and L. De Galan, *J. Chromatogr.* 205 (1981) 13.
11. P. J. Schoenmakers, A. Bartha, and H. A. H. Billiet, *J. Chromatogr.* 550 (1991) 425.
12. P. Schoenmakers, in *Handbook of HPLC*, E. Katz, R. Eksteen, P. Schoemnakers, and N. Miller, eds, Marcel Dekker, New York, 1998, pp. 218–220.
13. A. P. Schellinger and P. W. Carr, *J. Chromatogr.* 1109 (2006) 253.
14. J. W. Dolan, L. R. Snyder, N, M. Djordjevic, D. W. Hill, L. Van Heukelem, and T. J. Waeghe, *J. Chromatogr. A* 857 (1999) 41.
15. P. L. Zhu, J. W. Dolan, L. R. Snyder, D. W. Hill, L. Van Heukelem, and T. J. Waeghe, *J. Chromatogr. A* 756 (1996) 51.
16. W. Hancock, R. C. Chloupek, J. J. Kirkland, and L. R. Snyder, *J. Chromatogr. A* 686 (1994) 31, 45.
17. J. W. Dolan, L. R. Snyder, N, M. Djordjevic, D. W. Hill, L. Van Heukelem, and T. J. Waeghe, *J. Chromatogr. A* 857 (1999) 1.
18. J. W. Dolan, L. R. Snyder, T. Blanc, and L. Van Heukelem, *J. Chromatogr. A* 897 (2000) 37.
19. J. W. Dolan, *J. Chromatogr. A* 965 (2002) 195.
20. J. W. Dolan, L. R. Snyder, N. M. Djordjevic, D. W. Hill, D. L. Saunders, L. Van Heukelem, and T. J. Waeghe, *J. Chromatogr. A* 803 (1998) 1.
21. P. L. Zhu, J. W. Dolan, and L. R. Snyder, *J. Chromatogr. A* 756 (1996) 41.
22. K. D. Patel, A. D. Jerkovich, J. C. Link, and J. W. Jorgenson, *Anal. Chem.* 76 (2004) 5777.
23. H. Chen and Cs. Horváth, *J. Chromatogr. A* 705 (1995) 3.
24. S. K. Paliwal, M. de Frutos, and F. E. Regnier, in *Methods in Enzymology. Vol. 271. High Resolution Separation and Analysis of Biological Macromolecules*, W. S. Hancock and B. L. Karger, eds, Academic Press, Orlando, FL, 1996, p. 133.
25. J. J. Kirkland, F. A. Truszkowski, and R. D. Ricker, *J. Chromatogr. A* 965 (2002) 25.
26. L. Xiong, R. Zhang, and F. E. Regnier, *J. Chromatogr. A* 1030 (2004) 187.
27. M. Gilar, A. E. Daly, M. Kele, U. D. Neue, and J. C. Gebler, *J. Chromatogr. A* 1061 (2004) 183.
28. A. P. Schellinger, D. R. Stoll, and P. W. Carr, *J. Chromatogr. A* 1064 (2005) 143.
29. P. Brown and E. Grushka, eds, *Design of Rapid Gradient Methods for the Analysis of Combinatorial Chemistry Libraries and the Preparation of Pure Compounds*, Marcel Dekker, New York, 2001.
30. I. Molnar, *J. Chromatogr. A* 965 (2002) 175.
31. L. R. Snyder and L. Wrisley, in *HPLC Made to Measure: A Practical Handbook for Optimization*, S. Kromidas, ed., Wiley-VCH, New York, 2006, p. 535.

32. R. G. Wolcott, J. W. Dolan, and L. R. Snyder, *J. Chromatogr. A* 869 (2000) 3.
33. J. W. Dolan, L. R. Snyder, L. C. Sander, P. Haber, T. Baczek, and R. Kaliszan, *J. Chromatogr. A* 857 (1999) 41.
34. S. Heinisch, E. Lesellier, C. Podevin, J. L. Rocca, and A. Tschapla, *Chromatographia* 44 (1997) 529.
35. R. Cela and M. Lores, *Comput. Chem.* 20 (1996) 175.
36. J. L. Glajch and L. R. Snyder, eds, Computer-assisted chromatographic method development, *J. Chromatogr.* 485 (1989).
37. P. J. Schoenmakers, J. W. Dolan, L. R. Snyder, A. Poile, and A. Drouen, *LCGC* 9 (1991) 714.
38. A. Tchapla, *Analusis* 20, no. 7 (1992) 71.
39. T. Baczek, R. Kaliszan, H. A. Claessens, and M. A. van Straten, *LCGC Europe* June 2001, p. 304.
40. N. Lundell, *J. Chromatogr.* 639 (1993) 97.
41. P. L. Zhu, J. W. Dolan, L. R. Snyder, N. M. Djordjevic, D. W. Hill, J.-T. Lin, L. C. Sander, and L. Van Heukelem, *J. Chromatogr. A* 756 (1996) 63.
42. J. Pellett, P. Lukulay, Y. Mao, W. Bowen, R. Reed, M. Ma, R. C. Munger, J. W. Dolan, L. Wrisley, K. Medwid, N. P. Toltl, C. C. Chan, M. Skibic, K. Biswas, K. A. Wells, and L. R. Snyder, *J. Chromatogr. A* 1101 (2006) 122.
43. T. Jupille, L. Snyder, and I. Molnar, *LCGC Europe* 15 (2002) 596.
44. S. V. Galushko and A. A. Kamenchuk, *LCGC Int.* 8 (1995) 581.
45. S. V. Galushko, A. A. Kamenchuk, and G. L. Pit, *Am. Lab.* 27 (1995) 33G.
46. V. Concha-Herrera, G. Vivó-Trujols, J. R. Torres-Lapasió, and M. C. Garcia-Alvarez-Coque, *J. Chromatogr. A* 1063 (2005) 79.
47. J. W. Dolan, L. R. Snyder, N, M. Djordjevic, D. W. Hill, L. Van Heukelem, and T. J. Waeghe, *J. Chromatogr. A* 857 (1999) 21.
48. G. Vivó-Truyols, J. R. Torres-Lapasió, and M. C. Garcia-Alvarez-Coque, *J. Chromatogr. A* 876 (2000) 17.
49. J. D. Stuart, D. D Lisi, and L. R. Snyder, *J. Chromatogr.* 485 (1989) 657.
50. J. D. Stuart, D. D. Lisi, and L. R. Snyder, *J. Chromatogr.* 555(1991) 1.
51. B. F. D. Ghrist, B. S. Cooperman, and L. R. Snyder, *J. Chromatogr.* 459 (1989) 1,
52. J. W. Dolan, D. C. Lommen, and L. R. Snyder, *J. Chromatogr.* 485 (1989) 91,
53. T. Sasegawa, Y. Sakamoto, T. Hirose, T. Yoshida, Y. Kobayashi, and Y. Sato, *J. Chromatogr.* 485 (1989) 533,
54. R. G. Lehman and J. R. Miller, *J. Chromatogr.* 485 (1989) 581.
55. D. J. Thompson and W. D. Ellenson, *J. Chromatogr.* 485 (1989) 607.
56. I. Molnar, R. I. Boysen, and V. A. Erdmann, *Chromatographia* 28(1/2) (1989) 39.
57. W. Markowski, T. H. Dzido, and E. Soczewinski, *J. Chromatogr.* 523 (1990) 81.
58. B. F. D. Ghrist, B. S. Cooperman, and L. R. Snyder, *HPLC of Biological Macromolecules*, F. E. Regnier and K.M. Gooding, eds, Marcel Dekker, New York, 1990, p. 403.
59. N. G. Mellish, *LCGC* 9 (1991) 845.
60. T.-Y. Liu and A. Robbat, Jr, *J. Chromatogr.* 539 (1991) 1.
61. I. Molnar, K. H. Gober, and B. Christ, *J. Chromatogr.* 550 (1991) 39.
62. T. H. Dzido, E. Soczewiński, and J. Gudej, *J. Chromatogr.* 550 (1991) 71.
63. P. A. Ryan, B. A. Ewels, and J. L. Glajch, *J. Chromatogr.* 550 (1991) 549.
64. R. Däppen and I. Molnar, *J. Chromatogr.* 592 (1992) 133.
65. R. C. Chloupek, W. S. Hancock, and L. R. Snyder, *J. Chromatogr.* 594 (1992) 65.
66. A. J. J. M. Coenen, L. H. G. Henckens, Y. Mengerink, Sj van der Wal, P. J. L. M. Quaedifleig, L. H. Koole, and E.M. Meijer, *J. Chromatogr.* 596 (1992) 59.
67. L. Wrisley, *J. Chromatogr.* 628 (1993) 191.
68. H. Fritsch, I. Molnar, and M. Wurl, *J. Chromatogr. A* 684 (1994) 65.
69. R. Bonfichi, *J. Chromatogr. A* 678 (1994) 213.
70. L. R. Snyder, *New Methods in Peptide Mapping for the Characterization of Proteins*. W. Hancock, ed., CRC Press, Boca Raton, FL, 1996, p. 31.
71. H. W. Bilke, I. Molnar, and Ch. Gernet, *J. Chromatogr. A* 729 (1996) 189.
72. T. H. Dzido and A. Sory, *Chem. Anal. (Warsaw)* 41 (1996) 113 .
73. W. Markowski and K. L. Czapinska, *Chem. Anal. (Warsaw)* 42 (1997) 353.

74. P. Haber, T. Baczek, R. Kaliszan, L. R. Snyder, J. W. Dolan, and C. T. Wehr., *J. Chromatogr. Sci.* 38 (2000) 386.
75. T. H. Hoang, D. Cuerrier, S. McClintock, and M. Di Maso, *J. Chromatogr. A* 991 (2003) 281.
76. G. Xue, A. D. Bendick, R. Chen, and S. S. Sekulic, *J. Chromatogr. A* 1050 (2004) 159.
77. J. L. Glajch, M. A. Quarry, J. F. Vasta, and L. R. Snyder, *Anal. Chem.* 58 (1986) 280.
78. H. J. Issaq and K. L. McNitt, *J. Liq. Chromatogr.* 5 (1982) 1771.
79. J. K. Strasters, H. A. H. Billiet, L. de Galan, and B. G. M. Vandeginste, *J. Chromatogr.* 499 (1990) 499.
80. E. P. Lankmayr, W. Wegscheider, J. Daniel-Ivad, I. Kolossváry, G. Csonka, and M. Otto, *J. Chromatgr.* 485 (1989) 557.
81. M. Kele and G. Guiochon, *J. Chromatogr. A* 830 (1999) 41, 55.
82. U. D. Neue, E. Serowik, P. Iraneta, B. A. Alden, and T. H. Walter, *J. Chromatogr. A* 849 (1999) 87.
83. M. Kele and G. Guiochon, *J. Chromatogr. A* 855 (1999) 423; 869 (2000) 181; 913 (2001) 89.
84. J. W. Dolan, A. Maule, L. Wrisley, C. C. Chan, M. Angod, C. Lunte, R. Krisko, J. Winston, B. Homeierand, D. M. McCalley, and L. R. Snyder, *J. Chromatogr. A* 1057 (2004) 59.
85. J. W. Dolan, L. R. Snyder, and T. Blanc, *J. Chromatogr. A.* 897 (2000) 51.
86. J. W. Dolan, L. R. Snyder, T. H. Jupille, and N. S. Wilson, *J. Chromatogr. A* 960 (2002) 51.
87. J. W. Dolan and L. R. Snyder, *J. Chromatogr. A* 799 (1998) 21.
88. A. P. Schellinger and P. W. Carr, *J. Chromatogr. A* 1077 (2005) 110.
89. J. W. Dolan, L. R. Snyder, D. L. Saunders, and L. Van Heukelem, *J. Chromatogr.A* 803 (1998) 3.
90. J. W. Dolan, D. C. Lommen, and L. R. Snyder, *J. Chromatogr.* 535 (1990) 55.
91. L. R. Snyder, J. W. Dolan, and D. C. Lommen, *J. Chromatogr.* 535 (1990) 75.
92. P. Jandera, J. Churáček, and H. Colin, *J. Chromatogr.* 214 (1981) 35.
93. J. J. Kirkland and J. L. Glajch, *J. Chromatogr.* 255 (1983) 27.
94. H. Vuorela, P. Lehtonen, and R. Hiltunen, *J. Liq. Chromatogr.* 14 (1991) 3181.
95. P. Haber, T. Baczek, R. Kaliszan, L. R. Snyder, J. W. Dolan, and C. T. Wehr, *J. Chromatogr. Sci.* 38 (2000) 386.
96. L. R. Snyder, *J. Chromatogr. B* 689 (1997) 105.
97. N. S. Wilson, M. D. Nelson, J. W. Dolan, L. R. Snyder, and P. W. Carr, *J. Chromatogr. A* 961 (2002) 195.
98. L. R. Snyder, A. Maule, A. Heebsch, R. Cuellar, S. Paulson, J. Carrano, L. Wrisley, C. C. Chan, N. Pearson, J. W. Dolan, and J. Gilroy, *J. Chromatogr. A* 1057 (2004) 49.
99. D. Guo, C. T. Mant, and R. S. Hodges, *J. Chromatogr.* 386 (1987) 205.
100. C. Z. Chuang and F. A. Ragan, Jr, *J. Liq. Chromatogr.* 17 (1994) 2383.
101. E. Van Gyseghem, I. Crosiers, S. Gourvenec, D. L. Massart, and Y. Vander Heyden, *J. Chromatogr. A* 1026 (2004) 117.
102. B. D. Karcher, M. L. Davies, J. J. Venit, and E. J. Delaney, *Am. Pharm. Rev.* 7 (2004) 62.
103. G. Xue, A. D. Bendick, R. Chen, and S. S. Sekulic, *J. Chromatogr. A* 1050 (2004) 159.
104. E. Van Gyseghem, M. Jimidar, R. Sneyers, D. Redlich, E. Verhoeven, D. L. Massart, and Y. Vander Heyden, *J. Chromatogr. A* 1074 (2005) 117.
105. P. Forlay-Frick, E. Van Gyseghem, K. Héberger, and Y. Vander Heyden, *Anal. Chim. Acta* 539 (2005) 1.
106. N. Tanaka, H. Kimura, D. Tokuda, K. Hosoya, T. Ikegami, N. Ishizuka, H. Minakuchi, K. Nakanishi, Y. Shintani, M. Furuno, and K. Cabrera, *Anal. Chem.* 76 (2004) 1273.
107. M. Vollmer, R. Horth, and E. Nagele, *Anal. Chem.* 76 (2004) 5180.
108. D. R. Stoll and P. W. Carr, *J. Am. Chem. Soc.* 127 (2005) 5034.
109. D. R. Stoll, J. D. Cohen and P. W. Carr, *J. Chromatogr. A*, 1122 (2006) 123.

CHAPTER 4

GRADIENT EQUIPMENT

> Man is a tool-using animal ... Without tools he is nothing, with tools he is all.
> —Thomas Carlyle, *Sartor Resartus*

Equipment for gradient elution differs from that needed for isocratic separation in that it must be able to generate accurate and reproducible changes in mobile phase composition during the run. Many users of isocratic methods employ gradient equipment for on-line blending of isocratic mobile phases because of the convenience; for isocratic applications, equipment differences may not be important. When carrying out a gradient procedure, however, differences in equipment are more likely to affect the separation. This chapter discusses the basics of gradient equipment design (Section 4.1), the characteristics that are desirable when choosing a system (Section 4.2), and how to measure gradient performance (Section 4.3). Because of the importance of equipment *dwell volume* in the selection of equipment for a given gradient elution application, this topic is called out for special attention (Section 4.4).

4.1 GRADIENT SYSTEM DESIGN

Gradient-elution HPLC systems are available in two basic designs, low-pressure mixing and high-pressure mixing. As the names imply, these differ primarily in whether the mobile phase is blended before or after it passes through the pump. There are some hybrid designs which do not cleanly fall into the high-pressure or low-pressure mixing categories. For example, at least one manufacturer's pump (Varian) has proportioning valves mounted on the pump head, so that solvents are proportioned and mixed within the pump head. Only high- and low-pressure mixing systems are discussed here.

4.1.1 High-Pressure vs Low-Pressure Mixing

The layout of a typical high-pressure mixing system is shown in Figure 4.1. Two pumps are used to meter the mobile phase components, A and B, into the mixer at high pressure. The blend of solvents is controlled by the flow rates of the two

---

High-Performance Gradient Elution. By Lloyd R. Snyder and John W. Dolan
Copyright © 2007 John Wiley & Sons, Inc.

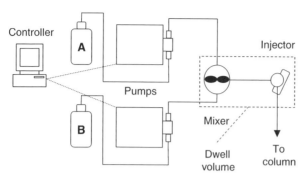

Figure 4.1 Diagram of a typical high-pressure mixing system.

pumps. For example, a 60:40 blend of A and B at 1 mL/min would be produced by pump A running at 0.6 mL/min and pump B at 0.4 mL/min. Gradients are formed by changing the relative flow rates of pumps A and B during the gradient while the total flow rate is kept constant.

High-pressure mixing systems usually are limited to two pumps, so only two solvents can be blended at a time. Low-pressure systems allow blending of as many as four solvents. Some high-pressure systems have an additional solvent selection valve that allows for a choice of solvents for each pump. For example, pump A might be set to use a buffer during sample analysis, then switch to water to flush buffer residues from the system when the sample batch is completed. Such a setup allows for selection of more than two solvents, but not online blending of more than two solvents.

A two-solvent, low-pressure mixing system is shown in Figure 4.2. In this layout, a single pump is used to deliver pre-blended solvent to the column. The blend of solvents is controlled by a set of proportioning valves that open and close one at a time under the command of the system controller. The flow rate is kept constant, while the cycle of the proportioning valves varies during gradient formation. For example, if the total valve cycle time were 100 ms and a gradient of

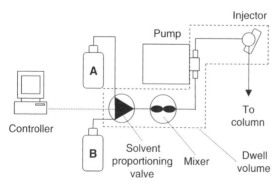

Figure 4.2 Diagram of a typical low-pressure mixing system.

5–95 percent B were selected, the gradient would start with the A solvent proportioning valve open for 95 ms followed by the B valve open for 5 ms (5 percent B). Gradually over the course of the gradient, the proportion of time the B valve was open would increase until, at the end of the gradient, the A valve would be open for 5 ms and the B valve for 95 ms of each cycle. The mobile phase components are mixed on the low-pressure side of the pump, which generally results in a larger dwell volume than the high-pressure counterpart. Because solvents are blended at atmospheric pressure, outgassing of the solvents upon mixing can be a major problem with low-pressure systems, so solvent degassing is required for reliable operation.

The dwell volume V_D (Section 2.3.6.1) includes the volumes of the gradient mixer plus the flow-path between the mixer and column inlet. Values of V_D tend to be smaller in high-pressure mixing systems than the equivalent low-pressure configuration. This can be important for applications such as LC-MS that use low-volume columns and low flow rates (Section 8.1), where dwell volumes of <0.5 mL are desired. The balance of the hardware is discussed in Section 4.1.3. See Chapter 5 for information about potential problems with gradient equipment.

4.1.2 Tradeoffs

The advantages and disadvantages of high- and low-pressure mixing systems are summarized in Table 4.1. The discussion below cites the general tradeoffs between high- and low-pressure mixing systems. However, the evolution of system design and the characteristics of specific equipment models will surely provide an exception for every one of the generalizations listed.

4.1.2.1 Dwell Volume The A and B solvents are mixed downstream from the pump with high-pressure mixing systems, so the dwell volume usually is smaller than for low-pressure mixing. For example, high-pressure mixing systems typically have dwell volumes in the 1–3 mL range, whereas low-pressure systems are typically 2–4 mL (but the newest low-pressure mixing systems can have smaller dwell volumes than older high-pressure systems). Small dwell volumes are important for small-volume applications, such as LC-MS (Section 8.1) or narrow-bore columns. It usually is easy to convert a high-pressure mixing system from a conventional

TABLE 4.1 Comparison of High-Pressure and Low-Pressure Mixing Systems

Feature	High pressure	Low pressure
Dwell volume	Smaller	Larger
Mixer conversion	Straightforward	May not be possible
Low-volume applications	Usually easy conversion	More difficult
Degassing	Usually recommended	Usually essential
Accuracy (%B)	±1 percent typical	±1 percent typical
Accuracy (compressibility)	More affected	Less affected
Complex mixtures	One pump required per solvent	Three or four solvents standard
Independent pump use	Easy	Not possible

dwell volume (e.g., 2.5 mL) to a small dwell volume (e.g., 200 μL) by disconnecting the standard mixer and replacing it with an after-market micromixer available from the system manufacturer or a fittings and parts supplier (e.g., Upchurch). This can be useful when changing a system from use with conventional 150 × 4.6 mm i.d. columns to use with short, narrow-bore columns, such as the 50 × 2.1 mm i.d. columns popular for LC-MS applications. (Note that mixer design and volume are chosen by the manufacturer to avoid compositional ripples, which can be visible under certain circumstances, such as TFA gradients at low-UV wavelengths. For this reason, if a manufacturer's mixer is replaced by an aftermarket device, one should be sure to check for adequate gradient performance, as described in Section 4.3.1.) The mixer and proportioning apparatus of low-pressure mixing systems is usually integral to the design of the system and seldom can be replaced with a smaller-volume unit.

Most newer gradient equipment has the capability of delaying the injection until after the gradient has started. Using this technique, one can time the injection to coincide with the time when the gradient reaches the head of the column, effectively reducing V_D to zero. See Sections 4.4, 5.2, and 5.3.1 for additional information on dwell volume and its practical consequences.

4.1.2.2 Degassing   Solvent outgassing (release of air bubbles) can occur whenever an aqueous and an organic solvent are mixed, unless the solvents are degassed in advance. When solvents are mixed at atmospheric pressure, as is the case with low-pressure mixing systems, bubbles often form in the mixer and may be drawn into the pump, which can result in poor pump performance and errors in solvent proportioning. On the other hand, when solvents are mixed under high pressure, the gas remains dissolved in the mobile phase mixture and does not cause bubble problems in the high-pressure portion of the system. However, when the mobile phase returns to atmospheric pressure after it leaves the column, outgassing in the detector flow cell can occur. It is a good idea to use a back-pressure device downstream from the detector to ensure that bubbles stay in solution until they leave the detector. One should be sure that the back pressure (e.g., 50–100 psi) does not exceed the pressure limits of the detector cell. Back-pressure restrictors are available from many HPLC parts and fittings vendors.

Different HPLC systems have varying susceptibility to bubble problems. This depends on the overall design (high- vs low-pressure mixing), as well as the manufacturer's specific design features. Some systems require continuous degassing for reliable operation, whereas others can tolerate degassing once a day, and still other systems may be able to operate without solvent degassing. Helium sparging historically has been the gold standard for degassing [1] and still is practiced widely as an effective degassing technique for any HPLC system. In recent years, in-line membrane degassers have proven to be reliable and easy to use. These in-line degassers are standard equipment on many HPLC systems and are configured such that each solvent supply line is independently degassed. As a general rule, every HPLC system will work more reliably if the mobile phase components are degassed. The almost universal practice of solvent degassing has reduced outgassing problems from a major problem to a minor irritant in most laboratories.

4.1.2.3 Accuracy High-pressure mixing systems rely on variation in the flow rate of the two pumps to form the gradient. For example, a 5–95 percent B gradient at 1 mL/min will start with the A solvent pump at 950 μL/min (95 percent of 1 mL/min) and the B pump at 50 μL/min. Over the course of the gradient, the flow rate of the A pump will decrease and the B pump will increase until the gradient ends with 50 μL/min of A and 950 μL/min of B. Older pumping systems were less reliable at the extremes of the gradient (e.g., <5 percent B or >95 percent B), because in these regions one pump was operating at a very low flow rate. This sometimes caused problems with respect to the linearity and reproducibility of the gradient [2]. However, modern high-pressure pumps, with piston volumes of ≤10 μL and piston step sizes of ≤10 nL, provide adequate accuracy (e.g., ±1 percent) throughout the gradient. Apart from column equilibration and buffer solubility, there are no compelling reasons for preferring a 5–95 percent gradient, instead of 0–100 percent, in terms of accuracy of the mobile phase composition.

Low-pressure mixing systems, on the other hand, use a single pump at a constant flow rate throughout the gradient, so flow rate accuracy should not affect gradient accuracy. The composition of the mobile phase is controlled by the valve cycle of the proportioning valves (see Section 4.1.1), which gives satisfactory performance throughout the gradient range. Typical specifications for accuracy of flow rate and gradient performance for both low- and high-pressure mixing systems are ±1 percent. It has long been known that the relationship between the proportioning valve timing, the pump piston cycle, and the mixing volume is critical to generation of high-quality gradients for low-pressure mixing systems [3, 4]. Today's systems are the result of years of research and development, thus any user modifications of the mixing system are unlikely to improve performance.

Shallow gradients (e.g., <1 percent change/min), as are required for separations of high-molecular-weight samples (Chapter 6), are more susceptible to small errors in mobile phase proportioning than are steeper gradients (e.g., >1 percent change/min). Some system designs may perform better than others with shallow gradients, so it is wise to check gradient performance for your system, so as to ensure that acceptable gradient accuracy can be obtained under the desired operating conditions. Tests of gradient system performance are described in Section 4.3. Note that pre-mixing the A and B solvents can serve to overcome narrowing the gradient range can serve to overcome problems due to shallow gradients (Section 5.5.4.2).

4.1.2.4 Solvent Volume Changes and Compressibility Three characteristics of mobile phase solvents contribute to non-ideal behavior of HPLC pumping systems:

- volume changes on mixing;
- compressibility;
- viscosity changes on mixing.

When two solvents are mixed, the *volume of the mixture* often is not the same as the combined component volumes. For example, at 60–65 percent methanol–water, the volume is reduced by ≈ 4 percent, whereas maximum changes for

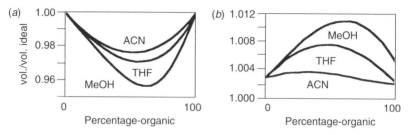

Figure 4.3 (a) Solvent volume change on mixing, and (b) compressibility of solvent mixtures. Conditions for same nominal flow rate that generates 2000 psi back pressure with 100 percent water (after Fig. 1 of [5]).

acetonitrile–water mixtures are ≈ 2.5 percent [5]. The volume change of solvents on mixing is illustrated in Figure 4.3(a). For low-pressure mixing systems, the mobile phase is blended upstream from the pump, and the pump operates at a constant flow rate, so the flow rate is not compromised by changes in the volume due to mixing. High-pressure mixing systems, on the other hand, blend the solvents after the pumps, so any change in volume due to mixing will be reflected in the flow rate (measured downstream from the mixer). Because the magnitude of the volume change depends on the proportion of solvents, the volume change is not constant throughout the gradient, and varies further with the organic solvent used as B solvent. Consequently, correcting for this effect would be difficult. However, any such volume changes will be consistent, so the gradient should be reproducible.

Solvent compressibility also contributes to nonideal behavior of mobile phase delivery. When a solvent compresses under pressure in the pump, it will generate a lower-than-expected flow rate within the column, but normal flow rate when measured at the end of the column, because the solvent expands to its original volume when it returns to atmospheric pressure. The compressibility of common mobile phase mixtures is generally <1 percent [5], as illustrated in Figure 4.3(b) (the plots of Fig. 4.3 are based on conditions of equal nominal flow rates that generate a pressure drop of 2000 psi for water). Because the compressibility effect is of opposite sign than volume changes, there is some canceling of these two effects.

Additional flow-rate errors can arise from *changes in solvent viscosity* upon mixing, over the course of a gradient [5]. All of these contributions can add up to a net change in flow rate of -2 to -5 percent over the course of the gradient (see further discussion in [2, 5, 6]); however, it is unlikely that such changes are of practical consequence, since they should be reproducible from one run to the next.

Most pump manufacturers specify some kind of "compressibility compensation" for their equipment. An adjustment is made for the difference in compressibility between the A and B solvents. It generally is not clear from the manufacturers' literature whether or not any additional compensation is made for changes in volume and/or compressibility during a gradient run. Gradient linearity generally is measured with the same solvent in both the A and B reservoirs (see Section 4.3.1.1), so nonideal effects due to solvent differences would not be noticed. Fortunately, any errors should be constant for a given system, so any variation in flow rate over the course of the run should be consistent. Furthermore, because the changes are due more to physical characteristics of the mobile phase than the hardware, it is expected that the errors will be of similar magnitude for different HPLC systems and therefore not likely to be a problem. From a practical standpoint, such errors are seldom the source of complaints about gradient performance. See the further discussion in Section 9.2.2.

Because most users ignore the use of compressibility compensation, there is one practical difference between high-pressure and low-pressure mixing systems in terms of flow rate (and thus retention). The flow rate for low-pressure mixing systems is always the set flow rate, because any change in solvent volume upon mixing takes place before the mobile phase is metered through the pump. On the other hand, the flow rate for high-pressure mixing systems is always less than the set flow rate, because the total volume delivered by the pumps is reduced by the volume change when the solvents are mixed. This means that, although the solvent composition is identical, the flow rate may be different between a high-pressure and low-pressure mixing system with otherwise identical settings. However, small changes in flow rate have very little effect on a gradient separation (Section 2.3.3 and Fig. 2.17).

4.1.2.5 Flexibility Low-pressure mixing systems with more than two solvents (often four are available) have an advantage of added flexibility over high-pressure mixing systems that typically are limited to two solvents. With four-solvent capability, it is possible to arrange the reservoirs so that several organic solvents can be used. This can be quite useful for unattended method development, such as when a series of experiments with different mobile phases are to be carried out overnight. Another interesting application of low-pressure mixing systems with more than two solvent reservoirs is the ability to run constant-buffer-strength gradients. In this configuration, one solvent reservoir (e.g., C) is programmed to deliver the buffer at a constant rate (e.g., one-tenth the flow rate of $10\times$ the desired buffer concentration), while the A and B solvents are blended to generate a solvent gradient. The availability of more than two solvents also allows the user to mix ternary or quaternary mobile phase blends on-line, which may be useful both for method development and routine analysis (Section 8.4).

In order to avoid possible buffer precipitation, it is best to flush the system with buffer-free mobile phase before flushing with 100 percent organic before system shut-down, especially when acetonitrile is used. (This assumes that the gradient used in the method stops at <100 percent B, because of buffer-solubility concerns.) The availability of an additional reservoir can be a useful tool for purging buffer from the system at the end of a series of runs, prior to system shut down (e.g., buffer in the A-reservoir, acetonitrile in the B-reservoir, and water in the C-reservoir). After a batch of samples has been run with buffer–acetonitrile gradients (A and B) in an unattended series of runs, the system can be programmed to flush the system with water/acetonitrile (C and B) so as to remove any buffer from the system before the pumps are turned off. See Section 3.2 for more information on buffer solubility.

Many manufacturers of high-pressure mixing systems provide the capability to add a third pump for ternary solvent use or even a fourth pump, but the additional cost of more pumps discourages this from being a popular option. An alternative offered for some high-pressure mixing systems is a solvent selection valve that enables the use of more than one solvent per pump, such as for the buffer-flushing example above, but not for generating ternary-solvent mobile phases.

4.1.2.6 Independent Module Use Low-pressure mixing systems generally are supplied as a complete HPLC unit, comprising the pump, autosampler, and column oven. The detector may or may not be an integral part of the unit. When constructed as a unit, the manufacturer may be able to reduce construction costs and also minimize the amount of bench space used. High-pressure mixing systems may be sold as individual modules that can be combined as desired by the user, or that can be stacked in a compact unit with a footprint similar to the low-pressure counterpart. Historically, users have argued that a high-pressure mixing design, because of its inherent modular nature, allows one to mix and match the best modules from each manufacturer so as to obtain a system that can outperform any individual manufacturer's system. This may have been true in the past, but is less likely today, because the software that controls one manufacturer's modules generally will not control another's, so mixed-manufacturer systems may not be practical. As a result, most users obtain the entire system from the same manufacturer, with the possible exceptions of the detector and data system, which may be easier to adapt from one vendor to another. For users who only occasionally need gradient capability, a high-pressure mixing system still may be advantageous, because when only isocratic methods are used, one pump can be removed to use on a second isocratic system. The independent nature of high-pressure mixing pumps makes them easier to combine for some column switching applications or parallel chromatography for improved LC-MS throughput (see Sections 8.1.5.2 and 8.1.6.8). Either a high-pressure or low-pressure mixing system will work satisfactorily as a convenience to avoid hand-mixing solvents, even if the system is used exclusively in the isocratic mode.

4.1.3 Other System Components

Whereas there are significant differences in terms of the solvent delivery system, the remaining components for both high- and low-pressure gradient systems, such as the autosampler, column oven, detector, and data system, are the same as those used in any isocratic HPLC system. These are discussed briefly below.

4.1.3.1 Autosampler Most HPLC systems include an autosampler instead of manual injector, so that unattended sample injection can be used. Autosamplers are designed to accommodate the needs of many types of users. For example, use with conventional sample vials or 96-well plates, refrigeration, and/or automatic plate changers may be features on various autosampler designs. The fixed-volume injector loop, common in the past, has largely given way to variable-volume injections using a partially filled injector loop. The precision of peak areas (or heights) is a reflection of autosampler performance, and is determined during system performance checks (Section 4.3.2.4).

4.1.3.2 Column For the most part, columns used in gradient elution are the same as those used for isocratic separations (Appendix III). Historic concerns about column instability with gradients can be forgotten for silica-based columns – improvements in column packing technology have resulted in very stable columns. For some polymeric supports, shrinking or swelling with a change in solvent composition may be a consideration, so it is wise to check the care-and-use instructions for any nonsilica-based column. Similarly, some protein-bonded columns, such as chiral phases, may be intolerant to high concentrations of organic, thereby limiting the allowable gradient range for these columns. Additional information on columns can be found in [7].

4.1.3.3 Detector Most detectors that are used for isocratic applications will also perform well for gradient elution. The exceptions are detectors that are sensitive to changes in the mobile phase composition. The most popular detectors are the variable-wavelength and photodiode array UV detectors. Many workers find the evaporative light scattering detector a suitable alternative for non-UV absorbing analytes. Fluorescence detectors and those based on mass spectrometry (MS) also are very popular for use with gradient elution. The refractive index (RI) detector relies on a differential change in the refractive index of the eluted sample relative to the background for detection. With gradients, the change in the mobile phase refractive index during the gradient is larger than the sample signal, making this detector useless for gradient elution.

The detector flow cell should be large enough for adequate detection, but not so large that dispersion within the cell causes excessive band broadening; otherwise, closely eluting peaks may overlap. For UV detectors used with conventional 150×4.6 mm i.d. columns, a typical detector cell is 10 mm long and 1 mm i.d., for about 8 µL of volume. Shorter cells are available for preparative applications, and narrower diameter cells are needed for applications which require maximum signal with minimum dispersion. One low-volume cell design (Waters' UPLC system) reduces dispersion by using a total-internal-reflectance cell wall and a small internal diameter (e.g., 0.5 mm i.d.) so as to simultaneously increase light transmission and reduce the cell volume, while maintaining the path length.

Some detectors have an adjustable time constant. The time constant is an electronic filter that serves to smooth the detector signal. However, if the time constant is too large, peak broadening and signal loss can occur. A rule of thumb is to set the time constant at less than or equal to one-tenth of the peak width at baseline. Thus, a 10 s wide peak could use a 1 s time constant without deleterious effects. Many workers prefer to set the time constant to a very small value (e.g., <0.1 s), or turn it off and rely on the data system for signal smoothing.

4.1.3.4 Data System The personal-computer-based data system has replaced other data reduction techniques in nearly every application. An electronic record of the sample is made and post-run processing allows the user to get the desired information from the chromatogram. To obtain high quality data that are representative of the true detector signal, a minimum of 10–20 data points need to be gathered across a peak. (The detector time constant and data system data rate both smooth the signal by averaging data points over time.) It is always a good idea to err on the side of gathering too much data, because post-run processing can always simplify the data set (combine data points), but can never expand it. Most data acquisition software can automatically select the appropriate acquisition settings based on the appearance of the chromatogram.

Although there are differences in the user interface between various data systems, they all perform the same basic functions. For specialty applications, such as preparative, microscale, or fast gradients, it is wise to check the data system capabilities. For example, one should make sure that the data system (and detector) is capable of taking data at a fast enough rate to adequately characterize a chromatogram that may elute in <20 s with a fast gradient separation.

4.1.3.5 Extra-Column Volume Any volume in those parts of the system, other than the column, through which the sample travels can contribute to peak broadening, that is, an increase in peak width and a decrease in sample resolution. The injection volume, connecting tubing and fittings, detector cell, and column all contribute to the observed peak volume [7]. The influence of the injection volume can be minimized by injecting small volumes or using a weak injection solvent to take advantage of on-column band compression [8–11 and Section 9.2.2]. The detector cell volume can be changed in some systems by substituting a larger or smaller detector cell, but sensitivity may suffer if the cell path length is changed. If fittings are assembled properly, their influence can be ignored when using conventional (e.g., 4.6 mm i.d.) columns. The tubing which connects the autosampler to the column and the column to the detector, is a system component that often is changed by the user; however, care should be taken when making plumbing changes. Extra-column band broadening prior to the column is relatively unimportant in gradient elution, because of gradient compression at the column inlet when the sample enters the weak initial mobile phase.

From a practical standpoint, the reduction in resolution due to extra-column band broadening from connecting tubing will seldom be noticed when small i.d. tubing is used conservatively for conventional separations. For example, with 3 μm columns of 2.1 mm i.d. or larger and 50 mm or more in length, 25 cm of 0.005 in. (∼0.13 mm) i.d. connecting tubing will cause ≤1 percent loss in resolution; with 0.007 in. (∼0.18 mm) i.d. tubing the loss will be ≤6 percent [12]. The effects of band broadening due to connecting tubing for larger-volume columns and/or larger particles will go unnoticed in all but the most demanding separations.

4.2 GENERAL CONSIDERATIONS IN SYSTEM SELECTION

A recurring question that arises in the authors' HPLC short courses is, "who makes the best gradient HPLC system?" The choice of a preferred HPLC brand and model was much more important (and difficult) 15 years ago than it is today. Design refinements, economics, and user feedback have resulted in a consistently high quality of available HPLC equipment; it can be truthfully stated that there are no "bad" HPLC systems being made today. Table 4.2 lists some of the more common manufacturers of HPLC systems and their website addresses. If equipment performance is

TABLE 4.2 HPLC System Manufacturers

Manufacturer	Web site
Agilent	www.agilent.com
Beckman	www.beckman.com
Bio-Rad	www.biorad.com
Bishoff	www.bischoff-chrom.de
Dionex	www.dionex.com
ESA	www.esainc.com
Gilson	www.gilson.com
Hitachi	www.hitachi-hta.com
Jasco	www.jascoinc.com
LC Packings	www.lcpackings.com
Michrom Bioresources	www.michrom.com
Micro-Tech	www.microlc.com
Perkin-Elmer	las.perkinelmer.com
Shimadzu	www.shimadzu.com
Spark Holland	www.spark.nl
Thermo Separations	www.thermo.com
Varian	www.varian.com
Waters	www.waters.com

roughly comparable, what other factors should be considered when buying a gradient HPLC system?

4.2.1 Which Vendor?

In the early days of HPLC it was common to purchase a pump from one vendor, a detector from another, and so forth to ensure that the "best" system components were obtained. Today's HPLC systems, with embedded microprocessors and system controllers, are seldom capable of communicating with modules from other manufacturers, so the multi-vendor approach is no longer practical. Detectors, ovens, and even autosamplers may be capable of functioning with other manufacturer's systems, but generally control is only via simple contact-closure switches or other primitive electronic communication (available with most systems for the control of external devices). It is best to purchase the entire system from a single vendor.

Should multiple systems in the same laboratory be of one brand and model or many? One might argue that one manufacturer may make the best system for ion chromatography, another for size exclusion, and another for general use, but there is little differentiation among the main-line HPLC systems today. On the other hand, there is a strong argument to stay with one manufacturer and one model of system in terms of parts, service, and training. For example, if the laboratory has five nominally identical systems, there is added justification to stock less-common parts, so downtime during maintenance or repair can be reduced. Whether service is done internally or via a service contract, multiple copies of the same system will simplify service. Workers in the laboratory and in-house service technicians will have more in-depth understanding of system repairs if they do not have to spread their knowledge over several different brands. A single service call by a manufacturer's personnel can accomplish annual maintenance tasks on several systems for service cost savings. From a very practical standpoint, workers tend to gravitate toward their favorite system and ignore units they do not like. This can reduce the flexibility of work assignments and thus decrease laboratory efficiency. Training laboratory staff on a single model of equipment can save training costs. Transfer of gradient methods from one system to another within a laboratory, or between laboratories, will be simpler if a common HPLC platform is used. All these factors strongly support the purchase of multiple instruments from the same manufacturer.

4.2.2 High-Pressure or Low-Pressure Mixing?

The choice of a high-pressure or low-pressure mixing system is a matter of personal preference, as much as anything. Most applications will work well on either design of system. Many manufacturers offer both high- and low-pressure mixing models, so you may not be restricted in choice by the vendor. For some applications, one design may have an advantage over another (see Section 4.1.2), but this tends to be the exception rather than the rule. See also the above discussion of Sections 4.1.1 and 4.1.2, as well as Table 4.1.

4.2.3 Who Will Fix It?

Service can be a major factor in the decision of which system to purchase. Whether service is done by in-house staff or under contract, sooner or later the vendor's service personnel will be needed. What is the expected response time? What is the experience of other users in your geographic area? What is the quality of telephone support that you can get? Can you get on-line support? Is there a service school to train your laboratory staff in the fundamentals of servicing a specific system? The answers to these questions can have important economic and downtime consequences. It is our belief that the quality of local service is one of the more important factors in a purchase decision.

4.2.4 Special Applications

There are special applications that can drive the choice of which instrument to buy. Short (≤ 50 mm), small diameter (≤ 2.1 mm i.d.), small particle (<5 μm) columns may push the limits of some HPLC systems, whereas others are designed specifically to work with miniaturized columns that generate very small peak volumes. Very fast gradient separations (1–2 percent B/s) will require (a) equipment that can generate the gradient accurately, (b) autosamplers that cycle quickly and can handle a large number of samples, and (c) detectors and data systems that can satisfactorily measure peak widths of <2 s. Columns for fast separations are necessarily short (e.g., ≤ 50 mm); small gradient volumes and a rapid return to initial conditions after a run require small equipment dwell volumes. There is growing interest in high-pressure applications (e.g., Waters UPLC system) with pressures of $>10,000$ psi, and <2 μm particles. Specially designed systems are required to meet these demands. At the other end of the scale, if semipreparative, or especially preparative, separations will be performed, the upper limit of flow rate of 10 mL/min for many conventional systems may prove inconvenient. Specialized preparative HPLC systems can pump hundreds of mL/min, use columns >10 cm i.d., and handle grams of sample or more per injection.

If your gradient application is in any way specialized, it would be wise to seriously consider an HPLC system designed to accommodate that need. In other words, if your application uses a conventional 150 × 4.6 mm i.d. column packed with 3 or 5 μm particles, it is likely that nearly any conventional gradient HPLC system will do the job. However, for preparative or microscale separations and other special applications, a system designed for the specific application is likely to give better performance.

4.3 MEASURING GRADIENT SYSTEM PERFORMANCE

Gradient elution requires on-line mixing of the mobile phase. This can introduce variables that may not be as easy to control as when hand-blending solvents for isocratic elution. Any variation from ideal behavior can make a gradient method hard to transfer. Method transfer can be especially difficult if you do not compensate for differences in the system dwell volume V_D (Sections 2.3.6.1 and 5.2.1). A system performance check can provide useful information on gradient linearity, accuracy, and dwell volume. The following discussion describes a system "test suite" from the laboratory of one of the authors (J.W.D.) as a periodic test of system performance. Similar tests have been recommended by others [13]. Tests of system performance are made semiannually or when deemed prudent following system maintenance. For laboratories that operate under regulatory guidelines (e.g., GLP, Good Laboratory Practice), this test may be used as part of the operational

TABLE 4.3 Typical System Performance Parameters

Parameter/test (section)	Typical specification	Typical acceptance criteria
Linearity (4.3.1.1)	±1 percent	Visual linearity 5–95 percent B
Dwell volume (4.3.1.2)	Varies	As measured
Step test (accuracy) (4.3.1.3)	±1 percent	±1 percent
GPV[a] test (4.3.1.4)	Not specified	≤2 percent plateau range[b]
Flow rate (collect 10 mL at 1 mL/min) (4.3.2.1)	±1 percent	±2 percent (±12 s on 10 mL at 1 mL/min)
Pressure bleed-down from 4000 psi (4.3.2.2)	Not specified	15 percent in 10 min
Retention reproducibility (4.3.2.3)	Not specified	±0.05 min
Area reproducibility (10 μL injection) (4.3.2.4)	±0.3 percent	±1 percent

[a]Gradient-proportioning-valve.

[b]Difference between highest plateau and lowest plateau when C or D solvent is blended with A or B solvent, as in Figure 4.6 or Figure 5.35; see Section 4.3.1.4.

qualification (OQ) test. A summary of typical system specifications and corresponding test acceptance criteria are listed in Table 4.3. Acceptance criteria are often more lenient than the system specifications, because measurement techniques and conditions typically are less controlled in the laboratory than at the manufacturer's test facility.

4.3.1 Gradient Performance Test

The most important part of the test suite for evaluating system performance is a pair of experiments to determine the linearity and accuracy of gradient formation, as well as measure the system dwell volume. The column is removed, and replaced with a piece of narrow-bore connecting tubing. For example, ~1 m of 0.005 in. (~0.13 mm) i.d. tubing can be used to connect the injector and detector. This provides sufficient back pressure to enable reliable operation of the pump check-valves and results in insignificant dead volume (~12 µL) or dispersion of the gradient. Next, water is placed in the A reservoir and water containing 0.1 percent acetone is placed in the B reservoir. (An alternative is to use methanol in A and 0.1 percent acetone in methanol in B, but one should be sure to use the same base solvent in both reservoirs.) The detector wavelength is set to 265 nm.

4.3.1.1 Gradient Linearity For the first test, the system is programmed to run a full-range gradient (0–100 percent B). A 20 min gradient time is recommended. The flow rate should be set such that the system generates sufficient pressure for reliable check-valve operation; generally 1–3 mL/min will be satisfactory. The autosampler should be in the inject mode, so that mobile phase is pumped through the loop. Because the injector loop is normally in the flow stream during a run, the loop volume contributes to the dwell volume (determined in Section 4.3.1.2). (If the system is usually run with the loop out of the flow stream, this test is run in the fill mode rather than the inject mode.) The test gradient is run and the data collected, which should appear as an S-shaped curve, as illustrated by the solid curve of

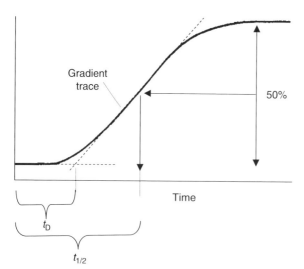

Figure 4.4 Dwell volume measurement from a blank linear gradient. See text for details.

Figure 4.4. This blank linear gradient can be used as a rough check of gradient linearity, and to measure system dwell volume as described below.

Gross deviations from gradient linearity can be checked by drawing a straight line that fits the middle of the gradient profile (dashed line in Fig. 4.4). The actual gradient (solid curve) should be smooth and not deviate from the line except for the slight "gradient rounding" at the beginning and end of the gradient. If deviations are seen, one should pay extra attention to the gradient step-test (Section 4.3.1.3). For low-pressure mixing systems, the gradient-proportioning-valve test (Section 4.3.1.4) will also help to isolate gradient linearity problems. If visible deviations from linearity are observed, these will often appear as a distinct shift or angle in the gradient plot (see Section 5.5.3.3, Fig. 5.33). It is wise to make additional step-test measurements in the region deviations were observed; e.g., 10, 25 and 30 percent vs the usual 20 and 30 %B steps (Section 4.3.1.3).

Most HPLC systems can be programmed to generate gradient rates of 10 percent/min or higher. If the system is to be used with steep gradients, it is a good idea to check gradient linearity by running the blank gradient test under the desired conditions. The 20 min gradient described earlier generates a 5 percent/min gradient; just adjust the gradient time to the desired steepness and rerun the test. Gradient rounding, which is usually not a significant problem for typical gradient conditions, tends to be more serious for very steep gradients (or smaller gradient volumes $t_G F$, where t_G is the gradient time and F is the flow rate; also see Sections 8.1.6.2 and 9.2.2). Usually gradient rounding can be reduced by reducing the system dwell volume (Section 4.4).

4.3.1.2 Dwell Volume Determination The gradient profile (as in Fig. 4.4) can be used to determine the system dwell volume. Dwell volume can be measured by means of one of two techniques. The first method is to extend the linearity test line (dashed line in Fig. 4.4) until it intersects the extended baseline. The time between this intersection and the start of the gradient is the dwell time, as shown in Figure 4.4. Dwell time t_D can be converted to dwell volume V_D by multiplying by the flow rate F: $V_D = t_D F$. This method to determine dwell volume is simple, but it is subject to any errors that result from inaccuracy in drawing the linearity test line through the gradient. It also may be inconvenient to make this measurement directly on a computer monitor from a data system output.

A second method for measuring the dwell volume is less error-prone and more convenient to perform on the computer monitor. This is shown graphically in Figure 4.4. Determine the detector response at the baseline (0 percent B) and at the top of the gradient (100 percent B). From these two values, locate the point on the plot at which the response has reached 50 percent B and note the time $t_{1/2}$ it took to reach this point. The dwell time t_D equals $t_{1/2}$ minus half the gradient time (e.g., 10 min for a 20 min gradient). The dwell volume equals $t_D F$.

4.3.1.3 Gradient Step-Test The second test, using the same system setup and A and B solvents, is a gradient step-test. This test determines the accuracy of solvent proportioning for selected mixtures. The system controller is set to deliver a series of solvent mixtures in a stair-step design. A good choice for this test is to use a

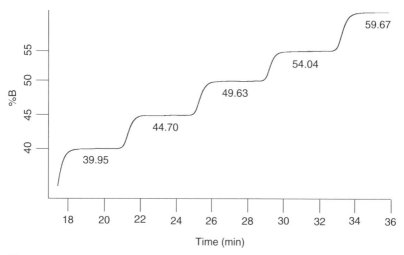

Figure 4.5 Gradient step-test showing steps for 40, 45, 50, 55, and 60 percent B. Actual composition is shown next to each step. See text for details.

10 percent step size so that mixtures of 0, 10, 20, ..., 80, 90, 100 percent B are formed for 3 min each. Problems are most commonly found near 50 percent B, so an additional step at 45 and 55 percent B should be added for a total of 13 steps. The remaining conditions are the same as those described for the gradient linearity test (Section 4.3.1.1). The results for the 40–60 percent B portion of this test for a well-behaved system are shown in Figure 4.5. For example, the 40 percent B setting actually delivered 39.95 percent B.

The %B for each step is calculated by measuring its height from the baseline (0 percent) and dividing by the difference between the 100 percent step and the 0 percent step. The %B for each step should compare favorably with the programmed value for the step. Typically, manufacturers specify accuracy of ± 1 percent B throughout the gradient. For example, in Figure 4.5, the 55 percent B step actually delivered 54.04 percent B – (barely) within the ± 1 percent criterion. For applications with gradient rates of ≥ 1 percent/min, accuracy of ± 1 percent is usually sufficient, and this is the acceptance criterion for the step test in the laboratory of one the authors (J.W.D.). When shallower gradients are used, smaller deviations may be required. It may be possible to improve proportioning accuracy by premixing solvents. For example, proportioning accuracy can improved 10-fold from e.g., ± 1 percent to ± 0.1 percent by replacing 100 percent aqueous solvent in reservoir A with hand-mixed 15:85 aqueous–organic and 100 percent organic in B with 25:75 aqueous–organic, and programming a 0–100 percent B gradient instead of 15–25 percent B. Examples of this technique are discussed in Sections 5.5.2.3 and 5.5.4.2.

4.3.1.4 Gradient-Proportioning-Valve Test A third test with a similar system setup as above is useful for low-pressure mixing systems, but does not apply to high-pressure mixing. This test checks the accuracy of the proportioning valve system and its associated control software. As an example, consider a four-solvent system (A, B, C, and D). The A and B inlet lines are placed in the water reservoir and

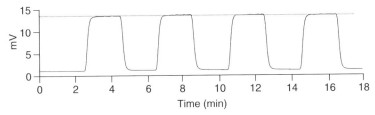

Figure 4.6 Gradient-proportioning-valve test. Valve sequence is 50 percent A + 50 percent B (baseline), 90 percent A + 10 percent C, 50 percent A + 50 percent B, 90 percent A + 10 percent D, 50 percent A + 50 percent B, 90 percent B + 10 percent C, 50 percent A + 50 percent B, 90 percent B + 10 percent D, and 50 percent A + 50 percent B. See text for details.

the C and D lines are placed in the water–acetone reservoir. The baseline is generated by pumping a 50 : 50 mixture of A and B. The various combinations of solvents are checked by blending 90 percent A or B with 10 percent of C or D. For example, the test results shown in Figure 4.6 (for an acceptable test result) are for the sequence shown in the caption. The height above baseline (50 : 50 A : B) of each 90 : 10 plateau is measured. The difference between the heights of the highest and lowest plateaus is divided by the average plateau height to determine the percentage range for the various proportioning valve combinations. In the laboratory of one of the authors (J.W.D.), a plateau range of ≤ 2 percent is considered acceptable, although ranges of ≤ 1 percent are common for well-behaved systems. (See Section 5.5.3.3 and Figure 5.35 for an example of a failed proportioning valve test.)

4.3.2 Additional System Checks

In addition to the accuracy and linearity of gradient formation, other factors affect the reliable operation of a gradient HPLC system. One of the authors (J.W.D.) uses the tests listed below on a semiannual basis to help ensure that the system is operating properly.

4.3.2.1 Flow Rate Check While a change in the flow rate F usually results in only minor changes in separation, large errors in F can be more serious. Consequently, a check of flow rate accuracy on a periodic basis is recommended. The second-by-second flow rate accuracy during a gradient is difficult to measure without specialized equipment (and is not very important), but a longer term volumetric check of flow rate can be made easily, by carrying out a timed collection of mobile phase into a 10 mL volumetric flask at a flow rate of 1 mL/min under isocratic conditions. For high-pressure mixing systems, flow rate can depend on solvent compressibility, so it is best to check the flow for representative solvents; for example, one should check the flow of 100 percent A with water and 100 percent B with acetonitrile or methanol. Typical manufacturer's system specifications are ± 1 percent for flow rate accuracy. A measured flow rate accuracy should fall within this range for routine operation. In Table 4.3, the acceptance criterion is set to ± 2 percent, because the combination of measurement errors (volumetric glassware) and the timing start/stop errors will add somewhat to the overall measured error. Besides, as illustrated in Figure 2.17, small changes in flow rate have little effect on retention or resolution.

4.3.2.2 Pressure Bleed-Down

Malfunctioning check valves and worn pump seals often show up as deviations in expected values of gradient linearity, accuracy, and flow rate. An additional test of the outlet check valves can be made with a pressure bleed-down test. For this test, the outlet tubing from the pump is blocked. The high-pressure shutoff limit for the pump is set near its maximum value, for example 5000 psi (350 bar) for a system capable of 6000 psi (400 bar). The pump is then turned on and allowed to shut off at the shutoff limit. The maximum pressure is recorded, and 10 min later the pressure is recorded again. A pressure drop of ≤ 15 percent indicates that the check valves are working properly. A larger drop suggests that the outlet check valve(s) should be cleaned or replaced or the pump seal(s) should be replaced. A pressure drop to atmospheric pressure over the 10 min test is more indicative of a leaky fitting.

4.3.2.3 Retention Reproducibility

A check of retention reproducibility is an overall check of gradient generation and pump performance. Although this check can be done with any sample, it is wise to use a sample that can be formulated easily under conditions that can be reproduced at any time. This allows one to check system performance independently of a specific method – a good tool for troubleshooting. Example test conditions are listed in Table 4.4. Be sure to use a sample concentration such that the peak is within the detector's linear range and is sufficiently large that baseline noise does not affect the precision of the measurements. For example, a sample that generates a peak height of 0.1–0.8 AU would be a good choice for UV. Retention time variation of no more than $\sim$0.05 min (1 SD) is

TABLE 4.4 Retention Time and Peak Area Test Conditions

Parameter	Value
Flow rate	1.5 mL/min
Column	C_{18}, 150 × 4.6 mm, 5 μm
Temperature	35°C
Detection	280 nm UV
Mobile phase A	Water
Mobile phase B	Methanol
Gradient/isocratic	5–95 percent B in 20 min; 80 percent B[a]
Equilibration time	$\geq$10 min
Sample	5 μg/mL 1-chloro-4-nitrobenzene[b]
Injection volume	10 μL
Retention (typical) gradient/isocratic	14 min/3 min

[a]Isocratic; adjust as necessary for $2 < k < 10$.
[b]Other nonpolar aromatic compounds (e.g., toluene, methyl benzoate) can be used.

acceptable for six replicate injections. Larger variations are an indication of problems. If the gradient performance tests are OK, but retention reproducibility is poor, one should look for leaks or air bubbles as the most likely problem sources.

With the system configured for gradient retention checks, it is easy to repeat the retention test using the isocratic conditions listed in Table 4.4. This will gather additional data on system performance. If the isocratic retention times are reproducible, but gradient retention is not, problems in gradient formation are the likely cause.

4.3.2.4 Peak Area Reproducibility The same chromatograms run for retention reproducibility can be used to determine peak area reproducibility, which is primarily a measure of injector (autosampler) performance. Variation in peak areas should be <1 percent RSD based on six injections. Most modern autosamplers will generate values of ≤0.5 percent RSD for injection volumes of ≥5 μL when injecting under standardized conditions. Poor area reproducibility usually can be traced to an autosampler problem. If the isocratic retention reproducibility test was run, check the area reproducibility for those data, as well. Similar peak area reproducibility should be obtained with both isocratic and gradient runs.

4.4 DWELL VOLUME CONSIDERATIONS

The dwell volume of the gradient HPLC system (Sections 2.3.6.1 and 4.3.1.2) can have a profound impact on chromatographic results. Retention, selectivity, run time, and the required time for between-run equilibration can be affected by dwell volume. Differences in dwell volume between HPLC systems can be a primary factor in problems associated with method transfer (Section 5.2 and [14]). A small dwell volume is preferred, because of shorter run times. Low-dwell-volume systems tend to be more flexible because additional dwell volume can be added, either physically or virtually (with an isocratic hold), so the same gradient can be generated (yielding an equivalent separation) as for a system with a larger dwell volume. In contrast, it may not be possible to reduce the dwell volume of a large dwell system. Many of the newer gradient HPLC systems allow for starting the gradient prior to sample injection, effectively reducing or eliminating the dwell volume (Section 5.2.1.1); however, this feature is not available on all instruments.

A practical rule of thumb is that the system dwell volume should be no more than about 10 percent of the gradient volume ($V_G = t_G F$), and smaller dwell volumes generally are desired. The gradient delay will then comprise no more than ~10 percent of the run time. Thus, a 20 min gradient run on a conventional 150 × 4.6 mm i.d. column at a flow rate of 1.5 mL/min would have a gradient volume of ~30 mL, so $V_D \leq 3$ mL is desired. For an LC-MS application, a 4 min gradient on a 50 × 2.1 mm i.d. column might be run at 0.5 mL/min, generating a 2 mL gradient, so ≤200 μL of dwell volume is then needed.

There is little opportunity to significantly reduce the dwell volume of a low-pressure mixing system, because all the components up to and including the pump, which comprise a majority of the dwell volume, are seldom user-replaceable. Of the components downstream from the pump, only the injector loop and (high-pressure) mixer volume are of any consequence in terms of dwell volume. A change in the loop volume will have a practical reduction in V_D/V_G only for small volume applications. With high-pressure mixing systems, dwell volume can be reduced by replacing the mixer. For example in the laboratory of one of the authors (J.W.D.), a stock mixer is replaced with a micromixer, reducing the dwell volume from 2.3 mL to 300 μL, including a 100 μL injector loop. This makes the system compatible with 50 × 2.1 mm i.d. columns used for LC-MS applications. Whenever system components (other than the injector loop and excess tubing) are replaced so as to reduce dwell volume, the user should be aware that detector noise may increase – manufacturers carefully design mixer characteristics to minimize detector noise due to incomplete mobile phase mixing.

For additional information on dwell volume see Sections 2.3.6.1 (basics), 4.3.1.2 (measurement), 5.2 (method transfer), and 5.2.1 (compensating for differences in V_D).

REFERENCES

1. J. N. Brown, M. Hewins, J. H. M. van der Linden, and R. J. Lynch, *J. Chromatogr.* 204 (1981) 115.
2. M. A. Quarry, R. L. Grob, and L. R. Snyder, *J. Chromatogr.* 285 (1984) 1.
3. H. A. H. Billiet, P. D. M. Keehnen, and L. de Galan, *J. Chromatogr.* 185 (1979) 515.
4. D. L. Saunders, *J. Chromatogr. Sci.* 15 (1977) 129.
5. J. P. Foley, J. A. Crow, B. A. Thomas, and M. Zamora, *J. Chromatogr.* 478 (1989) 287.
6. P. Jandera, J. Churáček, and L. Svoboda, *J. Chromatogr.* 192 (1980) 37.
7. L. R. Snyder, J. J. Kirkland, and J. L. Glajch, *Practical HPLC Method Development*, 2nd edn, Wiley-Interscience, New York, 1997.
8. R. P. W. Scott and P. Kucera, *J. Chromatogr.* 119 (1976) 467.
9. K. Slais, D. Kourilova, and M. Krejci, *J. Chromatogr.* 282 (1983) 363.
10. H. A. Claessens and M. A. J. Kuyken, *Chromatographia* 23 (1987) 331.
11. M. J. Mills, J. Maltas, and W. J. Lough, *J. Chromatogr. A.* 759 (1997) 1.
12. J. W. Dolan, *LCGC* 23 (2005) 130.
13. G. Hall and J. W. Dolan, *LCGC* 20 (2002) 842.
14. S. K. MacLeod, *J. Chromatogr.* 540 (1991) 373.

CHAPTER 5

SEPARATION ARTIFACTS AND TROUBLESHOOTING

> ... the worst thing you can do to an important problem is discuss it ... one of the more poignant fallacies of our zestfully overexplanatory age.
>
> —Simon Gray, *Otherwise Engaged*

Prior to 1985, gradient elution had a reputation for being unreliable and problematic. The equipment was still undergoing improvement, the technique was more complicated than isocratic elution, and few HPLC users were experienced in its use. Furthermore, gradient elution is subject to problems that either do not occur or are unimportant in isocratic elution. Finally, workers familiar with isocratic separation were often surprised at the results of different changes in conditions. When changing flow rate or column length in gradient elution, resulting changes in retention time and resolution were usually far smaller than in isocratic elution (Sections 2.3.2 and 2.3.3). This sometimes led to the conclusion that something was "just not right," when in fact normal gradient elution behavior was being observed.

The performance of today's gradient elution equipment (Chapter 4) is much improved, and Chapters 1–3 provide the foundation for a better understanding and use of gradient elution techniques. In this chapter, we will summarize our recommendations for avoiding potential problems in the use of gradient elution (Section 5.1); we will discuss a number of problems that are either unique to or more important for gradient elution (Sections 5.2–5.4); and we will present troubleshooting guidelines for solving gradient elution problems as they arise (Section 5.5).

Many gradient problems, symptoms, and solutions are scattered throughout this chapter as different topics are covered. These are summarized at the end of the chapter in Table 5.5, organized by the symptom with its associated cause and a cross-reference to where the problem is discussed.

High-Performance Gradient Elution. By Lloyd R. Snyder and John W. Dolan
Copyright © 2007 John Wiley & Sons, Inc.

CHAPTER 5 SEPARATION ARTIFACTS AND TROUBLESHOOTING

5.1 AVOIDING PROBLEMS

By adhering to certain practices, a number of potential problems in gradient elution can be avoided. A series of recommendations are outlined in the flow chart of Figure 5.1 for use in both routine analysis and method development. Before injecting samples, it is important to verify that the equipment is working

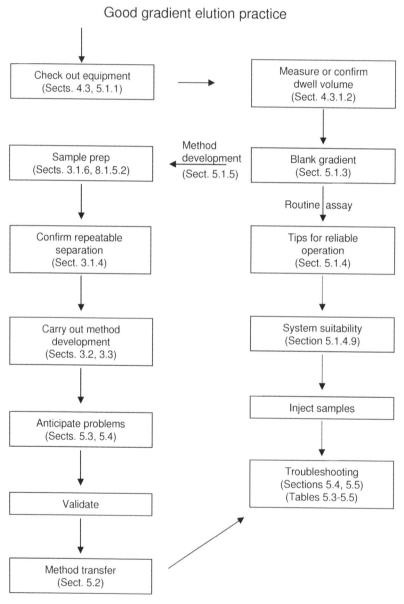

Figure 5.1 Recommendations for avoiding problems in gradient elution.

properly (Section 5.1.1). Because the dwell volume V_D of the system plays a major role in its performance, the operator should know the value of V_D for his or her equipment before proceeding further (Section 4.3.1.2). Before running actual samples, a blank gradient should be run in order to verify that the mobile phase is pure and the system is clean (Section 5.1.3).

At this point, further action depends on whether gradient separation involves routine sample analysis (Section 5.1.4) or method development (Section 5.1.5). For routine sample analysis, the operator should be following a detailed method procedure. In order to ensure that the gradient system is operating properly and that the method is being followed correctly, a system suitability test often is carried out prior to running samples (Section 5.1.4.8). If the results of this test are acceptable, the operator can begin injecting samples for analysis. Additional tips for routine analysis can be found in Section 5.1.4.

For method development, it is important to know that the samples that will be run are ready for injection. That is, no further pretreatment of the sample ("sample prep") is required (Sections 3.1.6 and 8.1.5.2). At this point, method development (Section 3.3) can begin. Some guidelines for reliable method development can be found in Section 5.1.5. It is important initially to confirm that each separation can be replicated; retention times should be reproducible within narrow limits (e.g., $\pm 0.01 - 0.02$ min or $\pm 0.1 - 0.2$ percent). Carrying out runs in duplicate should be the rule, until the operator is convinced that no problems with retention variability are likely. Repeatable gradient runs require adequate equilibration of the column with the starting mobile phase between each run. A 10-column-volume equilibration with the initial mobile phase usually is adequate, but shorter equilibration times often are acceptable (Section 5.3). Following these initial experiments, the recommendations of Chapter 3 can form the basis of further method development.

When conditions for a final method have been established, it is important to anticipate possible problems that might arise during the routine use of the method, and to carry out experiments that can alleviate those problems. Problems of this kind include the following:

- differences in equipment dwell volume;
- inadequate temperature control;
- equipment bias or incorrect separation conditions;
- batch-to-batch variations in the column.

Differences in the dwell volume for different gradient systems can result in unacceptable changes in separation. When it is anticipated that the routine use of a method will involve different equipment, adjustments may be required to correct for this (Section 5.2.1). The contraction of mobile phase upon mixing organic and water can result in a slight drop in flow (2–4 percent) in the middle of the gradient with high-pressure mixing systems when compared with low-pressure mixing (Section 4.1.2.4 and Fig. 4.3). It is unlikely that most users will notice this effect, but it may explain small differences in retention observed between low- and high-pressure-mixing systems. *Inadequate temperature control* can take either of two forms (Section 5.2.2.5), especially when a method is to be run at temperatures above ambient: (a) incomplete thermal equilibration of the solvent entering the column resulting in distorted peaks and poor resolution; and/or (b) inaccurate temperature calibration (bias) of the gradient system. *Equipment bias or errors* might involve errors in flow rate, temperature setting, mobile phase pH, and so on. These errors can be diagnosed more easily if illustrative chromatograms are supplied as part of the method procedure. For example, mobile pH may need to be controlled within ±0.05 units or less, which can be difficult to achieve routinely. By providing a diagnostic chromatogram where the pH is deliberately varied by 0.1–0.2 units, the operator can easily confirm an error in mobile phase pH. This general approach is illustrated in Figure 5.2 for a separation as "originally" developed (*a*), the same separation for an increase in dwell volume V_D from 1.0 to 5.0 mL (*b*), and the same sep-

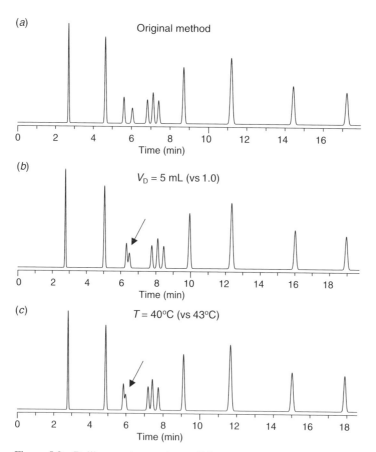

Figure 5.2 Deliberate changes in conditions as a means of diagnosing problems encountered for a routine gradient elution procedure. (*a*) Original separation described in the method procedure for the irregular sample of Table 1.3; conditions: 250 × 4.6 mm column, 25–50 percent B in 17 min, 2.0 mL/min, $V_D = 1.0$ mL; (*b*) same separation as in (*a*) for a dwell volume $V_D = 5.0$ mL [simulated by an isocratic hold of $(5 - 1)/2 = 2$ min]; (*c*) same separation as in (*a*) for a temperature of 40°C.

aration as in (a) except for a change in temperature from 43 to 40°C (c). In Figure 5.2(b) and (c), the characteristic changes in separation that are shown could be compared with an actual chromatogram which does not match the original separation [as shown in Fig. 5.2(a) and documented in the method procedure]. Note in Figure 5.2(b, c) that these two changes in conditions happen to result in superficially similar chromatograms for this particular separation, so that an incorrect diagnosis of the problem could easily result. However, a closer attention to changes in retention time for different peaks should avoid any misinterpretation. For critical peaks, the final method procedure should list the percentage change in retention per unit change in various conditions as a means of simplifying the identification of a changed condition for example, +1.2 percent change in t_R per 1°C increase in temperature for peak "X," +0.5 percent change for peak "Y," and −0.2 percent change for peak "Z." A comparison of observed changes in retention for a given method with corresponding documented changes as a result of change in temperature, pH, dwell volume, and so on, can usually identify the changed condition, as well as determine how much change in that condition has occurred. In addition to the convenience of this approach to method adjustment, inclusion of such data in the method documentation can justify method adjustment of regulated methods (e.g., operation under Good Laboratory Practice, GLP) without re-validation.

Finally (*batch-to-batch variations in the column*), the specific brand and model of column used in method development may not be reproducible from batch-to-batch. In method development, this possibility is usually investigated by carrying out the method on two or more different batches of the selected column. However, a method intended to be run over several years in laboratories around the world may eventually encounter a column batch which is no longer acceptable. One way of anticipating this possible problem is to identify one or more "equivalent" columns (e.g., different brand and/or model) which can be used interchangeably with the original column brand and model, that is, gives an equivalent separation for the same separation conditions. If at a later time the original column proves to be unusable because of changes in its selectivity (the most likely problem), one of the equivalent columns can be used instead. A general procedure for selecting columns of equivalent selectivity now exists [1], as verified in a collaborative study by several different laboratories [2]. See Appendix III for details.

5.1.1 Equipment Checkout

We cannot overemphasize the importance of ensuring that the gradient HPLC system operates in a reliable and reproducible manner. If the equipment does not operate properly, it will be impossible to obtain satisfactory results. For this reason, it is important to ensure that the system works well before running samples.

5.1.1.1 Installation Qualification, Operational Qualification, and Performance Qualification One way to demonstrate that a new HPLC system is functioning properly is to follow the practice of the pharmaceutical industry and perform installation qualification (IQ), operational qualification (OQ), and performance qualification (PQ) tests prior to releasing the system for routine work. The IQ test ensures that the instrument is installed according to the manufacturer's procedures. IQ is often done by the vendor if installation is included with the purchase of the system. The documentation accompanying the system will outline the IQ test. OQ demonstrates that the instrument meets the manufacturer's specifications, or some subset of them. The OQ test results may also be included in the documentation. Alternatively, the performance tests outlined in Section 4.3 can be followed and the results compared with the manufacturer's specifications (generally found in the back of one of the operator's manuals). The PQ test generally is user-designed and may range from extensive testing, such as the performance tests of Section 5.3, to simply running a standard curve plus a few mock samples, in order to show that the expected results can be obtained for a gradient method. Once these three tests have been performed, the system should be ready for routine use.

5.1.2 Dwell Volume

Previous sections have discussed the origin of system dwell volume (Section 4.1.1), its measurement (Section 4.3.1.2), and the effects of changes in dwell volume on separation (Section 2.3.6). As noted in the example of Figure 5.2(*b*), a change in dwell volume can lead to unacceptable changes in separation. Problems in transferring a gradient method can often be traced to differences in dwell volume for different gradient systems. For this reason, when beginning a series of gradient runs for either routine analysis or method development, it is important either to measure the value of V_D for the equipment to be used, or to verify that the equipment has not changed since a value of V_D was last measured. Once V_D has been determined, this value should be recorded in the notebook that tracks equipment maintenance.

A gradient method, once developed, may be carried out on different equipment having different dwell volumes. There are several approaches that can be used to deal with dwell-volume variability, as discussed in Section 5.2.1.

5.1.3 Blank Gradient

Gradient baselines can exhibit drift due to differences in absorbance of the A and B solvents (Section 5.4.1), and excessive noise if the solvents are not mixed properly (Section 5.4.2). In addition, gradients are susceptible to background or "ghost" peaks that arise from impure reagents or other sources (Section 5.4.3). To identify potential problems and avoid surprises, it is wise to run a blank gradient (normal gradient, but with no injection) prior to starting method development or routine sample analysis. This may not be required on a daily basis for a routine method, but the blank run is inexpensive insurance to help protect against loss of valuable sample data, or a need to repeat method development experiments. Consult Section 5.4.3 for details on running blank gradient tests.

5.1.4 Suggestions for Routine Applications

To obtain high quality data from routine gradient elution assays, the HPLC system must perform in a reliable and reproducible manner. This section lists some additional tips and techniques that will help improve the likelihood of high quality results.

5.1.4.1 Reagent Quality Gradient elution tends to concentrate nonpolar impurities in the A and B solvents at the head of the column, followed by their release as the gradient progresses. These impurities can show up as peaks in both blank and sample runs (Section 5.4.3). For this reason it is essential to use HPLC-grade reagents for gradient work. Lower quality reagents may be suitable for isocratic applications, but even the most minor impurities can cause problems with gradient elution. Aqueous reagents and buffers should be discarded frequently (e.g., weekly) to avoid contamination by microbial growth. Water impurities can be especially problematic (Section 5.4.3.1).

5.1.4.2 System Cleanliness Just as reagent quality is important in order to minimize interfering peaks, a clean instrument will also help avoid unwanted peaks. The system should be thoroughly flushed with strong solvent, such as acetonitrile, at the end of the day, or prior to shutting off the system. A system should not be shut off that contains buffers or salts – it should either be flushed with non-buffered mobile phase or the flow rate reduced to 0.1 mL/min. This will help to avoid the formation of buffer crystals in the equipment. Highly aqueous reagents (e.g., <10 percent organic) tend to support microbial growth over time, so it is a good idea to replace the water or buffer every week or so and to use clean reservoirs (Section 5.4.3.1). Spills, leaks, and other potential sources of contamination should be cleaned up. One should be sure to wash or replace solvent reservoirs on a regular basis (e.g., weekly).

5.1.4.3 Degassing Although some gradient systems will operate without degassing the mobile phase, every system will operate more reliably with degassed solvents (Section 4.1.2.2). Trapped air bubbles and solvent outgassing are quite common problems in gradient elution that can be largely avoided by solvent degassing. It is a good idea to purge the pump(s) and solvent inlet lines daily by opening the purge valve(s) and operating at an elevated flow rate (e.g., 5 mL/min) for a few minutes to remove any air bubbles.

5.1.4.4 Dedicated Columns Each analytical method should have a column dedicated to that method. We strongly advise against sharing columns between methods, because peaks that are not of concern in one method may cause interferences in a second method. Dedicated columns last longer, so fewer columns will need to be purchased over time if each method has its own column.

5.1.4.5 Equilibration Prior to each run, the column should be equilibrated to the same extent as the other runs in the run sequence. Complete equilibration may or may not be necessary (Section 5.3).

5.1.4.6 Priming Injections Some methods will give better results if several "priming" injections are made before the first sample is injected. These injections of standards or mock samples may help to load slowly equilibrating active sites on the column, so that more reproducible separations can be obtained. Sometimes the system suitability injections serve as priming injections.

5.1.4.7 Ignore the First Injection Because some methods require the priming process (Section 5.1.4.6) and the column is likely to be equilibrated differently for the first injection (relative to subsequent injections; Section 5.3.3), we advise setting up a routine method so that the first injection is not used for quantitative purposes. The second and subsequent runs will be more reliable than the first injection.

5.1.4.8 System Suitability Many methods that run under the oversight of regulatory agencies (FDA, EPA, OECD, USP, etc.) will require a system suitability test prior to sample analysis. System suitability serves as a confirmation that the equipment and analytical method are operating in a fashion that will produce reliable results. Requirements for system suitability tests vary, so the regulatory guidelines should be consulted to help select appropriate tests. Many workers use retention time and area reproducibility, peak response (detection sensitivity), peak width, peak tailing, resolution, and column back-pressure, either alone or in combination, as part of the system suitability test. The system suitability sample may be a diluted pure standard, a mock sample in extracted matrix, or some other sample selected to demonstrate system performance. The important concept is to select system suitability samples that test the ability of the method to perform its desired function. Whether or not a system suitability test is required, we strongly suggest running such a test prior to routine analysis, even if it is just an injection of a standard to see if the retention and peak size are as expected.

5.1.4.9 Standards and Calibrators For quantitative analysis, the response of unknown samples is compared with the response for standards of known concentration. The range of standard concentrations, number of replicates, and sequence of injection may depend on the specific application. Either external or internal standardization can be used. In any event, running at least one standard prior to running unknown samples will provide assurance (system suitability) that the analytical method is working properly before potentially valuable samples are injected.

5.1.5 Method Development

Method development in gradient elution often is cast in an aura of mystery. Most of the mystery can be eliminated by taking a systematic approach to method development (Chapter 3) and using many of the same precautions during method development that are used for routine analysis (Section 5.1.4). The key to minimizing problems with method development is to ensure that the process is in control at all times. Poorly controlled method development conditions often result in an unreliable method. Just like the computer adage "garbage-in, garbage-out," the quality of method development results will reflect the care that is taken during the process. We have listed below some key elements that need to be part of any gradient method development process.

5.1.5.1 Use a Clean and Stable Column Just as it is important to dedicate columns for routine analysis (Section 5.1.4.4), a new project should always be started with a new column so that contaminants remaining on the column from previous uses do not confuse the method development process. Preferably a type B, high-purity-silica column should be used; the less reproducible type A silica columns used in the past should be avoided. The manufacturer's column-care-and-use instructions should be read in order to define column usage (e.g., temperature, allowed mobile-phase pH range), so that conditions are selected for reliable operation. For some combinations of a new column, sample, and other conditions, retention may be found to change significantly during the first few injections (Section 5.1.4.6). A problem of this kind can be confirmed by carrying out the initial separation in duplicate. Even small changes in retention for initial runs should be regarded with suspicion; further replicate runs can be used to determine whether retention drift is likely to continue for several runs before leveling off.

5.1.5.2 Use Reasonable Mobile Phase Conditions
When selecting the mobile phase, it must be compatible with the column (and other system components). Particular attention needs to be paid to pH – most reversed-phase columns are stable in the pH range of 2.5–7.5, but specialty columns are available for use outside this range. Choose solvents which are stable and will not degrade the sample.

Close attention needs to be paid to the solubility of the buffer in the organic solvent. The solubility of a buffer or salt is a function not only of the bulk properties of the reagents, but also of the microenvironment in which the buffer and organic are mixed. This means that some buffers that appear to be soluble in bulk solutions of organic–water may precipitate when they are blended inside the HPLC system. System design also can play a role, so some brands and models of HPLC systems will do a better job of blending buffers and organic solvents. It should also be noted that buffer precipitation is often time-dependent, suggesting that fast gradients ($t_G < 5$ min) may be able to tolerate buffer concentrations that would cause problems with slower gradients.

Of particular concern is the solubility of the most popular buffer, phosphate, when used with acetonitrile. Buffers are much less soluble in ACN than in MeOH, which is less subject to precipitation problems. For example, one study [3] reported the solubility of potassium phosphate in 80 percent ACN as only 5 mM vs 15 mM in MeOH. A simple test for buffer solubility is to add the A solvent drop-wise to a test tube of B solvent and vice versa. If any cloudiness or precipitation occurs, it is likely that there will be problems with on-line mixing. It is common practice to improve mixing and reduce solubility problems by premixing a little A solvent into the B solvent, and B into the A solvent. For example, fewer problems with solubility are observed when 5 percent ACN is added to the buffer (A solvent) and 5 percent buffer to the ACN (B solvent), but check whether precipitation occurs for the latter mixture! The gradient program then is adjusted to deliver the desired absolute %B during the gradient. (In this case, gradients originally run from 5 to 95 percent B would now be programmed to run from 0 to 100 percent B to cover the same absolute 5 to 95 percent ACN range.) If the primary function of the phosphate is to achieve a low mobile-phase pH, phosphoric acid alone will provide sufficient buffering for most applications [4], while eliminating any solubility problems. It should be apparent that the common practice of using buffer as the A solvent and buffer-free organic as the B-solvent will allow a higher buffer concentration in the A solvent. Note, however, (a) a *reverse* gradient in buffer concentration will be run (may or may not matter), and (b) care should be taken to avoid buffer precipitation at high concentrations of the B-solvent.

The gradient must be allowed to reach a sufficiently strong B solvent composition (with added isocratic hold, if necessary, as in Fig. 3.8*b*) to completely elute all sample components from the column. Without sufficient column washing, analytes that are not eluted in one run may come out in a subsequent run and confuse interpretation of the chromatograms (Section 5.4.4.3), although this is much less of a problem with gradient elution than for isocratic methods. It is simple and straightforward to trim the gradient to eliminate wasted time at the beginning (Section 2.3.4) and end (Section 2.3.5) of the run while maintaining constant k^* (and thus selectivity). However, it is best to delay such adjustments until initial method development experiments are complete (Section 3.2.1.1). The use of higher %B values in the gradient to clean the column between sample injections also can be carried out with step gradients as a means of shortening run time (Section 3.3.4 and Fig. 3.8*b*).

5.1.5.3 Clean Samples Cleaner samples provide longer column lifetimes. During method development, the cleanest possible samples should be sought until method conditions are determined. Extraneous sample components (proteins, excipients, etc.) can contaminate the column and change its performance characteristics, leading to nonreproducible method development runs. A balance must be made between the cost and effort of developing extensive sample cleanup procedures (Sections 3.1.6, 8.1.5.2), the lifetime of the column, and reliability of the method.

5.1.5.4 Reproducible Runs It is very important, particularly when computer simulation is used to aid method development (Section 3.4), to make sure that retention times gathered during method development are repeatable. It is recommended to carry out duplicate runs under each gradient condition tested. If the same retention times are obtained from duplicate runs, more confidence can be placed in the quality of the results.

5.1.5.5 Sufficient Equilibration As with routine methods (Section 5.1.4.5), sufficient equilibration must be allowed between runs (Sections 3.3.7 and 5.3). It is better to err on the side of too much equilibration during method development, rather than too little; during method development, allow at least 10 column-volumes of the initial mobile phase for between-run column equilibration. Once the final method conditions are determined, experiments can be made to reduce the equilibration time and thus shorten the total run time (Section 5.3.3).

5.1.5.6 Reference Conditions The development of a gradient method will usually take several days to a week or more. Over this period of time, it is possible that the column may degrade, or some other change in the system may take place, such that results obtained are not equivalent to the same system settings on an earlier day. Any drift in retention or system performance makes it difficult to compare results, and the quality of the final method may suffer. One way around this is to pick a set of conditions to use as a reference. For example, the best separation from day 1 can be chosen and repeated every day to ensure that the system is working as expected. This provides a kind of system suitability test (Section 5.1.4.8) for use during method development to help guarantee high quality results.

5.1.5.7 Additional Tests Besides the tests mentioned above, some additional checks should be made, whether method development or routine analysis is the goal. These include basic instrument performance tests (Sections 4.3 and 5.1.1), dwell volume determination (Sections 4.3.1.2 and 5.1.2), and a blank gradient (Sections 5.1.3 and 5.4.3).

5.2 METHOD TRANSFER

When a gradient method is transferred between two laboratories, the separation in the second laboratory is sometimes found to differ in important respects. Method transfer also can be a problem for isocratic separations, but usually is less so than for gradient methods. Changes in separation during method transfer can arise from differences in the way the separation is carried out in each laboratory, that is, the use of a different column, mistakes in the selection of experimental conditions, or differences in equipment. The most common reason for method transfer problems that involve gradient elution is the use of different HPLC systems. Almost all gradient equipment has a significant dwell volume V_D, and values of V_D often vary between different systems. Differences in dwell volume result in differences in sample retention times, and in some cases significant changes in resolution (Section 2.3.6).

5.2.1 Compensating for Dwell Volume Differences

When a gradient separation is initiated, the system dwell volume results in a delay of the gradient reaching the column. The appearance of the chromatogram will change when a method is moved between HPLC systems with different dwell volumes. As discussed in Section 2.3.6, if peaks are well retained, a change in dwell volume primarily affects retention time, but early eluted peaks can undergo significant changes in peak spacing as well. In any case, a change in retention with a change in HPLC systems is generally not desirable.

To avoid problems created by dwell volume differences between equipment, four different approaches can be taken:

- injection delay;
- adjustment of the initial isocratic hold;
- use of maximum-dwell-volume methods;
- adjustment of the initial %B to simulate a reduction in dwell volume.

These options are discussed below. In each case, the original system will refer to the equipment from which the method is transferred and the new system will be the equipment to which the method is transferred.

5.2.1.1 Injection Delay In the usual operation of a gradient method, the injection (arrows pointed down in Fig. 5.3) is made at the same time the gradient is

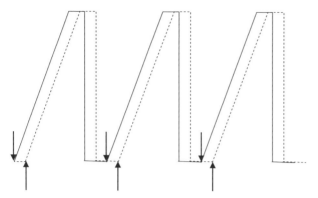

Figure 5.3 Compensation for dwell volume by injection delay; %B is plotted vs time. (——) Gradient as programmed (at mixer); (- - -) gradient at column inlet. See text for details.

started, resulting in a lag between sample injection and the time that the gradient reaches the head of the column (arrows pointed up); this lag is equivalent to the dwell time $t_D(=V_D/F)$. This is illustrated in Figure 5.3, where the programmed gradient (——) is delayed by the dwell time so that the actual gradient (- - -) reaches the head of the column t_D min later. Many of the newer HPLC systems incorporate a feature that allows one to delay the injection until after the gradient has started. The concept of a delayed gradient is not new [e.g., 5–8], but only recently has this feature become a common option for HPLC equipment. If the injection is delayed until the gradient reaches the column (arrows pointed up in Fig. 5.3), the dwell volume is effectively reduced to zero with the obvious benefit that dwell volume effects are eliminated. The "cost" of this delayed injection is equal to the initial delay only (for the first injection); there is no additional time penalty for subsequent injections (of course adequate inter-run equilibration is still required, Section 5.3). The use of a delayed injection to compensate for, or eliminate, dwell volume differences can simplify the transfer of methods between HPLC systems, *but delayed injection can be used only with systems that have this capability (and these are limited at present)*. An added bonus is that the user has more control over the mobile phase composition at the beginning of the gradient, so that the gradient can start immediately upon injection, if desired. See Section 5.3 and Figure 5.6 for a further discussion of the timing of the injection relative to arrival of the gradient at the head of the column.

5.2.1.2 Adjustment of the Initial Isocratic Hold When the dwell volume of the original system is larger than that of the new system, adjustment for dwell volume differences can be made in a straightforward manner. The new system will need to have the dwell volume increased to match the original system. Once the dwell volumes are matched, the separation should be the same on both systems. Although one could physically add dwell volume to the new system, such as by plumbing in an additional mixing chamber, this is impractical and unnecessary. Instead, the gradient program can be adjusted to include an isocratic hold that matches the required increase in dwell volume. For example, if the original system had a dwell volume of 3.5 mL and the flow rate was 2 mL/min, the gradient would reach the column $3.5/2 = 1.75$ min after the program started. If the new system had a dwell volume of 1 mL, the same gradient program and flow rate would cause the gradient to reach the head of the column $(3.5 - 1.0)/2 = 1.25$ min earlier than the original system. The addition of 1.25 min of isocratic hold at the beginning of the run would exactly compensate for this difference, and the chromatograms should be identical. It should be noted that this strategy works only for the case in which the new system has a smaller dwell volume than the original system. This often is not the case, because new methods typically are developed on the latest equipment, which often has a smaller dwell volume than equipment in the routine laboratory where the method will be transferred.

5.2.1.3 Use of Maximum-Dwell-Volume Methods Neither of the two preceding procedures may be applicable, because the "new" system (to which a method is to be transferred) will often be an older piece of equipment with a larger dwell volume and no provision for delayed injection. The maximum-dwell-volume technique can compensate for these shortcomings of the two preceding procedures (Sections 5.2.1.1 and 5.2.1.2). The concept is to develop the original method so that it can correct for the maximum dwell volume that will be encountered on any other system [9]. This technique has been used successfully in the laboratory of one of the authors (J.W.D.), so that gradient methods transferred to other laboratories exhibited no problems due to dwell volume differences.

The maximum-dwell-volume technique is simple, but takes advanced planning. First, the largest dwell volume that will be encountered in "new" systems is determined. Alternatively, the method can be specified to work for any system up to a certain dwell volume. Gradient conditions then are adjusted in the original laboratory so they correspond to this maximum dwell volume. For example, a multinational pharmaceutical company might have HPLC systems with dwell volumes as large as 4.5 mL for some older systems still in service. The development laboratory might have the latest equipment with a dwell volume of 0.5 mL. All methods would be developed such that they included an additional isocratic hold equivalent to 4.0 mL at the beginning of each gradient, so that the originating system had an *effective* dwell volume of 4.5 mL. The method document should be written to allow the end user to correct for dwell volume differences as follows. The difference in dwell volumes is calculated for the "new" vs "original" systems: δV_D mL. A gradient delay equal to $\Delta V_D/F$ min is added, where F is the flow rate in mL/min. For example, if the new system had $V_D = 3.5$ mL, $\Delta V_D = 4.5-3.5 = 1.0$ mL. For $F = 2.0$ mL/min, the gradient delay for the "new" system would be $1.0/2 = 0.5$ min. With a similar correction, all systems will be able to generate identical gradients and should give the same separation without further adjustment. This approach can largely reduce or eliminate problems encountered in transferring methods between HPLC equipment with different dwell volumes.

5.2.1.4 Adjustment of Initial Percentage B A final way of compensating for differences in equipment dwell volume is to adjust the initial %B value of the gradient. Consider first the separation of the irregular sample of Table 1.3 as shown in Figure 5.4(a). This optimized gradient (10–45 percent B in 53 min) for a gradient system with a dwell volume $V_D = 1.0$ mL results in the baseline separation of peaks 2 and 3 ($R_s = 1.9$). The transfer of the method to a second system with $V_D = 5$ mL results in a decrease in resolution for this sample (Fig. 5.4b, with $R_s = 0.8$), with an increase in run time of 4 min. In this case, it can be assumed that no provision for a larger dwell volume was considered during method development, and none of the preceding ways of compensating for differences in dwell volume were

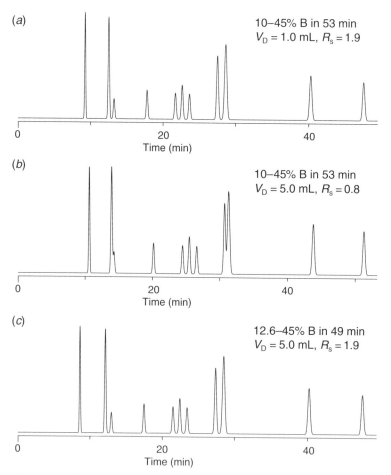

Figure 5.4 Gradient separation of the irregular sample of Table 1.3 with systems having different dwell volumes. Conditions: 15 × 4.6 mm C_{18} column, 44°C, 1.0 mL/min. (a) Optimized separation (10–45 percent B in 53 min) using first system with $V_D = 1.0$ mL; (b) separation with same conditions as in (a) using second system with $V_D = 5.0$ mL; (c) separation on second system ($V_D = 5.0$ mL) with gradient adjusted (12.6–45 percent B in 49 min) to compensate for larger dwell volume. See text for details.

applicable. However, in many cases this dwell-volume-related loss in resolution (and increase in run time) can be overcome in the following way.

First, observe the various gradients illustrated in Figure 5.5. Gradient *i* is the original gradient programmed into the system (10–70 percent B in 10 min). For a system with $V_D = 0$, this is also the gradient measured at the column inlet. Now assume that the separation is repeated on a second system, with $V_D = 3.5$ mL. Since the flow rate assumed in Figure 5.5 is 1.0 mL/min, the dwell time $t_D = 3.5$ min. The gradient measured at the column inlet is now delayed by 3.5 min, as illustrated by gradient *ii* in Figure 5.5. This gradient delay can affect the separation, as illustrated in Figure 5.4(*a* vs *b*).

Now assume that the initial %B of the gradient is increased from 10 to 31 percent B, to give the programmed gradient *iii* and the actual gradient at the column inlet *iv*. Note that at time t_D the initial %B at the column inlet for gradient *i* will be initial %B + $[(t_D/t_G)$(final %B $-$ initial %B$)] = 10 + (3.5/10)(60) = 31$ percent B. If a new gradient starts at this %B and maintains the same %B/min, gradient *iii* results. The gradient at the column inlet (for $V_D = 3.5$ mL) will then be given by gradient *iv*, which is seen to overlap the original programmed gradient *i* (except prior to 3.5 min). As a result, gradient *iv* for a system with a dwell volume $V_D = 3.5$ will more closely resemble the original gradient for a system with $V_D = 0$. This means that the separation for the second system with $V_D = 3.5$ should look more like the separation achieved with the first system ($V_D = 0$) and the original - gradient.

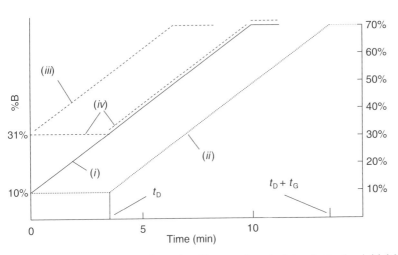

Figure 5.5 Hypothetical gradients that illustrate the principle of adjusting initial %B for the gradient in order to compensate for differences in equipment dwell volume. Curve *i* is the programmed gradient for an initial separation with zero dwell volume; curve *ii* is the resulting gradient at the column inlet for a second system with $V_D = 3.5$ mL and resulting dwell time to $t_D = 3.5$ min; curve *iii* is the proposed gradient program to compensate for the larger dwell volume of the second system; curve *iv* is the gradient at the column inlet for the second system using the gradient program *iii*. See text for details.

Now return to Figure 5.4, in order to see how we can adapt the above proposal to this particular example. In this case, the gradient is 10–45 percent B in 53 min, and $V_D = 1.0$ mL for the original system of Figure 5.4(a). Since the flow rate is 1.0 mL/min, the dwell time $t_D = 1.0$ min for the initial separation (a). For the separation on the second system (b) with $V_D = 5.0$ min and $t_D = 5.0$ min, the gradient in (b) will be offset from that in (a) by $5.0 - 1.0 = 4.0$ min. In that time, the programmed gradient will have increased from 10 percent B to 10 percent B + $(4.0/53)(45 - 10) = 12.6$ percent B. This will be the initial %B value used for the new gradient for the gradient system of Figure 5.4(b) (with $V_D = 5.0$ mL). Since the gradient steepness is to be maintained constant, the new gradient time must be reduced by the difference in t_D values, or by 4 min (i.e., to $53 - 4 = 49$ min). The proposed gradient for the second system, in order to maintain a similar separation as in (a), will therefore be 12.6–45 percent B in 49 min. The separation of this sample on the second system with the new gradient is shown in Figure 5.4(c). The original resolution ($R_s = 1.9$) for peaks 2 and 3 has been restored, and the retention times of all peaks except the first three are almost identical in the separations of (a) and (c).

To recap, the present procedure for compensating for differences in equipment dwell volume is as follows. First, calculate the difference in dwell times for the second system vs the first: $\Delta t_D = \Delta V_D / F$. Second, subtract this value (Δt_D) from the gradient time for the original separation ($t_{G,1}$), to give the gradient time $t_{G,2}$ to be used for the second system (with higher dwell volume). Note that if the second system has a *lower* dwell volume, the procedure of 5.2.1.2 (addition of an isocratic hold) should be used instead. Finally, calculate the new value of initial %B for the gradient, equal to (original value of initial %B) + $(\Delta t_D / t_{G,1})$[(final %B) − (initial %B)]$_{original}$.

The adjustment of initial %B for the gradient as above will tend to equalize the separations on two gradient systems which differ in dwell volume, including total run time. However, there may be some compression of the early part of the chromatogram after adjusting the gradient, so this approach to compensating for differences in dwell volume may not always work as expected. If the adjusted method is still unsatisfactory, further trial-and-error adjustments of initial %B and gradient time (maintaining gradient steepness constant) are recommended.

5.2.2 Other Sources of Method Transfer Problems

Although differences in dwell volume between gradient equipment represent the most common problem when transferring a gradient method from one HPLC system to another, other factors also can be important. Several of these factors are discussed below:

- gradient shape;
- gradient rounding;
- inter-run equilibration;
- column size;

- column temperature;
- interpretation of method instructions.

5.2.2.1 Gradient Shape We strongly recommend the use of linear gradients wherever possible. Many HPLC systems allow the use of curved gradients (e.g., Fig. 1.4*b* and *c*). Although curved gradients might seem a logical choice in order to to control peak spacing via a change in gradient steepness during a run, curved gradients are very hard to reproduce between HPLC systems of different manufacturers. A better choice is to use segmented gradients (e.g., Fig. 1.4*d*), where two or more linear segments of different slope are joined to form the total gradient. If segmented gradients are used, the number of segments should be limited to two or three. The more segments that are included in a gradient run, the more difficult it will be to transfer that method to another system. See Section 1.3 for additional discussion on gradient shape.

5.2.2.2 Gradient Rounding All gradient mixing systems are a compromise between efficient mixing and minimal volume, and all have some degree of rounding at the ends of a test gradient profile, as is illustrated in Figure 4.4. Differences in the amount of rounding between systems can result in different effective gradient shapes [5, 10], especially for small gradient volumes V_G, which may make gradient transfer more difficult. See also the discussion in Section 9.2.2.

5.2.2.3 Inter-Run Equilibration The degree of column equilibration between runs can affect the ease of transfer of gradient methods from one system to another. Although partial equilibration may generate reproducible results for a given method on a given set of hardware (Section 5.3.3), differences in dwell volume and mixer washout characteristics may make partially equilibrated methods more difficult to transfer than more fully equilibrated ones.

5.2.2.4 Column Size If column length or i.d. is changed when a method is transferred, the gradient conditions must be adjusted to maintain constant k^* [see Equation (3.3) and [11]]; otherwise, major changes in selectivity and separation can result (e.g., Fig. 2.16). A change in column size (volume) also will change the value of V_D/V_G. This can affect the chromatogram even when gradient time is varied to maintain k^* constant – especially for early eluting peaks (Section 4.4). Obviously, it is preferable not to change column size.

5.2.2.5 Column Temperature The column temperature can affect retention and selectivity in gradient methods, so temperature control is essential for reliable gradient methods. Column temperature errors can be of two types: (a) bias in the temperature settings of the gradient system; or (b) incomplete thermal equilibration of the solvent entering the column. If the original and new oven are not properly calibrated, the same temperature settings may produce different oven temperatures with a resulting difference in the chromatogram (as in Fig. 5.2*c*). Different types of oven design may have different heating efficiencies. For example, one study [12] showed that up to a 10°C difference in measured column temperature was observed at a

55°C nominal setting, depending on the column oven type (block heater vs air bath) and column mounting technique. Another study [13] showed a 4.5°C difference in the effective column temperature between a Peltier-heated oven and a block heater oven nominally set at 40°C. No matter which heating technique is used, it is necessary to have the column at the same temperature on the original and new gradient systems, so that the same separation can be obtained. Temperature differences on the order of $\pm 5°C$ can be corrected for by adjusting the column temperature of the new system until the retention times are the same as for the original system (assuming same column, mobile phase, and method settings) [13]. Although temperature calibration errors can be a problem, the uniformity in temperature along the column also is important. Poor control of the axial temperature of the column usually results from inadequate preheating of the mobile phase prior to entering the column; this can severely distort peak shapes [13]. As long as the solvent at the column inlet is within $\sim 5°C$ of the column temperature, the temperature should be sufficiently uniform. A simple preheater can be constructed from a length of small-i.d. stainless steel tubing (e.g., 50 cm of 0.005 in. i.d. for an air-bath oven, 10–20 cm for a block heater) [13].

It has long been known (e.g., [14, 15]) that, as the mobile phase passes through the column, frictional heating occurs, but this is of little practical consequence with conventional columns and equipment (e.g., ≥ 3 μm particles, flow rates for pressures ≤ 6000 psi). However, with increasing interest in <2 μm particles and $>10,000$ psi operation, special precautions may need to be taken in equipment design to avoid problems due to temperature gradients along the column from this source.

5.2.2.6 Interpretation of Method Instructions Some methods are quite sensitive to small changes in reagents or other variables, whereas other methods are robust to similar changes. If the expected results are not obtained, the method instructions should be inspected carefully for possible misinterpretation. For example, are buffers prepared by mixing equal-molar acid and base components to reach the desired pH, or is the basic component titrated with concentrated acid? Are organic–aqueous blends made by combining measured volumes or by adding one component to the other to obtain the desired volume? Are the A and B solvents unblended, with all mixing performed by the equipment, or preblended so that the A reservoir contains 95 percent A + 5 percent of the B solvent (as in Section 5.1.5.2)? When following a written procedure, shortcuts or "equivalent" ways of doing the same thing should be avoided.

5.3 COLUMN EQUILIBRATION

Incomplete column equilibration arises when the column has not been flushed by the initial mobile phase (of composition ϕ_0) for an adequate time before starting the next gradient run. Unless the column has been fully equilibrated by the starting mobile phase, the stationary phase of the column will contain an excess of B solvent from the preceding gradient, and this may result in (a) decreased retention times for solutes that elute soon after the gradient starts, and (b) chromatograms from successive gradient runs that are not reproducible. Consequently, it has been recommended in the past to "fully" equilibrate the column between successive gradient runs, for example, by flushing the column with 5–10 column volumes (the "equilibration volume" V_{eq}) of the starting mobile phase [16, p. 394]. As will be seen, the requirement for column "equilibration" differs for routine analysis vs method development. In routine analysis, each sample will be injected on a timed basis, such that the equilibration time for each run after the initial sample injection will be the same. In this case, incomplete equilibration may not be a problem, since every run after the first is treated identically. Also, for routine analysis there is a considerable incentive to keep the equilibration time between runs as small as possible, for shorter overall run times and maximum sample throughput.

In method development, where the time between experiments often varies, and where repeatable retention times are necessary, there is a greater incentive for a more complete equilibration of the column between runs. Also, the added time involved in column equilibration (e.g., 5–10 min for a 150 × 4.6 mm i.d. column) is less critical in method development, since usually the total number of runs required for gradient method development is not large (Section 3.3).

5.3.1 Primary Effects

When we consider column equilibration in greater detail, a somewhat complicated picture emerges (Fig. 5.6), one which it will prove worthwhile to examine in detail. Figure 5.6(a) shows a series of successive gradients, with the injection of samples 1, 2, and so on, shown by the vertical arrows. The injection of each sample occurs just as the previous gradient has ended, and the next gradient is starting (no inter-run column equilibration). The gradient profiles of Figure 5.6a are measured at the column inlet as a function of time after simultaneous gradient initiation and sample injection. In this idealized example (Fig. 5.6a), it is assumed that the dwell volume of the equipment is zero, there is no mixing or dispersion of the gradient by the equipment or column, and the mobile phase is in instantaneous equilibrium with the column; none of these assumptions are ever valid for "real" gradient separations.

The first complication to consider is the equipment hold-up or "dwell" volume V_D (Sections 2.3.6.1, 4.1.2.1, and 4.3.1.2) or dwell time $t_D = V_D/F$. Relative to the gradient that is selected, the gradient arriving at the sample injector and column inlet will be delayed by a time t_D (*dashed* gradients in Fig. 5.6b). Because of this dwell time, it is seen that the second sample injection occurs before the end of the first gradient has reached the column inlet. As a result, early peaks will be eluted by mobile phase that is too strong, with compression of these early peaks and resulting poor resolution. This is illustrated in Figure 5.7, where the initial sample injection (*a*) occurs prior to the arrival of the gradient at the column inlet, but the second injection (*b*) occurs before the first gradient has cleared the column (or even moved past the column inlet). Not only are early peaks eluted early, but also changes in relative retention and loss of sample resolution result for this "irregular" sample. In order to avoid the problem illustrated in Figure 5.7, it is necessary to introduce an inter-run column equilibration of duration $t_{eq} \geq t_D$ (Fig. 5.6c).

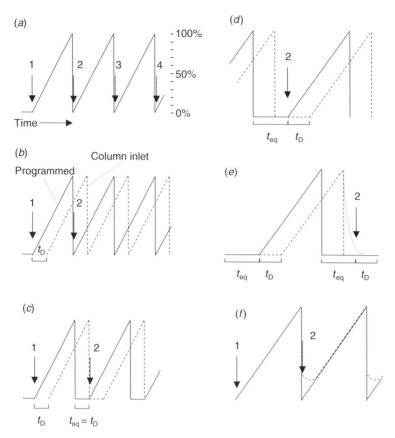

Figure 5.6 Gradient profiles (%B vs time) that illustrate different contributions to incomplete column equilibration between gradient runs. (*a*) "Ideal" gradient profiles assuming no problem with between-run column equilibration; (*b*) difference between selected gradient profiles (——) and the gradient profile at the column inlet (- - -) as a result of equipment dwell volume; (*c*) use of a minimum equilibration time $t_{eq} = t_D$ between successive gradient runs; (*d*) same as (*c*) with $t_{eq} > t_D$; (*e*) gradient distortion due to the equipment mixing volume V_M; (*f*) slow intra-column equilibration.

The use of column equilibration as in Figure 5.6(*c*) appears to avoid the injection of sample before the end of the preceding gradient, but it allows for no equilibration of the column (removal of excess B solvent from the column and equilibration of the column with mobile phase before the start of the next gradient). For this reason, it is necessary to provide an equilibration time $t_{eq} > t_D$ (Fig. 5.6*d*). However, we have to consider still another characteristic of the gradient equipment that affects column equilibration, namely gradient dispersion and distortion due to the equipment mixing volume V_M (Section 9.2.2). As a result of gradient distortion, at the time the gradient ends the concentration of B solvent does not return immediately to its value at the start of the gradient (Fig. 5.6*e*). If gradient distortion is severe enough, and if insufficient column-equilibration time is allowed, the start of the

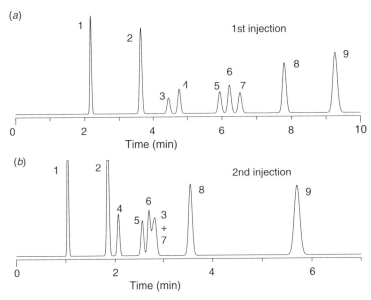

Figure 5.7 Effect of sample injection prior to the end of the gradient passing the column inlet; compounds 1–9 of "irregular" sample. (*a*) First gradient run; (*b*) second gradient run, assuming gradients as in Figure 5.6(*b*). See text for details.

following gradient will not return to its programmed initial value of %B. Therefore, the gradient equilibration time t_{eq} must be large enough to overcome any gradient distortion as in Figure 5.6(*e*), account for system dwell volume, and also restore the column to be in equilibrium with the mobile phase at the start of each new gradient run.

5.3.2 Slow Equilibration of Column and Mobile Phase

By column-mobile phase equilibration, we refer to flushing the column with sufficient starting mobile phase to create an equilibrium between mobile phase and column stationary phase at the start of the gradient. That is, the column is filled with mobile phase of the same composition (corresponding to the start of the gradient), and the stationary phase is in equilibrium with the mobile phase. Column-mobile phase equilibration, apart from the various related effects of Figure 5.6(*a–e*), is discussed further in Section 9.2.1.1. As a result of slow equilibration, the starting value of %B in the second and following gradients will be slightly greater than the value programmed into the system (regardless of gradient carryover), as illustrated in Figure 5.6(*f*) (dashed curve, highly exaggerated, similar to the effects of gradient distortion in Fig. 5.6*e*). Recent reports [17–18b] have shown that complete equilibration of the column when changing the mobile phase or running successive gradients can require a time varying from 1 to 10 h, which could definitely complicate the use of gradient elution. However, this potential problem is, for several reasons, of little practical importance.

First, actual changes in %B at the start of the gradient (as inferred from changes in retention as a function of equilibration time) appear to be quite small, typically resulting in changes in retention time of no more than 0.2 percent of the gradient time t_G. Second, retention shifts, as a result of slow column equilibration, decrease for later eluting peaks. Therefore, *adjacent* peaks with similar retention times will experience similar (small) shifts in retention time, so that their resolution will be little affected by a lack of complete column equilibrium (Section 9.2.1). Finally, it is only the first sample injection that experiences differences in retention time because of variable column equilibration, and it is recommended that this sample not be used for analysis (Section 5.1.4.7). Later injections experience the same retention time shifts (same degree of non-equilibration), so all runs after the first sample injection will show the same retention times – despite incomplete column equilibration.

A further observation from the study of [18] is that the addition of 1 percent propanol to the A and B solvents that form the gradient can allow *complete* inter-run equilibration with only 1–2 column volumes of the initial mobile phase. The use of propanol for this purpose was first suggested by Dorsey [19, 20]. Since the contribution of slow column equilibration to the required equilibration time t_{eq} is usually minor, the latter procedure is not commonly used.

5.3.3 Practical Considerations and Recommendations

Column equilibration in gradient elution is primarily affected by "gradient carryover" from the preceding gradient run, that is, as occurs in Figure 5.6(*b*) (due to dwell volume) and Figure 5.6(*e*) (gradient distortion). If a subsequent gradient is not significantly affected by gradient carryover (due to the provision of a column-equilibration time t_{eq}). There should be few differences between sequential gradient runs (including the first run). The extent of gradient distortion is determined by the mixing volume V_M of the gradient system (Section 9.2.2), which will usually be slightly less than the dwell volume V_D [5]. To avoid a major contribution from gradient carryover, it is recommended that $t_{eq} \geq 2t_D$ (assumes a conventional gradient system with sample injection occurring at the same time as gradient initiation). Note that the latter recommendation is expressed in time rather than (as previously) column volumes of the initial mobile phase (i.e., the required equilibration time will be proportionally shorter for higher flow rates). This emphasis on equilibration time rather than volume reflects a much greater significance of gradient carryover (Section 5.3.1) vs column equilibration (Section 5.3.2).

For method development, where the time spent for each experiment usually is less critical than for routine analysis, a larger value of t_{eq} is recommended. Gradient equipment used today usually has values of $V_D \leq 3.0$ mL, and we recommend a flow rate of 2.0 mL/min (Table 3.2). This suggests that the inter-run equilibration time t_{eq} should be at least 3 min. An actual equilibration time of 5–10 min represents a more than adequate safety margin that does not add much to the overall time required for each gradient experiment. The adequacy of the latter column equilibration can be verified by comparing retention times for replicate, sequential gradient runs (Section 5.1.5.4).

For routine separations, the equilibration time can comprise a significant fraction of the gradient time t_G, and for this reason it may be worthwhile to reduce the equilibration time t_{eq} as much as possible. The above discussion suggests a minimum equilibration time equal to $2t_D$. Very fast runs, with sample injection rates of one per minute or less, are becoming more common (Section 3.3.7). Clearly, such runs are incompatible with a 3 min equilibration time (for $V_D = 3$ mL) or even with a gradient system where $V_D \approx 1$ mL. There are several possibilities for reducing t_{eq} to less than a minute. First, some gradient systems sold today allow for the injection of the sample after gradient initiation (Section 5.2.1.1). From Figure 5.6(b) it can be seen that, if sample injection is delayed by a time equal to t_D, sample injection and the arrival of the gradient at the column inlet are now in phase, thereby eliminating any effect of the dwell time on column equilibration. The contribution of gradient distortion (Fig. 5.6e) to slow column equilibration can be similarly eliminated, by a modification of the gradient system that involves an additional valve and pump [18]. Finally, it is advantageous for very fast gradient elution to use short columns and flow rates >2 mL/min (Section 3.3.7); the required equilibration time t_{eq} is inversely proportional to F, allowing much smaller values of t_D for modern equipment with $V_D \leq 0.5$ mL.

The use of traditional ion-pair reagents (e.g., octane sulfonate) in gradient elution can result in slower column equilibration and longer equilibration times, especially if the concentration of the reagent in the mobile phase is small, and the reagent is strongly retained from the initial mobile phase (A solvent). We recommend against using gradient elution with such slowly equilibrating reagents. Nevertheless, by an appropriate choice of ion-pair reagent (smaller, less retained molecules preferred) and gradient conditions (higher %B in the A solvent), it is possible to reduce the volume of equilibration solvent to 1–3 column volumes [21]. However, ion-pair gradient elution, other than the use of trifluoroacetic acid for the separation of peptides and proteins (Section 6.2), is not generally recommended.

Slow inter-run equilibration also can be much more pronounced for normal-phase chromatography (NPC), especially separations on bare silica [22]. The presence of trace amounts of water in the nonaqueous mobile phases usually used for NPC can lead to column equilibration times measured in hours. However, the use of dried solvents with %B values >6 percent speeds up equilibration time and improves retention reproducibility [23]. Otherwise, reproducible gradient separations with NPC may not be practical. An alternative to the use of dried solvents with NPC is to use solvents partially saturated with water or those containing a trace of a polar solvent, such as propanol [16, p. 288]. See Section 9.2.1 for further details on slow inter-column equilibration. NPC using polar bonded phases (Section 8.3) is a viable alternate technique for gradient elution under normal-phase conditions.

5.4 SEPARATION ARTIFACTS

Separation artifacts include baseline problems, peaks in blank gradients, unexpected peaks in sample gradients, and unexpected peak shape. Various terms have been used to describe unexpected peaks in the chromatogram: artifact peaks, system peaks, pseudo peaks, vacancy peaks, eigen peaks, induced peaks, spurious peaks, and ghost peaks [24]. The term "ghost peak" appears to be the most frequently used term and will be used here to encompass any peak in the chromatogram that arises from the reagents or other nonsample-related sources.

We recommend running a blank gradient to help isolate the source of unexpected peaks in a method. Simply run the gradient program without making an injection, or, if the HPLC system will not allow this, make a minimum-volume injection (e.g., 1 μL) of the starting mobile phase.

5.4.1 Baseline Drift

With gradient elution methods that use UV detection, it is rare for the weaker A solvent and the stronger B solvent to have identical UV absorbance, especially for detection wavelengths <220 nm. This results in a drift in the chromatographic baseline, even when a blank gradient is run. In most reversed-phase methods, the aqueous A solvent has less UV absorbance than the B solvent (typically ACN, MeOH, or THF), so the baseline will drift upwards during a run. However, various baseline shapes are possible, depending on the nature of the solvents (see below).

As a general rule, drift is worse at lower wavelengths. An example is shown in Figure 5.8 [25] for a gradient of 5–80 percent methanol/phosphate buffer in 10 min. At 254 nm the absorbances of the buffer and methanol are nearly the same, so little, if any, gradient drift is observed. At 220 nm, however, methanol absorbs more strongly than the buffer and a positive baseline drift is observed (~0.03 absorbance units, AU). Gradient drift is common, but as long as the baseline remains on-scale over the course of the run, most data systems can accurately quantify all peaks. Whereas methanol and acetonitrile have sufficiently low UV absorbance to be useful for gradients with detection below 220 nm or even lower, other solvents, such as THF, have sufficient absorbance to preclude their use for full-range gradients (e.g., 5–100 percent B) at low wavelengths. As an example, a THF–water gradient is shown in Figure 5.9 [25], where the drift at 254 nm is more noticeable than with methanol (Fig. 5.8). At 215 nm, moreover,

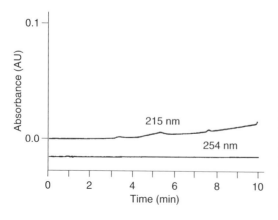

Figure 5.8 Baselines obtained using phosphate–methanol gradients. Solvent A: 10 mM potassium phosphate (pH 2.8); solvent B: methanol; gradient 5–80 percent B in 10 min. Adapted from [25].

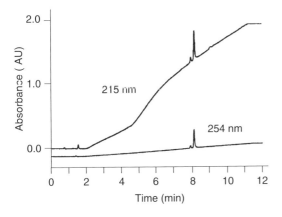

Figure 5.9 Baselines obtained using phosphate–tetrahydrofuran gradients. Solvent A: 10 mM potassium phosphate (pH 2.8); solvent B: THF; gradient 5–80 percent B in 10 min. Adapted from [25].

the drift for THF is >2 AU, which is too large for a practical gradient method. (Peaks at ~8 min in Fig. 5.9 are discussed in Section 5.4.3 below.) *Note when comparing Figures 5.8–5.12, that the y-axis varies from ~0.1 to 2.0 AU full-scale for different figures.*

Although it is much more common to observe a positive gradient drift, where the B solvent absorbs more strongly than the A solvent, negative drift can occur if the B solvent has lower UV absorbance than the A solvent. This is illustrated in Figure 5.10 [25] for a gradient of ammonium acetate (A) to methanol (B) (215 nm trace). Negative drift is more problematic from a practical standpoint, because many data systems will not record or integrate data if the baseline drops more than a certain amount (e.g., 10 percent) below the starting baseline value. Thus, in the case of the ammonium acetate–methanol gradient at 215 nm (Fig. 5.10), it would be difficult to gather data for more than the first 5 min of the gradient. (To obtain the 215 nm trace of Fig. 5.10 so that the full baseline is displayed, the initial baseline was set at approximately 1.2 AU and was not auto-zeroed when the gradient started. This would be impractical to do in routine analysis.)

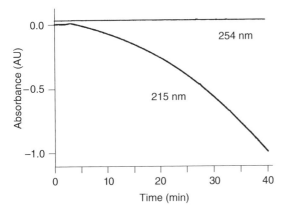

Figure 5.10 Baselines obtained using ammonium acetate–methanol gradients. Solvent A: 25 mM ammonium acetate (pH 4); solvent B: 80 percent methanol in water; gradient 5–100 percent B in 40 min. The absorbance scale at 215 nm is offset by −1.2 A. Adapted from [25].

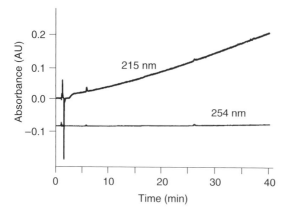

Figure 5.11 Baselines obtained using equimolar ammonium acetate–methanol gradients. Solvent A: 25 mM ammonium acetate (pH 4) in 5 percent methanol; solvent B: 25 mM ammonium acetate in 80 percent methanol; gradient 0–100 percent B in 40 min. The absorbance scale is relative, not absolute. Adapted from [25].

One often can compensate for positive or negative baseline drift by doping the less absorbant mobile phase with an absorbing compound. For example, an unretained UV-absorbing compound could be added to the less-absorbing mobile phase component (usually the A solvent). Inorganic ions (e.g., nitrate, nitrite, azide), small organic ions (e.g., formate, acetate), and hydrophilic, low-molecular-weight compounds (e.g., urea, thiourea, formamide) are possibilities [16, 26, 27]. It also may be possible to adjust the concentration of the existing buffers or additives to match the absorbances. In the example of Figure 5.11 [25], the negative drift of Figure 5.10 was reversed by adjusting the solvent composition so that *both* the A and B solvents contained (UV-absorbing) 25 mM ammonium acetate.

In still other cases, the absorbance of a mixture of A and B may have higher or lower absorbance than either A or B alone. This is seen in Figure 5.12 [25] for ammonium bicarbonate–methanol gradients. At 215 nm, the mixture of ammonium bicarbonate and methanol has less absorbance than either ammonium bicarbonate or methanol alone. Compensating for such drift may not be possible.

These examples of gradient drift are likely to appear (though usually to a lesser degree) in nearly every gradient, especially at wavelengths ≤220 nm. The difference

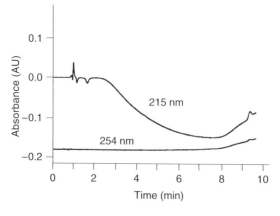

Figure 5.12 Baselines obtained using ammonium bicarbonate–methanol gradients. Solvent A: 50 mM ammonium bicarbonate (pH 9); solvent B: methanol; gradient 5–60 percent B in 10 min. The absorbance scale is relative, not absolute. Adapted from [25].

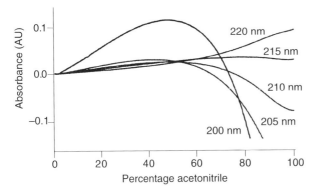

Figure 5.13 Baseline drift for TFA–ACN gradients at selected wavelengths. Gradient: A = 0.1 percent TFA in water; B = 0.1 percent TFA in acetonitrile; 0–100 percent B in 100 min; 1 mL/min; no column. The absorbance scale is relative, not absolute. Adapted from [25].

in UV absorbance of the A and B solvents is a physical property of the solvents that cannot be avoided. As long as the drift is smooth and unidirectional, it can be minimized by adjusting the absorbance of one solvent or the other. Once drift is minimized, the data system should have no problem accurately quantifying peaks on a mildly drifting baseline.

A special case for baseline drift occurs with TFA–ACN gradients. TFA is popular for use with biological molecules because it provides ion pairing capabilities that are compatible with gradient elution and has favorable UV absorbance at wavelengths <220 nm. Typically for such applications, 0.1 percent TFA is added to both the A solvent (water) and the B solvent (acetonitrile). The UV absorbance is dependent both on the detector wavelength and the water–ACN ratio, so baseline drift is observed with UV detection. Examples of TFA/ACN baselines at several wavelengths are shown in Figure 5.13 [28]. The preferred detection wavelength for minimum drift (over the complete range in %B) is 215 nm [28, 29]. Further reduction in baseline drift can be obtained by adding 15–20 percent more TFA to the A solvent than the B solvent (e.g., A = 0.115 percent TFA, B = 0.1 percent). See Section 5.5.2.9 for further recommendations on the use of TFA.

5.4.2 Baseline Noise

In addition to baseline drift, baseline noise is possible if two solvents of unequal absorbance are not thoroughly mixed. This is illustrated in the hypothetical example

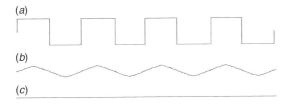

Figure 5.14 Hypothetical solvent blending. (a) Pulses of A and B solvents; (b) partial mixing; (c) complete mixing.

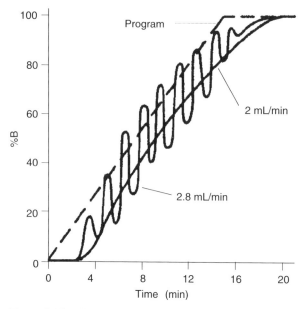

Figure 5.15 Illustration of beat frequency from mismatched solvent proportioning timing and pump piston cycles. Gradient program of water to 0.01 M potassium bromide in water over 15 min (dashed line). Well-matched system (straight solid line), 2 mL/min; mismatched system (wavy line), 2.8 mL/min. Adapted from [30].

of Figure 5.14, where Figure 5.14(a) portrays pulses of A and B sequentially metered into the mixer, as would happen with a low-pressure mixing system. As mixing takes place, the square wave of solvent pulses will begin to be rounded, as in Figure 5.14(b), until complete mixing takes place as in Figure 5.14(c). From a practical standpoint, no on-line mixer is perfect, so, if magnified sufficiently, the mixture over time will always look more like Figure 5.14(b) than Figure 5.14(c).

Poorly mixed mobile phase should not be confused with another artifact that can appear as baseline noise that is caused by a phenomenon called *beat frequency*. This is a particular problem with low-pressure mixing systems when the cycle time of the solvent proportioning valves is not coordinated properly with the pump piston cycles, such that the delivered concentration of a mobile phase component varies during a run, even under isocratic conditions. The problem has been recognized for many years [e.g., 30, 31] and should not be an issue with newer equipment. In the extreme, the beat frequency can appear as sinusoidal noise on the baseline, as illustrated in Figure 5.15 [30], where proper valve cycle timing (2 mL/min) gives a smooth mixing curve, but the beat frequency shows up when the timing is off (2.8 mL/min). Proper system design optimizes the valve and pump cycles to eliminate the beat frequency problem.

5.4.2.1 Baseline Noise: A Case Study
Some baseline problems are more complex than represented above, as illustrated by the following example [32].

5.4 SEPARATION ARTIFACTS   181

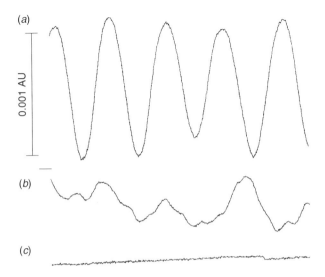

Figure 5.16 Baseline noise problems. (*a*) Typical baseline noise (~1.2 mAU peak-to-peak) without premixing of solvents, measured at 210 nm; (*b*) noise improvements (~0.4 mAU peak-to-peak) resulting from premixing the mobile phase, matching the A and B solvent absorbances, and changing to 220 nm; (*c*) baseline (~0.02 mAU peak-to-peak) with degassed mobile phase. Adapted from [32].

Figure 5.16 illustrates a problem in baseline noise that apparently was the result of several causes. The blank gradients shown were obtained for a very shallow reversed-phase gradient of 5–80 percent B over 75 min; solvent A was 0.1 percent TFA in water and solvent B was 0.1 percent TFA in acetonitrile. A two-pump high-pressure mixing system was used with a mechanically stirred mixer and a manual injection valve. The column and injector were maintained at 30°C and UV detection was at 210 nm. This system had worked well historically without solvent degassing. A back-pressure regulator (~50 psi) was used after the detector to prevent outgassing in the detector. Initially, the cyclic baseline seen in Figure 5.16(*a*) was observed. In this case, the baseline cycle roughly corresponded to the pump cycle, so pump problems were suspected. However, neither replacement of all four check-valves nor addition of a supplemental solvent mixer improved the baseline.

The baseline noise was reduced to that shown in Figure 5.16(*b*) following three changes. There is general agreement that TFA–water–ACN mixtures are difficult to blend, so the solvents were premixed such that the A mobile phase contained 5 percent ACN and the B mobile phase contained 20 percent water (80 percent ACN; see also Section 5.5.2.3). A second change was to match the absorbance of the A and B solvents (as discussed in Section 5.4.1) by using 0.115 percent TFA in A and 0.1 percent TFA in B. The detector wavelength also was changed to 220 nm so that the absolute absorbance of the mobile phase was reduced. These three changes decreased the amplitude of the noise by 3-fold and greatly diminished the sinusoidal nature of the noise. A further reduction in noise (Fig. 5.16c) was obtained by thoroughly degassing the mobile phase. (See Section 5.5.4.4 for additional discussion of this problem.)

This problem illustrates several artifacts related to baseline noise. First, pump problems can produce pulsating baselines, either from pressure fluctuations or pulses of solvent that are not thoroughly mixed. The pump(s), especially the check-valves and pump seals, must be in good working order. Second, premixing some of the A solvent into the B solvent and vice versa can improve the mixing characteristics of the solvents. This, in turn, leads to more thoroughly blended mobile phases and quieter baselines. Third, matching the absorbance of the A and B mobile phases can reduce baseline noise, as well as drift. Finally, degassing almost always improves the performance of gradient HPLC methods (Section 4.1.2.2). In the present example, degassing may represent the main contribution to the overall correction of the original problem.

5.4.3 Peaks in a Blank Gradient

When a blank gradient is run (no injection of sample), a smooth baseline with some drift (Section 5.4.1) and noise (Section 5.4.2) is expected. The appearance of distinct peaks (so-called "ghost" peaks) in a blank gradient may come as a surprise, such as is seen in the blank buffer–THF gradients of Figure 5.9 at about 8 min. Often the source of such peaks can be traced to the water used in the mobile phase (Section 5.4.3.1), as well as the organic solvent, reagents, equipment, or technique. It is important to run a noninjection blank gradient to ensure that no false peaks appear in the gradient which might interfere with analysis.

5.4.3.1 Mobile Phase Water or Organic Solvent Impurities If organic contaminants are present in either the A or B solvent, they tend to (a) concentrate at the head of the column during the between-run equilibration, and (b) elute during the following gradient. Such peaks may be indistinguishable from normal sample peaks. The problem of spurious background peaks in the chromatogram can range from insignificant to severe.

The blank gradients of Figure 5.17 [33] are an example of significant mobile phase contamination. The method was a stability-indicating assay for a pharmaceutical product. Such methods require quantification of any peaks ≥ 0.05 percent of the active ingredient. The method was run on a high-pressure mixing HPLC system with a C_{18} column and UV detection at 255 nm. Normally a 10 min equilibration was used between runs (Section 5.3). The blank gradient of Figure 5.17(a) was run first to check for the presence of any potentially interfering peaks (note that the y-axis scale is in units of 10^{-4} AU). The manufacturer's specifications for acetonitrile allowed for no peaks larger than 0.001 AU at 254 nm with a 100 percent water to 100 percent ACN gradient. Only the large peak at 11.5 min exceeds this specification. However, this stability-indicating assay required the quantification of peaks with absorbances >0.0005 AU, so reduction of the background to <0.0002 AU was necessary.

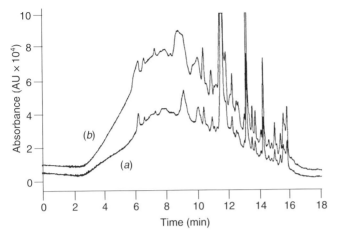

Figure 5.17 Blank gradient runs after (a) 10 min, and (b) 30 min equilibration; 150 × 4.6 mm i.d., 5 μm C_{18} column; mobile phase of 27 mM TFA (A, pH 3) and acetonitrile (B); flow 1.5 mL/min; gradient 5–83 percent B in 13 min plus 5 min isocratic hold at 83 percent; UV detection was at 255 nm. Adapted from [33].

It is more likely that spurious peaks in a gradient run arise from the A solvent. Solvent impurities concentrate at the head of the column during equilibration and are released during the gradient as normal peaks, when the mobile phase strength increases sufficiently. A simple way to verify the source of the extra peaks is to increase the equilibration time. In this manner, a higher concentration of the A-solvent contaminants accumulates at the head of the column and the gradient peaks should be correspondingly larger. For the present example, the equilibration time was increased from 10 min (a) to 30 min, resulting in the chromatogram of Figure 5.17(b). Although not obvious from this figure, the major peaks in Figure 5.17(b) are approximately three times larger than those in (a), implicating the A solvent as the primary source of the problem. In the process of troubleshooting this problem, the degassing process, column, TFA, and laboratory glassware were eliminated as the problem source (the troubleshooting process for a similar problem can be found in Section 5.5.4.3). When a different water source was used, the pattern of background peaks changed and their magnitude was significantly reduced, confirming that contaminated water was the main problem source. (Note that, in this example, the technique discussed below for Figure 5.18 was not used to determine if any of the problem peaks originated from the organic solvent.)

For the best baselines with gradient elution, only HPLC-grade water should be used in the mobile phase. Whereas adequate results often may be obtained for isocratic separations when deionized or distilled water is used, gradient background traces will be much better with commercially prepared or in-house purified HPLC-grade water. Aqueous, dilute organic (e.g., <10 percent organic), and buffer solutions tend to support microbial growth, which also can be the source of ghost peaks from these reagents (as well as column blockage). Such solutions must be changed regularly (e.g., expiration dates ≤1 week) and clean containers

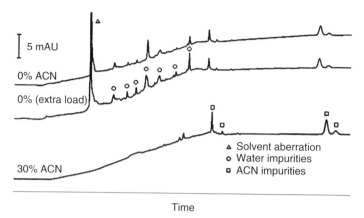

Figure 5.18 Determining the origin of ghost peaks caused by mobile phase impurities. Column: Zorbax XDB-C8, 150 × 4.6 mm, 5 μm; 40°C; 2 mL/min; UV 215 nm; A = water, B = acetonitrile. Gradient: 10 min equilibration at initial conditions followed by gradient to 100 percent B in 10 min and hold at 100 percent B for 10 min. Initial conditions 0 and 30 percent B as noted; all equilibration times are 10 min, except that "extra load" is longer equilibration at 0 percent B. Adapted from [24].

used to prevent re-contamination. In some cases, solvent reservoir filters and the interior walls of tubing and instrumentation can become coated with a "biofilm" of bacteria, spreading contamination. Periodic flushing of highly aqueous solvent lines with methanol or acetonitrile will help to keep these surfaces clean [24]. (See Section 5.5.4.3 for more information on water problems.)

Although it may seem illogical that impurities from organic solvents might build up on the column and be eluted as ghost peaks, it is important to remember that contaminants in the organic solvent are present in low-%B solutions during the beginning of the gradient; these contaminants therefore can concentrate at the head of the column, because the solvent is weak (i.e., values of k for the contaminants are large) [24]. A simple test to help identify the source of organic-solvent impurities has been described [34]. The standard extended-equilibration test described above will not distinguish between impurities that originate from the aqueous or organic mobile phase components, because extended equilibration allows materials from both sources to concentrate on the column (in all cases except where the A solvent is 100 percent aqueous). Usually the equilibration solvent is primarily A solvent, so its impurities will make up a much larger portion of the concentrated total than the B solvent impurities. However the source of the impurities may not be certain from just one equilibration condition. This problem is illustrated in Figure 5.18 [24] where the initial equilibration is 10 min at 0 or 30 percent acetonitrile. If only the equilibration time is varied ("0 percent" vs "0 percent-extra load," where the equilibration time in 0 percent B is extended), all the impurity peaks increase in size in proportion to the increased equilibration time (this also was seen in Fig. 5.17). In this case, with 0 percent ACN the peaks which increased in size (circles in Fig. 5.18) originated from the water, because the column was not exposed to ACN during the extended equilibration period. If the equilibration mobile phase contains B solvent and only the equilibration time is changed, the ratio of the impurity peaks stays the same, so it is not possible to determine if they originated from the A or B solvent (although, from a practical standpoint, the aqueous phase is generally the source of most contaminants, as in the case of Fig. 5.17). However, if the %B for initial equilibration is increased (with constant equilibration time), the proportion of ghost peaks originating from the B solvent will increase relative to the A solvent peaks. This can be seen for the increasing peak size for the acetonitrile impurities (marked with squares) in the 0 and 30 percent ACN equilibration runs of Figure 5.18. (The "solvent aberration" peak of Fig. 5.18 was attributed to a physical mixing effect or solvent outgassing because it did not change in size as the equilibration was increased [24].) *One should be cautious in quantitative interpretation of any of the extended-equilibration experiments – in the authors' experience the increase in peak size is only semi-quantitative and somewhat variable. This can be seen in the extended-equilibration experiments of Figures 5.17 and 5.18.*

5.4.3.2 Other Sources of Background Peaks The examples above illustrate problems arising from contaminated water or organic solvent, but solvents are not the only sources of potential background peaks in blank gradients. Similar problems can result from contaminated reagents, dirty equipment, or poor laboratory technique. Impurities from buffers will also concentrate at the head of the column with extended equilibration, just as the case for water impurities. Buffer impurities can be identified by comparing extended-equilibration experiments using A solvent with and without buffer. Improper use of laboratory equipment can contribute mobile phase contaminants, as can improperly cleaned labware. The case study of Section 5.5.4.3 gives an example of the isolation of some of these problems.

5.4.4 Extra Peaks for Injected Samples

In Section 5.4.3, blank-gradient peaks were discussed. Additional peaks in the chromatogram also can appear when injections are made of samples, blank matrix spiked with analyte, or reagent blanks. (A "reagent blank" sometimes is called a "sample-pretreatment blank," where all the sample preparation steps are followed without processing a sample.) The sources of some of these extra peaks are discussed here. Sample carryover may also be a source of extra peaks (Section 5.5.2.11).

5.4.4.1 t_0 Peaks It is common to observe a peak at the column dead time t_0 in both isocratic and gradient runs. Usually this can be attributed to unretained sample components or a detector response from a mismatch between the sample injection solvent and the mobile phase. This peak (or cluster of peaks) sometimes is referred to as the "garbage peak," "solvent front," or "injection disturbance." A typical t_0 disturbance for a "clean" sample is seen as a negative peak at ∼1 min in Figure 5.19(*a*) (arrow) [35]. This is most likely due to a change in the mobile phase refractive index as the sample solvent passes through the detector flow cell, resulting in scattering of the light passing through the flow cell (Schlieren effect). A much larger t_0 peak due

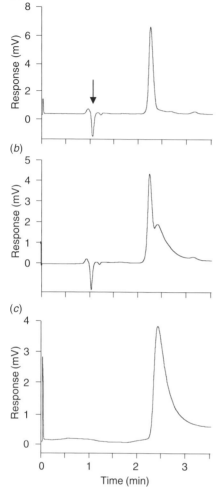

Figure 5.19 Air peak in a chromatogram. (*a*) Normal chromatogram for a 5 μL injection of sample with the analyte peak at 2.24 min; (*b*) analyte peak plus an interfering peak eluted just after the analyte; (*c*) chromatogram obtained from an injection of 5 μL of air. Isocratic separation using 35 : 65 25 mM phosphate buffer (pH 2.5)–methanol mobile phase at a flow rate of 1.0 mL/min with a 150 × 4.6 mm i.d., 5 μm C_{18} column operated at 30°C and UV detection at 230 nm. Adapted from [35].

to unretained sample components is illustrated as the first peak in Figure 1.2. In gradient elution, it usually is easier to separate the t_0 peak from the first peak of interest, compared with isocratic elution. With a well-designed gradient method, the first peak of interest will be eluted with a retention time of at least twice t_0, so this early peak usually can be ignored.

5.4.4.2 Air Peaks Although uncommon, it is occasionally possible to observe peaks in the chromatogram that are due to injection of air with the sample. This can arise from dissolved air in the sample or from an injected air bubble. These are real peaks – air may be retained in the column and may appear more or less like a sample peak. As long as air has a higher UV absorbance than the background, the peak will appear as a positive peak with a UV detector. An example of this is shown in Figure 5.19 [35]. Normally, this separation gave the chromatogram of

Figure 5.19(a), with a sample peak eluted at 2.24 min with acceptable peak shape. During routine method operation, chromatograms such as that of Figure 5.19(b) were observed, containing a broadly tailing peak on the back of the analyte peak. It was discovered that the autosampler needle depth had been set incorrectly, so air was drawn up with some of the samples. Injection of 5 µL of air gave the chromatogram of Figure 5.19(c), which corresponds to the extra peak in Figure 5.19(b) at 2.45 min. Adjustment of the autosampler corrected the problem. Note that this is an isocratic example, but similar results can be expected in gradient elution. An air peak typically will elute at 60–70 percent B in a gradient.

Air can arise from bubbles in the injector, as in the example above, or from dissolved air in the mobile phase. Although inconvenient, the samples can be degassed using vacuum or helium sparging to remove dissolved gas if the problem is persistent and interferes with obtaining acceptable results [35].

5.4.4.3 Late Peaks With isocratic methods, peaks that have not fully eluted prior to the end of one run can appear in the next (or later) chromatogram as a broad (relative to neighboring peaks), unexpected peak (arrow in Fig. 5.20 [36]). In isocratic runs, one way of correcting this problem is to extend the run time to exceed the retention time of the problem peak (shown in its normal position at 38.5 min in Fig. 5.20). Late eluted peaks are much less likely to be a problem with gradient elution, since a peak that is strongly retained in one gradient also will be strongly retained in the next run. To avoid buildup of strongly retained materials in gradient elution, extend the gradient to a higher %B or add a step-gradient flush at the end of the run, as in Figure 3.8b.

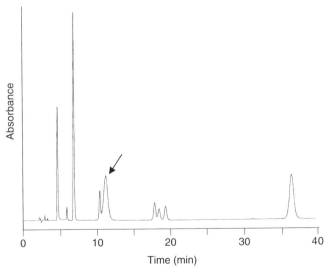

Figure 5.20 Late eluted peak normally eluted at 38.5 min appears at 12.0 min (arrow) in the next chromatogram of a shortened isocratic run. Adapted from [36].

5.4.5 Peak Shape Problems

5.4.5.1 Tailing and Fronting Peak fronting and tailing are present in many assays; if values of either the peak asymmetry factor or USP tailing factor fall within a range of ∼0.9–1.5, peak shape is considered normal and acceptable. Fronting or tailing peaks with values outside this range may be indicative of problems that require attention.

Peak fronting is fairly rare in reversed-phase HPLC. The most common cause of peak fronting is a "bad" column or guard column. This can occur when the column is poorly packed, or a void or channel occurs after extended use. Fronting can occur gradually over time or can be abrupt, with one chromatogram appearing normal and the next showing severe fronting. The best way to check for column failure is to replace the column and/or guard column. Peak fronting also can occur if the injection solvent is too strong relative to the starting mobile phase (Section 5.4.5.4). Another possible source of peak fronting can arise from sample decomposition (Section 5.4.5.5).

Peak tailing is much more common than peak fronting. The most common cause of peak tailing is overloading of ionized-silanol sites within the column by a protonated, basic sample compound. Tailing has become much less common with the introduction of columns made from high-purity, less-acidic, type B silica. We strongly recommend the use of type B silica columns when developing any new method or revising an existing method for a replacement column. Older, more-acidic, type A silica contains larger amounts of metal impurities and readily ionizable silanol groups, which are prone to peak-tailing for protonated basic compounds. Additives such as triethylamine have been used in the past to reduce tailing, but additives are seldom necessary for type B silica columns. For a list of common RP-LC columns characterized by silica type ("A" or "B"), as well as a further characterization of columns in terms of their tendency to give tailing peaks for basic samples, see Appendix III.

Tailing is somewhat less common with gradient elution than for isocratic separation, because of gradient compression (Section 9.1.2.1). In gradient elution, the tail of the band moves in a stronger solvent than the front, so the molecules in the band tail are "pushed" toward the band front. As a result, bands which tail in isocratic elution are often symmetrical in gradient elution (e.g., Fig. 1.3a vs b).

Sample overload is a special case of peak tailing, in which too large a sample mass is injected. Overloaded peaks tend to take on a right-triangle shape, with shortened retention times (e.g., Fig. 7.9). A simple test for overload is to reduce the injected mass 10-fold and examine the results. If the retention time increases and tailing is reduced, overload is the likely cause. For a further discussion of peak shape for overloaded peaks in gradient elution, see Section 7.3.

5.4.5.2 Excess Peak Broadening In isocratic separations, abnormal peak broadening can be checked by calculating the plate number N. If a value of N drops noticeably over time, column failure might be suspected. Unusually low values of N for just one of the peaks in a chromatogram might also signal some problem with that peak. In gradient elution, a value of N cannot be calculated in the same way as for isocratic elution, although it is possible to determine a value

TABLE 5.1 Relationship Between Peak Width, Plate Number, and Resolution in Gradient Elution

	Percentage change in	
Peak width (W)	Plate number (N)[a]	Resolution (R_s)[b]
+10	~−20	~−10
+20	~−30	~−20
+30	~−40	~−25
+40	~−50	~−30
+50	~−55	~−35

[a] Equation (2.5), N is inversely proportional to W^2.
[b] Equation (2.6), R_s is inversely proportional to W (if both peaks are wider by the same percentage).

of N in gradient elution [Equation (2.20)]. However, for purposes of determining abnormal peak broadening in gradient elution, it is usually more convenient to simply compare peak widths. The data of Table 5.1 can be used to help evaluate the impact of peak broadening on the separation. For example, in Table 5.1 an increase in peak width of 20 percent corresponds to a decrease of N by ~30 percent and loss of ~20 percent in R_s. For modest increases in peak width, the percentage loss in resolution is about the same as the percentage increase in peak width, so that peak width can be monitored as a surrogate for resolution to help determine when a column needs to be replaced.

If excess peak broadening is suspected, it is wise to change to isocratic conditions and repeat the column manufacturer's test for column plate number. If N is less than 70–80 percent of the column manufacturer's test value, a new column should be installed and the test repeated. If N is now OK, the original column has deteriorated; if N is still low, look for problems related to extra column effects or (especially) data system and time-constant problems.

Extracolumn effects are contributions to peak broadening that take place before and after the column. The most common problems are due to excessive lengths and diameters of connecting tubing, excessive volume in the detector cell, or (less frequently in gradient elution) injector plumbing. If peak broadening due to extracolumn effects is suspected, the tubing length can be reduced to the minimum convenient length. Where sample is transported between the injector and column, and between the column and detector, tubing no larger than 0.007 in. i.d. (~0.17 mm) should be used. Smaller i.d. tubing (0.005 in. i.d., ~0.12 mm) can be used, but it is more prone to blockage. Extracolumn effects prior to the column are less important in gradient elution, because all peaks (with the possible exception of some early-eluting peaks) are compressed at the column inlet so as to minimize any precolumn peak broadening. Therefore, if extra connecting tubing is required, it is better to have the additional length before the column with a short run between the column and detector, because on-column concentration of the sample upon injection often will cancel out any precolumn peak broadening.

Data system sampling rate. The rate at which the data system collects data can influence the peak shape (Section 4.1.3.4). As a general rule, the data rate should be set – either automatically or manually – such that at least 10–20 data points are collected across a peak. This is a good compromise that gives high-quality data without too much noise. It is better to err on the side of collecting at too high a data rate – the data set can be simplified by post-run data averaging, but it is not possible to create additional data points once the run is completed at an insufficient data rate.

Most detectors have *built-in time constants* that act as electronic filters to smooth out excessive noise in the detector signal (Section 4.1.3.3). Most users ignore the detector time constant and use the default setting, relying on the data system sampling rate to control the amount of peak smoothing achieved.

5.4.5.3 Split Peaks

Peak splitting (doubling) or distortion can occur for one or all peaks in a chromatogram. When the splitting affects all peaks in the same manner, the most likely cause is a blocked in-line filter or column-inlet frit, or a void in the column. We strongly recommend using an 0.5 μm porosity in-line filter between the autosampler and the column to catch any particulate matter that might otherwise block the column frit. A blocked frit often is accompanied by a rise in the system pressure. If a blocked frit is suspected, the best approach is to successively replace the in-line filter (if used) and column with a new one. Alternatively, backflushing the column may correct the problem (consult the column care and use instructions to see if backflushing is allowed). Frit blockage can arise from particulates originating from the sample, the mobile phase, or wear of pump seals and/or injector seals. Use of HPLC-grade reagents and solvents plus mobile phase filtration (~0.5 μm) should eliminate mobile phase particulates. Pump seals continually wear and should be replaced at least yearly, or more often if historic data suggest this (Section 5.5.2.5). Injector manufacturers quote rotor seal lifetimes of 20,000 injections or more, which may represent several years for most users. Rotor seal failure is often accompanied by leaks. Samples are the most common source of particulates. Some users filter all the samples, but this is expensive and time-consuming. Sample filtration may require validation to ensure that sample is not adsorbed onto the filter or that nothing leaches from the filter into the sample. Physical sample-volume loss during filtration can be a problem for very small samples. Centrifugation of the samples in a benchtop centrifuge (e.g., $\geq 1500g$ for 5 min) removes most particulates, is faster than filtration, and can be done batch-wise.

Peak splitting or distortion also can be caused by injecting the sample in a solvent that is not compatible with the mobile phase (see following Section 5.4.5.4). Generally, injection solvent effects are more apparent for early-eluting peaks than ones that are well retained (Section 5.4.5.4).

5.4.5.4 Injection Conditions

For improved separation in gradient elution, it is desirable to inject a very small volume of a sufficiently small sample mass dissolved in the initial mobile phase; otherwise, peak distortion may result. In many routine separations, however, it is impractical or impossible to achieve both a small volume and matched injection solvent and still get sufficient sample mass onto the column. If the injection solvent is too strong, peak distortion can occur, as is shown in Figure 5.21 for injections of 4 μL of a caffeine solution in 33 percent (Fig. 5.21*a*), 66 percent (*b*),

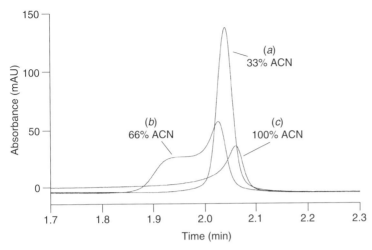

Figure 5.21 Elution profiles for caffeine dissolved in (*a*) 33 percent acetonitrile, (*b*) 66 percent acetonitrile, and (*c*) 100 percent acetonitrile. Caffeine at 0.75 mg/mL, 4 μL injection. Column: Luna 3 μm C_{18}-2, 50 × 2.0 mm, 1 mL/min. Gradient 5–95 percent acetonitrile (with 0.1 percent formic acid) in 7.5 min. Adapted from [37].

and 100 percent (*c*) ACN [37]. In each case, the injection solvent concentration is much greater than the initial gradient conditions (5 percent ACN–0.1 percent formic acid). It can be seen that the two higher injection-solvent concentrations produce significantly distorted peaks, whereas the peak for the 33 percent ACN injection is symmetric. So, although it is best to match the injection solvent with the starting mobile phase, some difference in solvent concentration can be tolerated.

In the example of Figure 5.22, the effect of injecting 1–4 μL of caffeine dissolved in 100 percent ACN is seen (same gradient conditions as in Fig. 5.21) [37]. Whereas a 4 μL injection in 100 percent ACN distorts the peak (same injection as Fig. 5.21*c*), reduction of the injection volume to 1 μL results in a symmetric peak. Here, too, although an injection solvent of 100 percent ACN is much stronger than the starting mobile phase, a sufficiently small volume can be tolerated with acceptable performance. As one might expect, the combination of injection solvent strength and injection volume is important. Additionally, peak distortion depends upon the position of the peaks in the gradient – earlier peaks are more affected than later-eluting ones (as is the case for the relatively early-eluting peak of Figs 5.20 and 5.21). This is illustrated in Figure 5.23 [37] for injection volumes of 1–10 μL of a sample mixture dissolved in 100 percent ACN. Larger volumes of strong injection solvent carry the effect later into the chromatogram. Thus, for 1 μL injection, only the first peak (thiourea, at 0.3 min) is distorted; with 2 μL, caffeine (at 2 min) is misshapen, and so forth for larger volumes and later peaks. The foregoing examples (Figs 5.21–5.23) are each based on the use of a low-volume (50 × 2 mm i.d.) column, which enhances the peak distortions shown. For the same injection volume, peak distortion will be less significant for a larger column (e.g., 150 × 4.6 mm i.d).

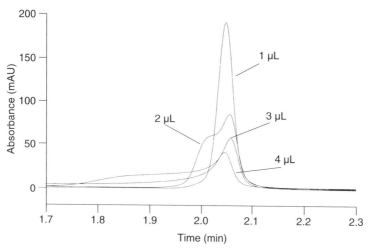

Figure 5.22 Effect of sample injection volume. Caffeine at 0.75 mg/mL in 100 percent acetonitrile. Other conditions as in Figure 5.21. Adapted from [37].

The examples of Figures 5.21–5.23 illustrate the problem of using too large a volume of an injection solvent that is too strong. The injection volume usually can be increased by using a sample solvent that is *weaker* than the starting mobile phase, so as to take advantage of on-column concentration (Section 7.3.3). In the weak injection solvent, solute molecules have a higher affinity for the stationary phase, so they concentrate in the stationary phase at the head of the column. Equation

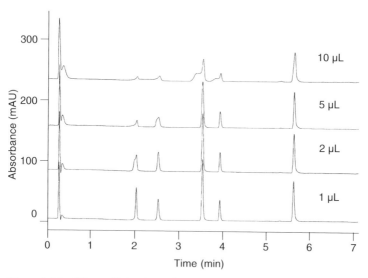

Figure 5.23 Effect of injection volume and analyte retention factor on peak distortion. Sample mixture of thiourea, caffeine, phenol, acetophenone, dimethylphthalate, and valerophenone (in retention order) in 100 percent acetonitrile. Other gradient conditions as in Figure 5.21. Adapted from [37].

(5.1) can be used to estimate how large a volume increase is allowed by the use of a more dilute injection solvent.

$$V_{\text{inj(new)}} \approx V_{\text{inj(norm)}}(k_{\text{new}} + 1)/(k + 1) \tag{5.1}$$

Here $V_{\text{inj(new)}}$ is the maximum allowed injection volume with a new (weaker) injection solvent, for no more than a 5 percent loss in resolution; $V_{\text{inj(norm)}}$ is the maximum allowed injection volume for a 5 percent loss in resolution using the original injection solvent; k_{new} is the isocratic k-value of the solute in the new injection solvent; and k is the k-value in the original mobile phase [38]. Equation (5.1) can be rearranged to

$$V_{\text{inj(new)}}/V_{\text{inj(norm)}} \approx (k_{\text{new}} + 1)/(k + 1) \tag{5.2}$$

which tells us that the factor by which the injection volume can be increased by the use of a weaker sample solvent (and larger value of k_{new}). We can use the *rule of three* (a 10 percent change in isocratic %B changes k by $\sim$3-fold; corresponds to $S \approx 5$), in order to estimate the effect of diluting the injection solvent. For example, if a solute had $k = 3$ under the initial conditions, it would elute early in the gradient and likely be subject to distortion from too large an injection or too strong a solvent, as discussed above. A reduction of the injection solvent concentration to 10 percent less than the initial mobile phase would give $k_{\text{new}} \approx 3 \times 3 = 9$. Equation (5.2) allows us to estimate that we could increase the injection volume 2.5-fold [(9 + 1)/(3 + 1)] with only a 5 percent loss in resolution under these conditions. *The use of the combination of simultaneously diluting the injection solvent and increasing the injection volume is a practical way to increase the mass of sample on column with little or no penalty in separation quality.*

5.4.5.5 Sample Decomposition

If a compound is not stable under the chromatographic conditions used for a method, decomposition can take place within the column. Sometimes this decomposition takes place at the head of the column, possibly catalyzed by the metal frit of the column, without further reaction of the sample during separation; as a result, the decomposed sample is chromatographed without further change in sample composition, yielding product and reactant peaks of normal appearance. However, if the rate of decomposition is slow, the sample may degrade while the sample transits through the column, resulting in a distorted peak. This is the result of two (or more) distinct molecular structures passing through the column, with the ratio of their concentrations changing as they chromatograph [39, 40]. Some molecules are converted from the parent compound to degradant at different places in the column, so the final chromatogram contains a distorted peak.

A good example of both fast and slow sample reaction is provided in Figure 5.24 [40, 41]. The gradient separation of tipredane (structure in Fig. 5.24) epimers is shown for different samples and conditions. In Figure 5.24(*a*), the *S*-epimer (ethylsulfone substitutent) was injected, and peaks for both the *R*- and *S*-epimers are observed in the chromatogram (i.e., reaction of *S*-epimer to *R*). Because the two peaks are sharp and well separated, the reaction of *R* to *S* must have occurred

194 CHAPTER 5 SEPARATION ARTIFACTS AND TROUBLESHOOTING

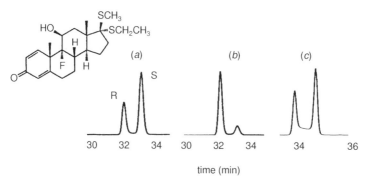

Figure 5.24 Separation of tipredane epimers. Conditions: (a, b) 100 × 4.6 mm Hypersil ODS column; 29–32–62 percent acetonitrile–pH 7.2 buffer at 0–10–20 min; 1.5 mL/min; 26°C; (c) 150 × 3.9 mm Resolve C_{18} column and similar, but not identical gradient conditions. (a) Injection of S-epimer; (b) injection of R-epimer; (c) injection of S-epimer, using a different column. Adapted from [40, 41].

prior to extensive elution through the column. The injection of the R-epimer (Fig. 5.24b) shows a similar, but reduced conversion to the alternate epimer; that is, the R-epimer seems to be the preferred species in an equilibrium mixture. The two separations of (a) and (b) were each carried out on a Hypersil ODS column. When the column was changed to Resolve C_{18}, the separation of Figure 5.24(c) was obtained. In this case, a characteristic "saddle" is observed between the two peaks, indicating that sample reaction has occurred more slowly *during* the separation, rather than primarily during sample injection.

If degradation is suspected, this can often be confirmed by changing the chromatographic conditions (temperature, pH, etc.) to speed or slow the rate of degradation. For example, increasing or decreasing temperature will usually speed or slow the rate of sample reaction, with a predictable effect on peak shape. For the sample of Figure 5.24, it was found that a higher temperature accelerates sample reaction, while a higher mobile phase pH slows the reaction.

5.5 TROUBLESHOOTING

The flow chart of Figure 5.25 may prove useful to help guide you through the troubleshooting process described in this section.

In a hurry? *If you have a problem to solve and want to get right to work, optionally read Section 5.5.1, then go right to Table 5.3 (at the end of the chapter) to classify the problem. Table 5.3 will lead you to more specific symptoms, causes, and solutions in Table 5.4, Table 5.5, and cross-references to specific sections in this chapter.*

A comprehensive guide to troubleshooting would be a book by itself [42], not to mention more than 250 installments of an HPLC troubleshooting-advice column by one of the authors [43]. A comparable treatment here is therefore beyond the scope of the present book. Instead we will present a general strategy for

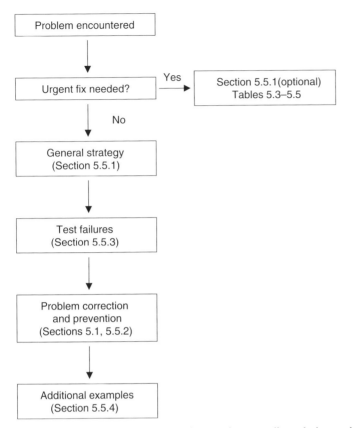

Figure 5.25 Guide for isolating and correcting a gradient elution problem.

troubleshooting problems common to gradient elution, and will not concentrate on more general problems (calibration, autosampler problems, detector problems, etc.) that are common to all HPLC modes. The topics in this section cover:

- problem isolation (Section 5.5.1);
- general practices for troubleshooting and preventive maintenance (Section 5.5.2);
- a detailed discussion of failure of the gradient performance checks of Section 4.3.1 (Section 5.5.3).

Finally (Section 5.5.4), we present several case studies that both highlight the troubleshooting process and give examples of problems not covered elsewhere in this chapter.

5.5.1 Problem Isolation

Two main questions arise when a problem occurs with an HPLC system. First, is the problem related to (a) the system hardware (pump, autosampler, detector, etc.), or

(b) the analytical method (instrument settings, mobile phase, column, sample preparation, etc.)? Second, how can we identify a specific failed component or other problem source? Two different, but complementary, approaches will help answer these questions:

- divide and conquer; and
- module substitution.

For the *divide-and-conquer* approach (Section 5.5.2.12), successive experiments should each eliminate a large portion of all possible problems, allowing one to quickly isolate the root cause of the problem. Whether the instrument or the method is responsible for the problem can be determined by running one or two tests. But before running any tests, a quick visual scan should be made of the system – are there any leaks, is the pressure too high or low, or is there anything obviously wrong? The source of these problems (Table 5.3) should be tracked down before going further.

Once any obvious problems have been eliminated, the system suitability test can be run for the method (Section 5.1.4.8). If system suitability passes, the hardware is capable of producing acceptable results, the hardware has been eliminated from further troubleshooting. Therefore, the problem must be specific to samples or with the implementation of the analytical procedure. Is there something unique about real samples vs the system suitability sample? Are the samples prepared properly? Are the standards at the right concentration and in the right injection solvent? Are the instrument settings correct? The written method procedure should be thoroughly checked to see that it has been followed properly.

If the system suitability test fails, the method or the instrument may be at fault. The next divide-and-conquer experiment (Table 5.3) is to determine if the instrument will work under ideal conditions. A new column is placed on the system and the column manufacturer's isocratic column test should be repeated. Results should be obtained that give retention times and peak widths that are within about 5 percent of those from the manufacturer's test chromatogram (a system, at its best, may never be able to exactly reproduce the column manufacturer's results obtained on a specialized column testing instrument). If the isocratic column test passes, gradient retention time and peak area reproducibility tests should be perfomed (Sections 4.3.2.3 and 4.3.2.4; Table 4.4). If all of these tests pass, the instrument is operating properly, so it must be the analytical method. If any of these tests do not pass, the problem source has to be the instrument, because the column test completely bypassed the method conditions. At this point, you should run the gradient performance check (Sections 4.3 and 5.5.3, Table 5.4). (All of these tests need to be run initially when the instrument and method are known to be operating properly, or there will be no reliable reference for comparison.)

Module substitution is the second technique to speed up problem isolation (Section 5.5.2.12), when the hardware is known to be at fault. Simply replace a suspect component (e.g., pump, column, mobile phase) with one known to be in good working condition. We do this without much thought when the column is suspect, but it can be a powerful tool to isolate problems throughout the system. As discussed in Section 4.2, the ability to have equivalent modules for this troubleshooting process is one argument for staying with one brand and model of HPLC system when multiple systems are purchased.

Finally, it can be assumed that a new problem with an HPLC system or method is the result of a single change that needs to be identified and corrected. This is not always the case (e.g., the example of Section 5.5.4.2), but it is a good place to start.

5.5.2 Troubleshooting and Maintenance Suggestions

This section contains a collection of tips and techniques for troubleshooting HPLC problems and for the maintenance of the HPLC system. These are individually cross-referenced throughout this chapter. Also see Section 5.1.4 for additional suggestions for the routine operation of gradient methods. Although troubleshooting and correcting problems are important, preventive maintenance also should be high on the HPLC operator's priority list. A review of the following topics will help identify preventive maintenance techniques, as well.

5.5.2.1 Removing Air from the Pump The internal parts of the HPLC pump and associated hardware have many small, often angular, passages that can trap air bubbles. Sometimes a sharp tap with a wooden or plastic object, such as a screwdriver handle, will help to dislodge bubbles. A system flush with a thoroughly degassed, low-viscosity, low-surface-tension solvent such as methanol will sometimes dissolve bubbles that resist displacement using other techniques. The use of degassed solvents on a routine basis will help to prevent accumulation of bubbles in the system, because the solvent will have an additional capacity for dissolved gas and will help to solubilize tiny bubbles before they become a problem. Every HPLC system will work more reliably if the mobile phase is degassed.

5.5.2.2 Solvent Siphon Test All HPLC systems will perform more reliably if the reservoirs are elevated relative to the pump, so that a slight siphon head-pressure helps to deliver mobile phase to the pump. To ensure a free flow of solvent to the mixer, it is important to check the solvent inlet-line frits occasionally. For both low- and high-pressure mixing systems, one should expect the reservoir to be able to deliver several times more solvent by siphon action than will be required by the pump. To test this, the solvent inlet line can be disconnected at the mixer (low-pressure mixing) or the pump inlet manifold (high-pressure mixing) and the solvent allowed to siphon through the tubing. A 10-fold excess of solvent is a good rule of thumb for adequate delivery. For example, if the typical operation of the system is 1 mL/min, at least 10 mL/min of solvent through the siphon should be expected. This will supply enough solvent that starvation of the pump will never be an issue. If the flow is lower than expected, a blocked solvent-inlet-line frit, a pinched inlet line, or a poorly vented reservoir should be checked for. A restricted solvent supply in low-pressure-mixing systems can cause mobile phase proportioning errors (Section 5.5.3.2).

5.5.2.3 Premixing to Improve Retention Reproducibility in Shallow Gradients Gradient slopes of <1 percent/min are often required for high molecular weight compounds, such as peptides and proteins, so that reasonable k^*-values can be obtained for these samples with large S-values (Section 6.1.1). In some

198 CHAPTER 5 SEPARATION ARTIFACTS AND TROUBLESHOOTING

cases, the HPLC equipment cannot generate shallow gradients with sufficient accuracy to obtain an acceptable separation. This was seen for a high-molecular-weight polystyrene standard (230 kDa) when using a gradient of 86–91 percent THF over 50 min with an 80 × 6.2 mm, 5 μm C_8 column [44, 45]. In a run to check the purity of this standard, the chromatogram of Figure 5.26(*a*) was obtained. At first glance, this separation looks very good, with the separation of several sample components. However, this sample was known to be a polystyrene with a very narrow molecular weight distribution, for which only a single peak was expected (the separation of individual oligomers in this sample is highly unlikely; Section 6.1.4.1). The gradient slope of this run is only 0.1 percent/min, which was generated by a low-pressure mixing system. Low-pressure mixers generate the gradient by mixing alternate pulses of the A and B solvents. Because mixing is never complete, a small residual variation in mobile phase composition (and strength) remains when the mobile phase reaches the column (see discussion of Figure 5.14 in Section 5.4.2). This is visualized in Figure 5.27 as the solid trace overlaid on the programmed gradient (dashed line). Thus, there is an oscillation of values of %B around the programmed gradient, throughout the separation. For this very shallow gradient, the value of %B

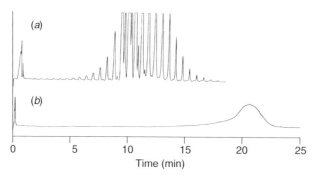

Figure 5.26 Chromatogram of a 230 kDa polystyrene sample. Mobile phase: (*a*) solvent A = 100 percent water and solvent B = 100 percent THF; (*b*) A = 86 percent THF–water and B = 91 percent THF–water; gradient 86–91 percent THF over 50 min. Adapted from [45].

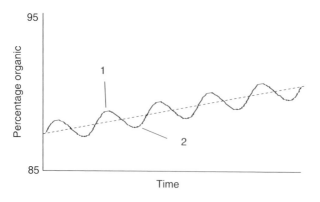

Figure 5.27 Theoretical shallow gradient profile (- - - -); profile actually observed (——). Mobile phase is stronger at 1 than at 2, even though point 1 occurs earlier in the gradient.

at point 1 is greater than at point 2, even though point 1 appears earlier in the gradient.

The gradient separation of the present polymer sample is expected to result in a fairly wide peak at elution. As this peak leaves the column, the small variations in %B, as shown in Figure 5.27, lead to corresponding large changes in k at elution (because of the large value of S for this high-molecular-weight sample). These oscillations in k, in turn, lead to segments of the sample band being either retained (large k) or released (small k) as the band leaves the column, giving the appearance as in Figure 5.26(a) of a separation of this single peak into several peaks. To correct this problem, the A solvent was premixed to 86 percent THF–water and the B solvent to 91 percent THF–water. Now instead of requiring the HPLC system to proportion an accurate slope of 0.1 percent/min, the effective program was 100 percent A to 100 percent B in 50 min, or 2 percent/min. This is within the normal performance specifications of a low-pressure gradient HPLC system. The resulting chromatogram under these conditions is seen in Figure 5.26(b). The sample now appears as a single broad peak, as expected for a high-molecular-weight polystyrene with a narrow molecular-weight distribution.

This example illustrates our ability to compensate for equipment limitations in mobile phase mixing by premixing the mobile phase. The requirement of generating a gradient of 0.1 percent/min was beyond the system capability when starting with pure A and B solvents (at least for this high-molecular-weight sample), but premixing allowed a reasonable gradient slope of 2 percent/min to be programmed to generate the actual 0.1 percent/min gradient. Another, less dramatic, example of the benefit of mobile phase premixing is presented in Section 5.5.4.2.

5.5.2.4 *Cleaning and Handling Check-Valves*
As an alternative to replacement of pump check valves with new parts, faulty check-valves often can be rejuvenated by sonication in alcohol. Note that the inlet and outlet check-valves are seldom interchangeable; they are not clearly marked, an identifying mark should be scribed on the check-valve body to indicate position and flow direction. Some check-valves will come apart when inverted, so it is wise to check for this prior to sonication or one may be surprised to find small parts in the sonicator after a cleaning attempt. We recommend placing each check-valve in a separate beaker with enough methanol or isopropanol to cover the check-valve, then sonicating for 5–10 min. In our experience, this will fix leaky check-valves most of the time, presumably by removing unwanted contaminants from the ball and/or seat of the seal. If a check-valve comes apart, the parts should be cleaned in alcohol, then carefully reassembled using forceps; contacting the internal valve parts with paper, cloth, or fingers should be avoided, because a small piece of fiber or a fingerprint can cause the valve seal to leak.

5.5.2.5 *Replacing Pump Seals and Pistons*
The replacement of pump piston seals is a simple, user-performed service operation; the pump operator's manual should be consulted for instructions for each specific pump. One begins by carefully removing the pump head, following which the old seal can be removed with a seal removal tool, if available. Alternatively, a brass wood screw can be used (operated

like a cork screw), taking care not to damage the pump head; then a new seal inserted. A seal that is compatible with the mobile phase must be used (some manufacturers make piston seals that are intended for use only with aqueous solvents). Clean the piston with a lint-free laboratory tissue and a few drops of alcohol. Stubborn deposits of buffer or seal residue can be removed by rubbing with a little toothpaste. The piston should be rinsed and inspected for scratches or chips. This is done easily by holding a pocket flashlight or laser pointer to the end of the sapphire piston so that it light-pipes (glows). Any scratches or chips will appear as dust or lines on the surface that cannot be wiped off. A broken piston will have a rough, irregular end. Replace a scratched, chipped, or broken piston. Prior to assembly, the piston and seal should be moistened with alcohol to lubricate it during reassembly. Many pumps have a seal-wash feature that will help extend pump seal lifetimes when mobile phases contain buffers or salts.

5.5.2.6 Leak Detection Mobile phase leaks may be obvious, or not. Drips, puddles, and leak alarms usually make location of the leak simple. When buffers are used, a white, crystalline deposit may show up on a slowly leaking fitting where no liquid is obvious. A simple leak detector for hard-to-find leaks can be made from a piece of thermal-printer paper (used by some electronic balance printers, bar code printers, etc.). A narrow, pointed strip of thermal paper is cut (e.g., 0.5 × 5 cm, pointed at one end) and the pointed end used to probe suspected fittings, seals, or other possible leak sources. The paper will turn black when it contacts organic solvent; this can be useful for locating leaks that are hard to detect by other means.

5.5.2.7 Repairing Fitting Leaks Correcting a leaky fitting may be as simple as tightening the fitting a quarter-turn to see if a leak can be stopped. If this does not fix the problem, the fitting should be disassembled, rinsed, and tried again or the ferrule replaced with a new one. A stainless steel fitting should never be over-tightened, because it can distort enough that the ferrule will "mushroom" out beyond the fitting threads, making the connection impossible to disassemble. For stubborn fitting leaks, PEEK (poly-ether-ether-ketone) ferrules often are superior to stainless steel because they deform sufficiently to seal with otherwise imperfect surfaces. When a leak is encountered with PEEK fittings and tubing, it is best to shut off the mobile phase flow, loosen the nut, reseat the tubing in the fitting body, and re-tighten the nut. Sometimes when a PEEK fitting is tightened with the flow on, the tube end can slip in the fitting, creating a small cavity at the tip of the tube, which in turn can cause unwanted band broadening (extracolumn effects, Section 5.4.5.2).

5.5.2.8 Cleaning Glassware Organic residues on "clean" glassware can be the source of ghost peaks in blank gradient runs (e.g., Section 5.5.4.3), so a thorough cleaning of the glassware is essential. Various techniques to clean glassware have been recommended [24]. To avoid inadvertent contamination of the mobile phase by glassware, only the cleanest possible glassware should be used and care should be taken that the cleaning process does not add contaminants to the glass surfaces. Extra rinsing (10 rinses with tap water followed by 10 rinses with deionized water) of glassware that had been washed with laboratory dishwashing detergent was found to be satisfactory in one study [46]. Other workers [47] were unable to remove detergent residues with multiple water washes or with dilute HCl, but this was suspected to be due to the "softness" of the water [24]. Still others avoid detergents altogether, preferring to rinse glassware used only for HPLC mobile phases (including reservoirs, pipettes, and graduated cylinders) with water and then a clean organic solvent [24].

5.5.2.9 For Best Results with TFA TFA is a widely used additive for gradient mobile phases. TFA is readily miscible, provides a low-pH mobile phase (0.1 percent TFA ≈ pH 1.9), acts as an ion-pairing reagent with biomolecules (Section 6.2.1.2), can be used at wavelengths <220 nm, and is sufficiently volatile to use with mass spectrometric or evaporative light scattering detectors. TFA is available in a highly purified form suitable for HPLC use, but degrades rapidly upon exposure to air. For best results, one should purchase HPLC-grade TFA (or equivalent spectral grade) in 1 mL ampoules and use the entire ampoule in a single use. TFA is available in larger containers (e.g., 25 mL) at a much lower cost per milliliter, but in the experience of one of the authors, it is impossible to prevent rapid degradation of the reagent once the bottle is opened, even when working in an inert atmosphere and carefully resealing the bottle. However, once mixed with water, resulting TFA solutions are fairly stable (e.g., 1 week). With UV detection at <220 nm, some drift with TFA–acetonitrile mobile phases may be observed (Section 5.4.1 and Fig. 5.13). To minimize baseline drift, add the same amount of TFA (e.g., 0.1 percent) to both the A and B solvents. If drift is still observed, add a little more TFA to the A solvent (e.g., 0.115 percent in A, 0.1 percent in B), in order to compensate for differences in TFA absorbance due to the solvent.

5.5.2.10 Improved Water Purity In the examples of Figure 5.17 (Section 5.4.3.1) and Figure 5.39 (Section 5.5.4.3), the problem of background contaminants from water and reagents is discussed. It is important to use the highest quality reagents in order to avoid unnecessary background peaks (Section 5.1.4.1). This usually means purchasing HPLC-grade reagents for all salts, buffers, and organic solvents. Most laboratories prepare their own HPLC-grade water with a water purification system, such as the Milli-Q system (Millipore). Such water purifiers combine physical filtration (≤ 0.2 μm), ion exchange, and carbon filters to remove organic contaminants. Sometimes ultraviolet photo-oxidation is carried out in order to kill bacteria and oxidize organic species [24].

A further cleanup of the water may be required for maximum removal of background peaks, for example, for applications such as stability-indicating pharmaceutical assays, for which peaks ≥ 0.05 percent of the parent peak area must be quantified. Cleanup can be performed by passing the water through a C_{18} HPLC column [34], but this is inconvenient. Another option for high-pressure mixing systems is to install a scrubber column between the A pump and the mixer [33, 34, 48]. The scrubber column traps organic materials before they reach the mixer and prevents them from entering the analytical column and producing background peaks. In one configuration [48], a 50 × 4.6 mm column was hand-packed with 40 μm

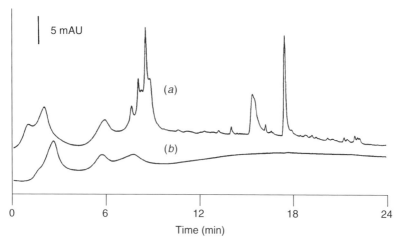

Figure 5.28 Comparison of baselines obtained (a) without and (b) with a scrubber column for cleanup of solvent A. Zorbax SB-C_{18}, 150 × 4.6 mm i.d. analytical column operated at 1 mL/min. Solvent A: 1 percent ACN–72 mM triethylamine phosphate (TEAP); solvent B: ACN. Gradient 5–5–80–80 percent B at 0–5–25–30 min. Backwash of scrubber column for 4 min between injections. Detection, 210 nm; ambient temperature. Adapted from [48].

C_{18} particles from a solid-phase-extraction (SPE) cleanup column. A six-port valve was used to enable flushing the scrubber column with strong solvent between each injection (but automated flushing is probably not required [33]). Figure 5.28 shows the ability of this setup to remove extraneous background peaks from a gradient run (especially peaks that appear later in the chromatogram). Cleanup was very reproducible, as shown in Figure 5.29 for five consecutive blank runs. Another approach [33] used a 10 × 10 mm semipreparative guard column hand-packed with C_{18} material from a used C_{18} analytical column. The guard column was placed between the A pump and the mixer. In this case, the capacity of the guard column was such that it required flushing only once every 3 weeks. The cleanup of contaminated mobile phase was dramatic (Fig. 5.30). A commercial C_{18} guard

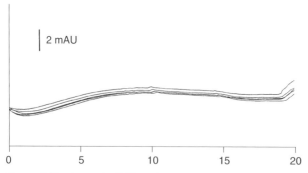

Figure 5.29 Reproducibility of baselines for five successive gradients. Same conditions as Figure 5.28, except detector scale as noted. Adapted from [48].

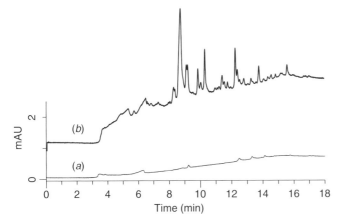

Figure 5.30 Blank gradient runs performed (*a*) with, and (*b*) without cleanup column between A pump and mixer. Column 150 × 4.6 mm C_{18}; 1.5 mL/min; 35°C; UV detection at 255 nm. Gradient: 0–83 percent ACN–water in 13 min with 5 min hold. Adapted from [33].

column could be used as an alternative, but would have a smaller volume and therefore a lower capacity for collecting contaminants, thus requiring more frequent replacement.

Other devices have been suggested for a final polishing of water prior to use. These include passing the water through a C_{18} SPE cartridge prior to use, using a low-back-pressure in-line C_{18} filter-cartridge before the mixer (low-pressure mixing) or A pump (high-pressure mixing) [24, 49], or pumping the water through a C_{18} column off-line prior to use [34].

5.5.2.11 Isolating Carryover Problems "Carryover" describes the repeated appearance of a peak in later chromatograms, when the sample is only injected in an initial run. That is, remnants of the sample remain after the first run and are somehow introduced into one or more subsequent blank runs. There are three main types of carryover:

- true carryover;
- carryover due to adsorption;
- incomplete elution.

Each of these carryover problems is described below, with some tips for distinguishing between them and correcting the problem.

True carryover is classic carryover, in which a small amount of sample is trapped in the sample-injection system and is unintentionally injected the next time the injector cycles. It is characterized by a constant percentage area of the carryover peak relative to the previous peak. For example, if an injection had an area of 100,000 area counts and 1 percent carryover was seen, a blank injection following a normal injection would have a peak 1 percent as large as the original (1000 area counts), but with the same retention time. An additional blank injection would have 1 percent of the 1 percent peak, or 10 area counts. This constant, serial dilution in subsequent blank injections characterizes "true" carryover. Because of the dilution effect, it is rare to have any carryover peak after two or three blank injections. Small, unintentional volumes are seen where sample might get trapped and then diluted out, such as in poorly assembled fittings. Ineffective autosampler flushing between samples also can be a problem source – the autosampler wash mechanism must be working properly.

Adsorptive carryover may appear to resemble true carryover, but it does not disappear as rapidly in subsequent injections. For example, the first blank may have 1 percent carryover from the original peak (e.g., 1000 area counts), but the second blank may also have a larger carryover than 1 percent of 1000 (e.g., $\gg$ 10 area counts). The source of such carryover is sample adsorption on surfaces within the system or column. Sample adsorption on the internal polymeric surfaces of the sample injector, the autosampler loop, and the inside of the injector needle are common sources of adsorption. Injection of a very hydrophobic sample dissolved in a polar solvent (e.g., water) is one common cause of adsorptive carryover. Addition of a few percent of organic to the injection solvent often will correct the problem. Also it may be useful to increase the strength and volume of the autosampler wash solvent or change the nature of the surfaces (e.g., replace a stainless steel loop with a PEEK one). Sometimes true carryover and adsorptive carryover can occur together. In such cases, a constant-fraction drop-off in peak size would be seen in the first and perhaps second blank injection, as the true carryover peak disappeared, but the adsorptive peak would persist for later injections.

Incomplete elution – with isocratic separation, if a sample is not fully eluted from the column during the run, it can elute in a following injection, but this is highly unlikely in gradient elution (Section 5.4.4.3).

5.5.2.12 Troubleshooting Rules of Thumb

The following list of rules of thumb can be a helpful guide in the process of troubleshooting:

- divide and conquer;
- easy vs powerful;
- change one thing at a time;
- only reproducible problems;
- module substitution;
- put it back.

Divide and conquer – this is the primary strategy of troubleshooting. Changes should be made that allow potential problems to be eliminated – the more, the better. A typical example is to run a new-column test to determine if a problem is related to the analytical method or the hardware. A new column can be installed and the manufacturer's column performance test repeated. If the same results (within $\sim$10%) are obtained as the column manufacturer, the HPLC system is working satisfactorily and the method is more likely to be the problem source. The column performance test checks isocratic performance; it may be necessary to supplement this with a gradient linearity or gradient-step test (Sections 4.3.1 and 5.5.3). See Section 5.5.1 for additional information.

Easy vs powerful – it is important to balance which tests are done first so as to make the best use of time. For example, if the problem is larger than normal retention times, a flow rate check is easy and fast. Although it may not be as effective at isolating a problem as making up a new batch of mobile phase, it may be chosen first for convenience and speed. Of course, common sense should lead you to focus on the most common problem areas, even if they are not as easy to troubleshoot.

Change one thing at a time – also called *the rule of one*, this reminds us to use the scientific method during troubleshooting. A change is made and the result evaluated. Sometimes it is faster to make several changes at a time, but this offers little insight into the real source of the problem, a knowledge of which can be used to (a) help design preventive maintenance procedures, or (b) solve similar problems in the future. The case study presented in Section 5.4.2.1 illustrates the problem of changing more than one thing at a time. The baseline of Figure 5.16(*a*) was improved compared with that seen in Figure 5.16(*b*) by making three *simultaneous* changes – unfortunately, there is no way to determine the individual effect of any of the changes.

Only reproducible problems – this is also called *The Rule of Two*, making sure the problem happens at least twice. Chromatographic problems that are not reproducible are difficult to troubleshoot and it is even more difficult to know that they have been corrected. The problem needing to be solved must be sufficiently reproducible for one to be confident that it has been corrected.

Module substitution – replacing a suspect part with a known good part, whether it is a column, check-valve, circuit board, detector, or other part, is one of the easiest and most powerful ways to isolate a problem. This strategy constitutes a good argument for having multiple copies of a given brand and model of HPLC system in a laboratory, so that there are more equivalent parts to interchange. Plenty of consumable items should always be kept on hand, such as filters, frits, guard columns, columns, tubing, and fittings so that they are available for substitution.

Put it back – this reminds one that if a known good part has been substituted for a suspect one and it does not correct the problem, the original part should be re-installed. This helps to avoid the accumulation of used parts of questionable quality. Of course, common sense should be used – it does not make sense to put the old seal back if replacing a pump seal did not solve the problem.

5.5.3 Gradient Performance Test Failures

In Section 4.3, a series of gradient test procedures and suggested acceptance criteria were described. The present section discusses reasons for the possible failure of these tests, and how to correct or (better) prevent the problem. *This section is organized in the same order as Section 4.3 and each heading is followed by a reference to the corresponding Chapter 4 section in parentheses. For example, Section 5.5.3.1 is titled "Linearity (4.3.1.1)," providing the cross reference to the Gradient Linearity test of Section 4.3.1.1.*

5.5.3.1 Linearity (4.3.1.1)

A direct test of gradient linearity is described in Section 4.3.1.1, based on a linear gradient from water (A) to water–acetone (B). Three different forms of gradient nonlinearity or deviations from ideal behavior can be recognized:

- rounding;
- nonlinear break;
- step-test failures.

Depending on the gradient method and the severity of the problem, gradient nonlinearity may or may not be a concern. In general, if a gradient method will be used only on a single instrument, linearity problems are of little consequence as long as the gradient profile is reproducible. If the method is to be used on more than one instrument, however, poor linearity can result in different gradient profiles for different instruments and thus generate irreproducible results.

Rounding – the gradient linearity test, as in Section 4.3.1.1, almost always will show some rounding at the ends of the gradient, as seen in Figure 4.4. Rounding is a result of the washout characteristics of the mobile phase mixer and the volume of the flow path between mixer and column inlet. For conventional conditions (0.5– 2 mL/min, 150 × 0.46 mm i.d. columns), minor rounding can be ignored. Severe rounding, where little or no linear region is observed in the middle of the gradient profile, is correspondingly more serious. Plumbing problems (all fittings between the mixer and the detector should be disconnected and checked) or running too steep a gradient (a shallower gradient should be used) may be the cause.

Nonlinear breaks – a gradient may show one or more offsets in the gradient profile (see example of Section 5.5.3.3 and Fig. 5.33). This generally is due to an equipment controller error or faulty proportioning valve and should be corrected before proceeding.

Step-test failures – the gradient step test of Section 4.3.1.3 may show irregular step sizes that are in excess of the system specifications; see Section 5.5.3.2.

5.5.3.2 Step Test (4.3.1.3)

As described in Section 4.1.1, the blend of mobile phase delivered to the column is controlled by relative pump speeds of the A and B pumps for a high-pressure mixing system or by the open–close sequence of the solvent proportioning valves in a low-pressure mixing system. Changes in the pump speed or proportioning valve timing must be made in a smooth and controlled manner in order to obtain a linear gradient. Each step of the step test demonstrates the accuracy of solvent blending for either system. *The following discussion uses case studies to illustrate possible causes of step-test failure and a related problem caused by pump starvation.*

In a first example, the operator was unable to obtain reproducible retention times [50]. A gradient step test and linearity test were run, with the obviously unacceptable results shown in Figure 5.31. In this case, the cause was suspected to be trapped air bubbles in the pumping system, because occasional pressure fluctuations were observed. Thorough purging of the system with degassed solvent accompanied by tapping on each component of the system with a screwdriver handle (to dislodge

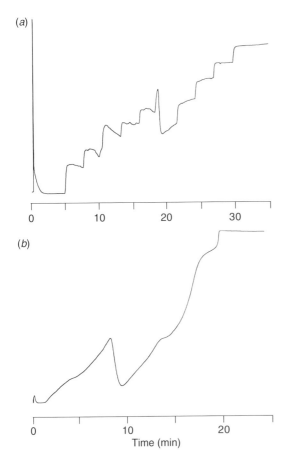

Figure 5.31 Results for an unacceptable (*a*) gradient step test, and (*b*) gradient linearity test. Adapted from [50].

bubbles adhering to internal surfaces) resulted in a series of bubbles in the waste stream. The tests were rerun and passed easily. (See Section 5.5.2.1 for hints on removing entrapped air.)

In a second example, the method worked well, with acceptable retention time, precision, accuracy, and resolution. However, when the gradient step test was run, the results of Figure 5.32 were obtained [50]. It can be seen that a small secondary step is located between each major step (note the small step marked by the arrow between the 10 and 20 percent B major steps). In the process of eliminating likely causes, the autosampler was replaced with a manual injector (module substitution, Section 5.51), at which time the problem disappeared. Replacement of two stainless steel frits within the autosampler corrected the problem. It was not clear why these blocked frits generated the secondary steps of Figure 5.32. In retrospect, it may be that the frits controlled flow through a flow bypass channel that is used in some autosampler designs to minimize pressure pulses to the column [51–53]. In such designs, part of the mobile phase flow bypasses the injection valve so that flow is not shut off when the injection valve is rotated (for additional information, see p. 238 of [42]). One of the authors has observed retention time and peak

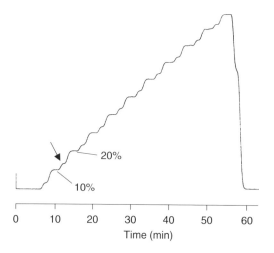

Figure 5.32 Unacceptable step test showing small secondary steps between each major step. Adapted from [50].

width problems when the flow through such a passage was disturbed, and one can appreciate that a disturbance in the gradient also is possible if such a partial blockage occurs.

The proportioning-valve system for low-pressure pumps has certain liabilities that can compromise system performance. If the seal on the proportioning valve leaks (due to a piece of particulate matter lodged in the seal), or if the solvent inlet-frit in the mobile phase reservoir is partly blocked, solvent proportioning can be compromised. For example, with the flow rate set at 1 mL/min, if the A solvent frit is partially blocked so as to prevent delivery of the 1 mL/min of A solvent required when the A valve opens, a slight vacuum will build in the mixing chamber. The A valve will then close and the B valve will open for the next step of the proportioning program. Because of the vacuum in the mixer, the B reservoir will deliver more than 1 mL/min to satisfy the vacuum, resulting in a mobile phase that is stronger in B than was requested by the controller. Solvent inlet-frit blockage can occur when unfiltered mobile phases are used, or when microbial growth takes place in the reservoir (e.g., when buffers are used past their expiration dates). When solvent proportioning problems are encountered, it is wise to check the condition of the inlet-line frits and run the solvent siphon test (Section 5.5.2.2). This can avoid the replacement of an expensive set of proportioning valves, when an inexpensive frit or blocked tube was the actual cause of the problem.

5.5.3.3 Gradient-Proportioning-Valve Test (4.3.1.4)

The GPV test of Section 4.3.1.4 tests various combinations of the proportioning valves used to control the mobile phase composition in low-pressure-mixing systems. One type of non-linearity that can be observed with low-pressure-mixing systems is shown in Figure 5.33. In this case [54], the proportioning-valve timing algorithm apparently changes at 25, 50, and 75 percent B; there is an error in the changeover at each of these points.

In the same set of experiments used to gather the data of Figure 5.33, the step-test results of the lower trace of Figure 5.34(a) were obtained. The acceptance criterion was that each step had to be within ±1 percent of the programmed value. All steps

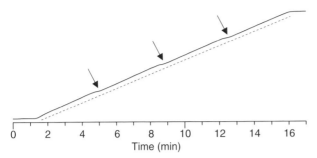

Figure 5.33 Plot of linear gradient with faulty proportioning valves. Arrows show deviations from linearity; dashed line drawn below plot for reference. Gradient 0–100 percent B in 15 min at 1 mL/min; A = water, B = 0.1 percent acetone in water; detection UV 265 nm. Adapted from [54].

passed this criterion, yet there was obviously something wrong with the gradient, as can be seen from the deviations from linearity in Figure 5.33. In order to more closely examine the gradient near these deviations, a step-test in 1 percent increments was run over the range 45–55 percent B, as shown in Figure 5.34(b). The arrow highlights the "short" step at the 50–51 percent B transition. The discovery of this error led to a more detailed examination of system performance. This low-pressure-mixing HPLC system was tested further using the gradient-proportioning-valve test. In this case, the inlet lines for solvents A and B were placed in the reservoir containing 100 percent water and the lines for C and D were placed in the water–0.1 percent acetone reservoir. A series of 2 min steps was run from 50:50 A:B to 90:10 A–C to 50:50 A:B to 90:10 A–D to 50:50 A:B to 90:10 B:C to 50:50 A:B, and so forth until all valve combinations were checked. The 50:50 A:B (100 percent water) serves as a baseline and each of the 90:10 steps should reach the same offset if the system is working properly. The results of this test are shown in Figure 5.35. To pass the test, the difference between the minimum and maximum plateau heights must be less than 5 percent of the average height. In this example, the maximum difference (between A/C and A/D) was 12.4 percent.

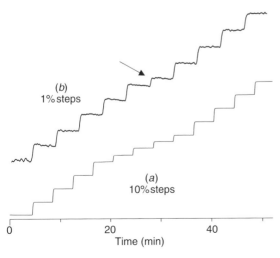

Figure 5.34 Gradient step test results for HPLC system of Figure 5.33. (a) Steps of 0, 10, 20, 30, 40, 45, 50, 55, 60, 70, 80, 90, and 100 percent B; (b) 45–55 percent in 1 percent steps. Arrow shows "short" step between 50 and 51 percent B. Adapted from [54].

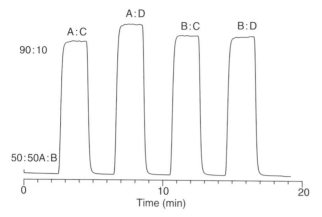

Figure 5.35 Gradient proportioning valve test results for conditions of Figure 5.33. Baseline is generated by 50:50 A:B; the remaining plateaus are 90:10 A:C, A:D, B:C, and B:D from left to right. A = B = water; C = D = 0.1 percent acetone in water. See text for details. Adapted from [54].

The latter HPLC system has a procedure to re-calibrate the proportioning valves based on the test results, but several tries gave little improvement in the test results. Several other possible problem sources were examined. A partially blocked inlet-line frit in the reservoir could restrict flow from one of the reservoirs, and cause insufficient delivery on one solvent to the proportioning manifold. The siphon test (Section 5.5.2.2) showed that all lines and frits were operating properly. Although less likely, air bubbles in the pump, poorly operating check-valves, or failed pump seals might contribute to the problem. To check these items, the mobile phase was degassed again, the check-valves were sonicated to ensure that they were clean, and the pump-seal replacement log was checked (the seals were nearly new). After it was determined that none of these simple fixes corrected the problem, the proportioning valve assembly was replaced and the step-test result was 0.9 percent variation, easily passing the 5 percent limit.

5.5.3.4 *Flow Rate (4.3.2.1)*

The flow rate test (Section 4.3.2.1) checks the flow rate over an extended period (e.g., 10 min), but may not identify short-term flow rate problems. Except for the case of a faulty pump setting or controller problem, the measured flow rate should never be greater than the set value. There are several possible causes for lower-than-expected flow rates discussed below:

- bubbles;
- check-valve or pump-seal problems;
- leaks;
- compressibility compensation errors.

Bubbles – one of the most common problems with both isocratic and gradient methods is the presence of bubbles in the mobile phase. This problem also is one of the easiest to eliminate – the mobile phase is simply degassed. Bubbles are particularly problematic under gradient conditions, because when reversed-phase solvents are blended, the solubility of air in the mixture is less than in the pure starting solvents. With insufficient degassing, the result is solvent outgassing during the gradient run. Bubble problems are more common with low-pressure mixing than high-pressure mixing, because in low-pressure mixing the solvents are mixed at atmospheric pressure and outgassing can occur in the mixer or the pump during the reduced pressure of the fill stroke of the piston. Pump pressure fluctuations are a common result. Sometimes a loose fitting or other leak source on the low-pressure side of the pump can be small enough to allow air to leak in without allowing liquid to leak out. If this is suspected, it should be ensured that all the low-pressure connections are tight.

With high-pressure mixing, the elevated pressure in the mixer usually keeps excess gas in solution. Also, the mixing takes place after the pumps, so pump problems due to bubbles are less common than with low-pressure mixing. Outgassing can occur at the detector when the system pressure is reduced to nearly atmospheric pressure. A pressure restrictor (e.g., 50–75 psi) after the detector often is sufficient to prevent outgassing in the detector for high-pressure mixing systems. In any event, more reliable HPLC system operation is observed with all equipment designs if the mobile phase components are degassed prior to use.

Check-valve problems – reliable check-valve operation is essential for smooth flow rates, constant pressure, and reproducible retention times. Pump-pressure fluctuations are one of the most common symptoms of faulty check-valves. Cleaning by sonication (Section 5.5.2.4) or replacement of the check-valves will usually correct such problems. Retention time changes also can be symptomatic of check-valve problems.

Pump seal problems – the piston pump seals will generally last 6–12 months. High-salt (e.g., >100 mM) mobile phases tend to increase pump seal wear by abrasion from the salts that evaporate behind the pump seal. If check-valve cleaning and/or replacement does not correct a flow rate problem, pump seal replacement is a good idea, especially if the seals have several months of use on them. Some seal replacement tips are presented in Section 5.5.2.5.

Leaks – leaks are an obvious reason why the measured flow rate might be less than expected. Some leaks are easy to find, such as when a fitting drips, but others can be very small and hard to detect; see Sections 5.5.2.6 and 5.5.2.7 for tips on locating and eliminating mobile phase leaks.

Compressibility-compensation errors – most HPLC systems have a mechanism to correct for differences in compressibility between different solvents (Section 4.1.2.4). The simplest of these merely increases the flow setting slightly, in order to compensate for the compressibility of the organic solvent. More sophisticated designs are able to correct for changes in mobile phase compressibility during the gradient as the mobile phase composition changes. If the compressibility setting is not correct, the delivered flow rate may not be what was selected at the system controller. For example, if the B pump of a high-pressure mixing system is adjusted to compensate for compressibility of methanol, measurement of the flow with water will result in a higher measured flow rate than selected. Many HPLC systems have adjustments for compressibility that can be made by the operator – the pump operator's manual should be checked for specific instructions. Before making compressibility adjustments, the flow rate for each pump should be checked under typical operating conditions (same pressure and solvent). From a practical standpoint, compressibility compensation errors are unlikely to cause chromatographic problems as long as a given analytical procedure is used on a single instrument, but small chromatographic differences may be seen when transferring the method to another instrument.

5.5.3.5 Pressure Bleed-Down (4.3.2.2) The pressure bleed-down test (Section 4.3.2.2) is performed by blocking the pump outlet, turning on the flow, allowing the pressure to rise to a high value (e.g., 5000 psi), and observing the pressure stability when the flow is turned off. A pressure decay of >15 percent in 10 min is indicative of a leak. First check for fitting leaks (Section 5.5.2.6), which usually will be accompanied by an obvious drip, puddle, or leak alarm. The fitting should be tightened or replaced to correct for leaks. If a leaky connection is not found, one or more check-valves is the likely problem. The suspect check-valve(s) can be sonicated for a few minutes in alcohol or replaced with new check-valve(s). (See Section 5.5.2.4 for tips on cleaning and handling check-valves.) The pump seal(s) should be replaced if the pressure bleed-down test fails after cleaning or replacing the check-valves, or if the system log book indicates that replacement is due (generally every 6–12 months for most applications). (See Section 5.5.2.5 for tips on replacing pump seals and pistons.)

5.5.3.6 Retention Reproducibility (4.3.2.3) The retention-time reproducibility test of Section 4.3.2.3 is a holistic test of gradient system performance – all the system components and test method conditions must be working together properly for the retention times to be reproducible. Generally retention times should be consistent within $\sim \pm 0.05$ min ($n = 6$, 1 SD) and/or compare favorably with historic data for the same conditions. Bubbles in the pump are the first suspect for retention variability, in which case one should degas the mobile phase and re-run the retention reproducibility test. Leaky check-valves or pump seals (Sections 5.5.2.4 and 5.5.2.5) are another common cause of retention irreproducibility. Finally, mobile phase proportioning (Sections 5.5.3.2 and 5.5.3.3) should be tested if the problem still persists. Retention reproducibility problems during system testing are usually accompanied by failure or near-failure of one of the other system tests (see discussion of other tests in Section 5.5.3).

5.5.3.7 Peak Area Reproducibility (4.3.2.4) The peak area reproducibility test (Section 4.3.2.4) primarily is a test of autosampler performance and assumes that the signal-to-noise ratio (S/N) is large (e.g., >100). Most autosamplers should give peak-area precision of ± 0.5 percent ($n = 6$, 1 SD) or better for injections of ≥ 5 μL. If larger imprecision is observed, a check should be made to be sure the same sample is used for all injections, the vial is properly vented, and the sample is homogeneous. Other problem sources can be a syringe fill-rate that is too fast (causing outgassing in the needle), air bubbles in the syringe or sample (causing irreproducible sample volumes), mechanical problems with the autosampler, or leaks in the connections or injector seal(s).

When the S/N is small (e.g., <10), peak-area precision can be limited by the S/N. The contribution of S/N to the coefficient of variation (CV) can be estimated [16, p. 71] as:

$$CV \approx 50/(S/N) \qquad (5.3)$$

Thus, a peak with S/N of 10 would have CV $\approx$ 5 percent. In this case, the autosampler precision would not play a major role in the overall precision. To test autosampler precision, a large S/N ratio (e.g., >100) is required.

5.5.4 Troubleshooting Case Studies

Many specific problems and their causes are described elsewhere in this chapter, but in this section we will illustrate the troubleshooting *process* with a few specific case studies. These case studies also provide additional examples of gradient problems and their solutions, using Tables 5.3–5.5 as a troubleshooting guide. With a new problem, it is easy to follow a step-wise troubleshooting strategy, such as was described in Section 5.5.1. With historic problems, such as those included below, the process may flow differently – as is often the case in real life. Nevertheless, the same troubleshooting principles apply, as we will attempt to show.

5.5.4.1 Retention Variation – Case Study 1 This first case involves retention-time variation between two consecutive runs of a system-suitability standard, on a reversed-phase gradient with a low-pressure-mixing system (Fig. 5.36). System suitability requirements for the method allowed a retention variation of no more than 0.1 min between runs. For the first and last peaks, the difference was slightly over this limit, and the variation of >0.4 min for the peak in the middle

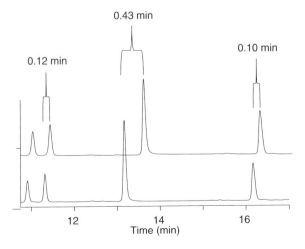

Figure 5.36 Overlay of chromatograms from two consecutive gradient runs showing larger errors for peaks near the gradient midpoint (13 min). Adapted from [55].

was excessive [55]. Additional runs (not shown) gave similar results, with the first and last peaks close to the limits and the middle peak varying by ±0.3 min from the average retention. The retention of all four peaks in a given run showed the same direction of shift (increase or decrease), but the direction of retention variation from run to run was random.

We should start at the top of troubleshooting guide of Table 5.3 and work our way down the table, eliminating or identifying symptoms that correspond to those we observe. There were no pressure problems or leaks, so the problem does not fall in the "visual checks" section of Table 5.3. The system suitability test failed (retention variation was too large), so Table 5.3 suggests a new-column test as the next step. The new-column test was not run, so we do not have the benefit of this information. The next item in Table 5.3 is a gradient performance check (Sections 4.3.1.3 and 5.5.3.2), which was conducted. We are directed to Table 5.4 for evaluation of the (failed) step-test results. Several possible causes of step-test failure are listed in Table 5.4 (discussed in more detail above), and we are directed to Section 5.5.3.2 for help on correcting the problem.

A full-range 0–100 percent step-test was run and the irregularities near the middle of the range were immediately apparent. The middle segments of the gradient step test are shown in Figure 5.37 [55] for the 5 percent steps from 40 to 60 percent B. All of the steps deviate from the nominal values by >1 percent, a greater change than is acceptable. Furthermore, notice that the steps for %B <50 percent are all below the nominal value, whereas those ≥50 percent are all higher than expected, resulting in a compositional change between 45 and 50 percent of not 5 percent, but 8.4 percent. This problem step occurs very near the mobile phase composition at which the 13 min peak of Figure 5.36 elutes.

Additional problems are seen with the shape of the 45–50 percent step, and, to a lesser extent, other steps in Figure 5.37. Irregular step shape can be a result of faulty check valves, bubbles in the pump, or other short-term discontinuities in flow. A well-performing test would have all steps shaped like those of Figure 4.5. The list of possible causes of step-test failures in Table 5.4 includes bubble and check-valve problems, so the mobile phase was degassed (Section 5.5.2.1) and the check valves were cleaned by sonication (Section 5.5.2.4). These actions corrected the step-shape problems (not shown), but the step size was still in error.

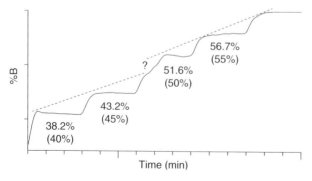

Figure 5.37 Results of proportioning step test performed near the midpoint of the gradient used for Figure 5.36. Theoretical values shown in parentheses. Dashed line shows approximate gradient slope (see text). Adapted from [55].

To highlight the step-size problem, the slope of the step-gradient is estimated with the dashed lines of Figure 5.37. The slope below the 50 percent point is approximately the same as that above 50 percent, but there is an offset between the two ("?" in Fig. 5.37). This discontinuity is at the 50 percent point, where there is often a change in the algorithm controlling the proportioning valve (see similar problem in Fig. 5.33 and discussion in Section 5.5.3.3). In this HPLC system, the proportioning-valve algorithm could be adjusted by the user; the GPV test of Sections 4.3.1.4 and 5.5.3.3 would have added valuable information, but this was not run. Recalibration of the mobile-phase proportioning valves corrected the step-size problem. Although it is a good first assumption that a chromatographic problem is the result of a single root cause, in this case it appears that proportioning-valve calibration *plus* bubbles and/or check-valve problems all contributed to the poor retention reproducibility.

5.5.4.2 Retention Variation – Case Study 2

In this case, a peptide separation had been developed on a high-pressure-mixing system with a C_{18} column using a 19–24 percent ACN–0.1 percent TFA gradient in 30 min [56]. The 15–20 min portion of the chromatograms for three consecutive runs are overlaid in Figure 5.38(*a*) (one peak for each run). It can be seen that the peptide peak retention varies by 2.1 min between three runs. For the same three runs, retention of the t_0 peak varied by only 0.014 min (not shown). The system suitability specification called for retention-time reproducibility of <0.2 min range for the injection of three consecutive standards. When the method was used a month earlier, it ran without problems.

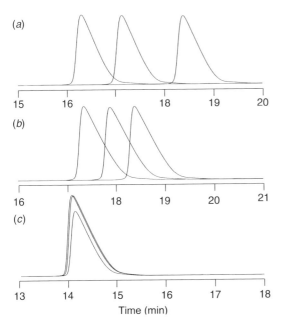

Figure 5.38 Expanded chromatograms from three consecutive injections of a peptide standard. Chromatograms generated (*a*) using the original system configuration (2.1 min retention range), (*b*) after replacing all check valves and pump seals (1.0 min range), and (*c*) using premixed mobile phase (<0.1 min range). Column: 250 × 4.6 mm, 5 μm C_{18} operated at 1.5 mL/min and 35°C with detection at 215 nm. Gradient: 19–24 percent ACN : 0.1 percent TFA in water over 30 min. Adapted from [56].

Table 5.3 can be used as a guide for solving this problem. The system suitability test failed, so the table suggests running a new column test to determine if the problem is related to the method or the hardware. The new-column test was not run, but it is likely that it would have passed (based on data presented below). Next, Table 5.3 suggests that the gradient performance tests be run; these were not run, either, but we suspect they would have passed, too. The last item in Table 5.3 directs us to look over the symptoms in Tables 5.4 and 5.5 to see if we can find one or more that correspond with our observations (this is the fall-back position if other tests are not run – just look in the tables for symptoms that match yours). In Table 5.4, under the retention reproducibility symptom (the problem in Fig. 5.38), the possible causes include bubbles, check valves, pump seals, and mobile phase proportioning problems. In Table 5.5, under retention time problems, retention variation between runs corresponds to our problem. The causes include the need to premix solvents as well as the problems listed from Table 5.4.

The common potential problem areas from Tables 5.4 and 5.5 are bubbles, check valve or pump seal problems, and mobile phase proportioning. Bubbles can be eliminated from consideration, because all mobile phases were thoroughly degassed. We will look at the remaining potential causes in light of the reported data.

With high-pressure mixing systems, a flow-rate problem directly affects retention, as well as mobile phase composition. A recent system check (Section 4.3) had passed the flow rate check within the manufacturer's specifications of ± 2 percent accuracy [56]. The standard flow rate check (Section 4.3.2.1) extends over 10 min, so it averages the flow rate over the collection time. Short-term flow rate problems could exist (and cause mobile phase proportioning irregularities), masked by the 10 min collection period. A check of shorter-term flow-rate performance can be made by examining the retention of the peak at t_0. Since the t_0 peak is unretained, it should be directly related to flow-rate variation with little or no influence from retention by the stationary phase. The retention time range of the t_0 peaks was 0.014 min, which exceeded the manufacturer's specifications of ± 0.3 percent precision for the pump flow rate (1.5 mL/min $\times \pm 0.3$ percent $= \pm 0.0045$ min). At this point, the rule of one (Section 5.5.2.12) was ignored and all eight check valves and four piston seals were replaced (two 2-headed pumps). A rerun of the system suitability check showed that the retention time variation had dropped by half ($\sim$1.0 min range, Fig. 5.38b), but was still greater than the acceptance criteria of <0.2 min. It appeared at this point (with replacement of valves and seals) that the pumps were working properly. This eliminates check valves and pump seals from our list of possible problem sources, leaving mobile phase proportioning as the primary suspect.

The method used a very shallow gradient, which covered a range of 5 percent B in 30 min, or $\sim$0.17 percent/min. This HPLC system had a specification for mobile phase proportioning accuracy of ± 1 percent. Under "normal" gradient conditions (e.g., as in Table 3.2), a gradient slope of $\sim$6 percent/min is typical; for such conditions, ± 1 percent proportioning accuracy is satisfactory, but it is inadequate for gradients of $\ll 1$ percent/min. A way around this problem is to premix the mobile phase (Section 5.5.2.3). The A solvent was changed from 0.1 percent TFA in water to 10:90 ACN:0.1 percent TFA–water; B was changed from 100 percent ACN to 30:70 ACN–0.1 percent TFA–water. A controller setting of 40–65 percent B in 30 min gave an effective gradient of 18–23 percent ACN:0.1 percent TFA in 30 min (vs 19–24 percent ACN in 30 min originally). This new gradient rate was ~0.83 percent/min, a 5-fold steeper gradient (in terms of the controller settings) – much more in line with the capability of the HPLC system. The results are shown in Figure 5.38c, where the retention range is <0.1 min, well within the system suitability limits of <0.2 min. (From the data presented in [56], we wonder how the validation initially passed without premixing.) An added benefit of premixing the solvents was a significant reduction in the baseline noise [56], lowering the detection limit by improving the signal-to-noise ratio. Premixing mobile phases to enhance the precision and accuracy of gradient formation is a simple technique that can be used whenever shallow (e.g., <1 percent/min) gradients are needed, as with large-molecule separations (Chapter 6). A more dramatic example of the effect of premixing mobile phases is presented in Section 5.5.2.3.

5.5.4.3 Contaminated Reagents – Case Study 3 In this example [46], numerous peaks were discovered when a blank gradient (Fig. 5.39a) was run as part of the method development process (Section 5.1.3) for a method using a C_{18} column and an ACN–buffer gradient. There were no leaks or pressure problems, and the system suitability test had not yet been developed, so the troubleshooting guide of Table 5.3 might seem of little use at first. However, we can use the "Other" category to guide us to Tables 5.4 and 5.5 to consider other symptoms. In Table 5.5, one of the baseline problems is peaks in a blank gradient. We are referred to Section 5.4.3, which tells us how to confirm contaminated water or reagents by extending the equilibration time. The results of this experiment are shown in Figure 5.39 for (a) a blank gradient with 10 min equilibration, and (b) a 30 min equilibration (note that the y-axis scale is in units of 10^{-3} AU). As in the example of Figure 5.17 (Section 5.4.3.1), an increase in peak size that is approximately proportional to the increase in equilibration time points us to contaminated mobile phase, usually the aqueous phase. HPLC-grade water should have no more than 1 mAU of extraneous peaks, and typically has much lower background – there are at least five peaks in Figure 5.19(a) that exceed this threshold. Contaminated buffer was one possible problem source, and its contribution was confirmed by running a blank gradient with water instead of buffer for the A solvent (compare the bottom trace in Fig. 5.40 with the buffer chromatograms). Buffer normally was prepared by blending 10 mM monobasic potassium phosphate with 10 mM dibasic potassium phosphate to achieve a buffer of pH 7.0, as measured with a pH meter. In an effort to find better quality buffer, mono- and dibasic phosphate salts were purchased from four different vendors and tested. The results of blank gradients using these different buffer sources are shown in the upper four traces of Figure 5.40. There are many more peaks in the buffer runs than in the chromatogram using water as the A solvent, so the problem is related to the buffer. However, since most of the peaks appear in all of the buffers, the primary cause does not appear to be a difference in contaminants between buffers from different suppliers. Because much cleaner baselines with buffer had been observed previously, it was

218 CHAPTER 5 SEPARATION ARTIFACTS AND TROUBLESHOOTING

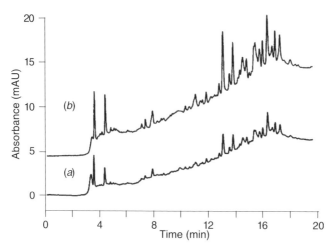

Figure 5.39 Blank gradient runs after (a) 10 min and (b) 30 min equilibration. C_{18} column; gradient 5–80 percent ACN–10 mM phosphate buffer (pH 7) in 15 min plus 5 min hold at 80 percent; UV detection at 215 nm. Adapted from [46].

suspected that contaminants had been introduced during the buffer preparation process. This suggests that the buffer itself may not be entirely responsible for the problem.

At this point, the divide-and-conquer strategy (Section 5.5.2.12) was used to break the potential problem sources into logical units that could be isolated. Several steps that had potential for contaminating the solvents were involved in preparing the buffer solutions, including glassware contact, microfiltration, pH adjustment,

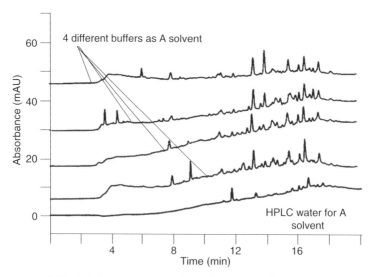

Figure 5.40 Blank gradients from four phosphate buffers (upper traces) and HPLC-grade water (lowest trace). All runs include 30 min equilibration. All other conditions as in Figure 5.39. Adapted from [46].

and degassing. To check each of these sources, buffer was replaced with water to simplify the troubleshooting process. Glassware contamination was checked by swirling HPLC-grade water in six 600 mL beakers to extract any surface contaminants. Filtered water was prepared by passing it through three separate 0.45 μm membrane filters. The pH adjustment step was checked by placing a stir bar in a beaker of HPLC-grade water and stirring it for 10 min with the pH meter probe immersed in the water. Degassing by helium sparging was used only for the degassing check. The chromatograms are shown in Figure 5.41 for each test, where exposure to the other sources was minimized (*a*, glassware) or eliminated (*b*, filter; *c*, pH; *d*, degassing). (These tests were run only once and should be compared only semiquantitatively.) There are peaks in each trace in the 14–18 min region of the chromatograms that are in common with the glassware contact test, and glassware exposure is the only exposure that was common to all four tests. This suggests that dirty glassware might be a source of at least part of the contamination. Normally, glassware was washed by hand with a commercial laboratory dish soap, rinsed six times with tap water, then six times with deionized water. Most glassware was baked dry in a laboratory oven. Modification of this procedure to add an additional six rinses with deionized water eliminated most of the spurious peaks in the 14–18 min region.

Once the glassware contaminants were minimized, the pH probe appeared to be the major remaining source of contamination (peaks in the 13–14 min region of Fig. 5.41*c*). The buffer preparation procedure was altered so that an aliquot of buffer was removed from the bulk solution and checked for pH, then discarded. This technique eliminated contact between the pH probe and the bulk buffer

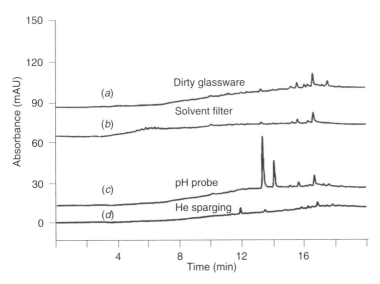

Figure 5.41 Chromatograms from suspected contaminated sources: (*a*) dirty glassware; (*b*) solvent filter; (*c*) pH probe; and (*d*) helium sparging apparatus. All other conditions as in Figure 5.39. Adapted from [46].

solution. Use of the extra-clean glassware and external adjustment of the pH resulted in a buffer that gave the acceptable blank gradient shown in Figure 5.42(*b*) (Fig. 5.42*a* is run under the same contditions as the original run of Fig. 5.39*a*). Additional experiments [46] identified the pH electrode filling solution (KCl) as the primary source of contaminants.

This case study is a good example of the systematic, step-wise checking and elimination of suspected problem sources that is often involved in troubleshooting a problem. A similar approach could be taken with other examples of this kind to identify the source of any spurious peaks in blank gradients. This example also suggests a general recommendation: avoid buffer contact with the pH meter probe when minimum baseline noise is required.

5.5.4.4 Baseline and Retention Problems – Case Study 4

Baseline problems that were encountered with a synthetic peptide sample were discussed in Section 5.4.2.1. Those problems were minimized by premixing the mobile phase components placed in the A and B reservoirs. The resulting method was a reversed-phase gradient of 5–80 percent ACN in 75 min, where A was 5:95 ACN:0.115 percent TFA–water and B was 80:20 ACN:0.1 percent TFA–water. The same sample exhibited the retention time reproducibility problems summarized in Table 5.2 [32]. The first two rows of Table 5.2 show the maximum retention time differences among three injections of the same sample containing six peaks run, in each of two batches of three runs. System suitability for the method allowed retention time variations of ≤ 0.1 min. It can be seen that some peaks in the two batches meet this criterion, but several of the peaks show

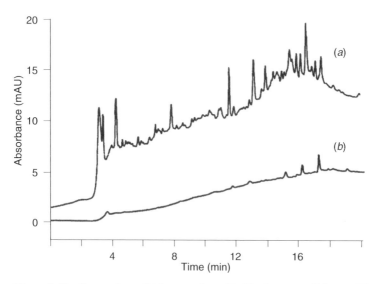

Figure 5.42 Comparison of (*a*) contaminated buffer (same conditions as Fig. 5.39*a*), and (*b*) buffer prepared with extra-clean glassware and no exposure to the pH probe. All other conditions as in Figure 5.39. Adapted from [46].

>0.5 min shifts in retention. Furthermore, the retention shifts were not consistent for a particular peak in both batches (e.g., peak 2 is bad in batch 1, but OK in batch 2).

At this point, the troubleshooting guide of Table 5.3 can be used to help isolate the problem. System suitability failed, so Table 5.3 suggests that a new-column test should be run. Instead of a new-column test, a batch of three runs containing six low-molecular-weight test probes was run with the same method, because the higher-molecular-weight peptide sample might be more sensitive to small variations in mobile phase %B (Section 6.1). This resulted in the data on the third line of Table 5.2, showing (barely) acceptable retention time variation for the low-molecular-weight samples, and confirming a possible problem with gradient reproducibility that is specific to high-molecular-weight samples. The additional demands for proportioning accuracy placed upon the hardware when shallow gradients are used to separate high-molecular-weight samples (see the example of Section 5.5.4.2), suggest that the problem may be related more to the hardware rather than the method itself. In this case, we are directed (Table 5.3, new column test fails) to run the gradient performance tests.

To check proportioning, a gradient step test (Section 4.3.1.3) was run. Irregular steps were observed in the 30–40 percent B region (not shown), so the step test was repeated in 1 percent steps in the 30–40 percent B range with the result shown in Figure 5.43. The regular nature of the step test was interrupted by negative spikes (arrows in Fig. 5.43), suggesting air bubbles. However, the mobile phase had been thoroughly degassed prior to the step test, so outgassing problems were unlikely. One or more faulty check valves could cause intermittent misproportioning of mobile phase, which would then result in small shifts in the baseline (or dips) and changes in retention due to short-term variation in mobile phase composition. All four check-valves in this two-pump high-pressure mixing system were replaced and the retention time variation stabilized to acceptable levels (≤ 0.1 min), as seen in the last line of Table 5.2.

This problem also illustrates the increased sensitivity of some compounds to problems with the gradient program. Peptides have significantly larger molecular

TABLE 5.2 Retention Time Reproducibility for a Peptide Sample [55]

		Retention-time range (min) for peak					
Run[a]	Sample	1	2	3	4	5	6
1	Six peptide polymers[b]	0.00	0.58	0.39	0.56	0.35	0.29
2	Six peptide polymers[b]	0.01	0.09	0.18	1.01	0.61	0.43
3	Six low-molecular-weight test probes[c]	0.02	0.02	0.05	0.08	0.10	0.07
4	Same as 1 and 2, after check valve replacement	0.02	0.02	0.05	0.05	0.04	0.05

[a] Manual injection for all runs.
[b] Duplicate sets of three runs each.
[c] Range for three consecutive runs.

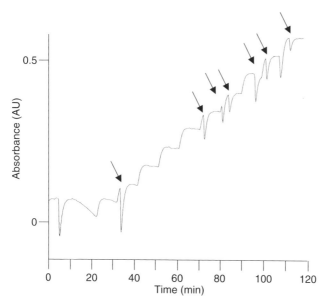

Figure 5.43 Gradient step test results for diagnosis of method of Section 5.5.4.4. Steps of 1 percent from 30 to 40 percent methanol spiked with acetone. Arrows highlight suspected bubble problems. See text for details. Adapted from [32].

weights than the small-molecular-weight test probes of Table 5.2. Higher-molecular-weight compounds in turn have larger S-values (Section 6.1.1) and are more sensitive to small changes in mobile phase composition than lower-molecular-weight samples.

... Common sense is not so common ...

—Voltaire, *Dictionnaire Philosophique*

TABLE 5.3 Divide and Conquer Troubleshooting Experiments

Symptom	Cause	Solution (see Section)
Visual Checks		
Pressure Problem		
High	Blockage	Step-wise loosen fittings until isolated, clear blockage
	Wrong flow rate setting	Reset to proper value
Low	Leak	Locate and fix
	Wrong flow rate setting	Reset to proper value
Cycles	Long term (once per run)	Normally due to viscosity changes during gradient
	Pulses (with pump cycle)	Bubbles, leaks, check valves, pump seals
	Pulses (with injector cycle)	Injector blockage, clean or replace tubing, loop, rotor seal
Leaks		
	At seals	Replace piston seal, injector, rotor seal
	At fittings	Tighten or replace
	With overpressure	Clear blockage first
Test Results		
System suitability[a] passes	Method problem	Table 5.5
System suitability[a] fails		
New isocratic column test passes *and* gradient retention and area reproducibilty test pass	Hardware OK, method problem	Table 5.5
New isocratic column test fails *or* gradient retention and area reproducibility tests fail	Hardware problem	Run gradient performance (Sections 4.3 and 5.5.3)
Failed gradient performance check	Hardware problem	Table 5.4 and Section 5.5.3
Other	Various	See symptoms of Tables 5.4 and 5.5

[a] If a system suitability test is not defined, make three to six replicate injections of a reference standard or other well-characterized sample.

TABLE 5.4 Failed Gradient Performance Checks

Symptom	Cause	Solution (see Section)
Nonlinearity	Mixing or solvent proportioning	5.5.3.1
Step test failure	Mixing or solvent proportioning; pump starvation; air bubbles in pump; blocked injector bypass; blocked mobile phase inlet line frit; check valve problems	5.5.3.2
	Proportioning valve failure	5.5.3.3
Flow rate too high	Improper settings; controller problems	5.5.3.4
Flow rate too low	Improper settings; bubbles in pump; compressibility compensation errors	5.5.3.4
	Check valve problems	5.5.2.4, 5.5.3.4
	Bad pump seals	5.5.2.5, 5.5.3.4
	Leaks	5.5.2.6, 5.5.2.7, 5.5.3.4
Pressure bleed-down	Check valve problems	5.5.2.4, 5.5.3.4
	Bad pump seals	5.5.2.5, 5.5.3.4
	Leaks	5.5.2.6, 5.5.2.7, 5.5.3.4
Retention reproducibility	Bubbles	5.5.2.1, 5.5.3.6
	Check valve problems	5.5.2.4, 5.5.3.6
	Bad pump seals	5.5.2.5, 5.5.3.6
	Proportioning problems	5.5.3.2, 5.5.3.3, 5.5.3.6
	Flow rate; column temperature; column dimensions; column degradation; mobile phase composition changes	5.5.4.1, 5.5.4.2
	Case study examples	5.5.4.1
Peak area reproducibility	Autosampler problems; lack of sample homogeneity; air bubbles; leaks	5.5.3.7

5.5 TROUBLESHOOTING

TABLE 5.5 Gradient Problem Symptoms, Causes, and Solutions

Symptom	Cause	Solution (see Section)
Retention Time Problems		
Change in retention of all peaks in same direction when move to another system	Dwell volume difference	5.2.1
Retention in run 1 does not match retention in subsequent runs	Insufficient inter-run equilibration	5.3.3
	Need to "dope" column with priming injection(s)	5.1.4.6
	Need to premix solvents	5.5.2.3, 5.5.4.2
Retention time varies between runs	Flow rate changes; bubbles; pump problems	5.5.4.1
Retention time varies over several hours	Poor temperature control	Use column heater; 5.2.2.5
	Solvent composition change	Make new mobile phase
	Column chemistry change	Replace column
Baseline Problems		
Baseline drift up or down	Difference in absorbance of A and B solvents (normal) or insufficient UV doping of solvents	5.5.2.9, 5.4.1
Excessive baseline noise	Inadequate mixing; bubbles in pump; beat frequency	5.4.2
Peaks in blank gradient	Contaminated water or reagents	5.4.3, 5.5.4.3
Extra Peaks		
Peaks in blank gradient	Mobile phase impurities	5.4.3, 5.5.2.8, 5.5.2.10, 5.5.4.3
Extra (unexpected) peaks	Impurities eluting at t_0	5.4.4.1
	Air peaks	5.4.4.2
	Late peaks; carryover	5.4.4.3, 5.5.2.11
Peak Shape Problems		
Fronting peaks	Decomposition on column; bad column	5.4.5.1, 5.4.5.5
Tailing peaks	Silanol interactions	5.4.5.1
Strong tailing peaks with retention loss	Overload	5.4.5.1
Broad peaks	Extra-column effects, data system acquisition rate too slow; detector time-constant too large	5.4.5.2
	Injection effects	5.4.5.4

Split or doubled peaks	Column frit blockage; column void	5.4.5.3
	Injection effects	5.4.5.4
Distorted peaks, especially later-eluting peaks	Temperature gradient along column	5.2.2.5
Distorted peaks, especially early-eluting peaks	Injection solvent too strong; too large injection volume	5.4.5.4

REFERENCES

1. L. R. Snyder, J. W. Dolan, and P. W. Carr, *J. Chromatogr. A* 1060 (2004) 77.
2. J. W. Dolan, A. Maule, L. Wrisley, C. C. Chan, M. Angod, C. Lunte, R. Krisko, J. Winston, B. Homeier, D. M. McCalley, and L. R. Snyder, *J. Chromatogr. A* 1057 (2004) 59.
3. A. P. Schellinger and P.W. Carr, *LCGC* 22 (2004) 544.
4. G. W. Tindall, *LCGC* 20 (2002) 1114.
5. M. A. Quarry, R. L. Grob, and L. R. Snyder, *J. Chromatogr.* 285 (1984) 1.
6. P. Jandera, J. Churacek, and L. Svoboda, *J. Chromatogr.* 192 (1980) 37.
7. J. W. Dolan, *LC/GC*, 6 (1988) 388–390.
8. H. Nakamura, T. Konishi, and M. Kamada, *Anal. Sci.* 4 (1988) 655.
9. S. K. MacLeod, *J. Chromatogr.* 540 (1991) 373.
10. R. A. Henry and D. S. Bell, *LCGC* 23 (2005) 496.
11. J. W. Dolan and L. R. Snyder, *J. Chromatogr. A* 799 (1998) 21.
12. P.-L. Zhu and J. W. Dolan, *LCGC* 14 (1996) 944.
13. R. G. Wolcott, J. W. Dolan, L. R. Snyder, S. R. Bakalyar, M. A. Arnold, and J. A. Nichols, *J. Chromatogr. A* 869 (2000) 211.
14. H. Poppe, J. C. Kraak, J. F. K. Huber, and J. H. M. Van den Berg, *Chromatographia* 14 (1981) 515.
15. H.-J. Lin and C. Horváth, *Chem. Engng Sci.* 36 (1981) 47.
16. L. R. Snyder, J. J. Kirkland, and J. L. Glajch, *Practical HPLC Method Development*, 2nd edn, Wiley-Interscience, New York, 1997.
17. D. H. Marchand, L. A. Williams, J. W. Dolan, and L. R. Snyder, *J. Chromatogr. A* 1015 (2003) 53.
18. A. P. Schellinger, D. R. Stoll, and P. W. Carr, *J. Chromatogr. A* 1064 (2005) 143.
18a. A. P. Schellinger, D. R. Stoll, and P. W. Carr, *J. Chromatogr. A*, in press (Part I).
18b. A. P. Schellinger, D. R. Stoll, and P. W. Carr, *J. Chromatogr. A*, in press (Part II).
19. L. A. Cole and J. G. Dorsey, *Anal. Chem.* 62 (1990) 16.
20. D. L. Warner and J. G. Dorsey, *LCGC* 15 (1997) 254.
21. M. Patthy, *J. Chromatogr.* 592 (1992) 143.
22. P. Jandera, *Adv. Chromatogr.* 43 (2004) 1.
23. P. Jandera, private communication.
24. S. J. Williams, *J. Chromatogr. A* 1052 (2004) 1.
25. D. N. S. Wilson, R. Morrison, and J. W. Dolan, *LCGC* 19 (2001) 590.
26. V. V. Berry, *J. Chromatogr.* 236 (1982) 279.
27. Sj. van der Wal and L. R. Snyder, *J. Chromatogr.* 255 (1983) 463.
28. C. T. Mant and R. S. Hodges, *High-Performance Liquid Chromatography of Peptides and Proteins: Separation, Analysis, and Conformation*, CRC Press, Boca Raton, FL, 1991, p. 90.
29. G. Winkler, P. Briza, and C. Kunz, *J. Chromatogr.* 361 (1986) 191.
30. H. A. H. Billiet, P. D. M. Keehnen, and L. de Galan, *J. Chromatogr.* 185 (1979) 515.
31. D. L. Saunders, *J. Chromatogr. Sci.* 15 (1977) 129.

32. J. W. Dolan, *LCGC* 7 (1989) 18.
33. J. W. Dolan, J. R. Kern, and T. Culley, *LCGC* 14 (1996) 202.
34. D. W. Bristol, *J. Chromatogr.* 188 (1980) 193.
35. J. W. Dolan, D. H. Marchand, and S. A. Cahill, *LCGC*, 15 (1997) 328.
36. J. W. Dolan, *LCGC* 7 (1989) 822.
37. J. Layne, T. Farcas, I. Rustamov, and F. Ahmed, *J. Chromatogr. A* 913 (2001) 233.
38. M. J. Mills, J. Maltas, and W. J. Lough, *J. Chromatogr. A* 759 (1997) 1.
39. W. R. Melander, H.-J. Lin, and Cs. Horváth, *J. Phys. Chem.* 88 (1984) 4527.
40. M. R. Euerby, C. M. Johnson, I. D. Rushin, and D. A. S. Sakunthala Tennekoon, *J. Chromatogr. A* 705 (1995) 229.
41. M. R. Euerby, C. M. Johnson, I. D. Rushin, and D. A. S. Sakunthala Tennekoon, *J. Chromatogr. A* 705 (1995) 219.
42. J. W. Dolan and L. R. Snyder, *Troubleshooting LC Systems*, Humana Press, Totowa, NJ, 1989.
43. J. W. Dolan, "LC Troubleshooting" a monthly column in *LCGC*, Advanstar Communications, Cleveland, OH, 1983–present.
44. J. W. Dolan, *LCGC* 5 (1987) 24.
45. M. Stadalius, personal communication (1986).
46. M. D. Nelson and J. W. Dolan, *LCGC* 16 (1998) 992.
47. C. K. Cheung and R. Swaminathan, *Clin. Chem.* 33 (1987) 202.
48. P.-L. Zhu, L. R. Snyder, and J. W. Dolan, *J. Chromatogr. A* 718 (1995) 429.
49. J. W. Dolan, *LCGC* 11 (1993) 640.
50. T. Culley and J. W. Dolan, *LCGC* 13 (1995) 456.
51. J. W. Dolan, *LCGC* 13 (1995) 940.
52. U. D. Neue, personal communication (1995).
53. T. Eidenberger, personal communication (1995).
54. J. J. Gilroy and J. W. Dolan, *LCGC* 22 (2004) 982.
55. J. W. Dolan, *LCGC* 14 (1996) 294.
56. D. H. Marchand, P.-L. Zhu, and J. W. Dolan, *LCGC* 14 (1996) 1028.

CHAPTER 6

SEPARATION OF LARGE MOLECULES

> "I weep for you," the Walrus said:
> "I deeply sympathize."
> With sobs and tears he sorted out
> Those of the largest size.
> —Lewis Carroll, *Alice's Adventures in Wonderland* (about oysters)

6.1 GENERAL CONSIDERATIONS

"Large" molecules include a variety of natural or synthetic polymers with molecular weights $M > 1000$ Da, especially compounds with $M > 10,000$. While separations of both large and small molecules by RP-LC are generally similar, there are some important differences that will be examined in this chapter. These differences in the separation of large vs small molecules can be traced to certain sample characteristics that vary with molecular size [1]:

- values of S; the rate of decrease in isocratic retention k as %B increases (Section 6.1.1);
- values of the plate number N (Section 6.1.2);
- conformational state (Section 6.1.3).

Apart from differences in these sample characteristics and their effects on separation, *the same general theory applies for the gradient separation of both large and small molecules*. Some workers have questioned this conclusion in the past, suggesting instead that the gradient separation of large molecules is based on a fundamentally different retention process than applies for samples with $M < 1000$ (Section 6.1.5). A preponderance of evidence, however, now suggests that *gradient separation usually takes place in the same fundamental way for molecules of all sizes*. Therefore, a similar approach can be used for understanding and controlling the separation of large molecules as was presented in Chapter 3 for samples with $M < 1000$.

High-Performance Gradient Elution. By Lloyd R. Snyder and John W. Dolan
Copyright © 2007 John Wiley & Sons, Inc.

6.1 GENERAL CONSIDERATIONS

For those readers with primarily a practical interest in the separation of large biomolecules (peptides, proteins, etc.), we suggest that you start with Sections 6.1.1–6.1.3, then proceed directly to Section 6.2.

6.1.1 Values of *S* for Large Molecules

Figure 6.1 illustrates the dependence of isocratic retention k on mobile phase %B for benzene and several peptides and proteins of varying molecular weight M (here, B is ACN). These plots of log k vs %B are approximately linear, as is usually the case for separations by RP-LC. The data of Figure 6.1 for each compound can be described quantitatively by an empirical equation that has been used in preceding chapters for

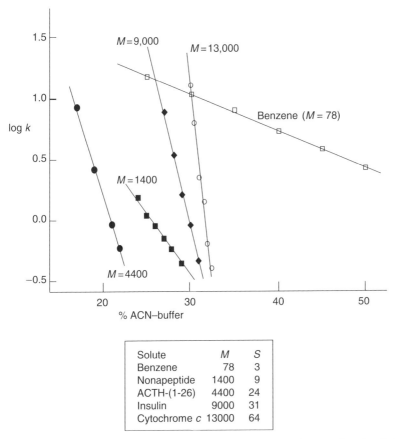

Solute	M	S
Benzene	78	3
Nonapeptide	1400	9
ACTH-(1-26)	4400	24
Insulin	9000	31
Cytochrome c	13000	64

Figure 6.1 Change in isocratic retention k with change in %B for a small molecule (benzene) and several peptides or proteins. Conditions: 100 × 4.6 mm C_{18} column; mobile phase, acetonitrile–pH-3.0 phosphate buffer; ambient temperature. Compounds are benzene ($M = 78$), a nonapeptide ($M \approx 1400$); ACTH-(1–26) ($M \approx 4400$); insulin ($M \approx 9000$); cytochrome c ($M \approx 13,000$). The figure is adapted from [4].

RP-LC retention:

$$\log k = \log k_w - S\phi \tag{6.1}$$

where k_w and S are constants for a given compound when only the volume fraction ϕ of the B solvent is varied ($\phi = 0.01$ %B). It is seen in Figure 6.1 that, as sample molecular weight increases, the slopes [$S = -\mathrm{d}\log(k)/\mathrm{d}\phi$] of these plots of $\log k$ vs %B become steeper. That is, larger molecules exhibit a faster change in retention for a given change in %B (values of S increase with M; see box at bottom of Fig. 6.1).

A relationship similar to Equation (6.1) for isocratic separation also applies for gradient elution (Section 9.1.1.1):

$$\log k^* = \log k_w - S\phi^* \tag{6.2}$$

where k^* is the median value of k during gradient elution (when the band has reached the column mid-point), and ϕ^* is the corresponding value of ϕ. Values of $\log k_w$ and S in Equations (6.1) and (6.2) have the same values for a given compound in "corresponding" isocratic and gradient separations (Sections 2.2.1.1 and 9.1.1.1), where experimental conditions differ only in whether %B is fixed (isocratic) or varies (gradient).

For RP-LC with water/organic mobile phases, molecules with molecular weights of 100–400 have values of $S \approx 4$ [2]. However, as sample molecular weight increases, values of S likewise increase [3–13], as in Figure 6.1. This is further illustrated in the log–log plots of Figure 6.2 for two series of compounds: peptides plus proteins (□, △) and polystyrenes (●). The data of Figure 6.2 can be represented approximately by the dashed curve in Figure 6.2,

$$S \approx 0.25(\text{molecular weight})^{1/2} \tag{6.3}$$

despite the use of different B solvents (ACN and tetrahydrofuran) for these diverse analytes. Equation (6.3) appears to be a general relationship for alkylsilica columns and water–organic mobile phases, where the organic can be ACN or THF. While data for larger molecules with methanol/water as mobile phases are lacking, Equation (6.3) provides reasonable predictions of S for small molecules with methanol as B solvent. Note also the scatter of data in Figure 6.2 (SD = 0.11); that is, values of S for compounds of similar molecular weight vary by an average $10^{0.11} = \pm 30$ percent.

Values of the sample parameter S affect the gradient retention factor k^* (Section 2.2.1):

$$k^* = t_G F/(1.15 V_m \Delta \phi S) \tag{6.4}$$

where t_G is gradient time (min), F is the mobile phase flow rate (mL/min), V_m is the column dead volume in mL (proportional to column length and diameter-squared), and $\Delta \phi$ is the change in ϕ during the gradient, that is, $\Delta \phi = 1$ for a 0–100 percent B gradient. Values of k^* in gradient elution should fall approximately within a range of $1 \leq k^* \leq 10$, similar to the preferred range of $1 \leq k \leq 10$ for isocratic separation.

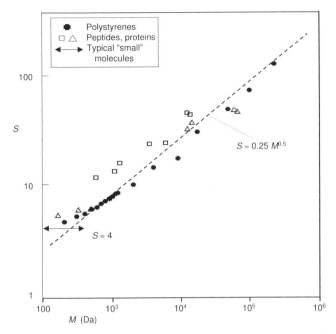

Figure 6.2 Experimental values of S for peptides and proteins ($\square$ [6], $\triangle$ [9]) and polystyrenes ($\bullet$) [5, 7] as a function of sample molecular weight (log–log plot); acetonitrile–water and THF–water mobile phases, respectively.

The control of k^* in gradient elution is usually achieved by the choice of gradient time t_G; larger values of t_G result in proportionately larger values of k^* [Equation (6.4)]. Equation (6.4) can also be expressed as

$$k^* = t_G/(1.15\, t_0 \Delta\phi S) \qquad (6.4a)$$

since the column dead time $t_0 = V_m/F$.

For isocratic retention, large values of S result in values of k that change significantly with small changes in %B, for example, for an S-value of 100 [$M \approx 160{,}000$ from Equation (6.3)], a change of 0.1 percent B in the mobile phase changes k by 25 percent. Consequently, the achievement of repeatable retention times from day to day for the isocratic separation of large molecules can prove inconvenient or even unattainable. More important, when a mixture of large molecules is separated isocratically, their values of k usually cover a range of several orders of magnitude. For such samples, it is not possible to select a mobile phase composition that provides acceptable values of k for all peaks (e.g., $1 \leq k \leq 10$). Consequently, the RP-LC separation of large molecules is usually carried out by means of gradient elution.

Experimental gradient elution conditions (t_G, F, column size V_m, $\Delta\phi$) that provide preferred values of k^* for typical "small-molecule" samples (e.g., Table 3.2), will result in much smaller values of k^* for large molecules with large values of S [Equation (6.4)]. Very small values of k^* in gradient elution in turn lead to poor separation (see discussion of Fig. 6.3 below). When separating macromolecular samples, therefore, rather different gradient conditions are needed for acceptable values of k^*: longer gradient times t_G (i.e., a shallower gradient), shorter columns, or (if possible) a decrease in gradient range $\Delta\phi$ – as suggested by Equation (6.4). The gradient time required for a given value of k^* can be obtained from a rearrangement of Equation (6.4):

$$t_G = 1.15 V_m \Delta\phi S k^* / F \qquad (6.4b)$$

Another way of illustrating the effect of the value of S on gradient separation is provided by the calculated plots of Figure 6.3, based upon preferred gradient conditions for the separation of a small molecule (similar to the recommended conditions in Table 3.2). Figures 6.3(a–c) describe the migration of a sample peak through the column, in the form of distance x vs time t plots (solid curves). In each of Figure 6.3(a–c), a value of $k_w = k_0 = 100$ is assumed for a gradient of 0–100 percent B in 15 min (2 mL/min, 150 × 4.6 mm column). Figure 6.3(d–f) shows the corresponding separations of two compounds for which the separation factor $\alpha = 1.2$ ($k_0 = 100$ and 120, respectively), for gradient times t_G equal 15 and 150 min and values of S are 4, 25, and 100, respectively.

Figure 6.3(a), for a low-molecular-weight compound ($S = 4$), shows the fractional migration x of the compound as a function of time ($\square$ marks $x = 1$ and elution from the column); instantaneous values of k (k_i) vs time (dashed curve, y-values on right) are superimposed onto this figure. A similar behavior is observed to that seen earlier in Figure 1.7, that is, continuously faster migration (dx/dt) of the band with time, because of decreasing values of k with time. When $x = 0.5$ (the column midpoint, marked by ●), $k_i \equiv k^* = 4$ (see dotted line in figure). A value of $k^* > 1$ means that sample resolution will not be seriously compromised because of insufficient sample retention [Equation (6.6) below]. A hypothetical example of the separation of two compounds with $S = 4$ and a separation factor $\alpha = 1.2$ is illustrated in Figure 6.3(d) (plate number $N \equiv N^* = 10{,}000$), for t_G equal to either 15 [$k^* = 4$; Equation (6.4a)] or 150 min ($k^* = 40$). Only a modest increase in resolution is observed for the longer gradient, because resolution is proportional to [$k^*/(k^* + 1)$] [Equation (6.6)], which increases in this example only by a factor of 1.2 (from 0.80 to 0.98).

Figure 6.3(b) shows a similar plot for a compound with a molecular weight of 10,000 and $S \approx 25$, where k^* is now only 0.7. This relatively small value of k^* will result in a significant loss in sample resolution, compared with the use of a gradient with $k^* \approx 4$ as in Figure 6.3(a). A hypothetical example of the separation of two compounds with $S = 25$ and a separation factor $\alpha = 1.2$ is illustrated in Figure 6.3(e) ($N^* = 3000$; a lower value than in Fig. 6.3d, due to a general decrease in N^* when the solute molecular weight increases; see Section 6.1.2 below). While the separation for a gradient time $t_G = 15$ is unsatisfactory, the use of a longer gradient ($t_G = 150$) leads to acceptable resolution of the two peaks, due to an increase in k^* from a value of 0.7 to 7 {and an increase in [$k^*/(k^* + 1)$] and R_s by a factor of 2.1}.

6.1 GENERAL CONSIDERATIONS 233

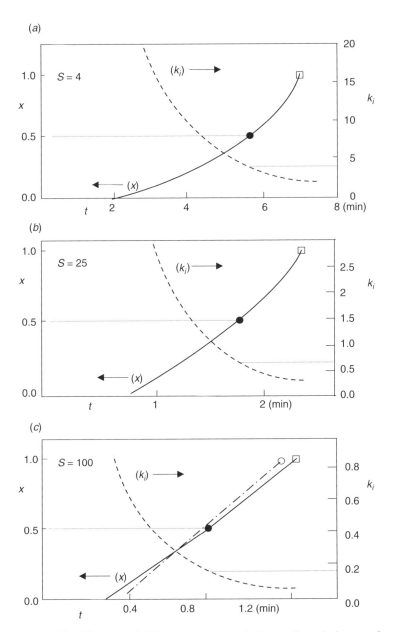

Figure 6.3 Migration of different compounds during gradient elution as a function of their molecular size or value of S and time t. Hypothetical examples based on identical conditions for each separation: gradient of 0–100 percent B in 15 min at 2 mL/min (150 × 4.6 mm column). The sample has S equal to 4 in (a, d), 25 in (b, e), and 100 in (c, f); $k_0 = 100$ in (a–c), and 100 and 120 for the two peaks in (d–f). (———) Fractional band migration x through the column; (– – –) instantaneous value of k (k_i) for a band at time t; (–·–·–) "on–off" behavior in (c). (●) Designates k^*; (□) designates $x = 1.00$. See text for details.

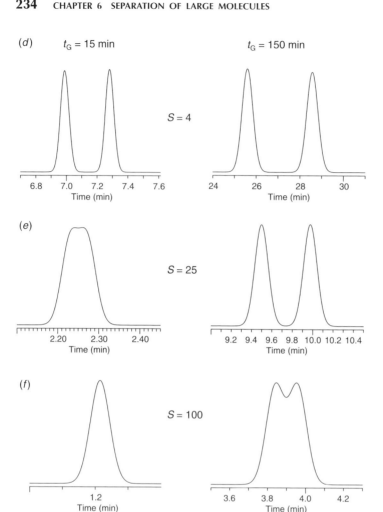

Figure 6.3 *Continued.*

Figure 6.3(c) provides an extreme example for the separation of a compound with a molecular weight of 160,000 and $S \approx 100$; $k^* \approx 0.2$. In this case, after band migration begins, the compound is essentially washed through the column with little retention, as well as reduced separation from adjacent peaks. A hypothetical example of the separation of two compounds with $S = 100$ and a separation factor $\alpha = 1.2$ is illustrated in Figure 6.3(f) ($N^* = 1000$; see above). For a gradient time of 15 min, there is little separation of the two peaks. However, for a longer gradient ($t_G = 150$, and $k^* = 2$), some resolution is achieved ($R_s = 0.7$), which could be improved by a further increase in t_G and k^*. However, the low value of $N = 1000$ for these separations, combined with $\alpha = 1.2$, means that the limiting possible resolution for a further increase in gradient time would be only $R_s \approx 1$. Because the retention times in Figure 6.3(d–f) for the second peak with $t_G = 150$ are only

4–30 min, the run times for these separations could each be shortened considerably by stopping the gradient after the last peak leaves the column.

Note also the dash–dotted plot in Figure 6.3(c), which corresponds to "on–off" elution behavior (Section 6.1.5); in the latter case, the sample is simply washed through the column, with no retention at all after band migration begins. Retention times (corresponding to $x = 1.00$) are indicated in Figure 6.3(c) for $S = 100$ (□; 1.22 min) and for on–off elution with $S \gg 100$ (○; 1.15 min). For these gradient conditions, we see that a sample with $S = 100$ is eluted with essentially "on–off" behavior.

6.1.2 Values of N^* for Large Molecules

Values of the plate number N (isocratic) $\equiv N^*$(gradient) as a function of separation conditions are discussed in Section 9.5. For small molecules (e.g., molecular weights of ≤ 500), typical separation conditions (e.g., as in Table 3.2) result in plate numbers of 5000–20,000. However, as sample molecular size increases, resulting plate numbers become smaller (assuming no change in the column or separation conditions) due to slower solute diffusion. For the separation of both large and small molecules, column plate number and sample resolution increase for longer columns, smaller particles, lower flow rates, and higher temperatures. It should be noted, however, that changes in column length or flow rate alone in gradient elution affect both N^* and k^* [Equation (6.4)]; when both N^* and k^* change, the resulting effect on resolution is more complex and less predictable [see Equation (6.6) below]. Therefore, *when changing column length or flow rate for the purpose of increasing N^* and resolution in gradient elution, it is advisable to vary gradient time t_G at the same time, so as to maintain values of k^* constant* [Equation 6.4(b)]; an identical recommendation was made in Section 3.3.5 for small-molecule separations.

There is usually a decrease in N^* of 2- to 3-fold for each 10-fold increase in sample molecular weight. Unless wide-pore columns (diameter >15 nm) are used, still smaller values of N^* can result for very large molecules, because of the slower diffusion of large molecules within small-diameter pores [14, 15]. Larger-pore columns (e.g., pore diameters of 30 nm instead of the usual 8–12 nm) are recommended for proteins and other samples with molecular weights >10,000, both to avoid slow diffusion of the sample within narrow pores (and smaller N^* values), and to provide complete access of the sample to the interior of column particles (for potentially increased column capacity and larger weights of injected sample in preparative separations; Section 7.2.1.2). Even larger values of N^* can be achieved by means of small, nonporous ("pellicular") particles, in place of the usual porous particles [16]; pellicular columns are especially effective for large sample molecules because of the elimination of (slow) diffusion within the pores. Finally, N^* also increases at higher temperatures, which means potentially faster and/or better separations of peptides and proteins at elevated temperatures [17], as in the examples of Figure 3.13. Because of the large number of peaks present in many biochemical samples, as well as generally lower values of N^* (because of larger M), the baseline resolution of all peaks of interest by a single separation is often impractical. For this reason, two-dimensional gradient HPLC is being used increasingly for a more complete separation and analysis of complex biochemical samples (Section 6.2.5).

Whereas a change in flow rate may have a relatively small effect on N^* and resolution for the separation of a small molecule (e.g., Fig. 3.11a, b), N^* becomes a stronger function of flow rate F for large molecules (for which $N^*F \approx$ constant). Consequently, large-molecule separations often can be improved significantly by reducing the flow rate, while maintaining k^* constant by a proportionate increase in t_G [Equation (6.4)]. While flow rates of 1–2 mL/min are typical for the separation of small molecules (4.6 mm column I.D.), lower flow rates (e.g., 0.5–1.0 mL/min) may prove a better choice for large molecules. Inasmuch as gradient time must be further increased for large molecules because of their large values of S (Section 6.1.1), the combination of lower flow, large S, and $1 \leq k^* \leq 20$ can require very long run times [Equation 6.4(b)]. Unfortunately, some proteins exhibit significant sample loss when very long gradients are used (Section 6.2.3). In order to reduce run time and improve sample recovery without affecting sample resolution, the gradient range $\Delta \phi$ should be adjusted so that peak migration starts soon after the gradient begins, and the gradient ends soon after the last peak leaves the column (this minimizes gradient time t_G, while keeping $\Delta \phi / t_G$ and k^* constant).

6.1.3 Conformational State

The conformation of a molecule is defined by the relative positions of its different parts (atoms, structural groups, etc.) with respect to each other, for example, a compact vs an expanded shape for a flexible molecule. In this chapter, protein conformation is mainly of interest, because of its potentially large effect on RP-LC separations of protein-containing samples. The primary structure of a peptide or protein is defined by the sequence of connected amino acids in the polypeptide chain. Because of strong intramolecular interactions due to hydrogen bonding (as well as covalent disulfide linkages), the polypeptide chain can assume a more compact ("tertiary") structure, referred to as the "native" protein (the preferred conformation in nature). Separate polypeptide units can further associate into a complex (quaternary structure); however, quaternary structure is almost always lost during reversed-phase separation, because of a dissociation of the parts that comprise the complex. Loss of tertiary structure also usually occurs during RP-LC separation, leading to a less compact, "denatured" protein.

The use of a low-pH mobile phase (usually preferred for RP-LC separations of proteins; Section 6.2.1.2) in combination with the hydrophobic nature of reversed-phase columns usually leads to a rapid denaturation of the protein following its injection into the column. Denaturation involves the opening up of the protein molecule, thus exposing its hydrophobic interior for maximum interaction with the hydrophobic stationary phase of the column. If protein denaturation does not proceed to completion soon after the protein first enters the column, slow changes in tertiary structure (i.e., partial protein denaturation) *during* reversed-phase separation can lead to the appearance of more than one peak for a given protein [18]. Some aspects of this process are visualized in Figure 6.4; in (a) the sequential denaturation of a protein is visualized, from native protein (N), to partially denatured protein

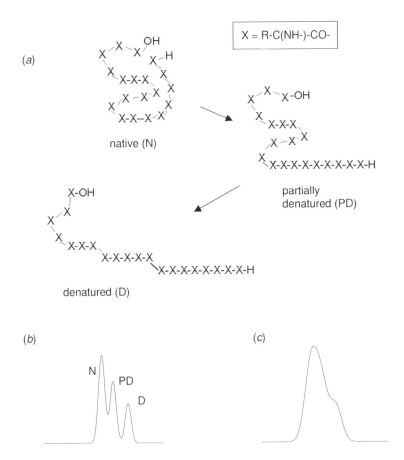

Figure 6.4 Hypothetical representation of protein denaturation during RP-LC separation. (*a*) Possible protein conformations; (*b*) gradient separation of a sample that is partially denatured prior to separation; (*c*) same as (*b*), except for slow denaturation during separation. See text for details.

(PD), to fully denatured protein (D). In (*b*), the gradient separation of the protein is shown, assuming that partial denaturation occurs *prior* to injection, so that individual species are formed and partly resolved as the sample migrates through the column; it is assumed in (*b*) that further denaturation during separation is relatively slow. In (*c*) slow denaturation *during* the separation is assumed, with the successive formation of partially denatured and denatured protein as the sample migrates through the column. Consequently, the three peaks (N, PD, D) leaving the column overlap to form a single, apparently misshapen peak. However, a distinction between the two processes illustrated in (*b*) and (*c*) may not be obvious in practice. It should also be noted that many *experimental* separations as in Figure 6.4(*b*, *c*) appear to involve only the native (N) and fully denatured (D) species; partially denatured protein (PD) is seldom observed as a distinct peak.

Multiple peaks can also occur during the separation of peptides that contain the amino acid proline, due to a slow interconversion of *cis* and *trans* configurations of the proline group [19] – especially at lower temperatures. The slow interconversion of different peptide or protein conformers can also cause increased peak broadening (lower *apparent* values of N^*) [20], without the appearance of multiple peaks or peak distortion as in Figure 6.4(*b*, *c*). Finally, changes or differences in molecular conformation during the separation of peptides or proteins can result in other, less important anomalies [21–23], when compared with theory-based predictions. A number of examples have been reported of degraded separation as a result of the partial denaturation of protein samples during separation (see [1, 24] and Refs 4–18 of [25]). If separation conditions are selected that favor the denaturation of protein samples *prior* to their migration through the column, these adverse effects of partial denaturation can be minimized (Section 6.2.1.2).

The native form of a protein molecule does not allow the (more hydrophobic) interior of the molecule to interact directly with the stationary phase of a RP-LC column. As a result, the interactions of native proteins with the column are reduced vs that of denatured proteins, and the native protein generally elutes before the denatured species – as illustrated by the hypothetical example of Figure 6.4(*b*). Values of S for the native (most compact) conformation will usually be considerably smaller than for the denatured (more expanded) conformation [25].

Conformational effects can also arise in the RP-LC separation of very large synthetic polymers ($M > 100,000$). For example, it appears that the A or B solvent can be preferentially sequestered within a folded or more compact polymer molecule. The time required to equilibrate the polymer molecule with mobile phase of changing percent-B during gradient elution can then be slow, with various (usually adverse) consequences for gradient separation. Similar complications can arise as a result of polymer crystallinity [26]. See [10, 13, 27] for further details.

6.1.4 Homo-Oligomeric Samples

We will define "homo-oligomeric" samples (referred to here simply as "oligomers") as mixtures of compounds that arise from the linear combination of n identical subunits X to form molecules X_n. More generally, a functionally different subunit Y can be present at the end of the molecule, to form compounds X_nY. Homo-oligomeric samples include mixtures of homologs such as the *n*-alkylcarboxylic acids, synthetic polymers such as polystyrene or polyethyleneglycol, and some polysaccharides – but *not* peptides or oligonucleotides that are formed from *different* amino acid or nucleic acid base building blocks (Section 6.2). Synthetic homo-oligomeric samples typically comprise a mixture of compounds that cover some range in n (e.g., from $n = 5$ to $n = 20$), with all intermediate values of n present (e.g., $5 \leq n \leq 20$).

Homo-oligomeric samples can be regarded as prototypically "regular" (Section 1.6), with values of S and k_w that increase continuously with oligomer molecular weight. The separation of an oligomeric sample is illustrated in Figure 6.5 for

6.1 GENERAL CONSIDERATIONS **239**

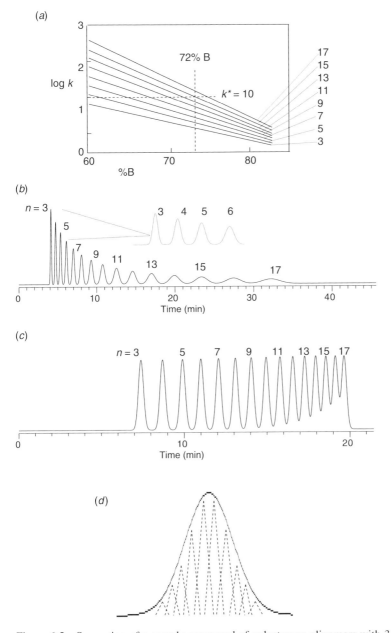

Figure 6.5 Separation of a sample composed of polystyrene oligomers with $3 \leq n \leq 17$. Conditions are tetrahydrofuran–water mobile phase, 35°C, 2.0 mL/min and a 250 × 4.6 mm C_{18} column. (*a*) Plot of log k vs %B (tetrahydrofuran) for odd-numbered oligomers only; dashed lines describe isocratic retention (72 percent B) and gradient retention ($k^* = 10$); (*b*) isocratic separation of the sample with 72 percent B; (*c*) gradient separation with 60–100 percent B in 60 min ($k^* = 10$); (*d*) hypothetical illustration of peak broadening for a higher-molecular-weight synthetic polymer sample (see text). Computer simulations based on data of [5].

a mixture of polystyrenes with $3 \leq n \leq 17$, using a C_{18} column and tetrahydrofuran/ water as mobile phase. Figure 6.5(a) shows plots of log k vs %B for some of the 15 sample components (those with odd values of n). Note that the slopes S of these plots increase continuously from $S = 5$ for $n = 3$ to $S = 11$ for $n = 17$.

Figure 6.5(b) shows the *isocratic* separation of this polystyrene sample with 72 percent B as mobile phase. To understand how resolution varies with n in such separations, consider the usual resolution equation:

$$\text{(isocratic)} \quad R_s = (1/4)[k/(1+k)](\alpha - 1)N^{1/2} \quad (6.5)$$

Values of k increase with n, so that $[k/(1+k)]$ also increases with n – but not by much when $k \geq 3$ for the initial peak in the chromatogram, as is the case in Figure 6.5(b). The value of α for adjacent peaks in an oligomeric sample will be approximately constant (the Martin rule [28]), as can be seen also in Figure 6.5(a) for isocratic separation with 72 percent B (vertical dashed line). That is, here the change in log k for adjacent oligomers is a constant 0.08 log units for 72 percent B, or $\alpha = 10^{0.08} = 1.20$ for each adjacent pair of oligomer bands. The major contribution to the decrease in resolution of later peaks in Figure 6.5(b) is a 2-fold decrease in N on going from $n = 3$ to $n = 17$ (Section 6.1.2), which results in a 1.4-fold reduction in resolution from initial to final peaks in this chromatogram.

Figure 6.5(c) shows the separation of the same sample with a gradient from 60 to 100 percent B in 60 min ($k^* \approx 10$). Unlike the similar spacing of peaks in the isocratic separation of Figure 6.5(b), the gradient separation of Figure 6.5(c) shows bunching of peaks at the end of the chromatogram, with greatly reduced resolution. The latter gradient separation is illustrated in Figure 6.5(a) by the dashed horizontal line (labeled "$k^* = 10$"). Resolution in gradient elution is given by a similar equation as for isocratic separation [Equation (6.5)]:

$$\text{(gradient)} \quad R_s = (1/4)[k^*/(1+k^*)](\alpha^* - 1)N^{*1/2} \quad (6.6)$$

except that values of k^* for different peaks in the chromatogram do not change as much as in isocratic separation. The main difference between the separations of Figure 6.5(b) and (c) arises from differences in values of α^* for gradient elution vs values of α in isocratic separation. Whereas α is approximately constant for all peak pairs in Figure 6.5(b), α^* decreases by a factor of 2 from the beginning to the end of the gradient separation of Figure 6.5(c) (note the spacing of peaks for the $k^* = 10$ line of Fig. 6.5a). The decrease in α^* (aggravated by decrease in N^*) as n increases leads to the reduced resolution and bunching of later peaks observed in Figure 6.5(c), as predicted by Equation (6.6); *this is a characteristic feature of the gradient separation of synthetic-polymer samples.*

In the past, it has been suggested that linear-gradient separations such as that of Figure 6.5(c) can be improved by the substitution of a *convex* gradient. Because gradient steepness decreases during a convex gradient, larger values of k^* at the end of the gradient should result, giving a better resolution of later peaks (as seen in Fig. 2.26c). In practice, however, the use of convex gradients usually provides somewhat limited improvement in the separation of oligomeric samples. A simpler and similarly effective approach is the use of flatter gradients, combined with larger values of initial %B for the gradient [29].

The average molecular weight and molecular-weight distribution of oligomeric samples are the result of the synthetic process *per se*, as well as any fractionation of the original sample into narrower-molecular-weight fractions. The analysis of narrow polymer fractions of this kind (for $n > 50$) by means of gradient elution usually results in a single peak which *resembles* the usual Gaussian peak for a single compound. A visualization of such a peak (composed of overlapping peaks of varying n) is shown in Figure 6.5(d). The observed width W of the composite peak (shown in Fig. 6.5d) is seen to greatly exceed that of any of the individual peaks making up the sample; in fact, the observed peak width is largely determined by the difference in retention times of the initial and final peak components – *with little contribution from N^* or the related width of individual oligomer peaks*. As a result, changes in column length or flow rate (while maintaining values of k^* constant) which are intended to increase N^*, reduce peak width, and improve resolution may have little effect on the separation of one oligomer fraction from another, despite any change in the widths of individual polymer bands. It should be noted that the example of Figure 6.5(d) is only illustrative; actual peaks for narrow polymer fractions tend to front, rather than be symmetrical, as suggested by the example of Figure 2.26(b).

Peak broadening due to overlapping multiple peaks as in Figure 6.5(d), which may not be much affected by an increase in N^*, can also occur for certain proteins as a result of their partial denaturation during separation (Section 6.1.3). Similar peak broadening due to the presence of more than one distinguishable species for nominally the same protein could also occur because of the *cis–trans* isomerization of prolines within a peptide, disulfide mismatches, and/or variable protein glycosylation (with relatively minor effect on RP-LC retention), and so on. All of these processes become increasingly likely as the size of the protein molecule increases (especially for $M > 20{,}000$), making the RP-LC separation of larger proteins less promising.

6.1.4.1 Separation of Large Homopolymers *This section is primarily of academic interest and can be skipped by readers interested mainly in the practical separation of large molecules.*

In the gradient separation of adjacent homo-oligomeric peaks, the decrease in α^* and resolution for larger n (as in Fig. 6.5c) is a general phenomenon that can result in the almost complete overlap of adjacent peaks when their molecular weights are large. This should be obvious from the example of Figure 6.5, but can be further illustrated as follows. Assume a polymer molecule composed of $n = 100$ sub-units, a value of $k^* = 5$, and $N^* = 1000$ (even smaller values of N^* are possible for such samples). The equilibrium constant for retention $K = k^*/\psi$, where ψ is the phase ratio. Further assume (a) a phase ratio $\psi = 0.05$, estimated from $\psi = w_s/V_m$, where w_s is the column saturation capacity (Section 7.2.1.2), equal to about 0.4 mg/m$^2 \approx 0.4$ μL/m^2 [30], (b) the surface area for a 250 × 4.6 mm column ≈ 300 m^2, and (c) $V_m \approx 2.5$ mL. We then have $K = k^*/\psi \approx 5/0.05 \approx 100$. The interaction of each oligomer sub-unit [e.g., a —CH$_2(\phi)$CH$_2$— group in the case of polystyrene, where $\phi =$ phenyl] with the column will be roughly constant, so that in isocratic elution the value of α^* for two adjacent bands (n and $n+1$) is also constant. Finally, from the Martin equation [28], $K \approx \alpha^{100} \approx 100$, which assumes that K for an oligomer with $n = 1$ equals α. Therefore, $\log \alpha^* \approx \log(100)/100 \approx 0.02$, and $\alpha^* = 1.05$ for adjacent polymers with $n = 100$ and 101. Equation (6.6) then yields $R_s = (1/4)(5/6)(1.05 - 1)1000^{1/2} = 0.3$ for the latter adjacent polymer molecules. The exact value of R_s will vary for different samples and conditions, as well as with n, but this example should reinforce our observation that there is a continuous decrease in resolution as n increases for adjacent oligomer peaks in linear gradient elution. When n exceeds a value such that $R_s < 0.5$, all later peaks will merge into a single, broad peak as in Figure 6.5(d). The preceding discussion applies only qualitatively to polymers with end groups whose retention differs greatly from that of the repeating unit; also, the assumption that the Martin equation applies is only a crude approximation for large values of n.

The above discussion further suggests that small changes in a macromolecule should have only a minor effect on its relative retention and resolution from adjacent peaks. However, this is often not the case for the separation of peptides and proteins (examples cited in p. 56 of [31]); also see [25, 32, 33]). For example, the oxidation of a single methionine in either human IGF-1 [34] or interleukin-2 [25] leads to a sufficient change in retention to allow the separation of oxidized from unoxidized protein. Large changes in retention for minor changes in the structure of a large peptide or protein can be explained either by resulting differences in molecular conformation (e.g., secondary structure) or by the preferential involvement of only a part of the sample molecule in the retention process.

6.1.5 Proposed Models for the Gradient Separation of Large Molecules

Prior investigations of gradient elution with high-molecular-weight samples have suggested several different retention processes that might contribute to these separations:

- conventional gradient elution as described by the present linear-solvent-strength (LSS) model;
- precipitation chromatography;
- size exclusion;
- "on–off" chromatography;
- "critical elution" behavior.

Conventional gradient elution in terms of the LSS model is described in Chapters 2, 3, and 9, primarily for "small" molecules with $M < 1000$. This treatment, which is derived from a combination of empirical observation [Equation (6.2)] and the general theory of chromatography (Chapter 9), provides a consistent and reliable description of both isocratic and gradient elution, for sample molecules of any size, and for both natural and synthetic polymers. We believe that most separations of biomolecules and synthetic polymers are best explained by means of conventional gradient elution theory.

Precipitation chromatography ("precipitation–redissolution") [35–40] refers to gradient separations of synthetic polymers which depend on compound solubility, independent of interactions of sample molecules with the stationary phase (as in conventional RP-LC gradient elution). In theory, precipitation chromatography proceeds as follows. A column packing with small pores is used, the B solvent is a good solvent for the sample, and the A solvent is a poor solvent ("good" vs "poor" solvents correspond to greater and lesser polymer solubility, respectively; also referred to by polymer chemists as "solvents" and "nonsolvents," respectively). At the start of the gradient (low %B, poor solvent), the injected sample will be insoluble in the mobile phase; it is therefore held initially at the column inlet as a precipitate. With the passage of time, the concentration of B solvent will increase sufficiently (thereby becoming a "better" solvent) so as to allow dissolution of part of the sample, that is, the more soluble components in a polymer blend or copolymer (usually those molecules with lower M). This dissolved sample fraction will therefore begin to move through the column without interacting with or being retained by the stationary phase (due to the choice of column and A and B solvents, $k \approx 0$). During the passage of mobile phase through the column, however, the small molecules of B solvent have access to the narrow pores of the column packing, while the large sample molecules do not (size exclusion). As a result, the dissolved sample fraction moves through the column faster than do molecules of the B solvent, and into mobile phase of lower %B. The sample fraction is then re-precipitated onto the column packing from this "less good" solvent. The concentration of B solvent at that point in the column continues to increase with time, however, so that the sample fraction later re-dissolves in the mobile phase and again begins moving through the column – until it again outruns the gradient and re-precipitates. Following many repetitions of this precipitation–redissolution process, various sample components are separated from each other on the basis of their solubility in the mobile phase, leading to the later elution of less-soluble sample molecules. Separation by means of precipitation chromatography should be unaffected by the nature of the column, except for the importance of column pore size. Gradient RP-LC separations of synthetic polymers can change from "conventional gradient elution" to precipitation chromatography when large enough sample weights are injected [38], because relative solubility (determined by sample–mobile phase interactions) becomes more important than the interaction of the sample and column in determining retention.

Some workers have assumed a precipitation–solubilization process for the separation of synthetic polymers, when the evidence instead suggests conventional gradient elution. Other workers have recognized that, even when precipitation–solubilization may be occurring, simultaneous retention of the sample by the stationary phase is also possible. It should be kept in mind that an increase in polymer solubility as a result of increase in %B also results in proportionally smaller values of k for conventional gradient elution [38]. That is, an increase in %B results in a decrease in retention for *both* precipitation chromatography and conventional gradient elution. For this reason, comparisons of retention time vs solubility as a function of %B are not reliable as a means for differentiating these two retention processes; a better means is the study of separation as a function of sample weight [38]. The emphasis in this chapter is on biomacromolecules, which normally do not precipitate during RP gradient elution; so little more will be said here about precipitation chromatography.

Size exclusion has sometimes been cited as a contribution to sample retention in gradient separations of large molecules [41, 42], whether natural or synthetic polymers. It is assumed that sample molecules of a certain size (measured by the Stokes or hydrodynamic diameter of the compound) cannot enter pores of smaller diameter. This is indeed the case for "true" size-exclusion chromatography, where conditions are intentionally selected to minimize the retention of sample molecules by the stationary phase. However, under conditions where strong retention of a large molecule by the stationary phase is possible, size exclusion effects become much less important – at least for molecules with $M \leq 50,000$ and columns with pore-diameters ≥ 6 nm [5, 43]. Under these circumstances, the sample molecule can unravel and be retained (even in narrow pores), as long as the minimum cross-section of the unraveled molecule allows its entry into the pore ("minimum cross section" is always much smaller than the Stokes diameter of a synthetic-polymer molecule or of a denatured protein), and k is sufficiently large. Conversely, if gradient conditions are selected to yield $k^* \ll 1$, size exclusion can play a more important role, as reported for the separation of synthetic polymers with $M > 200,000$ using 10 nm-pore columns [13]. It should be noted that size exclusion results in the preferential retention of smaller molecules (usually the opposite of RP-LC retention). Thus, when size exclusion effects are significant, they can compromise the separation of synthetic-polymer samples.

"*On–off*" chromatography refers to the initial, strong retention of a sample compound followed by its rapid release at a certain point in the gradient. That is, the value of k for the compound varies strongly with %B (corresponding to a very large value of S), so that retention k goes from a very large value to almost zero, for a very small change in ϕ. "On–off" chromatography was illustrated in the example of Figure 6.3(*c*), for gradient conditions that result in a very small value of k^*. In such cases, after band migration begins, the compound is then washed through the column with little retention or separation from closely adjacent peaks (small values of α). As a result, there is little acceleration of band movement during migration, and the plot of x vs time (solid curve) in Figure 6.3(*c*) is almost a straight line. This behavior for $S = 100$ can be compared with the almost identical dash–dotted plot in this figure, corresponding to complete "on–off" elution behavior; for the latter case, a compound stays at the column inlet (after injection and the start of the gradient) until mobile phase of a certain %B arrives at the inlet, at which point the value of k^* for the band immediately changes from essentially infinite to zero (log $k_w = S = \infty$). "On–off" separation can be expected for both natural and synthetic polymers whenever experimental conditions are selected [Equation (6.4)] that result in $k^* \approx 0$. In most cases, "on–off" separations are unsuitable for samples usually encountered in biochemical analysis, where the resolution of molecules of similar structure and properties is often desired. However, when gradient elution is used to separate synthetic-polymer samples with large M, the resolution of individual polymer molecules becomes impractical (Section 6.1.4.1), and "on–off" separation may then be preferred in some cases.

"*Critical elution behavior*" was suggested in the early 1980s as a means of explaining certain unexpected results that were encountered in the gradient separation of large molecules of both natural and synthetic origin [44, 45]. Later, an entirely different kind of "critical chromatography" was employed for the separation and characterization of synthetic polymers, referred to as *liquid chromatography under critical conditions* (LCCC) [46]. There is no connection between these two processes; gradient separation that is related to LCCC ("pseudo-critical" gradient elution) is discussed separately in Section 6.3.2. Returning to "critical elution behavior," it was proposed for molecules with large enough M that the dependence of retention k on %B is so steep as to be effectively infinite. As a result, isocratic retention of synthetic polymer fractions becomes impossible for sample molecular weights M above a certain value. Similarly, the gradient separation of such samples is restricted to "on–off" behavior, *regardless* of experimental conditions. Note that on–off behavior can occur for molecules of any size, when the gradient is sufficiently steep so as to result in values of $k^* \leq 0.2$ (as in Fig. 6.3c). That is, for certain experimental conditions, on–off behavior is predicted by conventional gradient elution theory. However, conventional theory also predicts that on–off behavior will disappear with sufficiently flat gradients.

A number of other workers have reported other "unusual" results for the gradient separation of mixtures of proteins (discussed below). Together these observations suggest that the theory of conventional gradient elution does not apply for the separation of compounds whose size exceeds some maximum value of M. *This is an important issue, in as much as an incorrect theory of these separations can lead to the misinterpretation of experimental separations and/or an incorrect approach to method development for the gradient separation of high-molecular-weight samples.* In retrospect, the various observations in support of "critical elution behavior" can be attributed to a variety of misconceptions, as summarized in Sections 6.1.5.1 and 6.1.5.2. Although much of the evidence that contradicts "critical elution behavior" comes from separations of synthetic polymers, the question of "critical" vs "conventional" gradient elution theory is actually more important for separations of large biomolecules.

6.1.5.1 "Critical Elution Behavior": Synthetic Polymers *The reader may prefer to skip the remainder of this section and following Section 6.1.5.2, if he or she accepts that conventional gradient elution theory and the LSS model describe these separations adequately. These two sections are directed at readers who (a) have a more theoretical interest in large-molecule separations and/or (b) have followed the literature on "critical elution behavior" since 1984.*

"Critical elution behavior" assumes values of S for large molecules that are so large as to be unmeasurable. As a result, there exists a critical value of ϕ (ϕ_c) such that $k \approx \infty$ for $\phi < \phi_c$, and $k \approx 0$ for $\phi > \phi_c$; that is, a plot of log k vs ϕ would be essentially vertical. Consequently, attempts at the isocratic elution of a large molecule or polymer fraction will yield either no elution (k too large) or elution at t_0 ($k = 0$). Gradient separations under "critical" conditions result in "on–off" behavior, with ϕ at elution (ϕ_e) equal to ϕ_c (which does not change with separation conditions), as well as other unusual or unexpected features. A detailed refutation of each of these and other arguments in favor of "critical elution behavior" was reported in 1986 [47] and further supported by subsequent research [10, 13, 46, 48–51].

A key observation that led originally to the proposal of "critical elution behavior" involved failed attempts to achieve the isocratic elution of large-M polymer fractions, except with $k = 0$ or the complete retention of the sample. For large values of S, isocratic elution with, for example, $0.5 \leq k \leq 20$ is only possible over a very narrow range of %B, which can be of the order of 0.5 percent B or less. Therefore, attempts at locating the latter range in %B might easily be overlooked in trial-and-error experiments. A second difficulty in carrying out isocratic separations of polymers with large M is the ability of the polymer to sequester "good" (i.e., less polar) solvent within the molecule, in turn requiring adequate equilibration of sample and mobile phase prior to the column [10]. Without adequate mixing of sample and mobile phase for a sufficient time to allow equilibration, erratic experimental results can be expected. Finally, even "narrow" polymer fractions typically comprise molecules that span a wide range in M, so that resulting values of k for different oligomers (at any value of ϕ) can vary by several orders of magnitude. A wide range in M can preclude the observation of a distinct peak with $0 < k < \infty$ in isocratic elution, because compounds of lower M will elute with $k \approx 0$, while (for the same experiment) compounds with higher M will have $k \gg 20$ [47]. Consequently, a failure to observe isocratic elution for high-molecular-weight polymer samples does not necessarily contradict conventional chromatographic theory.

Despite the difficulty in observing isocratic elution for polymers of large M, several experimental studies have been successful in this regard: $M \leq 50,000$ [7]; $M \leq 2.5 \times 10^5$ [13]; $M \leq 2.8 \times 10^6$ [10]. Also, measurements of S for polymer fractions with very large M are possible by means of gradient elution (Section 9.3.4), although very accurate gradient retention data are required for reliable results (Section 6.1.5.3).

6.1.5.2 *"Critical Elution Behavior": Biopolymers*

At the same time that "critical elution behavior" was being invoked as an explanation for *apparently* atypical behavior in the gradient separation of synthetic polymers, a number of studies with similar results for the separations of proteins were also reported [52–59]. The most common observation was that sample resolution did not change when column length (only) was increased or shortened [52–56, 58]. Similarly, resolution did not improve when flow rate (only) was reduced [54, 59]. We have seen previously (Sections 2.3.2 and 2.3.3) that, when column length or flow rate are changed, with no change in other separation conditions, resulting changes in gradient separation bear little resemblance to corresponding changes for isocratic elution. The main reason for this difference in isocratic vs gradient elution is that k does not change when column length or flow rate are changed in isocratic elution, but in gradient elution k^* is a function of column size and flow rate [Equation (6.4)]. When column length is increased, N^* increases, but k^* decreases, resulting in offsetting effects on resolution. Similarly, when flow rate is increased, N^* decreases but k^* increases.

The application of conventional theory to a quantitative interpretation of peptide and protein separations by gradient elution was underway by 1983, and subsequent studies were able to confirm a general agreement between experimental data and predictions based on theory [1, 25, 49, 60, 61]. Thus, for peptide and protein samples with $600 \leq M \leq 14{,}000$, gradient experiments have been carried out as a function of column length, flow rate, gradient time, and gradient range ($\Delta \phi$). Experimental retention times, peak widths, and peak heights agreed closely with predictions from theory. Finally, for separations carried out with reasonable values of k^*, the migration of a protein through the column should accelerate with time, as in Figure 6.3(a) (concave plot of x vs t). The use of glass-walled columns with colored proteins confirmed this behavior [62], thus demonstrating the applicability of conventional gradient elution theory for the separation of protein samples. Critical elution behavior or on–off separation should have resulted in a linear plot of x vs t (as illustrated in Fig. 6.3c by the plot marked $-\cdot-\cdot-$).

As one group of workers concluded in 2005 [50], "it is generally accepted that the mechanisms involved [i.e., retention process in the gradient separation of synthetic polymers with large M] will depend on the sample, the concentration of sample injected onto the column, on the choice of mobile phase and on the strength of the interaction between the sample and the stationary phase." With this in mind, the evidence reported so far supports "conventional chromatography" as a preferred explanation for gradient separations of both synthetic polymers and biopolymers by RP-LC with water-containing mobile phases. Separations of synthetic polymers based on nonaqueous RP-LC or normal-phase chromatography also appear to be generally consistent with a conventional retention process, with but few exceptions, for example, differences in values of S measured by isocratic or gradient elution [63] (but note that the determination of values of S in such cases requires extremely careful experimental measurements; see Section 6.1.5.3 and [64]). Size-exclusion effects represent a minor complication for very large molecules separated on small-pore columns [10], but with little consequence for the application of conventional gradient elution theory. While the basis of retention may still be ambiguous for a few cases that involve the separation of large-M samples with steep gradients, resulting deviations of experimental data from predictions based on conventional chromatography are often within experimental error and almost never have any practical significance. The foregoing observations should resolve any doubt that "large molecule" gradient elution obeys the same rules as "small molecule" separations.

6.1.5.3 Measurement of LSS Parameters for Large Molecules Values of $\log k_w$ and S for a given compound and experimental conditions can be measured directly from a best fit of isocratic retention data to Equation (6.1). Similarly, values of these two parameters also can be determined from two gradient runs where only t_G is varied (Section 9.3.3). Resulting values of $\log k_w$ and S determined in either way can be quite accurate (and therefore in agreement with each other) for samples with molecular weights $<10{,}000$. However, as M increases further for the compounds in a sample, the accuracy of these derived parameter values begins to decrease, and for $M > 100{,}000$ special care must be taken in order to obtain repeatable values from either isocratic or gradient data. Thus, for a measurement of S from isocratic values of k for two different values of ϕ, $S = \log (k_2/k_1)/(\phi_1 - \phi_2)$. For large M (and therefore large S) *and* a given value of $\log (k_2/k_1)$, the value of

($\phi_1 - \phi_2$) will be small. Errors in ϕ will therefore have a larger effect on derived values of S when M is larger. For gradient elution, values of S are related to differences in %B at elution for two runs with varying t_G (Section 9.3.3). As M increases, these differences in %B become smaller and smaller, and their measurement correspondingly less accurate. For a review of ways in which the determination of values of log k_w and S for large molecules can be improved, see [7, 10, 50].

6.2 BIOMOLECULES

Biomolecules include peptides, proteins, oligonucleotides, nucleic acids, and polysaccharides. Each of these sample types can be regarded as a chain formed from repeating subunits, similar to the "homo-oligomers" of Section 6.1.4 or the small protein of Figure 6.4. With few exceptions; however, the nature of the repeating unit generally varies within a given biomolecule. Peptides and proteins are composed of various combinations of 20 different amino acids, oligonucleotides and nucleic acids are formed from four different nucleotides, and the building blocks for polysaccharides can comprise a variety of simple sugar molecules. Mixtures of biomacromolecules therefore tend to be "irregular," in contrast to the "regularity" of synthetic polymers and oligomeric mixtures (Section 6.1.4). As a consequence, potentially useful changes in retention order or selectivity are often seen for these samples when gradient time (and values of k^*) is varied. Because of larger values of S for large peptides and proteins, isocratic separation is rarely useful for samples containing these compounds. However, an exception can be noted for the preparative isolation of a single peptide or small protein, where isocratic separation of the desired product can be followed by a step-gradient to purge more strongly retained peaks from the column [65]; see also Section 7.3.2.1.

6.2.1 Peptides and Proteins

Various goals can be defined for the gradient separation of peptide or protein samples. An *assay procedure* usually aims at the measurement of the concentration of target compounds in related samples ("*quantitation*"). If UV detection is used, the baseline separation of every component is preferred; however, this is not always possible for complex mixtures that contain 15 or more components. *Generic procedures* are intended for the partial separation of different samples (e.g., a single gradient elution procedure for the separation of digests from *any* protein) and are often used with MS detection (Section 8.1). Generic separations are also used to characterize or "fingerprint" the proteins present in different plant species or varieties, for example, in the measurement of cereal proteins [66]. *Preparative separations* aim at maximum resolution for one or more compounds in the sample, in order to increase the weight of a purified compound that can be isolated in a single run (Chapter 7). Low-pH conditions which favor protein denaturation during RP-LC separation are preferred, but proteins with $M \leq 20{,}000$ can usually be restored to their native conformation by dissolving the recovered protein in a suitable buffer (Section 6.2.3).

Neutral-pH mobile phases can be used to minimize denaturation during RP-LC separation; however, peak shape and resolution tend to deteriorate under these conditions, especially for proteins with $M > 10{,}000$.

The analysis of peptide and protein samples is most often accomplished by reversed-phase gradient elution. Except for Section 6.2.2, our discussion of peptide and protein separations will be limited to assay procedures by means of gradient RP-LC.

6.2.1.1 Sample Characteristics The importance of peptide and protein conformation in RP-LC separation was discussed above (Section 6.1.3). Usually it is desirable to achieve sample denaturation prior to separation. Samples containing peptides and/or proteins tend to be "irregular," as defined in Section 1.6. This is illustrated by the scattered plots of S vs log k_w in Figure 6.6 for (*a*) the peptides in a digest of recombinant human growth hormone (*rhGH*), and (*b*) the 30S ribosomal proteins from *Escheria coli* Q13 (compare the plots of Fig. 6.6 with the similar plot of Fig. 1.11*b* for an irregular, "small molecule" sample). It has also been noted (Section 6.1.1) that values of S increase for larger molecules, with a corresponding effect on their separation by gradient elution [Equation (6.4)]. An average value of $S \approx 25$ can be assumed for the peptides in a protein digest when using acetonitrile–buffer gradients (Fig. 6.6*a*), while $S \approx 40$ for small proteins (average $S = 43$ in Fig. 6.6*b*, for proteins with molecular weights that cover the range $8K \leq M \leq 26K$). It can be assumed that the various peptides and proteins represented in Figure 6.6 are in each case denatured by the low-pH mobile phase used for these separations.

Values of $S > 10$ are observed for all of the compounds in Figure 6.6, whereas typical small molecules have $S \approx 4$. Compared with recommended separation conditions for small molecules (Table 3.2), flatter gradients will be required for peptide and protein samples in order to achieve comparable values of k^* and resolution (see Table 6.1). Because of the "irregularity" of peptide and protein samples (Fig. 6.6), changes in conditions that affect k^* [gradient time, column length, flow rate; Equation (6.4)] can also be used to control peak spacing and maximize the resolution of either individual peaks or the entire chromatogram. Changes in gradient time are usually preferred both for the control of k^* and for initial attempts to vary separation selectivity α^*.

6.2.1.2 Conditions for an Initial Gradient Run Unless appropriate experimental conditions are selected, separations of peptides and especially proteins can result in wide, tailing peaks and/or poor recoveries of the injected sample. These problems can be minimized by the use of certain preferred separation conditions (column type, temperature, mobile phase B solvent, pH, and buffer [67, 68]). Today, most such separations by RP-LC are carried out with conditions similar to those of Table 6.1. While these conditions are generally recommended for the separation of different peptide and protein samples, certain peptides and proteins may favor the use of other conditions. A discussion of the special requirements for separating individual proteins or protein types is outside the scope of the present book; for details, see [69–72]. Preferred conditions for preparative separations or

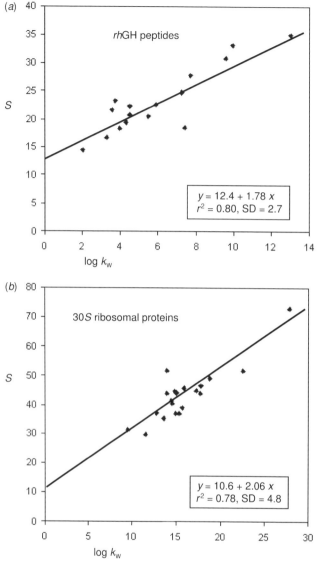

Figure 6.6 Plots of S vs log k_w for a mixture of peptides (*a*) and proteins (*b*), illustrating the "irregularity" of such samples. Data from [81, 82]; compare these plots with those of Figure 1.11.

the use of mass spectrometric detection are discussed in Sections 7.3.2 and 7.3.3, respectively.

Table 6.1 suggests initial conditions for peptide or protein samples of varying molecular weight. The small values of k^* for these different samples in the initial run ($k^* = 1–2$) are chosen for shorter run times, assuming that gradient time can be increased as necessary. On the basis of the initial run, "best" values of initial and

TABLE 6.1 Recommended Conditions for the Initial Gradient Separation of Peptide and Protein Samples. UV Detection at 210 nm is Assumed, but Similar Conditions Apply for LC-MS (See Section 8.1 for any Differences)

	Values for different samples		
		Proteins	
Condition	Peptides, digests $1 < M < 5$ kDa	$5 < M < 20$ kDa	$M > 20$ kDa
Sample treatment prior to injection	None	Add 8 M urea, store for 30 min	Add 8 M urea, store for 30 min
Column[a]	150 × 4.6 mm, type B C_{18} (8–12 nm pore diameter), 5 μm particles	150 × 4.6 mm, type B C_8 (12–30 nm pore diameter), 5 μm particles	50 × 4.6 mm, type-B C_4 ($\geq$30 nm), 5 μm particles
Solvent A	0.12% TFA[b]–water	0.12% TFA–water	0.12% TFA–water
Solvent B	0.10% TFA[b]–ACN	0.10% TFA–ACN	0.10% TFA–ACN
Gradient range	0–60% B	5–100% B	5–100% B
Temperature	30–35°C	30–35°C[c]	30–35°C[c]
Flow rate (mL/min)	2.0	1.0	0.5
Gradient time (min)	25	50	50
k^*	2	1	1
%B/min	2.4	1.2	1.2
Value of S assumed	25	40	70

[a]Columns should have reduced silanol acidity (Appendix III and [87]) and be stable at low pH and temperatures $\leq$60°C; other column lengths, diameters, and particle sizes can be used, in which case gradient time and flow rate should be adjusted to maintain similar values of k^* with acceptable pressure drop; for example, for >20 kDa proteins, 150 × 4.6 mm columns can be used with gradient times of 150 min (other conditions the same). The choice of ligand length (C_8, C_{18}) seems less critical when using recently introduced, type B columns (with low silanol acidity; e.g., C[2.8] $\leq$ 0.00 in Appendix III).

[b]A recent reference has suggested that 0.3 percent TFA in both the A and B solvent may be preferable for some protein digests [73], especially those containing more basic peptides; an increase in TFA increases the retention of these highly polar compounds and avoids their early elution; it also improves peak shape, possibly as a result of minimizing silanol interactions.

[c]Higher temperatures (e.g., 60–70°C) can be desirable for some protein samples [67], especially those with $M > 20,000$; column stability for these conditions should be verified before the use of >60°C and pH < 2.5.

final %B values can be chosen for the gradient, allowing larger values of k^* in subsequent runs (because of reduced $\Delta\phi$), without a proportional increase in run time. Separations of peptide samples can usually be carried out with a gradient of 0–60 percent B or 5–60 percent B, although strongly retained peptides may require a higher %B for complete elution of the sample. Proteins may benefit from higher %B-values at both the beginning and end of the gradient. In method development for small-molecule samples, it was recommended (Section 3.3.3, Fig. 3.1) to delay adjustments in the gradient range (initial and final %B values) until after peak spacing (values of α^*) has been optimized. Because separations of proteins often require long run times (sometimes accompanied by lower sample recoveries), and because these compounds tend to leave the column at similar values of %B (regardless of gradient conditions), narrowing the gradient range should be carried out following the initial gradient run, so as to reduce the time spent on subsequent experiments.

Protein samples (especially those with $M > 10,000$) may benefit from an initial denaturation of the sample prior to injection. This can be accomplished by adding an equal volume of 8 M urea to the sample, followed by mixing and incubation at ambient temperature for 30 min. The use of a higher *separation* temperature can further contribute to initial protein denaturation, especially for large proteins ($M > 20,000$). *It should be noted that proteins with $M > 50,000$ often exhibit problems when separated by RP-LC, and for this reason other HPLC modes (ion exchange, HIC, etc.) may prove more useful.* Peptide samples are usually separated with columns having 3–5 μm particles and pore diameters between 8 and 12 nm. Protein samples require a larger pore diameter; the use of 30 nm pores is common for all but very large protein molecules, for which commercial columns with 100 nm pores are available. The use of smaller-particle and/or pellicular columns is especially attractive for fast, analytical separations of peptides or proteins [73].

For various reasons, ACN is usually the preferred B solvent for peptide separations, while either acetonitrile or isopropanol is the best choice for separating proteins. A low-pH mobile phase provides generally better separations of proteins. The addition of ~0.1 percent TFA to both the A and B solvents is the most common means of providing a low pH (pH ≈ 1.9), although 0.3 percent TFA has been recommended for very basic peptides [74] (see Section 5.5.2.9 for some practical details concerning the use of TFA). Varying the concentration of TFA can also provide a means of changing selectivity [75], because of the ion-pairing of TFA with positively charged peptides; compared with other peptides in the sample, more positively charged peptides are preferentially retained with higher TFA concentrations. Similar, but more pronounced changes in selectivity are possible when using perchlorate as ion-pair reagent [76]. Formic acid or ammonium formate [77] is recommended as mobile phase additive when mass spectrometric detection is used (Section 8.1), and isopropanol is sometimes preferred over acetonitrile for the separation of proteins with $M > 20,000$.

RP-LC columns for peptide and protein separations must meet certain requirements [67], especially low silanol acidity. While polymeric columns do not contain silanols, and are therefore suitable for peptide and protein separations, somewhat lower values of N^* are typical of these columns. Most present separations of peptides and proteins use alkylsilica columns, as specified in Table 6.1. Alkylsilica columns can be classified according to silanol acidity by means of their cation-exchange activity **C** (see discussion of Appendix III); columns with lower values of **C** are preferred for peptide and protein separations. Some RP-LC columns are less stable at low pH, especially for temperatures >30°C. Therefore, some thought should be given to the selection of stable columns for peptide or protein separations. So-called "sterically protected" columns (sold under the label "StableBond" by Agilent) are thermally stable at low pH. The hybrid Xterra columns sold by

Waters are thermally stable at both low and high pH. (Columns stable at high or low pH are available from other vendors, as well.)

Many protein digests contain peptides that are weakly retained with 0 percent B as mobile phase, while most peptides elute before 60 percent B. This suggests the initial use of 0–60 percent B gradients for peptide samples, but keep in mind the occasional possibility of peak elution after the gradient ends, as illustrated in Figure 2.22(c). Proteins are less likely to be weakly retained with a mobile phase of 0 percent B, and they are more likely to elute after 60 percent B, so a 5–100 percent B gradient is recommended initially for protein samples. Small-molecule samples can be separated effectively with flow rates of 1–2 mL/min, but higher-molecular-weight samples benefit from lower flow rates – as suggested in Table 6.1 for peptides and proteins of varying molecular weight. The gradient times shown in Table 6.1 (25–50 min) reflect target values of $k^* \approx 1$–2 for each separation, corresponding to values of S estimated at 25, 40, and 70, respectively, for these three sample types (peptides, small proteins, large proteins).

6.2.1.3 Method Development
Method development for samples that contain large biomolecules proceeds in similar fashion as for small molecules (Section 3.3). An initial separation of the sample is carried out according to Table 6.1, following which conditions are varied to achieve an acceptable separation. We will illustrate this procedure for both a peptide and a protein mixture (protease digest of *recombinant human* growth hormone [*rh*GH] and a mixture of cereal proteins, respectively).

Figure 6.7(a) shows the initial separation of the *rhGH digest*. Nineteen peptides are present in the sample, but only peaks 6–14 are difficult to separate (Fig. 6.7b); further chromatograms (Fig. 6.8) will emphasize peaks 6–14, unless noted otherwise. The initial separation of Figure 6.7 results in a resolution of $R_s = 1.2$, which might be adequate for some applications. If improved resolution of this sample is required (e.g., baseline separation), the preferred approach is to carry out additional separations where both temperature and gradient time are varied (as in the case of small-molecule samples; Section 3.3). Beginning with the initial separation, gradient time should be increased by 2- to 3-fold, and temperature should be increased by 15–20°C, to give the four runs of Figure 6.8(a–d).

Visual inspection of separations as in Figure 6.8(a–d) may suffice for an evaluation of the preferred temperature and gradient time for maximum resolution and/or minimum run time. However, the use of computer simulation (Section 3.4) will usually prove quicker and more effective for optimizing the separation of peptide and protein samples [78–85]. We assume initially that computer simulation is *not* available. The choice of best conditions for this sample is complicated by the fact that several peak pairs can have marginal resolution for different conditions: 6–7 (b), 10–11 (c, d), 11–12 (a, b), and 13–14 (at intermediate temperatures, due to a peak reversal with change in temperature from 30 to 50°C). For a gradient time of 25 min (a, c), an increase in temperature is seen to increase the resolution of the critical peak pair 11–12, but with a decrease in resolution for peaks 10 and 11, and with complete overlap of peaks 13 and 14 at intermediate temperatures. The best (trial-and-error) separation for a 25 min gradient gives $R_s = 1.6$ for 45°C.

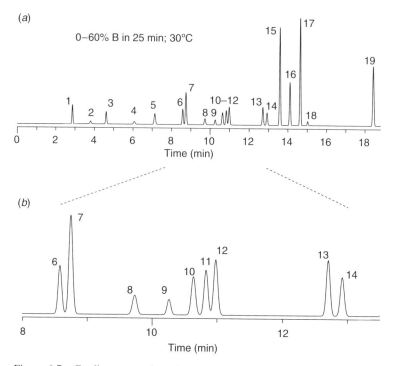

Figure 6.7 Gradient separation of peptide digest of *rh*GH. Conditions of Table 6.1 ("peptides"), temperature = 30°C. (*a*) Full chromatogram; (*b*) separation of peaks 6–14 only. Simulations based on data from [42].

Alternatively, for a gradient time of 75 min, the same trends with temperature are observed as for the 25 min gradient, but a better resolution can be obtained: $R_s = 2.1$ at 45°C. The latter separation is shown in Figure 6.8(*e, f*). Other changes in experimental conditions can be used for further improvements in peak spacing and separation; examples have been reported that involve different columns (C_4 vs C_{18}) or other B solvents (methanol and isopropanol) [84]. A somewhat different approach ("primitive grid search") for the optimized separation of peptide mixtures has been described by Lundell et al. [86, 87], varying gradient time, mobile phase pH, and different amounts of isopropanol or TFA added to the mobile phase.

The use of computer simulation can improve resolution slightly for this sample (with no need for exploratory experiments beyond the four runs of Fig. 6.8*a–d*): $R_s = 2.3$ for a 71 min gradient at 43°C, or $R_s = 2.6$ for a 91 min gradient at 63°C. Computer simulation can also facilitate the further improvement of this separation for reduced run time, as in Figure 6.8(*g*), where a segmented gradient is employed to compress the latter part of the chromatogram without a critical loss in resolution (see Section 3.3.4 for the similar use of segmented gradients for small-molecule samples). Other protein digests can contain many more peptides ($\gg 19$), and their baseline separation is then more difficult. An example of such a separation has been described [82] for the protein digest of recombinant tissue

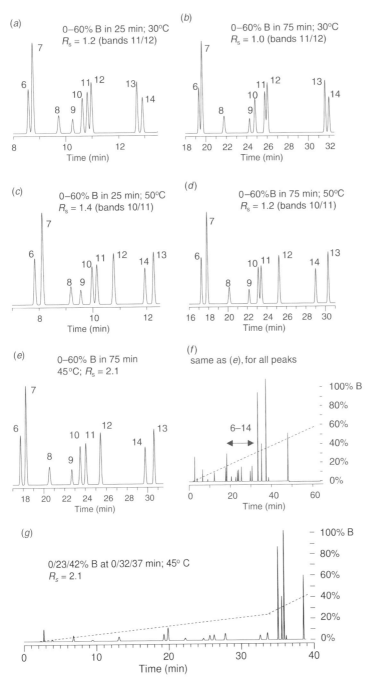

Figure 6.8 Separation of *rh*GH digest as in Figure 6.7, using different gradient times and temperatures in order to optimize selectivity and maximize resolution. See figure for conditions for each separation; consult the text for additional comment. Simulations based on data from [42].

plasminogen activator (rt-PA). This sample contains 37 major peptides, and the best separation that could be achieved by varying temperature and gradient time resulted in a resolution of only $R_s = 0.5$. Alternatively, the baseline separation of any one peptide from a digest is usually possible, using the approach described at the end of the present section (see discussion of Fig. 6.12).

Consider next the separation of a *cereal storage protein* sample (5–20 kDa). The initial separation with the conditions of Table 6.1 is shown in Figure 6.9(a) (only major peaks shown), except that a higher starting temperature (50°C) is used. Note that the first peak elutes at 29 min (well after the gradient starts), and the last peak leaves the column at 45 min) just before the gradient ends. Because values of k for proteins change markedly with %B, the value of %B at elution changes very little when k^* or gradient conditions are varied [Equation (2.15) and Section 2.3.1]. As a result, beginning the gradient just before the appearance of the first peak (~52 percent B) and ending it just after the last peak (~82 percent B) will save time in subsequent method development experiments [Equation

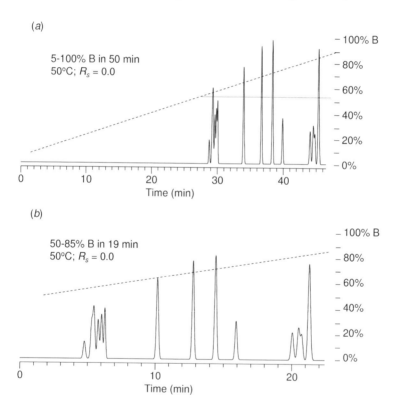

Figure 6.9 Separation of a cereal protein sample, beginning with similar conditions as recommended in Table 6.1, except for the use of a higher starting temperature. Conditions: 150 × 4.6 mm C_{18} column (30 nm pores); acetonitrile–water gradients (0.1 percent TFA added to each solvent); 1.0 mL/min; 50°C. Gradient range and time shown in figure. (a) Initial experiment as in Table 6.1; (b) revised conditions to shorten run time by compressing the gradient range. Simulations based on data from [82].

(2.15) provides the relation between retention time and %B at elution]. In this case, a 50–85 percent B gradient was selected, which requires a decrease in gradient time by a factor of $(85 - 50)/(100 - 5) = 0.37$. To maintain k^* constant (recommended in this second experiment), gradient time must be reduced by the same factor, to a new value of $t_G = 50 \times 0.37 = 19$ min (Fig. 6.9b).

Having established a desirable gradient range (50–85 percent B) for the cereal protein sample, four runs with gradient time and temperature varied were carried out next (Fig. 6.10). The higher temperature range (50–70°C) was at the option of the chromatographer, and required a large-pore column that was stable at these temperatures and low pH (Agilent Zorbax StableBond 300A C_{18}). The experiment of Figure 6.10(c) (40 min gradient, 70°C) gives the best separation ($R_s = 0.6$), but a further improvement in resolution is desirable. A comparison of the separation of critical bands 12–14 at 70°C as a function of gradient time (Fig. 6.10c, d) suggests that an intermediate gradient time may provide better resolution (several additional experimental runs might be required to pin down the best gradient time and temperature). Computer simulation (Section 3.4), based on data from just the four runs of Figure 6.10, can provide a resolution map (Fig. 6.11a), which allows the

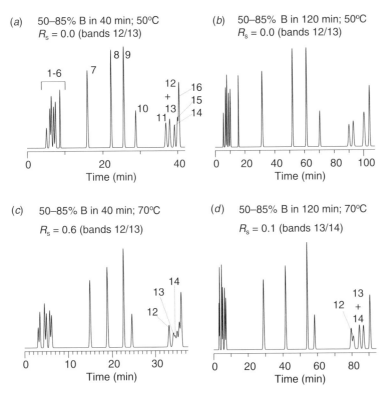

Figure 6.10 Separation of cereal protein sample as in Figure 6.9, using different gradient times and temperatures to optimize selectivity and maximize resolution. See figure for conditions and text for additional comment. Simulations based on data from [84].

easy selection of best values of temperature and gradient time (indicated by cross-hairs). The predicted separation ($t_G = 112$ min, 67.5°C, with $R_s = 1.1$) is shown in Figure 6.11(b).

Further changes in selectivity for peptide samples can be obtained by a change in B solvent (either propanol or methanol) or a change in column (C_4 vs C_{18}) [84], as well as by change in the concentration of TFA in the mobile phase [75]. Following any of the latter changes in conditions, it is suggested that four more runs with varying gradient time and temperature should be carried out.

Sometimes the goal of separation is the isolation of one or more sample compounds in pure form. As discussed in Section 7.2.2.1, the best conditions for

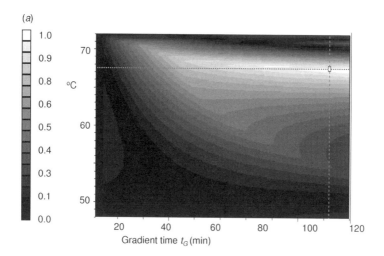

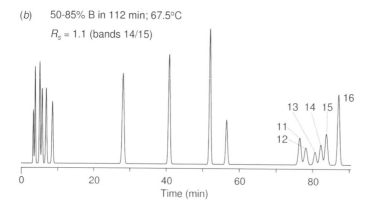

Figure 6.11 Selection of best conditions for the separation of the cereal protein sample using computer simulation. (a) Resolution map based on experiments of Figure 6.10; (b) separation with maximum resolution from (a).

achieving the preparative separation of a particular peak should maximize its resolution from adjacent peaks. As an example, consider protein 14 in Figure 6.10. None of the separations shown in Figures 6.9 and 6.10 provide $R_s > 1.1$ for this peak, but conditions *can* be selected for a much better resolution of peak 14 from adjacent peaks 13 and 15. This could be done by trial-and-error, while ignoring the separation of any peak except 14. Computer simulation provides a much more efficient approach by allowing the user to create a new resolution map (different from that in Fig. 6.11*a*, but requiring no new experimental data), based on selecting one or more peaks of interest, while ignoring the separation of other peaks from each other (but *not* ignoring their possible overlap onto peaks of interest). Figure 6.12 shows the results of such a computer simulation, where only the resolution of peak 14 is important. The resolution map of Figure 6.12(*a*) shows that a maximum resolution of peak 14 occurs for a temperature of 48°C and a gradient time of 37 min (cross-hairs, $R_s = 1.8$). The resulting separation of protein 14 from neighboring peaks is shown in Figure 6.12(*b*). The latter separation can also be achieved in shorter time with a gradient from 77 to 83 percent B in 13 min, although early peaks 1–8 are then bunched together at the front of the gradient and are therefore unresolved. However, this would be of no consequence when only the isolation of peak 14 is of interest.

6.2.1.4 Segmented Gradients As discussed in Section 3.3.4, segmented gradients can be used for different reasons:

- to clean the column between sample injections;
- to shorten run time;
- to improve separation by adjusting selectivity for different parts of the chromatogram.

Concerning *column cleaning*, samples of biochemical origin, such as peptides and proteins, often contain material that is of no interest to the analyst, but which is strongly retained and therefore can foul the column. For gradients that end short of 100 percent B, it is therefore common to add a steep ("column-cleaning") gradient segment that ends at 100 percent B (as in the example of Fig. 3.8*b*). The use of a segmented gradient for *shortening run time* was demonstrated above in Figure 6.8(*g*) for the separation of a protein digest.

Improving selectivity by means of segmented gradients was illustrated earlier for a small-molecule sample (Fig. 3.10), where early peaks were better separated with a flatter gradient and later peaks preferred a steeper gradient. In the case of small-molecule samples, the use of segmented gradients in this way is usually of limited value, because the separation of later peaks (with a different gradient steepness) is affected by the relative steepness of the initial gradient segment. For this reason, the separation of small molecules using a segmented gradient can never achieve a resolution of later peaks that is good as could have been obtained with a linear gradient whose steepness is optimized just for these later peaks.

In the case of protein samples, however, the use of segmented gradients as a means of optimizing selectivity and resolution is more promising, because the (a)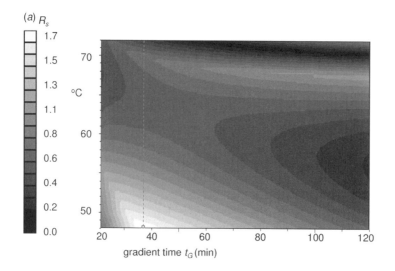

(b)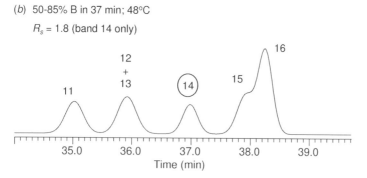

Figure 6.12 Selection of best conditions for the separation of peak 14 of the cereal protein sample using computer simulation. (a) Resolution map for peak 14, based on the experiments of Figure 6.10; (b) separation with maximum resolution of peak 14 based on optimum conditions from (a) (indicated by ●).

separation of later peaks is almost entirely determined by the steepness of the second segment – regardless of the steepness of the initial segment (due to large values of S, with resulting minimal migration of later peaks during the initial gradient segment). An example of this use of segmented gradients has been reported for the 30S ribosomal proteins [80], where a three-segment gradient allowed the first reported separation of all 20 proteins in this complex sample. The similar separation of the 50S ribosomal proteins using a four-segment gradient was able to resolve 31 of the 32 proteins as distinct peaks [80]. In each of the latter two examples, the experimental separation agreed closely with predictions from computer simulation (which was used to optimize these multi-segment gradients).

6.2.2 Other Separation Modes and Samples

Other biological samples include oligonucleotides, nucleic acids, viruses, and carbohydrates. Each of these compound types is relatively hydrophilic, and is therefore not often separated by means of RP-LC (mainly because of inadequate retention), except with the aid of ion-pairing [88]. As an example of the use of gradient ion-pair RP-LC for the separation of double-stranded nucleic acid fragments (strictly according to size), see Figure 6.13. The oligonucleotides from this DNA digest were first separated by other means, then recombined into the seven samples of Figure 6.13(a). The retention time of each oligonucleotide is plotted vs the number of base pairs in each fragment in Figure 6.13(b).

A variety of HPLC modes other than RP-LC have been employed for the separation of hydrophilic biochemical macromolecules: ion-exchange chromatography, hydrophilic interaction chromatography (HILIC), metal chelate chromatography, and some other less common procedures. The separation of very large and/or

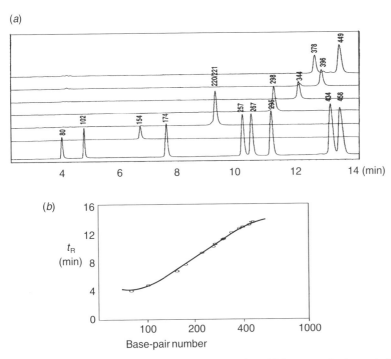

Figure 6.13 Separation of double-stranded nucleic acid fragments by ion-pair RP-LC. Sample: enzymatic digest of a DNA sample. Conditions: 50 × 4.6 mm nonporous alkylated-polystyrene column (DNAsep; Transgenomic, Palo Alto, CA, USA); 30–50–65 percent B at 0–3–15 min; the A solvent is 0.1 M triethylamine acetate (pH 7.0); the B solvent is solvent A plus 25 percent acetonitrile; 1.0 mL/min. (a) Separation of different oligonucleotides from DNA digest; (b) plot of retention time for individual oligonucleotides vs number of base pairs in the molecule. Adapted from [90].

hydrophobic proteins by RP-LC can also prove unsatisfactory: low sample recoveries, broad misshaped peaks, and nonreversible protein denaturation during separation. For such samples, hydrophobic interaction chromatography (HIC) can represent a preferred alternative to RP-LC. Separation procedures other than RP-LC can also be useful for a change in selectivity and for use in two-dimensional separation. Combinations of RP-LC and other HPLC procedures are commonly used for the isolation and purification of proteins from natural sources. Finally, many of the above HPLC procedures involve columns and conditions that minimize protein denaturation, whereas protein denaturation is favored by RP-LC.

6.2.2.1 Hydrophobic Interaction Chromatography HIC has been applied primarily to the separation of proteins, viruses, and (less often) nucleic acids [71, 90–92]. Columns for HIC are similar to those used for RP-LC, except that the bonded phase is less hydrophobic by virtue of (a) hydrophilic groups (e.g., ether) that are incorporated into short alkyl ligands, and (b) a less dense bonding. The mobile phase is usually an aqueous solution of an antichaotropic (nondenaturing) salt such as ammonium sulfate plus a buffer to control pH (usually, $6 \leq \text{pH} \leq 8$). HIC separations typically consist of inverse gradients from high to low concentration of ammonium sulfate. The combination of a less hydrophobic packing with purely aqueous mobile phases minimizes protein denaturation and usually allows the recovery of native (undenatured) proteins from separated fractions, especially for separations carried out at near-ambient temperature.

The dependence of retention k on the concentration of ammonium sulfate C_{AS} determines the applicability of the LSS model of gradient elution for HIC. As seen in Figure 6.14 for several proteins, plots of $\log k$ vs C_{AS} in HIC are typically linear and can be expressed as

$$(\text{HIC}) \quad \log k = \log k_0 + A_{\text{HIC}} \, C_{AS} \quad (6.7)$$

Here, k_0 refers to the value of k for $C_{AS} = 0$, and A_{HIC} is the slope of a plot as in Figure 6.14 [equal to $d(\log k)/d(C_{AS})$]. Note that k *increases* with increased concentration of ammonium sulfate (C_{AS}), so that C_{AS} must decrease during the gradient. Equation (6.7) can be transformed into the same form as Equation (6.2) for LSS gradients:

$$(\text{HIC}) \quad \log k = \log k_{2.5} - S_{\text{HIC}} \, \phi_{\text{HIC}} \quad (6.8)$$

where $k_{2.5}$ is the value of k for 2.5 M ammonium sulfate, and ϕ_{HIC} will normally vary from 0 to 1, corresponding to an inverse linear gradient from 2.5 to 0.0 M ammonium sulfate (the most popular HIC gradient). Here, S_{HIC} is arbitrarily set equal to $-2.5 \, A_{\text{HIC}}$, and ϕ_{HIC} is defined as $-(C_{AS} - 2.5)/2.5$. For example, values of C_{AS} equal to 2.5, 1.25, and 0.0 M correspond to ϕ_{HIC} equal 0.00, 0.50, and 1.00, respectively. Similar values of A_{HIC} and S_{HIC} are found for salts other than ammonium sulfate [92], but these other salts are less often used.

Since Equation (6.8) for HIC is equivalent to Equation (6.2) for RP-LC, *gradient separations based on either of these two separation modes are governed by the LSS model and follow similar qualitative and quantitative rules* (as in previous chapters for RP-LC gradient elution); for example, Equation (6.4) is

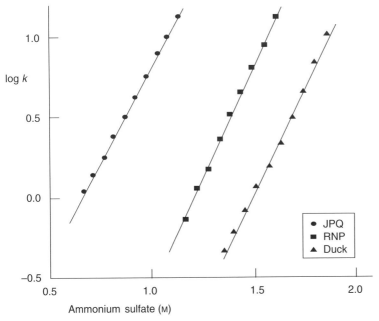

Figure 6.14 Isocratic retention in HIC as a function of the concentration of ammonium sulfate in the mobile phase (log–linear plot). Samples: various bird lysozymes (JPQ, Japanese quail; RNP, ring-necked pheasant); the column is a TSKgel Phenyl-5PW; the mobile phase is aqueous ammonium sulfate buffered with 10 mM phosphate at pH 7.0. Figure adapted from [93].

applicable for HIC, if S is replaced by S_{HIC}:

$$\text{(HIC)} \quad k^* = t_G F / (1.15 V_m \Delta \phi S_{HIC}) \tag{6.9}$$

Karger et al. [93] and others [94] have shown that retention and separation in HIC gradient elution are in quantitative agreement with the LSS model. This similarity of HIC and RP-LC greatly simplifies the treatment of gradient separations of proteins by HIC; that is, the various relationships developed for RP-LC gradient elution can be applied directly to separations by HIC.

An important difference between HIC and RP-LC separations of proteins concerns the respective values of S_{HIC} and S for each separation mode as a function of molecular weight. For proteins with $10^4 \leq M \leq 10^5$, RP-LC separations have values of S that range from roughly 25 to 80 (see Fig. 6.2). Figure 6.15 shows a log–log plot of S_{HIC} vs M, corresponding to values of S_{HIC} of 4–9 for this same protein molecular weight range ($10^4 \leq M \leq 10^5$). Thus, values of S_{HIC} for proteins are almost an order of magnitude smaller than corresponding values of S for gradient RP-LC separations, *which means that for a desired range in k^* of 1–10, steeper (shorter) gradients can be used in HIC separations of proteins, compared with RP-LC.*

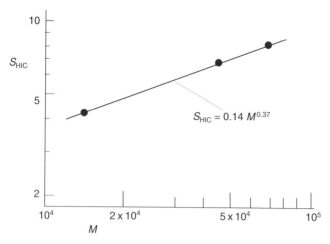

Figure 6.15 Variation of S in HIC (S_{HIC}) with protein molecular weight (log–log plot). Samples are lysozyme ($M = 14{,}000$), ovalbumin ($M = 44{,}000$) and conalbumin ($M = 68{,}000$). Values calculated from data from [94].

Kato et al. [95] have described several HIC separations of proteins where different gradient or column conditions were varied. Changes in initial %B (C_{AS}), column length, or flow rate were made *without* change in other conditions, so that k^* varied during these experiments. The results obtained appear to be in qualitative agreement with similar changes in conditions for RP-LC.

6.2.2.2 Ion Exchange Chromatography

Ion exchange chromatography (IEC) is widely used for the gradient separation of peptides, proteins, oligonucleotides, and nucleic acids [72, 89, 91, 96–99]. IEC columns have charge-bearing groups incorporated into the stationary phase, thereby rendering the column less hydrophobic and capable of ion exchange: negative groups for cation exchange and positive groups for anion exchange. The mobile phase is an aqueous solution of a salt such as sodium chloride, plus a buffer to control pH; occasionally acetonitrile or methanol are added to the mobile phase as a means of reducing peak width and increasing resolution [88], while urea is sometimes added in order to solubilize proteins that are difficult to dissolve. IEC typically uses linear salt gradients that start with a low salt concentration, for example, 0.005–0.50 M. The combination of a (less hydrophobic) IEC packing with an aqueous mobile phase usually prevents protein denaturation in IEC.

Retention in IEC is governed by the competition of sample and salt ions (e.g., X^- and Cl^-, respectively) for interaction with the charged groups (e.g., $-N[CH_3]_3^+$) in the stationary phase (Section 8.2). As a result, retention is given (for the usual choice of a salt gradient with a mono-valent counter-ion) by

$$\log k = C_{iec} - m \log C \qquad (6.10)$$

Here, C_{iec} is a constant for a given sample compound and experimental condition (equal to log k for $C = 1$), m is the absolute value of the effective charge on the sample compound, and C is the concentration of the salt counter-ion (assuming a univalent counter-ion). Note that the "effective" charge on a protein molecule is often smaller than the actual charge, because only part of the protein molecule is able to interact with the stationary phase [97].

At higher salt concentrations (C), Equation (6.10) may fail [98] due to the onset of HIC retention; that is, as salt concentration increases, IEC retention decreases, and HIC retention increases. Note also that the ionization of any molecule (and its value of m) can vary with pH. The *effective* value of m in Equation (6.10) is reduced for polyvalent counter-ions in proportion to the charge on the counter-ion; for example, for bivalent Ca^{2+} as counter-ion in place of Na^+, the value of m would equal the charge on the sample molecule divided by 2.

Equation (6.10) for IEC is seen to be of different form (log–log) than Equation (6.2) for RP-LC (log–linear); gradient retention is therefore described by somewhat different equations (Section 8.2):

$$\text{(IEC)} \quad b = \{V_m m \log([C]_f/[C]_0)\}/(t_G F) \tag{6.11}$$

and

$$\text{(IEC)} \quad k^* = 1/1.15b = t_G F/1.15\{V_m m \log([C]_f/[C]_0)\} \tag{6.12}$$

Here, $[C]_0$ and $[C]_f$ refer, respectively, to the value of C at the beginning and end of the gradient. The form of Equation (6.12) for k^* in gradient IEC is quite similar to Equation (6.4) for gradient RP-LC, except that $S\Delta\phi$ in Equation (6.4) is replaced by $m \log([C]_f/[C]_0)$ in Equation (6.12); m in IEC corresponds to S in RP-LC, and $\log([C]_f/[C]_0)$ corresponds to $\Delta\phi$. Changes in any of the conditions summarized in Equation (6.12) for IEC will result in analogous changes in separation as for gradient RP-LC. Thus, much of our preceding discussion of RP-LC gradient elution applies equally for IEC.

The required gradient time for some value of k^* can be derived from Equation (6.12):

$$\text{(IEC)} \quad t_G = 1.15 k^* V_m m \log([C]_f/[C]_0)/F \tag{6.13}$$

For example, assuming a 150 × 4.6 mm IEC column ($V_m \approx 1.5$), an average value of $m \approx 5$ for a protein sample, a flow rate of 1.0 mL/min, and a gradient from 0.005 to 0.50 M for a monovalent salt, the estimated gradient time for $k^* \approx 3$ would be $t_G = 1.15 \times 3 \times 1.5 \times 5 = 51$ min (notably shorter than for the RP-LC separation of a protein sample with $k^* = 3$). Resolution in gradient IEC can be predicted by means of Equation (6.6), using values of k^* from Equation (6.12). Note that the counter-ion concentration C is the sum of salt plus buffer concentrations (Section 8.2).

Proteins, oligonucleotides, and polysaccharides often have values of $m > 3$ [100], in which case Equation (6.10) for IEC can be approximated by Equation (6.2) for RP-LC, with ϕ replaced by C [96]. This means that IEC separations with $m > 3$ can be described accurately by corresponding equations for RP-LC [e.g., Equation (6.4) with ϕ replaced by C], and the interpretation and optimization of these IEC separations can be carried out in much the same way as described previously for RP-LC gradient elution. Several studies have reported the application of the LSS model [Equation (6.2)] to an interpretation of retention [1, 96, 100, 101] and band width in gradient IEC [1, 15]; good agreement is usually found between experimental and predicted data when $m \geq 3$. Kato et al. [102] have described several IEC separations of a protein sample where different gradient or column conditions were varied. Changes in gradient time and flow rate were made without change in other conditions, so that k^* varied during these experiments. However, the results obtained again appear in qualitative agreement with similar changes in conditions for RP-LC.

6.2.2.3 Hydrophilic Interaction Chromatography

Hydrophilic interaction chromatography (HILIC) separations are carried out with a hydrophilic column and gradients of acetonitrile–water (increasing water during the gradient [88]). Retention in HILIC decreases for a more polar mobile phase (one containing more water), just as in the case of normal-phase chromatography; indeed, HILIC can be regarded as a form of normal-phase chromatography. HILIC has been applied to the separation of carbohydrates, peptides, proteins, and oligonucleotides. See Section 8.3.3 for a discussion of the experimental aspects of HILIC and examples of HILIC separation.

Retention in HILIC is given by

$$\log k = \log k_{H_2O} - m_{HILIC} \log \phi_{H_2O} \tag{6.14}$$

Here, k_{H_2O} is the value of k for water as mobile phase, ϕ_{H_2O} is the volume-fraction of water in the mobile phase, and m_{HILIC} is the slope of plots of $\log k$ vs $\log \phi_{H_2O}$. The similarity of Equation (6.14) for HILIC and Equation (6.10) for IEC leads to similar equations for retention. Thus,

$$(\text{HILIC}) \quad k^* = t_G F / 1.15 \{ V_m \, m_{HILIC} \log ([C]_f / [C]_0) \} \tag{6.15}$$

$$= t_G F / 1.15 \{ V_m \, m_{HILIC} \log (\phi_{H_2O,f} / \phi_{H_2O,0}) \} \tag{6.15a}$$

where $\phi_{H_2O,f}$ and $\phi_{H_2O,0}$ refer to values of ϕ_{H_2O} at the end ("f") and beginning ("0") of the gradient. The value for t_G is the same as given by Equation (6.13), if m_{HILIC} is substituted for m, and $\log ([C]_f / [C]_0)$ is replaced by $\log (\phi_{H_2O,f} / \phi_{H_2O,0})$. The value of m_{HILIC} hence plays a similar role in HILIC separation as does m in IEC or S in RP-LC. Values of m_{HILIC} increase with the number of polar substituents in the solute molecule, as illustrated in Figure 6.16 for several peptide solutes. Here, m_{HILIC} is plotted vs the number n of amino acid groups in each peptide. The data of Figure 6.16 can be represented as

$$m_{HILIC} = 1.0 \times 10^{0.62n} \tag{6.16}$$

That is, m_{HILIC} increases with molecular size or the number of polar (e.g., amide) groups in the molecule. Similarly, for the HILIC separation of a series of oligoglycosides where n represents the number of sugar groups in the molecule (data of [104]),

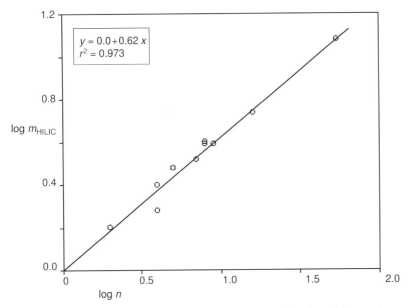

Figure 6.16 Values of the HILIC parameter m_{HILIC} as a function of the number n of amino acid groups in each peptide (log–log plot). Data from [104].

Equation (6.16) is obeyed with

$$m_{HILIC} \approx 2 \times 10^{0.6n} \qquad (6.16a)$$

We can conclude that $m_{HILIC} > 3$ for typical large biomolecules (peptides, proteins, oligonucleotides, polysaccharides), so that Equation (6.1) will be a good approximation for HILIC (with ϕ replaced by ϕ_{H_2O}), as in the case of protein separations by IEC (when $z > 3$). The LSS model will therefore apply quantitatively for the HILIC separation of these samples, and the various equations presented previously for RP-LC separation should be applicable.

6.2.2.4 Separation of Viruses[†]

Beginning in about 1995, interest in recombinant viruses for use in gene therapy motivated the development of liquid chromatography for the purification, analysis, and characterization of viruses. The initial driving force was a need for larger amounts of higher-purity material, compared with what could be obtained by classical purification methods such as CsCl density gradient centrifugation. As a result, the chromatographic purification of viruses is now a preferred approach, and corresponding analytical methods have also been developed. Anion-exchange gradient elution is most commonly employed for both preparative and analytical applications, although other liquid chromatography procedures have also proved useful. Recombinant adenoviruses (rAd) are widely used for gene therapy trials, but other viruses have also been purified and/or

[†]With Carl Scandella, Paul Shabram, and Gary Vellekamp.

analyzed by chromatography [e.g., adeno-associated virus AAV, and lentivirus [105]). While virus chromatography appears well established, chromatographic conditions may need to be varied for different virus samples. For additional background on the structure and properties of adenoviruses, see Appendix VI.

From a chromatographic standpoint, the most important feature of viruses is their huge molecular size, for example, 167×10^6 Da for an adenovirus. As a result, values of S (RP-LC) or m (IEC) for virus samples are expected to be much larger than for most other compounds of biochemical interest, meaning that quite small changes in mobile phase composition should have a very large effect on retention. Consequently, most virus separations are likely carried out in an "on–off" mode (Section 6.1.5), where a particular virus is first retained quite strongly, but at a certain point in the gradient retention quickly falls to zero (so that $k^* \approx 0$). While "on–off" separation normally leads to reduced resolution [Equation (6.6)], this is likely to be less true for the separation of large molecules (or particles in the case of viruses) other than synthetic polymers, because of very large differences in their k_0 values (not the case for synthetic-polymer molecules). Large values of S or m also mean that the isocratic separation of virus samples should be completely impractical, except by means of size-exclusion chromatography [106].

The large size of virus particles also results in their slow diffusion, for example, $D_m = 5 \times 10^{-8}$ cm^2/s for the virus Ad5, or about 10-fold slower than for larger proteins. Slow diffusion typically results in smaller values of N^* and wide elution peaks (Section 9.5). However, peak width is *decreased* for small values of k^* [Equation (9.34)], which should offset peak broadening due to a small plate number, and result in peak widths not much greater than observed in the gradient separation of proteins. Finally, the molecular (hydrodynamic) diameter of a virus is also quite large (of the order of 100 nm), which likely restricts penetration of the molecule into the pores of the column packing. This in turn reduces the available surface for retention of the virus; compared with protein chromatography, the amount of virus that can be separated (without column overload) is lower by a factor of 20–50 [106].

Virus purification – surprisingly good results for adenovirus have been obtained with gradient elution based on anion exchange, hydrophobic interaction, and metal chelate chromatography [107], despite the use of columns intended for the separation of proteins and small molecules. A typical result is shown in Figure 6.17 for the separation of p53Ad (ACN53) by anion-exchange gradient elution. The two major peaks in the chromatogram were identified as virus ($t_R \approx 19$ min) and DNA ($t_R \approx 28$ min) by absorbance ratios at 260 and 280 nm, as well as by other properties.

Hydrophobic interaction chromatography and metal chelate chromatography have also been evaluated for virus purification [107]. Initially yields were lower vs anion exchange; for example, 15–30 percent for HIC [107], but yields in the range of 70–85 percent can be achieved with technical refinements (G. Vellekamp, unpublished results). A zinc chelate column removed minor impurities that were not removed by anion exchange when the rAd peak from the Fractogel DEAE 650M column was rechromatographed on a zinc chelate column (Fig. 6.18). After trying several combinations of different columns [107], it was recommended that anion

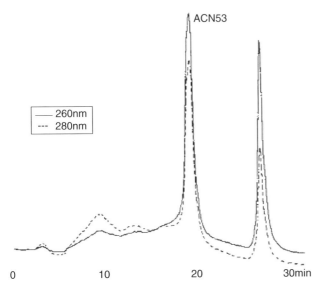

Figure 6.17 Separation of adenovirus by anion-exchange chromatography. Sample preparation: 293 cells were infected with p53Ad (ACN53), harvested 3–4 days after infection by centrifugation, lysed by freeze–thaw to liberate the virus, and clarified by centrifugation and filtration. Separation conditions: column, 6.6 × 50 mm Fractogel DEAE-650M; gradient 300–600 mM NaCl (50 mM Tris, pH 8.0 plus 2 mM $MgCl_2$ and 2 percent sucrose) in 10 min. Reprinted with permission from [107].

exchange should be followed by zinc chelate chromatography for the purification of adenovirus.

One of the more difficult challenges for preparative chromatography is the removal of empty capsids (the outer shell of the virus) [106, 108] (see Appendix VI for a description of virus structure and capsids), because capsids closely resemble the virus in terms of size and surface charge. Partial success at removal of empty capsids has been achieved by anion exchange, hydrophobic interaction, and metal chelate chromatograph, but there is room for improvement in this area [108].

Virus analysis – the measurement of recombinant adenovirus concentrations by anion exchange chromatography has been found to be superior to other assay procedures. Anion exchange is rapid, convenient, accurate, and sensitive, applicable to crude as well as purified samples, and able to distinguish aggregated and disrupted forms from intact virus [107, 109]. The measurement of adenovirus particle number (concentration) by anion exchange gradient elution is illustrated in Figure 6.19, using diode-array detection. In (*b*), a three-axis plot (absorbance, time, and UV wavelength) is shown for elution of a purified virus sample with diode array detection, compared in (*a*) with the spectrum of virus purified by density gradient centrifugation [112]. The absorption spectrum of the eluting virus peak (*b*)

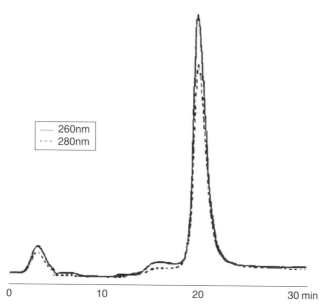

Figure 6.18 Further purification of the ACN53 peak from Figure 6.17 by metal-chelate chromatography. Conditions: column, 50 × 6.6 mm TosoHaas AF chelate 650 M; two successive gradients (a and b) were applied, with elution of ACN53 in gradient b: (a) 450–150 NaCl (50 mM HEPES, pH 7.5 plus 2 mM $MgCl_2$ and 2 percent sucrose; (b) 0–500 mM glycine (plus 150 mM NaCl) in 30 min. Reprinted with permission from [107].

compares closely with the spectrum of purified virus in (a); that is, the peak eluting at 12 min is adenovirus. Several batches of virus purified by density gradient centrifugation were assayed in this way, and the results were compared with assays by UV absorption at 260 nm and an infectivity assay. Good agreement was found among these three assay methods.

The latter anion-exchange separation of purified virus was extended to the assay of crude virus samples. The similar separation of an infected cell lysate containing adenovirus exhibited a more complex chromatogram (Fig. 6.20 [109]), but the large virus peak at 40 min was well resolved from adjacent impurity peaks. Spiking experiments were carried out to confirm peak identities. The ability of anion exchange gradient elution to resolve viruses from nucleic acid and protein contaminants, and to quantify virus particles present in unpurified and partially purified mixtures has proved to be a powerful tool for research, development and manufacturing of virus vectors for gene therapy [106, 110].

Separations by RP-LC, which can resolve individual proteins from adenovirus particles, are inherently denaturing. Early separations of rAd by means of RP-LC tended to be irreproducible, unless separation was carried out at >40°C. A higher temperature probably facilitates disruption of the virus capsid, favoring the release of proteins from the column matrix and viral DNA.

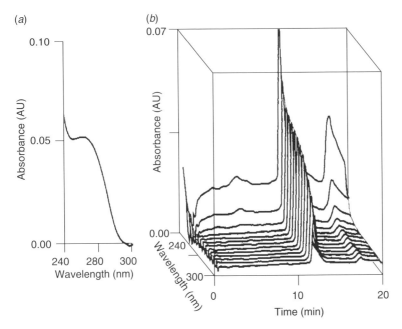

Figure 6.19 Assay of virus rAd5 by anion-exchange chromatography. (*a*) UV spectrum of rAd5 purified by CsCl density gradient centrifugation in 0.1 percent SDS buffer; (*b*) a three-axis plot of absorbance, wavelength, and retention time showing the elution of a peak of purified rAd5. Conditions: 1 mL Resource Q column (GE Healthcare); 300–600 mM NaCl (50 mM HEPES, pH 7.5). Adapted from [109].

6.2.3 Separation Problems

Gradient separations of peptides and proteins are subject to many of the same problems encountered in the RP-LC separation of small molecules (Chapter 5). In addition, some of the peculiarities of peptides and especially proteins can lead to additional sources of difficulty. The use of columns introduced prior to 1990 (so-called "type A" column packings with more active silanols [88]) often resulted in tailing peaks and poor recovery of peptides and proteins. Today a large number of less active ("type B") columns are available in a variety of ligand lengths and pore sizes suitable for protein separation (see Appendix III for columns with different pore sizes and low values of C, i.e., type B). Similarly, any major departure from the preferred conditions of Table 6.1 can lead to marginal protein separations.

Other problems are associated with incomplete protein denaturation at the beginning of separation (Section 6.1.3). The conditions recommended in Section 6.2.1.2 can be used to minimize these various possibilities (peak splitting, "ghost" peaks, broad peaks, low recovery of sample). While prior denaturation of the sample appears to preclude the recovery of proteins in their native form, the renaturation of medium-size proteins after RP-LC separation is usually possible, by

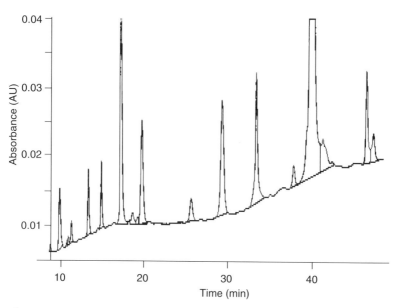

Figure 6.20 Separation of adenovirus proteins by RP-LC. The following method was adapted from [110]. A Jupiter C4 column (150 × 2 mm; 300 Å pore) run at 50°C with a three-part gradient. Solvent A was 0.1 percent TFA; solvent B was 0.1 percent TFA in acetonitrile/water (90 : 10, v/v). The gradient was 25–40–44–60 percent B at 0–10–25–40 min. Column regeneration was 2 min at 100 percent B followed by equilibration at 25 percent solvent B. Flow rate was 0.2 mL/min). Adapted from [111].

removing the mobile phase from collected fractions and dissolving the protein in an aqueous buffer. For many proteins, however, the protein chain must first be dissolved and fully denatured in an environment such as 6 M guanidine hydrochloride or 8 M urea, following which the protein is allowed to refold by slowly removing the denaturant in the presence of a redox buffer to allow proper formation of disulfide bonds [113].

Hydrophobic and/or higher-molecular-weight proteins often exhibit sample loss during gradient separation, which is usually reduced by the use of shorter gradients (smaller t_G), smaller columns (or a higher weight-ratio of sample to column packing), and higher separation temperatures. One study [68] found that the recovery of larger proteins was maximized at 60°C. An example of the effect of temperature on the recovery of a recombinant protein is illustrated by the example of Figure 6.21, where maximum recovery is observed for temperatures of 70–80°C. In confirmation of the role of temperature as a determinant of recovery in the separations of Figure 6.21, it was observed that a 30 min isocratic hold at 90°C (followed by gradient separation at 90°C) further reduced protein recovery by two-thirds (to a value of about 20 percent). Some hydrophobic proteins may be insufficiently water-soluble to allow their dissolution in a predominantly aqueous solvent. In such cases, surfactants are sometimes used to increase solubility [1, 114]; the surfactant may be added to both the sample and the mobile phase.

A problem encountered in the gradient separation of synthetic polymers is partial elution of the sample at t_0, that is, "break through" of the sample. This can result from the use of a chromatographically strong solution (e.g., the B solvent) to dissolve the sample initially, followed by incomplete mixing of the sample solution with the mobile phase at the start of separation. As a result, some of the sample never mixes with the initial mobile phase, is therefore unretained by the column, and elutes at t_0. A detailed analysis of this problem has been carried out, and steps to minimize sample break-through have been described [27].

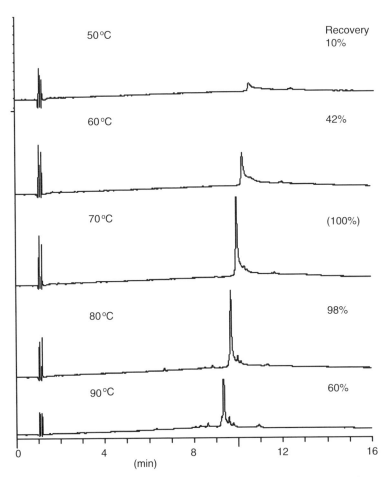

Figure 6.21 Recovery of a recombinant protein as a function of separation temperature. Conditions: 150 × 4.6 mm Zorbax 300SB-C_{18} column; 10–80 percent acetonitrile–TFA buffer in 15 min; 1.5 mL/min; temperatures shown in figure. Chromatograms courtesy of D.H. Marchand.

Sample breakthrough has also been observed for the gradient RP-LC separation of proteins. Here, the problem is associated with large weights of injected sample, with two possible causes [115]. First, if the gradient starts with water or a low %B value, more hydrophobic RP-LC columns (those with larger values of **H**; Appendix III) will not be wetted sufficiently, and as a result the sample will be unretained or only partially retained. The solution to this problem is to start the gradient with a somewhat higher %B and dissolve the sample in a higher %B solvent that corresponds to %B at the start of the gradient, so that the column is properly wetted at the start of the gradient. A second cause of sample breakthrough seems to be a displacement of sorbed B solvent from the column by the sample, with a resulting increase in %B leading to rapid elution of part of the sample. In this case, it is better to start the gradient with a lower %B value. Thus, either a higher or lower %B value at the start of the gradient is likely to reduce the problem of sample breakthrough.

6.2.4 Fast Separations of Peptides and Proteins

Some examples of fast gradient elution for the RP-LC separation of peptides and proteins were presented in Section 3.3.7 (Fig. 3.13), with some general comments on the requirements for separations that require only a few minutes or less. A similar example of fast separation using gradient ion-exchange chromatography is shown in Figure 8.15. Several other examples of this kind have been reported [116–119a]. Fast separation is favored by the use of short columns with small (preferably pellicular) particles, higher flow rates, and elevated temperatures, as well as by large values of α. Suitable gradient equipment is also necessary (Section 4.3.2.1).

6.2.5 Two-Dimensional Separations of Peptides and Proteins

Peptide and protein samples often contain a large number of individual components. The baseline resolution of such samples by means of a single HPLC separation is therefore often unlikely, especially in view of the generally smaller values of N^* that are observed for large molecules. There is an increasing interest in the separation and analysis of complex peptide and protein samples, especially in support of proteomics studies, which has in turn led to the increasing use of 2-D separation (Section 3.7.1). One approach is to combine an initial separation by gradient ion-exchange chromatography with a second separation by gradient RP-LC, with on-line transfer of fractions between the two separations, followed by mass spectrometric analysis of the effluent from the RP-LC separation [120] (but note the qualifications of [121]). A generic 2-D separation is generally desired; one that can be applied to samples composed of different peptides or proteins. Each of the two separations can be optimized for maximum peak capacity (Section 2.2.4), which in turn guarantees that a maximum number of separated peaks will be possible (e.g., 1000 peaks in 100 min [120], presumably for a sample that contains >1000 components).

6.3 SYNTHETIC POLYMERS

The separation of synthetic polymer samples by means of reversed-phase gradient elution has received considerable attention [35, 122–128]. Apart from their practical value, these polymer separations allow additional insight into the fundamental basis of analogous separations of proteins and other large biomolecules. Separations of proteins can be complicated by a number of phenomena (variable denaturation, aggregation, strong silanol-solute interactions, etc.) which are less likely for corresponding separations of synthetic polymers. Consequently, it can be argued that a study of gradient elution with synthetic polymers as samples is able to provide a clearer understanding of the gradient separation of large biomolecules, so far as the effects of molecular size *per se* are concerned (i.e., see above Section 6.1.5.1).

Synthetic polymer samples typically comprise a large number of individual compounds, the discernible resolution of which is unlikely when n is larger than about 50. Consequently, the separation of individual compounds in samples with $n > 50$ is usually not a goal of the chromatographer. However, an understanding of polymer gradient elution can facilitate other applications of the technique, such as

- determination of molecular weight distribution;
- determination of chemical composition distribution.

One approach to the interpretation of synthetic-polymer chromatograms is based on predictions of retention for each compound (of molecular weight M) as a function of experimental conditions [48, 50], in order to guide the selection of conditions for the achievement of a given goal (e.g., to measure a molecular-weight distribution for a given sample). Many of the "practical" applications of RP-LC for characterizing synthetic polymers use gradients with ACN as the A solvent and THF as the B solvent. A relationship similar to Equation (6.2) for water–organic mobile phases applies for gradient separation [50] with THF–ACN as mobile phase

$$\log k^* = \log k_{\text{ACN}} - S\phi^* \qquad (6.17)$$

where k_{ACN} refers to the value of k for pure ACN as mobile phase and ϕ^* is now the volume-fraction of THF in the mobile phase; S has its usual meaning. Values of S as a function of M for polystyrenes and ACN–THF mobile phases can be obtained from the isocratic data of [50]:

$$(\text{ACN} - \text{THF}) \quad S = 0.08 M^{0.56}$$
$$(r^2 = 0.97,\ 1.7\text{K} \leq M \leq 325\text{KDa}) \qquad (6.18)$$

Values of S for polystyrenes with ACN–THF mobile phases are about half as large as for water–THF mobile phases and increase approximately as $M^{0.5}$, similar to the behavior with water–THF as mobile phase (Fig. 6.2).

A second useful relationship relates the values of the sample parameters $\log k_w$ and S for water–organic mobile phases and various organic solvents B, where both parameters increase with M [129, 130]:

$$\text{(organic–water)} \quad S = p + q \log k_w \tag{6.19}$$

Here, p and q are constants for a given polymeric series and fixed experimental conditions other than ϕ [note the similar application of Equation (6.19) for "regular" small molecules in Fig. 1.11a; i.e., homologs are "regular" samples]. A variation of Equation (6.19) also applies for the nonaqueous HPLC separation of polystyrenes with gradients of THF in acetonitrile (Fig. 6.22), where k_{ACN} is the retention factor in pure acetonitrile:

$$\text{(THF–ACN)} \quad S = p + q \log k_{ACN} \tag{6.19a}$$

Once values of p and q have been determined for a given synthetic polymer (e.g., polystyrene), THF–ACN gradients, and some set of remaining conditions (column, temperature), it is possible to predict the elution curve (peak) for each compound of molecular weight M and to sum all the individual peaks into a final chromatogram. An example for the separation of several polystyrene standards is shown in Figure 6.23, where the predicted (a) and experimental (b) separations are compared. For mixtures of two different polymers (e.g., polystyrene and polymethylmethacrylate), calculations of the individual oligomer peaks can be combined so as to facilitate the measurement of each polymer type in the total sample.

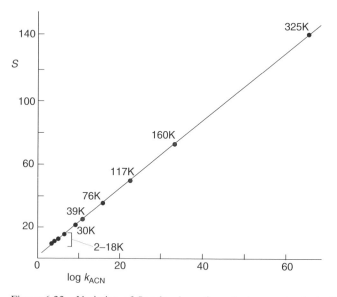

Figure 6.22 Variation of S vs $\log k_{ACN}$ for polystyrene standards (linear–log plot), Discovery C_{18} column (18 nm pores) and tetrahydrofuran–acetonitrile mobile phase. Numbers for data points are values of M for each standard. Adapted from [50].

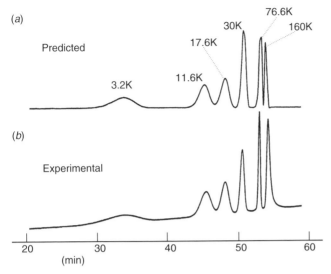

Figure 6.23 Comparison of predicted (*a*), and experimental (*b*) chromatograms for a mixture of several polystyrene standards (values of M indicated in figure). Discovery C_{18} column (18 nm pores), 5–95 percent tetrahydrofuran–acetonitrile in 60 min. Adapted from [48].

6.3.1 Determination of Molecular Weight Distribution

Synthetic-polymer samples comprise a mixture of compounds of varying M (Section 6.1.4). A molecular weight distribution can be expressed as a tabulation of percentage weight for each compound in the sample, but since individual oligomers cannot usually be resolved, a plot of normalized weight fraction vs log M is more practical. Such data can then be expressed in simpler terms (molecular weight average, polydispersity index, etc.). In the simplest case, the sample will be a homopolymer, and its molecular weight distribution can be determined by SEC [131]. Although SEC is a simple and reliable technique for this purpose, it can be limited by its relatively low resolution. One study [49] showed that RP-LC with THF–ACN gradients has about twice the resolution of SEC for the separation of high-molecular-weight oligomer fractions. Similar improvements in RP-LC gradient elution vs SEC have been claimed for the use of methylene chloride–ACN gradients [132]. While SEC remains the method of choice for carrying out the molecular-weight analysis of most synthetic polymer samples, gradient RP-LC can be more useful for isolating narrow fractions for subsequent analysis or use.

6.3.2 Determination of Chemical Composition

Homopolymers result from a synthesis that starts with a specific monomer, for example, styrene or methylmethacrylate, to form polystyrene (PS) or polymethylmethacrylate (PMMA), respectively. Such homopolymers have a given, constant chemical composition; the analysis of such synthetic polymers in terms of molecular weight distribution was discussed in Section 6.3.1. The characterization of polymers with a distribution in chemical composition is obviously more challenging; such samples can consist either of (a) mixtures of two (or more) different polymers ("blends"), or (b) a mixture of molecules which have been formed from two (or more) monomers used in the original synthesis ("copolymers").

The characterization of a blend of two polymers (e.g., PS plus PMMA) in terms of polymer type (not molecular weight distribution) can be achieved by means of gradient HPLC, if conditions can be selected that minimize the influence of M on sample retention [133], so-called "pseudocritical" conditions. The application of pseudocritical gradient elution to a mixture of two different polymers should result ideally in two sharp peaks, for example, a peak for PS and a peak for PMMA. If a molecular weight distribution also is desired for each polymer type, the two peaks can be collected separately and re-analyzed for their molecular-weight distributions by SEC (2-D separation). For copolymers, whose molecules contain both monomer types, a separation of the sample into each polymer type is obviously not possible. In this case, however, the retention time of the polymer peak may often be used to determine the fraction of each monomer type in the total sample.

Conditions for "pseudocritical" separation can be inferred as follows [134]. Insertion of S from Equation (6.19a) into Equation (6.17) gives

$$\log k = (1 - q\phi) \log k_{\text{ACN}} - p\phi \qquad (6.20)$$

Values of k for all oligomers of a given type are seen to become equal, for a *critical mobile phase composition* $\phi_{\text{crit}} = 1/q$. For higher-molecular-weight polymers, the range in values of ϕ at elution (ϕ_e) will be very small, because of the large values of S. For different polymers (and co-polymers of different composition), the critical compositions are usually quite different. RP-LC gradient elution offers a practical way to select conditions which result in separated (relatively narrow) peaks for each polymer type (or composition), by first measuring values of log k_{ACN} and S (Section 9.3.3). The value of $k^* = k_{\text{crit}}$ (corresponding to $\phi^* = \phi_{\text{crit}}$) can be obtained from Equations (6.17) and (6.20):

$$\log k_{\text{crit}} = -p/q \qquad (6.21)$$

Sometimes $q < p$, so that Equation (6.21) yields $\phi_{\text{crit}} > 1.00$, which means that the critical concentration cannot be adjusted within a "real" mobile-phase-composition range.

Several studies have been reported of the application of pseudo-critical gradient elution to various kinds of polymer samples [48, 125, 134–137].

REFERENCES

1. L. R. Snyder and M. A. Stadalius, in *High-Performance Liquid Chromatography. Advances and Perspectives*, Vol. 4, Cs. Horváth, ed., Academic Press, New York, 1986, p. 195.
2. L. R. Snyder and J. W. Dolan, *J. Chromatogr. A* 721 (1996) 3.
3. A. G. H. Lea, *J. Chromatogr.* 194 (1980) 62.
4. S. Terabe, S. Nishi, and T. Ando, *J. Chromatogr.* 212 (1981) 295.
5. J. P. Larmann, J. J. DeStefano, A. P. Goldberg, R. W. Stout, L. R. Snyder, and M. A. Stadalius, *J. Chromatogr.* 255 (1983) 163.
6. M. A. Stadalius, H. S. Gold, and L. R. Snyder, *J. Chromatogr.* 296 (1984) 31.
7. M. A. Quarry, R. L. Grob, and L. R. Snyder, *Anal. Chem.* 58 (1986) 907.
8. X. Geng and F. E. Regnier, *J. Chromatogr.* 296 (1984) 15.
9. G. Jilge, R. Janzen, H. Giesche, K. K. Unger, J. N. Kinkel, and M. T. W. Hearn, *J. Chromatogr.* 397 (1987) 71.
10. C. H. Lochmüller and M. B. McGranaghan, *Anal. Chem.* 61 (1989) 2449.
11. R. S. Hodges and C. T. Mant, *High-Performance Liquid Chromatogrpahy of Peptides and Proteins*, C. T. Mant and R. S. Hodges, eds, CRC Press, Boca Raton, FL, 1991, p. 3.
12. J. Koyama, J. Nomura, Y. Shiojima, Y. Ohtsu, and I. Horii, *J. Chromatogr.* 625 (1992) 217.
13. C. H. Lochmüller and C. Jiang, *J. Chromatogr. Sci.* 33 (1995) 561.
14. B. F. D. Ghrist, M. A. Stadalius, and L. R. Snyder, *J. Chromatogr.* 387 (1987) 1.
15. M. A. Stadalius, B. F. D. Ghrist, and L. R. Snyder, *J. Chromatogr.* 387 (1987) 21.
16. J. J. Kirkland, F. A. Truszkowski, and R. D. Ricker, *J. Chromatogr. A* 965 (2002) 25.
17. H. Chen and Cs. Horváth, *J. Chromatogr. A* 705 (1995) 3.
18. S. A. Cohen, K. P. Benedek, S. Dong, Y. Tapui, and B. L. Karger, *Anal. Chem.* 56 (1984) 217.
19. W. R. Melander, J. Jacobson, and Cs. Horváth, *J. Chromatogr.* 234 (1982) 269.
19a. D. R. Stoll, J. D. Cohen, and P. W. Carr, *J. Chromatogr. A*, 1122 (2006) 123.
20. M. T. W. Hearn and B. Grego, *J. Chromatogr.* 296 (1984) 61.
21. M. T. W. Hearn and M. I. Aguilar, *J. Chromatogr.* 397 (1987) 47.
22. A. W. Purcell, G. L. Zhao, M. I. Aguilar, and M. T. W. Hearn, *J. Chromatogr. A* 852 (1999) 43.
23. K. Buttner, C. Pinilla, J. R. Appel, and R. A. Houghten, *J. Chromatogr.* 625 (1992) 191.
24. K. L. Richards, M. I. Aguilar, and M. T. W. Hearn, *J. Chromatogr. A* 676 (1994) 33.
25. M. Kunitani, D. Johnson, and L. Snyder, *J. Chromatogr.* 371 (1986) 313.
26. H. J. A. Philipsen, M. Oestreich, B. Klumperman, and A. L. German, *J. Chromatogr. A* 775 (1997) 157.
27. X. Jiang, A. van der Orst, and P. J. Schoenmakers, *J. Chromatogr. A* 982 (2002) 55.
28. L. R. Snyder, in *Chromatography. Part A. Fundamentals and Techniques*, E. Heftmann, ed., 2nd edn, Elsevier, Amsterdam, 1992, p. A1.
29. B. F. D. Ghrist and L. R. Snyder, *J. Chromatogr.* 459 (1989) 43.
30. G. B. Cox and L. R. Snyder, *J. Chromatogr.* 483 (1989) 95.
31. W. S. Hancock and J. T. Sparrow, in *High-performance Liquid Chromatography. Advances and Perspectives*, Vol. 3, Cs. Horváth, ed., Academic Press, New York, 1983, p. 49.
32. B. de Collongue-Poyet, C. Vidal-Madjar, B. Sebille, and K. K. Unger, *J. Chromatogr. B* 664 (1995) 155.
33. T. J. Serada, C. T. Mant, and R. S. Hodges, *J. Chromatogr. A*, 695 (1995) 205.
34. C. Scandella, private communication.
35. G. Glöckner, *Polymer Characterization by Liquid Chromatography*, Elsevier, Amsterdam, 1987.
36. G. Glöckner and J. H. M. van den Berg, *Chromatographia* 19 (1984) 55.
37. G. Glöckner and J. H. M. van den Berg, *J. Chromatogr.* 317 (1984) 615.
38. M. A. Quarry, M. A. Stadalius, T. H. Mourey, and L. R. Snyder, *J. Chromatogr.* 358 (1986) 1.
39. R. A. Shalliker, P. E. Kavanagh, and I. M. Russell, *J. Chromatogr.* 543 (1991) 157.
40. M. Petro, F. Svec, I. Gitsov, and J. M. J. Frechet, *Anal. Chem.* 68 (1996) 315.
41. R. Van der Zee and G. W. Welling, *J. Chromatogr.* 244 (1982) 134.
42. R. Schultz and H. Engelhardt, *Chromatographia* 29 (1990) 205.
43. A. Alhedai, R. E. Boehm, and D. E. Martire, *Chromatographia* 29 (1990) 313.

44. K. H. Bui, D. W. Armstrong, and R. E. Boehm, *J. Chromatogr.* 288 (1984) 15.
45. D. W. Armstrong and R. E. Boehm, *J. Chromatogr. Sci.* 22 (1984) 378.
46. P. Schoenmakers, F. Fitzpatrick, and R. Grothey, *J. Chromatogr. A* 965 (2002) 93.
47. M. A. Stadalius, M. A. Quarry, T. H. Mourey, and L. R. Snyder, *J. Chromatogr.* 358 (1986) 17.
48. F. Fitzpatrick, R. Edam, and P. Schoenmakers, *J. Chromatogr. A* 988 (2003) 53.
49. F. Fitzpatrick, H. Boelens, and P. Schoenmakers, *J. Chromatogr. A* 1041 (2004) 43.
50. F. Fitzpatrick, B. Staal, and P. Schoenmakers, *J. Chromatogr. A* 1065 (2005) 219.
51. C. H. Lochmüller, C. Jiang, Q. Liu, and V. Antonucci, *Crit. Rev. Anal. Chem.* 26 (1996) 29.
52. G. Vanecek and F. E. Regnier, *Anal. Biochem.* 109 (1980) 345.
53. F. E. Regnier, *Science* 222 (1983) 245.
54. G. Lindgren, B. Lundström, I. Källman, and K.-A. Hansson, *J. Chromatogr.* 296 (1984) 83.
55. R. M. Moore and R. R. Walters, *J. Chromatogr.* 317 (1984) 119.
56. H. Engelhardt and H. Müeller, *Chromatographia* 19 (1984) 77.
57. N. Nimura, H. Itoh, and T. Kinoshita, *J. Chromatogr.* 585 (1991) 207.
58. Y.-B. Yang, K. Harrison, D. Carr, and G. Guiochon, *J. Chromatogr.* 590 (1992) 35.
59. W. Kopaciewicz, E. Kellard, and G. B. Cox, *J. Chromatogr. A* 690 (1995) 9.
60. M. A. Stadalius, H. S. Gold, and L. R. Snyder, *J. Chromatogr.* 296 (1984) 31.
61. M. A. Stadalius, H. S. Gold, and L. R. Snyder, *J. Chromatogr.* 327 (1985) 27.
62. J. M. DiBussolo and J. R. Gant, *J. Chromatogr.* 327 (1985) 67.
63. R. A. Shalliker, P. E. Kavanagh, and I. M. Russell, *J. Chromatogr. A* 679 (1994) 105.
64. Vivó-Truyols, G., Torres-Lapasió, and J.R., García-Alvarez-Coque, M.C., *J. Chromatogr. A* 1018 (2003) 169.
65. J. W. Dolan and L. R. Snyder, *LCGC* 17 (1999) S17 (April supplement).
66. B. S. Marchylo, D. W. Hatcher, and J.E. Kruger, *Cereal Chem.* 65 (1988) 28.
67. K. Nugent, W. Burton, and L. R. Snyder, *J. Chromatogr.* 443 (1988) 363.
68. K. Nugent, W. Burton, and L. R. Snyder, *J. Chromatogr.* 443 (1988) 381.
69. W. S. Hancock, ed., *High Performance Liquid Chromatography in Biotechnology*, John Wiley, New York, 1990.
70. C. T. Mant and R. S. Hodges, *High-performance Liquid Chromatography of Peptides and Proteins*, CRC Press, Boca Raton, FL, 1991.
71. R. L. Cunico, K. M. Gooding, and T. Wehr, *Basic HPLC and CE of Biomolecules*, Bay Bioanalytical Laboratory, Richmond, CA, 1998.
72. F. E. Regnier and K. M. Gooding, eds. *HPLC of Biological Macromolecules*, 2nd edn, Marcel Dekker, New York, 2002.
73. J. J. Kirkland, F. A. Truszkowski, and R. D. Ricker, *J. Chromatogr. A* 965 (2002) 25.
74. Y. Chen, A. R. Mehok, C. T. Mant, and R. S. Hodges, *J. Chromatogr.* 1043 (2004) 9.
75. D. Guo, C. T. Mant, and R. S. Hodges, *J. Chromatogr.* 386 (1987) 205.
76. M. Shibue, C. T. Mant, and R. S. Hodges, *J. Chromatogr. A* 1080 (2005) 49.
77. D. V. McCalley, *LCGC*, 23 (2005) 162.
78. L. R. Snyder, J. W. Dolan, and D. C. Lommen, *J. Chromatogr.* 485 (1989) 91.
79. I. Molnar, R. I. Boysen, and V. A. Erdmann, *Chromatographia* 28 (1989) 39.
80. B. F. D. Ghrist, B. S. Cooperman, and L. R. Snyder, in *HPLC of Biological Macromolecules*, F. E. Regnier, and K. M. Gooding, eds, Marcel Dekker, New York, 1990, p. 403.
81. R. C. Chloupek, W. S. Hancock, and L. R. Snyder, *J. Chromatogr.* 594 (1992) 65.
82. W. Hancock, R. C. Chloupek, J. J. Kirkland, and L. R. Snyder, *J. Chromatogr. A* 686 (1994) 31.
83. R. C. Chloupek, W. S. Hancock, B. A. Marchylo, J. J. Kirkland, B. Boyes, and L. R. Snyder, *J. Chromatogr. A* 686 (1994) 45.
84. M.-I. Aguilar and M. T. W. Hearn, in *Methods in Enzymology. Vol. 271. High Resolution Separation and Analysis of Biological Macromolecules* W. S. Hancock and B. L. Karger, eds, Academic Press, Orlando, FL, 1996, p. 3.
85. L. R. Snyder, in *New Methods in Peptide Mapping for the Characterization of Proteins*, W. Hancock, ed, CRC Press, Boca Raton, FL, 1996, p. 31.
86. N. Lundell, *J. Chromatogr.* 639 (1993) 97.
87. N. Lundell and K. Markides, *J. Chromatogr.* 639 (1993) 117.
88. L. R. Snyder, J. J. Kirkland, and J. L. Glajch, *Practical HPLC Method Development*, 2nd edn, Wiley-Interscience, New York, 1997.

REFERENCES 281

89. M. J. Dickman, *J. Chromatogr. A* 1076 (2005) 83.
90. S.-L. Wu and B. L. Karger, in *Methods in Enzymology. Vol. 271. High Resolution Separation and Analysis of Biological Macromolecules*, W. S. Hancock and B. L. Karger, eds, Academic Press, Orlando, FL, 1996, p. 27.
91. M. T. W. Hearn, in *HPLC of Biological Macromolecules*, K. M. Gooding and F. E. Regnier, eds, 2nd edn, Marcel Dekker, New York, 2002, p. 99.
92. J. L. Fausnaugh and F. E. Regnier, *J. Chromatogr.* 359 (1986) 131.
93. N. T. Miller and B. L. Karger, *J. Chromatogr.* 326 (1985) 45.
94. G. Ripple, Á. Bede, and L. Szepesy, *J. Chromatogr. A* 697 (1995) 17.
95. Y. Kato, T. Kitamura, and T. Hashimoto, *J. Chromatogr.* 298 (1984) 407.
96. M. A. Quarry, R. L. Grob, and L. R. Snyder, *Anal. Chem.* 58 (1986) 907.
97. W. Kopaciewicz, M. A. Rounds, J. Fausnaugh, and F. E. Regnier, *J. Chromatogr.* 266 (1984) 3.
98. G. Choudhary and Cs. Horváth, in *Methods in Enzymology. Vol. 271. High Resolution Separation and Analysis of Biological Macromolecules*, W. S. Hancock and B. L. Karger, eds, Academic Press, Orlando, FL, 1996, p. 47.
99. F. Regnier and K. M. Gooding, in *HPLC of Biological Macromolecules*, K. M. Gooding and F. E. Regnier, eds, 2nd edn, Marcel Dekker, New York, 2002, p. 81.
100. T. Sasagawa, Y. Sakamoto, T. Hirose, T. Yoshida, Y. Kobayashi, and Y. Sato, *J. Chromatogr.* 485 (1989) 533.
101. R. W. Stout, S. I. Sivakoff, R. D. Ricker, and L. R. Snyder, *J. Chromatogr.* 353 (1986) 439.
102. Y. Kato, K. Komiya, and T. Hishimoto, *J. Chromatogr.* 246 (1982) 13.
103. T. Yoshida, *J. Chromatogr. A* 811 (1998) 61.
104. A. Alpert, *J. Chromatogr.* 499 (1990) 177.
105. V. Slepushkin, N. C., R. Cohen, Y. Gan, B. Jiang, E. Deausen, D. Gerlinger, G. Ginder, K. Andre, L. Humeau, and B. Dropulic, *BioProcessing J.* (2003) p. 89.
106. N. E. Altaras, J. G. Aunin, R. K. Evans, A. Kamen, J. O. Konz, and J. J. Wolf, *Adv. Biochem. Engng/Biotechnol.* 99 (2005) 193.
107. B. G. Huyghe, X. L., S. Sujipto, B. J. Sugarman, M. T. Horn, H. M. Shepard, C. J. Scandella, and P. Shabram, *Hum. Gene Ther.* 6 (1995) 1403.
108. G. Vellekamp, F. W. Porter, S. Sujipto, C. Cutler, L. Bondoc, Y. H. Liu, D. Wylie, S. Cannon-Carlson, J. T. Tang, A. Frei, M. Voloch, and S. Zhuang, *Hum. Gene Ther.* 12 (2001) 1923.
109. P. W. Shabram, D. D. Giroux, A. M. Goudreau, R. J. Gregory, M. T. Horn, B. G. Huyghe, X. Liu, M. H. Nunnally, B. J. Sugarman, and S. Sutjipto, *Hum. Gene Ther.* 8 (1997) 453.
110. E. Lehmberg, P. Shabram, T. Schluep, S. F. Wen, B. Sugarman, M. Croyle, M. Koehl, E. Bonfils, D. Malarme, G. Sharpe, H. Nesbit, F. Borellini, A. Kamen, B. Hutchins, and G. Vellekamp, *BioProcessing J.* 2 (2003) 50.
111. G. Vellekamp, unpublished results.
112. J. V. J. Maizel, D. O. White, and M. D. Scharff, *Virology* 36 (1968) 115.
113. C. Scandella, private communication.
114. G. W. Welling, R. van der Zee, and S. Welling-Wester, in *HPLC of Biological Macromolecules*, F. E. Regnier and K. M. Gooding, eds, 2nd edn, Marcel Dekker, New York, 2002, p. 513.
115. G. B. Cox and L. R. Snyder, *J. Chromatogr.* 590 (1992) 17.
116. G. Lindgren, B. Lundström, I. Küllman, and K.-A. Hansson, *J. Chromatogr.* 296 (1984) 83.
117. S. K. Paliwal, M. de Frutos, and F. E. Regnier, in *Methods in Enzymology. Vol. 271. High Resolution Separation and Analysis of Biological Macromolecules*, W. S. Hancock and B. L. Karger, eds, Academic Press, Orlando, FL, 1996, p. 133.
118. U. D. Neue, J. L. Carmody, Y. F. Cheng, Z. Lu, C. H. Phoebe, and T. E. Wheat, in *Design of Rapid Gradient Methods for the Analysis of combinatorial Chemistry Libraries and the Preparation of Pure Compounds*, P. Brown and E. Grushka, eds, Marcel Dekker, New York, 2001, p. 93.
119. M. Gilar, A. E. Daly, M. Kele, U. D. Neue, and J. C. Gebler, *J. Chromatogr. A* 1061 (2004) 183.
119a. D. R. Stoll, J. D. Cohen and P. W. Carr, *J. Chromatogr. A*, 1122 (2006) 123.
120. K. Wagner, T. Miliotis, G. Marko-Varga, R. Bischoff, and K. K. Unger, *Anal. Chem.* 74 (2002) 809.
121. K.M. Gooding, B. Hodge, and R.K. Julian Jr, *LCGC* 22 (2004) 354.
122. T. C. Schunk, *J. Chromatogr. A*, 656 (1993) 591.
123. H. G. Barth and J. W. Mays, eds, *Modern Methods of Polymer Characterization*, Wiley, New York, 1991.

124. B. Klumperman and H. A. J. Philipsen, *LCGC Int.* 11 (1998) 18.
125. Y. Brun and P. Alden, *J. Chromatogr. A* 966 (2002) 25.
126. A. van der Horst and P. J. Schoenmakers, *J. Chromatogr. A* 1000 (2003) 693.
127. H. J. A. Philipsen, *J. Chromatogr. A* 1037 (2004) 329.
128. F. Fitzpatrick, Interactive chromatography for the characterization of polymers, Thesis for the University of Amsterdam, 3 May 2004.
129. P. J. Schoenmakers, H. A. H. Billiet, and L. De Galan, *J. Chromatogr.* 185 (1979) 179.
130. P. Jandera, *J. Chromatogr.* 314 (1984) 13.
131. W. W. Yau, J. J. Kirkland, and D. D. Bly, *Modern Size-exclusion Chromatography*, Wiley, New York, 1979.
132. R. A. Shalliker, P. E. Kavanagh, and I. M. Russell, *J. Chromatogr. A* 558 (1991) 440.
133. G. Glöckner, *Gradient HPLC of Copolymers and Chromatographic Cross-fractionation*, Springer, Berlin, 1991.
134. P. Jandera, M. Holčapek, and L. Kolářová, *Int. J. Polym. Anal. Charact.* 6 (2001) 261.
135. A. van der Horst and P. J. Schoemakers, *J. Chromatogr. A* 1000 (2003) 693.
136. F. Fitzpatrick, H.-J. Ramaker, P. Schoenmakers, R. Beerends, M. Verheggen, and H. Philipsen, *J. Chromatogr. A*, 1043 (2004) 239.
137. L. Kolarova, P. Jandera, E. C. Vonk, and H. A. Claessens, *Chromatographia* 59 (2004) 579.

CHAPTER 7

PREPARATIVE SEPARATIONS

> ... this is the question which I ask of you. If I had put within this bottle two pints, one of wine and the other of water, thoroughly and exactly mingled together, how would you unmix them? After what manner would you go about to sever them, and separate the one liquor from the other ... that you render ... the wine pure?
>
> —François Rabelais, *Gargantua and Pantagruel*, Chapter 52

7.1 INTRODUCTION

Previous chapters dealing with gradient separation have assumed that a small weight of sample is injected (e.g., <100 µg), as is generally the case for sample analysis. Preparative separation by liquid chromatography ("prep-LC") aims at the recovery of one or more purified sample components; this usually involves the injection of a larger weight of sample, in order to recover as much purified product as possible from each run. As sample size is increased, peaks within the chromatogram eventually become wider, accompanied by a decrease in sample resolution. Prep-LC can involve one of three strategies, as illustrated in Figure 7.1 for recovery of a purified product peak (3, labeled "*"). In Figure 7.1(*a*), the injection of a small sample is assumed, resulting in symmetrical, narrow peaks ("*nonoverloaded*" separation). The product peak can be isolated in pure form from such a separation, but only in small amounts: <100 µg for nonionized compounds, even less for ionized compounds (assuming a typical analytical column, e.g., 150 × 4.6 mm). Nonoverloaded separation as in (*a*) does not require special equipment, columns, or procedures.

In prep-LC, larger amounts of sample are usually available (≫100 µg), and the usual goal is to recover as much purified product as possible in each run, in the shortest time and with the least cost and effort. This can be accomplished most directly by increasing sample size to the point where peaks broaden and become distorted: so-called *column overload*. One approach is to increase sample size until the product peak widens enough to touch one of the two adjacent peaks – "*touching-peak separation*" as illustrated in Figure 7.1(*b*) for peaks 2 and 3. Touching-peak (T-P) separation allows a maximum recovery of purified product in each run, with minimum labor and with little additional method development

---

High-Performance Gradient Elution. By Lloyd R. Snyder and John W. Dolan
Copyright © 2007 John Wiley & Sons, Inc.

284 CHAPTER 7 PREPARATIVE SEPARATIONS

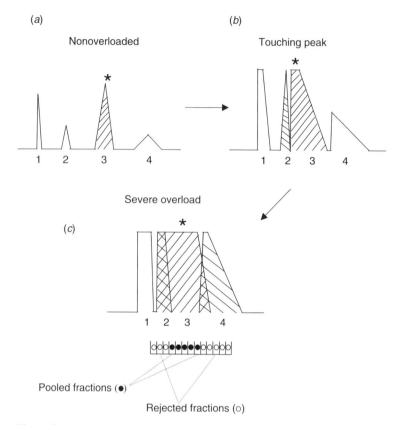

Figure 7.1 Hypothetical separations for different weights of injected sample. (*a*) Small sample; (*b*) touching-peak sample weight; (*c*) larger sample weight (severe column overload).

beyond that required for the small-sample separation in Figure 7.1(*a*). T-P separations correspond to ~100 percent recovery of injected product with ~100 percent purity; they can be used for the recovery of milligram to gram quantities of purified product (if the size of the column is increased). The present chapter will deal mainly with T-P separation as in (*b*).

If sample size is increased further for the same column ("*severe overload*"), the product peak begins to overlap one or more adjacent impurity peaks, as in Figure 7.1(*c*) where peaks 2 and 4 overlap the product peak 3. In this separation mode, small fractions are collected across the product peak, each fraction is analyzed for purity (e.g., by HPLC, as in Fig. 7.1*a*), and fractions which meet the purity requirement (typically ≥98 percent pure) are pooled. Larger quantities of purified product can be recovered in each run for such severely overloaded separations (other factors equal), but the total recovery (yield) of purified product from an individual separation is usually ≪100 percent. However, collected fractions which fail the purity requirement are often pooled and re-separated by prep-LC, in order to increase the overall yield of product. There is more work involved for each separation as in Figure 7.1(c), and a greater method-development effort will be required in order to optimize the final separation. However, severely overloaded prep-LC is usually preferred when large amounts of purified product are required (e.g., more than a few grams), because of lower cost. The practice, and especially theory, of severely overloaded prep-LC is complicated, and a detailed discussion of such separations is beyond the scope of the present book; see Section 7.4 for a practical overview.

A number of useful equations are presented in following sections for approximate predictions of T-P separation as a function of sample size and experimental conditions. These relationships can help guide method development, as well as add to our general understanding of how prep-LC works. While method development for T-P separation usually proceeds by trial and error, *a better grasp of the theory of prep-LC separation can help us select the right experiments as we proceed*; this should result in the best possible result, with the least time and effort spent on method development. As the scale of separation increases, a stronger reliance on basic theory to guide experiments becomes progressively more important.

In previous chapters, we have established the general similarity of isocratic and gradient separation (Section 2.2.1). Thus, when the same ("corresponding") conditions are used – except for fixed %B in isocratic separation vs varying %B in gradient elution – similar resolution can be expected when isocratic k-values are equal to gradient k^*-values. The advantage of this relationship is that general conclusions and procedures that are more easily developed and understood for (simpler) isocratic elution can also be applied to (more complex) gradient separations. The interpretation of gradient elution in terms of "corresponding" isocratic separation becomes even more useful when dealing with prep-LC, because of the combined complexity of gradient vs isocratic elution, *plus* preparative vs analytical separation. We will therefore first review the theory and practice of isocratic prep-LC (Section 7.2), followed by an analogous treatment of gradient elution in a preparative mode (Section 7.3). This is the same approach as followed earlier for the development of assay procedures by gradient elution (Chapter 3); that is, we first discussed (simpler) isocratic method development (Section 2.1), then we applied the same general principles to the development of (more complex) gradient methods.

7.1.1 Equipment for Preparative Separation

The same equipment used for sample analysis is often used for small-scale prep-LC, that is, columns with diameters of 4–8 mm, and equipment that can provide flow rates as great as 5–10 mL/min (Chapter 4). Such columns and equipment can allow the purification of a few milligrams or more of purified product in each run; repetitive runs, with pooling of purified product fractions, can then readily provide tens of milligrams or more of product. When several repetitive runs are required for the purification of a product, a fraction collector becomes necessary. Fraction collectors are sold by a number of companies that deal in HPLC equipment.

When the required amount of purified product is >50 mg, columns and equipment that are designed for prep-LC may be needed. Scale-up from analytical columns and equipment is discussed in Section 7.2.2.3, with a summary of the required column sizes and equipment flow rates.

7.2 ISOCRATIC SEPARATION

Figure 7.2(*a*) illustrates the result when different weights of a single compound are injected sequentially, and the resulting chromatograms are then overlaid on the same time axis. Peak 1 represents the injection of a small weight w_x of compound "X," such that no overloading of the column occurs, and a symmetrical peak results. Note the expanded insert on the right of this figure, showing the overlapped injections of small, but varying weights of X. For small weights of X, the height of peak 1 will increase as sample weight increases, but no increase in peak width will occur. That is, the plate number N [Equation (2.5)] does not change significantly for sample weights w_x below some maximum value.

Peaks 2 and 3 in Figure 7.2(*a*) correspond to the injection of successively larger weights of X, with further broadening of the resulting sample peak. We see that, when sample weight w_x becomes large enough, the peak begins to broaden and becomes asymmetrical, for example, peak 2 in Figure 7.2(*a*). Further increases

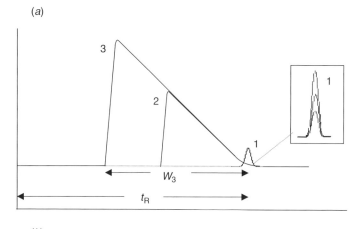

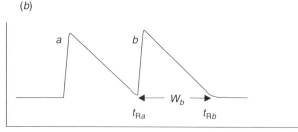

Figure 7.2 Isocratic peak shape as a function of sample size. (*a*) Overlapping peaks from the hypothetical separation of different weights of a single compound (sample weight increases in order of 1 < 2 < 3); (*b*) touching peak separation of two peaks. See text for details.

in sample weight lead to additional peak broadening, as for peak 3. At the same time, the resulting overloaded peaks can be described approximately as "overlapping right-triangles." It will prove convenient to define the retention time t_R for an overloaded peak as shown in Figure 7.2(a); t_R in prep-LC is equal to the retention time for a small sample and *does not change with sample size*. That is, t_R for peak 3 is the same as for peak 1, and is *not* defined by the highest point on peak 3 (which corresponds to $\ll t_R$). On the other hand, the baseline peak width W in prep-LC will be defined as shown in Figure 7.2(a) (W_3 for peak 3, measured from the start of the peak to t_R, that is, similar to the definition of peak width W for a small sample (Fig. 2.1).

7.2.1 Touching-Peak Separation

Touching-peak separation, which we recommend for prep-LC on a laboratory scale (e.g., requiring <1 g of purified product), is further illustrated in Figure 7.2(b) for two adjacent peaks *a* and *b*. The width of the second peak W_b is equal to the difference in small-sample retention times t_R for the two peaks ($t_{R,a}$ and $t_{R,b}$)

$$W_b = (t_{R,b} - t_{R,a}) \tag{7.1}$$

We can determine values of $t_{R,b}$ and $t_{R,a}$ from an initial analytical-scale (small sample) separation of the two compounds (a first prerequisite of prep-LC method development). If we can predict values of W_b as a function of sample size and experimental conditions, we can then estimate the weight of sample to inject for the desired T-P separation.

7.2.1.1 Theory The width W of an overloaded peak X, as in Figure 7.2(a), can be related to the weight of injected X and separation conditions [Appendix V, Equation (V.11)]:

$$W^2 = W_0^2 + W_{th}^2 \tag{7.2}$$
$$\equiv (16/N_0)t_0^2(1+k)^2 + 4t_0^2\, k^2(w_x/w_s) \tag{7.2a}$$

Here, W_0 is the peak width for injection of a small sample [Equation (9.27)], and W_{th} is the added contribution to peak width as a result of a larger sample weight w_x. N_0 is the column plate-number for the injection of a small weight of X (other conditions the same), t_0 is the column dead time, k is the retention factor for the injection of a small weight of X (previously referred to as "k_0" in [1]; in this book k_0 refers solely to the (small sample) value of k at the start of the gradient), w_x is the weight of injected compound X, and w_s is the column saturation capacity [maximum possible uptake of X by the column (stationary phase), for continued infusion of the column by an excess of concentrated X; referred to below simply as "column capacity"]. Section 7.2.1.2 below discusses column capacity w_s in detail. For a detailed justification of Equation (7.2), see Section V.1.1 in Appendix V.

W^2 is seen to be the sum of two terms, W_0^2 plus W_{th}^2. W_0 represents the normal peak broadening of a small sample, corresponding to the usual definition of the plate number N_0 (for a small sample):

$$N_0 = 16(t_R/W_0)^2 \qquad (7.3)$$

or

$$W_0 = 4N_0^{-1/2} t_R \qquad (7.3a)$$

Often in preliminary experiments where column overloading is significant, it can be assumed that $W \approx W_{th}$, so that peak width W then grows in proportion to $w_x^{1/2}$ [Equations (7.2) and (7.2a)]. The latter is a useful relationship for adjusting sample weight w_x during T-P method development (see Fig. 7.7 and related text). Since W_{th} is not a function of the column plate number N_0, W becomes less affected by W_0 (and the value of N_0) as the sample weight w_x increases. The column plate-number N_0 therefore often plays a less important role in prep-LC than in sample analysis (but note that this is not the case for severely overloaded separations; Section 7.4).

For T-P separation, Appendix V shows that

$$w_x/w_s \approx (1/6)([\alpha_0 - 1]/\alpha_0)^2 \qquad (7.4)$$

where $\alpha_0 = k_b/k_a$ is the separation factor for the small-sample separation of solutes a and b (k_b and k_a are values of k for peak b and a, respectively, for a small injection weight of each). Thus, the maximum weight of sample for T-P separation [w_x in Equation (7.4)] is determined by (a) the separation factor α for a small-sample separation, and (b) column capacity w_s (which is proportional to column volume). An effective strategy for prep-LC method development will aim first at finding separation conditions for a maximum value of α_0, because the maximum sample weight w_x increases approximately as $(\alpha_0 - 1)^2$, for example, a ~4-fold increase in w_x for an increase in α_0 from 1.05 to 1.10. Once a maximum value of α_0 has been determined, a further increase in sample size requires the use of a larger column volume (larger value of w_s for scale-up).

The peak width W for T-P separation is determined by sample size, column capacity, N_0, k, and t_0 [Equation (7.2a)]. Consequently, for a given separation with specified values of k, α_0, and t_0, T-P separation corresponds to some value of w_x/w_s, and requires some minimum value of N_0. Table 7.1 summarizes recommended values of N_0 as a function of values of α_0 and k for a T-P prep-LC separation. Table 7.1 also illustrates the large payoff, in terms of w_x, that is achieved when α_0 is increased. Larger values of N_0 than are shown in Table 7.1 are allowed; however, larger N_0 generally requires a longer run time than is necessary, accompanied by only a small increase in allowed values of w_x/w_s for T-P separation. Much smaller values of N_0 than are recommended in Table 7.1 can lead to large reductions in the allowed sample weight w_x for T-P separation (*very undesirable!*).

The values of w_x shown in the right-hand column of Table 7.1 for a 150 × 4.6 mm column are very approximate, because a value of w_s for the column is assumed based on Equation (7.5a) in the following section. See Section 7.2.1.2 for a further discussion of column capacity as a function of the column, separation conditions, and the sample. Note also that Table 7.1 is not useful for

TABLE 7.1 Sample Size w_x and Minimum Plate Number N_0 for T-P Separation as a Function of α_0 and k (isocratic) or k^* (gradient). See Appendix V for Details

	N_0			Approximate w_x for
				150 × 4.6 mm, 10 nm
α_0	$k, k^* = 0.5$	$k, k^* = 1.0$	$k, k^* \geq 3.0$	pore column[a]
1.05	190,000	85,000	38,000	0.05 mg
1.10	46,000	20,000	9,000	0.2 mg
1.15	23,000	10,000	4,500	0.4 mg
1.20	14,000	6,000	2,700	0.6 mg
1.3	6,900	3,100	1,400	1 mg
1.5	3,000	1,300	590	3 mg
2.0	1,100	500	220	6 mg
3.0	490	220	100	10 mg

[a] Assumes w_s = 140 mg; w_s is proportional to the reciprocal of pore diameter.

severely overloaded separations, because the required values of N_0 are then considerably larger than the values in Table 7.1 (Section 7.4.5), and values of w_x are also larger.

7.2.1.2 Column Saturation Capacity

It is convenient to visualize the uptake of sample molecules by the column as a filling of a flat surface. This is visualized in Figure 7.3(a) for the uptake of benzene as sample by the stationary phase surface. The maximum weight of a sample compound that can be taken up by a column (w_s) depends on several factors:

- the surface area of the stationary phase within the column;
- the charge (if any) on the retained sample molecule;
- to a lesser extent, other features of the sample molecule, column, and separation conditions.

The *surface area SA within the column* can be determined from the surface area (m² per gram) σ_g of the packing and the weight of packing in the column. The column surface area (SA in m²) can also be estimated as

$$SA \approx 0.0005 \, L \, d_c^2 \, \sigma_g \tag{7.5}$$

where L is column length (mm), and d_c is column i.d. (mm); see Section V.1.2 of Appendix V for details. The value of σ_g can vary from one column to another, and σ_g usually decreases in proportion to the pore-diameter of the column packing; Equation (7.5) assumes a pore diameter of 10 nm, while values of SA should be about a third as large for an otherwise similar 30 nm pore column. Smaller pore columns are therefore preferred for prep-LC, except for large-molecule separations, where larger pores permit better access of the sample.

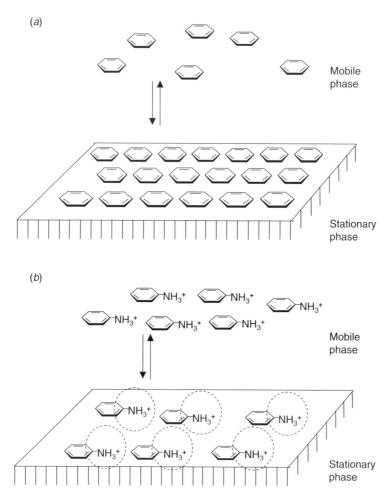

Figure 7.3 Hypothetical view of stationary-phase saturation by the sample. (*a*) Neutral sample compound; (*b*) ionized sample compound. See text for details.

For a variety of nonionized small molecules, it has been found [2] that column capacity can be estimated by

$$w_s = 0.4 \; SA \tag{7.5a}$$

(sample weight in milligrams and the column surface area *SA* in m^2). Since column surface area is proportional to column volume (or column length times diameter^2), column capacity w_s also increases in proportion to column volume. Actual values of w_s for neutral molecules can vary by ± 50 percent from predictions by Equation (7.5a), and w_s can be an order-of-magnitude or more smaller for ionized sample molecules – because of a repulsion of like charges within the stationary phase [2–4]. The latter effect is illustrated in Figure 7.3(*b*), for protonated aniline as sample. Fewer *charged molecules* can be accommodated within the stationary phase, because of the mutual repulsion of charges (either positive or negative) on adjacent molecules. This effect becomes even more important for multiply-charged molecules [5]. For ionizable compounds, the charge on the sample molecule will vary with mobile phase pH; maximum values of w_s therefore correspond to a pH which minimizes the charge on the sample molecule.

An example of the effect of mobile phase pH on column capacity and the allowed injection weight for T-P separation is illustrated in Figure 7.4, for the separation of three basic compounds (peaks D, O, T) as a function of sample weight and mobile phase pH (other conditions the same). At pH 3.8, the sample compounds are completely ionized, and T-P separation occurs for a sample size of about 0.5 mg. At

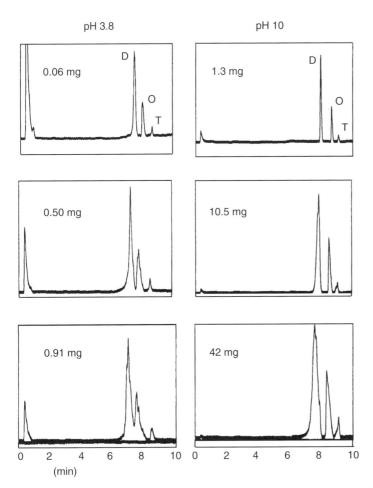

Figure 7.4 An example of the effect of mobile phase pH and sample ionization on the separation of a mixture of three basic compounds as a function of mobile phase pH and sample weight. Sample: diphenhydramine (D), oxybutinin (O), terfenadine (T). Conditions: 50.0 × 19 mm C_{18} column; 5–5–95 percent acetonitrile–buffer at 0–7–12 min; 30 mL/min. Adapted from [4], see text for details.

pH 10, the sample is largely nonionized, and T-P separation allows the injection of more than 42 mg of sample. Because $(\alpha_0 - 1)$ for peaks D and O is ~1.4-fold larger at pH 10, a 2- to 3-fold larger column capacity would be expected at pH 10, other factors being equal (ignoring the effect of sample ionization). However, after correcting for the latter difference in values of α_0, the value of w_s at pH 10 is at least 30-fold greater than at pH 3.8. For a further discussion of the effect of mobile phase pH on column capacity, see [4].

Separation conditions and other features of the column and sample molecule can also affect values of w_s, so that Equations (7.5) and (7.5a) represent only a rough approximation for a given separation. However, even rough estimates of w_s during the development of a prep-LC method can be quite useful, as illustrated in Section 7.2.2.2 below. If a more accurate value of w_s is needed, it can be calculated from an experimental value of W and Equation (7.2a).

7.2.1.3 Sample-Volume Overload

The volume of the sample can also affect peak width in prep-LC, aside from the effect of sample weight. A rough rule from the study of [6] is that sample volume has little effect on values of W in prep-LC, until the sample volume exceeds one-half the volume of the (overloaded) peak of interest. Thus, given a value of W for an overloaded peak as in Figure 7.2(a) and described by Equation 7.2(a) (small-volume injection), the maximum sample volume V_s (mL) should be limited to

$$V_s < 0.5 \; W F$$
$$< 0.5(t_{R,b} - t_{R,a}) \; F \qquad (7.6)$$

where W is in time units (min), and F is flow rate (mL/min); $t_{R,a}$ and $t_{R,b}$ refer to the retention times (min) of the product peak and its nearest neighbor in the chromatogram (as in Fig. 7.2b). Equation (7.6) assumes that the sample is dissolved in the mobile phase; Equation (7.6) has been confirmed by computer simulation [1].

7.2.2 Method Development for Isocratic Touching-Peak Separation

A systematic approach to the development of a prep-LC separation is outlined in Figure 7.5 and summarized in Table 7.2 (for gradient elution only). Prior to carrying out the initial method-development experiment (see Section 7.2.2.1 below), some attention should be given to the starting choice of column and separation conditions. The usual approach to method development for prep-LC begins with an analytical-scale column for initial experiments (e.g., 150 × 4.6 mm C_{18} column with 5 μm particles). Small-scale experiments are more economical and convenient than an initial use of prep-scale columns and equipment for method development.

If the quantity of purified product that will eventually be needed exceeds about 50 mg, it is likely that the separation developed on the analytical-scale column will require scale-up: the use of a larger column with possibly bigger particles, as well as prep-LC equipment. Where a rapid, trouble-free, and successful scale-up is critical, especially when ≫1 g of purified product will be required, it

7.2 ISOCRATIC SEPARATION   293

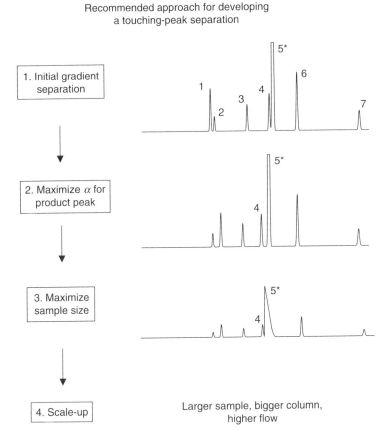

Figure 7.5 Recommended approach for developing an isocratic, touching peak separation. Product peak is 5 (see asterisk).

is best to use the identical packing material (i.e., same stationary phase, same selectivity) in both the analytical-scale column on which the method is developed and in the preparative column used for scale-up. Note that it may be difficult for some column manufacturers to guarantee identical selectivity for column packings of different particle sizes; if in doubt, method development should be begun with the *identical* packing (including particle size) that will be used in the final prep-LC separation.

If the compound to be purified by preparative RP-LC is an acid or base, two considerations should be kept in mind. First, a mobile-phase buffer will usually be required, and a volatile buffer can more easily be removed from the purified product. Commonly used volatile buffers include trifluoroacetic acid, formic acid, acetic acid, and their ammonium salts (or ammonia alone for a mobile phase pH > 9). Because a large sample will require more buffering, the pH of the sample solution should be adjusted to equal that of the mobile phase (but note the qualification of Section 7.2.2.4 concerning sample solubility). For the same reason, the

TABLE 7.2 Outline for the Development of a Routine Touching-Peak Separation by Gradient RP-LC (Compare with Fig. 7.5). Separations by Normal-Phase or Ion-Exchange Chromatography (Sections 8.2, 8.3) Proceed Similarly

Step (numbered as in Fig. 7.5)	Comment
1. Initial separation (small sample, analytical scale)	See recommended gradient conditions of Table 3.2 for "small" molecules, or Table 6.1 for large biomolecules, i.e., analytical-scale separation with a sample weight < 0.1 mg
2. Maximize α_0 for product peak; shorten run time (small sample, analytical scale)	(a) Vary gradient time or isocratic %B to see if an adequate value of α_0 can be achieved (Table 7.1) (b) If a further increase in α_0 is needed, change additional conditions that affect α_0; see Table 3.4 ("small" molecules) or Table 6.1 (peptides and proteins) (c) shorten run time by starting the gradient at the highest possible value of %B, while maintaining maximum resolution and α_0 by holding k^* constant; also, end the gradient as soon as the product peak leaves the column (use a steep gradient segment to remove later-eluting impurities and clean the column, if necessary)
3. Maximize sample size (analytical scale)	(a) Estimate sample weight for T-P separation from the value of α_0 for the product peak (Table 7.1, adjusting for a column size different than 150 × 4.6 mm, if necessary); for gradient elution, α_0 can be estimated from Equation (7.9) (b) Carry out a separation with this sample weight and the optimized conditions from step 2 (c) Consult Figure 7.7 and related text for further experiments to determine final sample weight for T-P separation; note that peak width will increase approximately as $w_x^{1/2}$, that is, (sample weight)$^{1/2}$ (d) Check for sample solubility (Section 7.2.2.4)
4. Scale-up	(a) Compare the weight of purified product that can be obtained from the final T-P separation of step 3 with the required weight of purified product (b) Determine the necessary column size that will provide the required amount of purified product in some reasonable number of replicate injections (Table 7.3). Increase flow rate in proportion to column volume so as to maintain k^* constant [Table 7.3 and Equation (2.13)] (c) If the necessary column and/or equipment are not available for scale-up, an alternative is further experiments aimed at increasing α_0 as in step 2 (small sample, analytical scale) (d) If the required column plate number from Table 7.1 is much smaller than the actual value, consider the use of a larger-particle column and increased flow rate

mobile-phase buffer concentration should be increased for samples that are partially ionized, for example ≥50 mM.

Second, it is desirable to select a mobile phase pH which favors the neutral (nonionized) molecule. The reason is that column capacity w_s can be much smaller for a fully ionized compound than for the corresponding less-ionized molecule (Section 7.2.1.2, Fig. 7.3). Thus, much larger weights of sample can be injected for T-P separation when acids are separated at low pH and bases are separated at high pH (as in Fig. 7.4). However, because the nonionized molecule is preferentially retained in RP-LC (by a factor of 10 or more), partial ionization of the sample compound may not greatly reduce column capacity. Thus, the separation of an acid or base at a pH that is close to its pK_a value (corresponding to ~50 percent ionization of the product) may still provide near-maximum values of column capacity. If a mobile phase pH > 7 is needed to suppress the ionization of a basic compound, a column that is stable at this higher pH will be required; some (but not all) manufacturers sell column packings that are adequately stable at pH > 7. Again, it is wise to verify that corresponding (i.e., same product designation) prep-scale columns are available before starting the initial, analytical scale development – if more than a few milligrams of purified product are required.

When a volatile buffer cannot be used, the product can be recovered from collected fractions by any of several techniques. The product solubility in the organic–aqueous mobile phase can often be reduced by a suitable adjustment of the pH, thus allowing its precipitation and recovery by filtration. Product solubility may also be reduced by partial evaporation of the fraction, thus reducing the concentration of the (more volatile and more solubilizing) organic component in the mobile phase. When the product solubility does not allow a complete precipitation of the product from a predominantly aqueous fraction, the remainder of the product may be recovered by means of one-stage or continuous extraction into a volatile organic solvent. In some cases, salts can be removed from the product by dilution of the sample solution with water, followed by loading the sample onto a second reversed-phase column. Mobile phase conditions can be chosen such that the product is strongly retained (large k), so that salts can be flushed from the column by a relatively small volume of the mobile phase. The product is later recovered as a concentrated solution by back-flushing the column with a strong solvent such as methanol or acetonitrile, followed by removal of the solvent by evaporation or distillation.

While prep-LC by means of RP-LC can be recommended for many samples, normal-phase HPLC is also widely used at present – usually for less polar samples – because of ease of recovery of the product from the (volatile) mobile phase, as well as for other reasons. The present chapter assumes separation by RP-LC, but most of the principles described apply equally to normal phase HPLC (NP-LC). For a further discussion of method development based on NPLC, see [7].

7.2.2.1 Optimizing Separation Conditions Initial prep-LC experiments for method development are generally similar to those used for the development of an assay procedure (Section 3.1), namely the trial-and-error improvement of separation by systematic changes in experimental conditions. For the development of either an isocratic or gradient elution method, it is recommended to start with a gradient run as defined in Table 3.2 for a low-molecular-weight sample or Table 6.1 for large biomolecules. This first run can then be used to confirm that isocratic elution is appropriate, and to estimate a value of %B for a second isocratic separation. For either small or large samples, the main emphasis should be on the control of separation selectivity α_0, so as to obtain the best possible resolution of the most difficult-to-separate peak pair; this can be especially rewarding when developing a prep-LC separation, because of a large increase in the weight of injected sample for T-P separation when α_0 is increased [Equation (7.4)]. In prep-LC, however, there is a major difference in separation goals when compared with analytical separation. This is illustrated by the two isocratic separations of Figure 7.6, for the separation of compounds 1–9 of the "irregular" sample of Table 1.3. In this example, temperature and mobile phase %B were varied as a means of changing both α_0 and resolution. The separation of Figure 7.6(a) (20 percent B, 45°C) provides the best resolution for the *total* sample ($R_s = 1.9$), and this would be an appropriate separation for the analysis of all nine compounds in

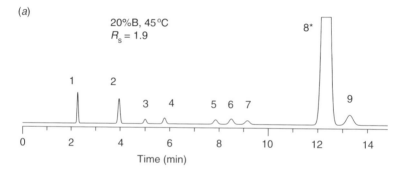

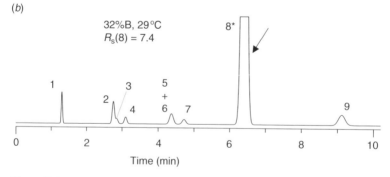

Figure 7.6 Analytical vs preparative conditions for the optimum separation of compounds 1–9 of the irregular sample of Table 1.3. Conditions: 150 × 4.6 mm C_{18} column; A-solvent is aqueous pH 2.6 buffer, B solvent is acetonitrile; flow rate 2.0 mL/min. Other conditions noted in the figure. (a) Optimum conditions for separation and analysis of all compounds in the sample; (b) optimum conditions for preparative purification of compound 8 (marked by asterisk). See text for details.

the sample. On the other hand, if the preparative isolation of peak 8 is desired, the goal is now a maximum resolution of this peak from adjacent peaks in the chromatogram. The separation of Figure 7.6(*b*) (32 percent B, 29°C) represents a much better choice for the preparative separation of peak 8 (*); its resolution from surrounding peaks is $R_s = 7.4$. A further examination of these two separations shows that $\alpha = 1.08$ in (*a*) vs 1.48 in (*b*); Equation (7.4) tells us that the weight of sample that can be injected for T-P separation in (*b*) is therefore about 20 times greater than in (*a*). That is, the time, effort, and cost required to obtain a certain weight of purified compound 8 can be reduced by as much as 20-fold, using the conditions of (*b*) instead of (*a*). Depending on the amount of purified product required, a separation as in Figure 7.6(*b*) might even avoid the need for prep-LC columns or equipment (scale-up; Section 7.2.2.3).

The use of higher temperatures in prep-LC as a means of optimizing α_0 and separation may be restricted by the available equipment and columns. This will be especially true when there is a need for scale-up (Section 7.2.2.3).

7.2.2.2 Selecting a Sample Weight for Touching-Peak Separation

The next step in method development (Fig. 7.5, step 3) is to increase sample size until T-P separation is achieved. Since α_0 is about 1.5 for the separation in Figure 7.6(*b*) of peak 8 from its nearest neighbor (peak 7), Table 7.1 suggests that about 3 mg of compound 8 can be injected, assuming (*a*) a column size of 150 × 4.6 mm, and (*b*) a neutral (nonionized) sample. In the present example, compound 8 is a slightly ionized acid under these separation conditions (pH 2.6), which suggests that a somewhat smaller sample weight *may* be necessary for T-P separation (because of smaller column capacities w_s for ionized compounds). We will ignore this possible complication for the moment.

Table 7.1 also gives the recommended (minimum) value of N_0 for the separation of Figure 7.6(*b*). In this case ($\alpha_0 = 1.5$, $k = 7.5$), a value of $N_0 = 590$ is suggested, which is far exceeded by the observed value of $N_0 = 8000$ in Figure 7.6(*b*). Therefore, column efficiency is more than adequate for T-P separation. Later, during scale-up, it will be possible to select conditions for smaller N_0 and a faster run time, so as to achieve a higher production rate of purified product (see Section 7.2.2.3 below and [8]).

Given an *estimate* (3 mg) of the required sample size for T-P separation, the next experiment should confirm this sample size by injecting a 3 mg sample for the conditions of Figure 7.6(*b*). An illustration of various possible results from the latter 3 mg separation is provided by Figure 7.7. The separation of peak 8 from peak 7 in the small-sample separation is shown in (*a*). T-P separation is shown in (*b*); if this is the actual result for a 3 mg injection, no further adjustment of sample size is needed (i.e., T-P separation has been achieved). Typically, the resulting separation will give a peak width W for the product peak that is either too small (*c*) or too large (*d*) for T-P separation. Based on Equation (7.2a) and $W \approx W_{th}$, a useful approximation is that w_x will be proportional to W^2. Since W for T-P separation is given by Equation (7.1), w_x (for T-P separation) in the examples of (*c*) or (*d*) should be changed by the factor $[(t_{R,b} - t_{R,a})/W]^2 = [(6.4 - 4.8)/W]^2 = (1.6/W)^2$ [W now refers to the value in (*c*) or (*d*) for the

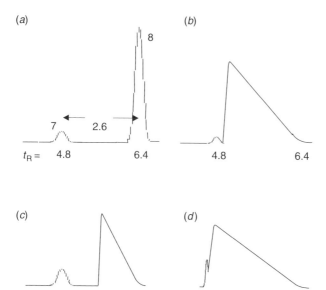

Figure 7.7 Hypothetical examples of the separation of peak 8 from peak 7 in Figure 7.6(b) for increasing sample size. Conditions of Figure 7.6(b) assumed, 3 mg sample injected in (b)–(d); (a) is a small-sample separation as in Figure 7.6(b). See text for details.

second (product) peak]. In (c), $W = (6.4 - 5.6) = 0.8$ min, so sample size should be increased by the factor $(1.6/0.8)^2 = 4$-fold (12 mg). In (d), $W = (6.4 - 3.5) = 2.9$ min, so sample size should be reduced by the factor $(1.6/2.9)^2 = 0.3$-fold (0.9 mg). One or two further experiments may be needed to fine-tune the final value of w_x for T-P separation.

7.2.2.3 Scale-Up

At this point, we will have determined how much purified product we can obtain from each run with our 150 × 4.6 mm column (3 mg for the present example, assuming separation as in Fig. 7.7b). Knowing how much purified product is needed, we can then calculate how many replicate runs with this column will be required, and how much time will be involved. If the effort required for this approach seems excessive, the simplest alternative is to use a larger column ("scale-up"); for example, increase column diameter by some factor x, while increasing sample weight by x^2 (or in proportion to column volume, when both column length and diameter are changed). Flow rate must also be increased by x^2, in order to maintain F/V_m and k^* constant [Equation (2.13)] for no change in separation selectivity and resolution. Columns for preparative RP-LC separation come in a variety of different dimensions (Table 7.3). Particles of 5–10 μm are typically used for T-P prep-LC separation, although larger particles are also available. Table 7.3 provides relative values of flow rate and sample size as a function of column dimensions, for either isocratic or gradient elution. If a larger column

7.2 ISOCRATIC SEPARATION

TABLE 7.3 Scale-Up Factor for Columns of Different Dimensions. Note that if F/V_m is Changed for Gradient Elution, α_0 May Change (Usually Undesirable)

Column internal diameter (mm)	Scale-up factor (increase in sample weight)[a]				Flow rate[b]
	Column length[a]				
	50 mm	150 mm	250 mm	500 mm	
4.6	0.3	(1.0)	1.7	3.3	(1)
7	0.8	2.3	3.9	7.7	2.3
10	1.6	4.7	7.9	15.8	4.7
21	6.9	20.8	34.7	69.5	20.8
25	9.8	29.5	49.2	98.5	29.5

[a]Factor-increase in w_x and column weight, relative to 150 × 4.6 mm column.
[b]Factor-increase in flow rate, relative to 4.6 mm diameter column (for constant pressure drop, if column length is unchanged); if both column length and diameter are changed, maintain flow rate/column volume constant to maintain the same separation selectivity.

and/or the associated prep-LC equipment is not available, or if the number of repetitive runs required is considered excessive, two alternatives are available. First, try to increase α_0 further by other changes in conditions (Table 3.4). If a significant increase in α_0 is possible, a much larger sample weight can be separated for a given size column (Table 7.1). Second, investigate the possibility of severely overloaded separation as in Figure 7.1(c) (Section 7.4).

If column length and/or particle size are changed during scale-up, the resulting plate number must be taken into account. Maintaining the same value of N_0 should give a very similar separation to that obtained with the original, small-scale column. However, a decrease in N_0 will not affect prep-LC separation much, as long as the new value of N_0 exceeds the required value for T-P separation shown in Table 7.1 for a given value of α and k ($N_0 = 590$ in the present example). A further refinement of the latter preparative separation is possible by adjusting column length, particle size, and flow rate so as to maximize the recovery of purified product per unit time. In many cases, such an improvement represents an unnecessary expenditure of effort for the collection of <1 g of purified product. However, when a major reduction in the time and/or cost for purifying the product is needed, the following approach may be attractive.

As noted above, the recommended value of $N_0 = 590$ from Table 7.1 suggests that we can reduce the original column plate number ($N_0 \approx 8000$) significantly without much affecting the separation (a slight decrease in w_x/w_s can be expected, however). A reduction in the required value of N_0 can in turn allow for a much faster run time. In prep-LC, a reduction in run time and N_0 is most conveniently achieved by the use of larger particles and higher flow rates, for example, 10 μm particles instead of the original 5 μm particles, plus an increase in flow rate by 4-fold, resulting in a 4-fold reduction in run time without increasing the column pressure (pressure drop is inversely proportional to particle diameter squared). Alternatively, the main consideration might be to reduce column pressure drop, because of pressure limitations imposed by either the equipment or column when working with much larger-diameter columns (compared with the small-scale separation). Section 9.5 can be consulted for guidance in the trade-offs involved for N_0 as a function of pressure, particle size, and column length (see especially Fig. 9.11). An experimental example of prep-LC method development which incorporates some of the above considerations is described in [8].

7.2.2.4 Sample Solubility

The above approach for the development of a prep-LC separation seems straightforward, but it assumes that the necessary weight of injected sample can be dissolved in a volume of mobile phase that is smaller than 0.5 $(t_{R,b} - t_{R,a})$ F [Equation (7.6) and Fig. (7.2b)]. In many cases, the sample will not be adequately soluble in this (relatively small) volume of mobile phase, whereas the injection of larger volumes of sample can significantly compromise the separation. Sample solubility can be improved in various ways. Samples that are inadequately soluble in water (or solvent-mixtures containing a large proportion of water) can usually be dissolved in various pure organic solvents. Because sample retention k decreases in pure organics, however, the injection of significant volumes of the sample dissolved in an organic solvent can lead to reduced initial retention. As a result, the sample peak can be further broadened and distorted, with increased overlap of the product peak by adjacent impurity peaks; this could lead to a major reduction in the amount of purified product recovered.

A similar problem can arise in the separation of acidic or basic samples. Often the solubility of such samples increases with their increasing ionization, but at the same time retention decreases, as does column capacity (Section 7.2.1.2). Therefore, if a mobile phase pH is selected to favor the neutral molecule, it is inadvisable to inject the sample in a solvent that causes near-complete ionization.

Sample solubility is often a major consideration in the design of a prep-LC separation, and consideration should be given to this possible problem at an early stage of method development. Over the years, a number of different techniques have been suggested to deal with difficultly soluble samples. In some cases, sample solubility may be limited by other constituents of the sample, rather than by the product. In these cases, it may prove useful to extract the sample into the mobile phase and discard the less soluble residue. In the case of normal-phase HPLC, it is often possible to dissolve the sample in a different solvent mixture, one that provides similar (or greater) retention *and* greater solubility than the mobile phase [9]. More often, workers deliberately use either larger volumes of the sample, or a sample solvent that provides reduced retention, thereby sacrificing some resolution of the final mixture. See also Section 7.3.3.

A recent proposal [10] provides a potential solution for the problem of limited sample solubility. This so-called "*at-column dilution*" technique works as follows. Means are created to introduce the sample (dissolved in a solvent that favors good solubility) into the mobile phase at the juncture where the mobile phase enters the column, thereby providing mixing of sample and mobile phase just before entry into column. For example, a basic sample dissolved in a lightly buffered, low-pH buffer (that promotes higher solubility) is injected into a more highly buffered, high-pH mobile phase (which favors reduced solubility). After mixing is achieved, the sample is now contained in a high-pH medium which favors strong retention (good), higher column capacity (good), but poor sample solubility (bad). However, sample precipitation under these conditions is not an instantaneous process, so that retention of the sample near the column inlet (but past the frit) can occur before precipitation advances to the point of clogging the frit or adjacent column packing. Similarly, a sample dissolved in pure organic might be injected into a water-containing mobile phase, again with increased retention and decreased solubility, but with a similar result (retention of the sample before its precipitation in the frit or column). Practical experience has shown that large sample loads can be accomplished successfully in many cases, using the at-column dilution technique. Occasionally, the sample precipitates too early, and the resulting pressure may switch off the gradient pump. However, the sample pump continues to feed organic solvent to the column at a slow flow rate, which can often redissolve the sample and prevent blockage. Nevertheless, it should be emphasized that special conditions and equipment are needed to make this approach work successfully; countless blocked frits, columns, and equipment have resulted from the careless use of sample solvents that have a composition different from that of the mobile phase.

7.2.3 Beyond Touching-Peak Separation

"Severely overloaded" separation refers here to injection of a sample mass in excess of that required for T-P separation (e.g., as in Fig. 7.1c). As a result, significant overlap will occur of the product peak and an adjacent impurity peak. This is illustrated in Figure 7.8 by some computer simulations of a model separation as a function of sample weight (note that such computer simulations have been shown repeatedly to give results that agree with experiment). In each example of Figure 7.8, separate peaks are shown for sample components A and B for the injection of indicated mixtures of the two compounds.

Figure 7.8(a and b) shows the injection of a small sample of 1 : 10 and 10 : 1 mixtures of A and B. Because the column is not yet overloaded by these small samples, Gaussian-shaped peaks are observed. Figure 7.8(c and d) shows corresponding separations for a sample large enough to create T-P separation in each case (0.5 g for the smaller peak, 5 g for the larger peak). There is a very small overlap of the two peaks, and the major peak in each case can be recovered in $\sim$99 percent or better purity and 99 percent or better yield. Note also that the retention time of the minor peak is unchanged by the presence of the larger peak.

When a still larger sample is injected [Fig. 7.8e and f, where the sample sizes of (c) and (d) are doubled], a more significant overlap of the two peaks occurs, that is, severe column overload. When a small peak (A) precedes the larger product peak (B) (Fig. 7.8e), the small peak is displaced with a reduction in its retention time, accompanied by some overlap of the following larger peak. When the small peak (B) follows the larger peak (A) (as in Fig. 7.8f), a so-called "tag-along" effect is observed. The smaller, later peak is dragged under the preceding overloaded peak, again with a reduction in its retention time. While in both cases of severe

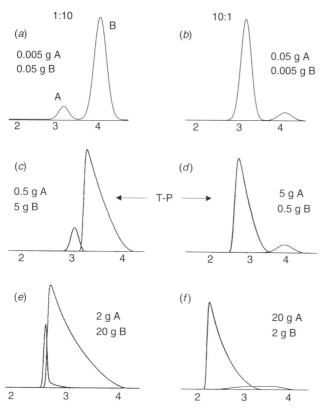

Figure 7.8 Isocratic separation of a two-component sample as a function of sample size. Computer simulations based on the Langmuir isotherm; Conditions: 250 × 50 mm column, 7 μm particles, 210 mL/min flow rate, $N = 800$; $k = 1$ and 1.5, respectively. Sample weights: (a, b), small sample; (c) 0.5 g A, 5 g B; (d) 5 g A, 0.5 g B; (e) 2 g A, 20 g B; (f) 20 g A, 2 g B. Adapted from [7], see text for details.

overloading (e and f) there is extensive peak overlap, it is usually possible to recover a much greater weight of purified product per injection than in the touching-peak separations of (c) and (d). See Section 7.4 for a further discussion of severely overloaded separation.

7.3 GRADIENT SEPARATION

The relative complexity of peak broadening and separation in isocratic ("overloaded") chromatography is further increased for overloaded gradient elution (more variables!). However, the LSS model provides an essentially complete parallelism between isocratic and gradient prep-LC for "corresponding" isocratic and gradient separation, when $k^* = k$ [11] (by definition, k^* does not change with sample size in gradient prep-LC – just as for the case of k in isocratic prep-LC). Thus, when gradient conditions are selected for $k^* = k$, the resolution of two adjacent peaks will vary in the same way when sample size is increased to the point of column overloading (for an example of this, see Fig. 7.11). Similarly, the sample weight w_x required for T-P separation will be the same in both isocratic and gradient elution, and the preferred value of N_0 will be the same (when α_0 and k or k^* are the same). Thus, method development for preparative gradient elution can proceed in very similar fashion as for isocratic separation (Fig. 7.5 and Table 7.2); likewise, the data of Tables 7.1 and 7.3 apply equally for both gradient and isocratic separation. While this similarity of isocratic and gradient prep-LC may appear reasonable to the reader (on the basis of what has so far been discussed in this book), otherwise knowledgeable workers in the field of prep-LC have occasionally appeared unaware of this relationship between isocratic and gradient elution [14] – as a result drawing incorrect conclusions on the expected effect of separation conditions on gradient prep-LC.

Peak shape changes in a similar way when sample weight is varied, for both gradient and isocratic elution. This can be seen by comparing the overlaid gradient peaks in Figure 7.9 with corresponding isocratic peaks in Figure 7.2(*a*). In each case, peaks for different sample weights are overlapped on the same time scale, resulting in what has been called "overlapping right-triangles." Overloaded gradient peaks (as in Fig. 7.9) tend to be more rounded ("shark-fin" shape), but essentially similar behavior is observed in both figures. The steeper gradient in Figure 7.9(*b*) vs (*a*), corresponding to a smaller value of k^*, shows narrower peaks – the same as for a "corresponding" isocratic separation with a smaller value of k.

For two adjacent peaks, a similar behavior is observed for either isocratic or gradient elution when sample size is increased beyond T-P separation. This is

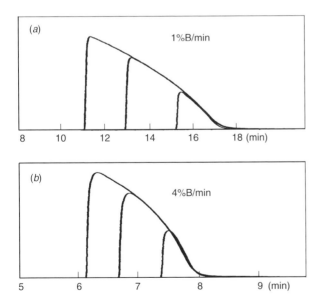

Figure 7.9 Peak width and shape in gradient elution as a function of sample size and gradient time. The sample is 2-phenylethanol. Adapted from [12].

illustrated by the computer simulations of Figure 7.10 for severely overloaded gradient elution, which can be compared with the simulations in Figure 7.8(c) for severely overloaded isocratic elution. In Figure 7.10(a), the small-sample separation of peaks A and B is shown. In (b) and (c), the overlapping chromatograms for a larger sample weight are shown, where the curves labeled A or B (– – –) are for injection of the individual compounds, and the remaining solid curves (A', B') are for the mixture of the two compound (but displaying individual peaks for each compound). The 10:1 relative weights of the two compounds are reversed in (b) vs (c). The phenomena of "displacement" (b) and "tag-along" (c) seen in Figure 7.10(b and c) for gradient elution are also seen in Figure 7.8(e and f) for isocratic elution. (note that the separations of Figs 7.8 and 7.10 are *not* corresponding separations, so only *general* similarities between isocratic and gradient separation can be observed).

Finally, Figure 7.11 compares the experimental separation of compounds A and B by isocratic elution (a) and gradient elution (b); these are "corresponding"

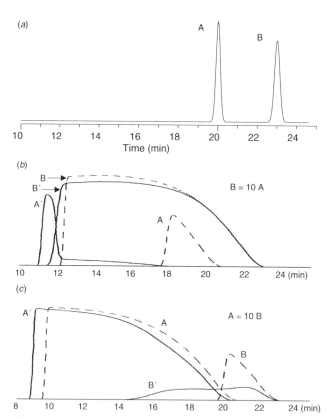

Figure 7.10 Severely overloaded gradient separation of a two-component sample as a function of sample size. Computer simulations based on the Langmuir isotherm. Peaks labeled A and B (– – –) are for the injection of samples of pure of A or B; peaks labeled A' and B' (———) are for the separation of mixtures of A and B. Adapted from [13], see text for details.

separations, with $k^* = k$. In each case, chromatograms for the injections of the individual compounds (A, B) are overlaid onto separations of the mixture (A′, B′) – as in Figure 7.10. Figure 7.11(a) shows the isocratic separation of peaks A and B for different injected weights of these two compounds: 2.5 mg of A and 2.5 mg of B in the first chromatogram, 2.5 mg of A and 10 mg of B in the second chromatogram, and so on [see sample weights for both (a) and (b) in the middle of Fig. 7.11]. The basic similarity of prep-LC using either isocratic or gradient elution is obvious from this example, and has been further affirmed in other studies based on computer simulation [15–17]; production rates (grams/hr of purified product), resolution as in Figure 7.11, and so on, are all similar for isocratic and gradient separations.

To summarize, the chromatograms of Figure 7.11 illustrate that separation as a function of sample weight is essentially equivalent for "corresponding" isocratic and gradient elution (with $k^* = k$). However, separation time often differs for gradient vs isocratic elution. Thus, the column must be equilibrated between gradient runs (Section 5.3), which usually increases run time. On the other hand, if the sample

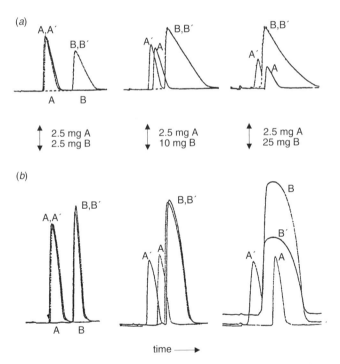

Figure 7.11 Similar effects of column overload in (a) isocratic, and (b) gradient elution. Separation of two xanthines [β-hydroxyethyltheophylline (A) and 7β-hydroxypropyltheophylline (B)] with k (isocratic) equal k^* (gradient). Sample weights shown in figure. Peaks labeled A and B are for the injection of samples of pure of A or B; peaks labeled A′ and B′ are for the separation of mixtures of A and B. Adapted from [11].

contains components which are more strongly retained than the product to be purified, isocratic run-times can be much longer vs gradient elution (unless a strong-solvent flush is used). Peak widths in gradient prep-LC are also narrower than in "corresponding" isocratic separations (Section V.2.1 in Appendix V).

7.3.1 Touching-Peak Separation

In the following discussion, we assume "corresponding" isocratic and gradient separation, where the sample is assumed to be the same, and separation conditions differ only in that %B is fixed in isocratic elution and varies in gradient elution. A further condition for "corresponding" separations is k^* (gradient) = k (isocratic) for two adjacent peaks of interest and injections of a small weight of sample. In gradient elution, peak width W for overloaded separation is given by Equation (7.2):

$$W^2 = W_0^2 + W_{th}^2 \qquad (7.2)$$

just as for isocratic separation. For "corresponding" small-sample conditions, where $W_0 \gg W_{th}$, the resolution of two adjacent peaks in both isocratic and gradient elution will be approximately equal (Section 2.2.3). As we will next show (and has been illustrated in Fig. 7.11), this is also the case for prep-LC with large samples ($W_0 \ll W_{th}$). Thus, from Appendix V, we have for gradient elution,

$$\text{(gradient)} \quad W_{th} \approx 2t_0 k_e (w_x/w_s)^{1/2} \qquad (7.7)$$

where k_e is the value of k at elution. Equation (7.7) is equivalent to W_{th} in isocratic elution [Equations (7.2) and (7.2a)]

$$\text{(isocratic)} \quad W_{th} \approx 2t_0 k (w_x/w_s)^{1/2} \qquad (7.8)$$

because k_e in gradient elution is equivalent to k in isocratic elution at the time the peak leaves the column. While this similarity of isocratic and gradient elution exists for peak width in prep-LC, resolution in gradient prep-LC is determined by k^*, rather than k at elution (k_e). Because $k_e = k^*/2$, peaks in small-sample and prep-LC gradient elution will be narrower than in "corresponding" isocratic separation, where k^* (gradient) = k (isocratic).

7.3.2 Method Development for Gradient Touching-Peak Separation

With minor exceptions, the same approach described in Section 7.2.2 for developing an isocratic T-P separation can be followed for T-P gradient elution. Initial conditions for the first gradient experiment should adhere to the same guidelines as for isocratic prep-LC (Section 7.1). For the separation of ionizable samples, a volatile buffer is recommended, and a mobile phase pH should be selected that limits the ionization of the desired product and adjacent impurities. If the pK_a value of the product is known, the mobile phase should have a pH that is 1 unit (or more) higher for a basic compound, and 1 unit (or more) lower for an acidic compound. A specific column packing should then be chosen that is (a) compatible with the planned mobile phase pH, and (b) available in columns of varying size for both preparative- and analytical-scale separation (depending on the weight of purified product that is required).

Figure 7.5 and Table 7.2 outline the overall approach for developing a gradient prep-LC separation. Steps 1 and 2 are carried out in the same way as for the development of an assay procedure (Section 3.3), but with two important changes. First, an assay procedure needs to separate all peaks of interest to baseline. However, *only the separation of the product peak(s) is required in prep-LC* (Fig. 7.6 and related discussion). Second, resolution in an assay procedure need not be greater than $R_s = 2$, and resolution in excess of $R_s = 2$ means a run time which is too long. In prep-LC, however, greater resolution means a larger value of α_0, which means that a larger weight of sample can be separated in each run; *usually this is of critical importance in prep-LC*. In prep-LC method development, it is often worthwhile to carry out additional experiments for the optimization of α_0 (more than would be the case for the development of an assay procedure), because of the direct effect of α_0 on the possible weight of recovered product in each run.

Returning to Figure 7.5 and Table 7.2, initial experiments (step 1) will be carried out on an analytical-scale column with sample weights <0.1 mg, while focusing on the product peak 5 (asterisk in Fig. 7.5). Next, separation conditions will be determined that provide a large enough value of α_0 for the product peak (step 2). In step 3, sample weight will be increased to provide T-P separation, while in step 4 the separation will be scaled-up if necessary. The amount of effort that is required in method development is related to (a) the amount of purified product that will be required, and (b) the difficulty of the separation. For smaller amounts of purified product and an easier separation (i.e., where a large value of α_0 is easily obtained), scale-up may be avoided altogether by means of repetitive injections with an analytical-scale column. By keeping in mind the amount of purified product that will ultimately be required, other opportunities may emerge for reducing the time spent on method development.

The following example will illustrate some of the options available during the method development process, as well as the value of assessing results and prospects at each stage in method development. It will be assumed that the isolation of an unknown component from an algal pigment extract is desired. More commonly, prep-LC is used for the purification of a crude product, but the recovery of minor components of a sample may be needed for their further characterization or other use. A major difference between the purification of a major constituent of the sample and the isolation of a minor component, is that the approximate weights of individual sample components may not be known. Therefore, a rough estimate will be required of the weight of critical sample components (either the product or a larger adjacent peak); such an estimate might be provided by relative peak size or (more accurately) by analysis of the sample. *Note that it is mainly the weight of the product peak (or a larger adjacent peak) that determines peak broadening and the value of w_x for T-P separation; the weights of other sample components can be ignored.*

Gradient separations of small weights of the algal pigment had been carried out previously, with the development of the preliminary separation shown in

Figure 7.12(a). The unknown peak of interest ("b^*") is indicated by an arrow. Compound b^* was originally an unknown sample constituent, whose isolation would permit its molecular characterization. The conditions for the separation of Figure 7.12(a) are as follows: 250 × 4.6 mm C_{18} column, flow rate of 2 mL/min, 70–100 percent MeOH in 15 min, 50°C. The chromatograms of Figure 7.12 are computer simulations based on the experimental data of [18]. A portion of the chromatogram in (a) is expanded in (b) to better show the relative separation of peak b^* from adjacent peaks. Clearly, the separation of b^* with these experimental conditions is inadequate.

The next step is to vary those conditions that affect selectivity, as a means of maximizing the resolution (and α_0) of b^* from its neighboring peaks. The same general procedure will be followed as recommended in Section 3.3.2 for the analysis of the total sample. The main difference is that we will focus on just that part of the chromatogram which contains the peak of interest, for example, as in Figure 7.12(b). Various means of changing selectivity can be used, as summarized in Sections 3.3.2 and 3.6. In the present example, we will vary gradient time t_G and temperature T. The latter two separation conditions have been found to be a good first step in attempting to optimize separation selectivity and resolution for analytical separations. However, the use of above-ambient temperatures for scale-up (Section 7.2.2.3) may be limited for some prep-LC equipment and columns, so this possibility should be kept in mind – unless it is anticipated that the required amount of purified product can be produced with the equipment used for the initial small-scale studies. In the present case, it had already been established that the column used for the separations of Figure 7.12 was temperature-stable for these separation conditions.

Figure 7.12(b–e) shows the results of four successive experiments where t_G and T were varied: (b) 15 min, 50°C; (c) 45 min, 50°C; (d) 15 min, 60°C; (e) 45 min, 60°C. A reasonable separation of the desired compound b^* is obtained only in (c), with $R_s = 2.1$. At this point, further experiments can be carried out to further improve this separation and increase the value of α_0 for a maximum weight of injected sample for T-P separation [Equation (7.4) and Table 7.1)]. However, before proceeding in this way, it is useful to estimate the amount of purified b^* that can be obtained from a single run with the conditions of (c). This requires (a) an estimate of α_0 for the separation of b^* from adjacent impurities c and a, and (b) an estimate of the concentration of desired product in the sample. From Equation (V.28) of Appendix V we have

$$\log \alpha_0 = (\Delta \phi S / t_G)(t_{R,b} - t_{R,a}) \tag{7.9}$$

For small-molecule samples (with molecular weights between 100 and 400), the sample parameter S is usually about 4, which allows an initial estimate of α_0 from the difference in retention times of the product and adjacent peaks ($t_{R,b} - t_{R,a}$). For large-molecule samples (e.g., peptides, proteins), a better estimate of S as a function of molecular weight M is (Section 6.1)

$$S \approx 0.25 \text{ (molecular weight)}^{1/2} \tag{7.10}$$

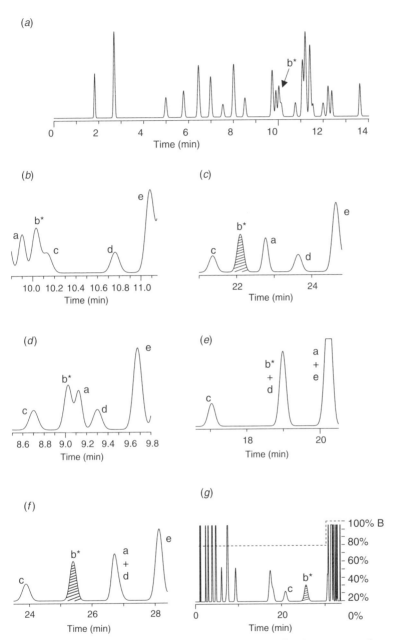

Figure 7.12 Example of prep-LC method development for the recovery of an purified component from an algal pigment extract. Conditions: 250 × 3.2 mm C_{18} column; 70–100 percent methanol–buffer gradients except for (g); 2.0 mL/min; gradient time and temperature vary. (a) Total chromatogram; 15 min gradient, 50°C; (b–f) partial chromatograms in vicinity of desired product b^*; (b) 15 min gradient, 50°C; (c) 45 min gradient, 50°C; (d) 15 min gradient, 60°C; (e) 45 min gradient, 60°C; (f) 57 min gradient, 52°C; (g) step-gradient; 79–79–100 percent MeOH at 0–28–33 min; 49°C. Peak identifications: (a, b^*) unknowns; (c) antheraxanthin; (d) alloxanthin; (e) diatoxanthin. Computer simulations based on data of [18]. See text for details.

The algal pigment sample has a molecular weight of about 500, so Equation (7.10) suggests a value of S equal to about 6.

Alternatively, it is possible to determine values of the gradient steepness b for compound b^* from the retention times for the two gradients and experimental conditions (Section 9.3.3). A value of S for the compound can then be obtained from Equation (2.11):

$$S = t_G F b / (V_m \Delta \phi) \qquad (7.11)$$

or

$$S = t_G b / (t_0 \Delta \phi) \qquad (7.11a)$$

In this case, S for b^* was determined equal to 13.3, an unexpectedly large number (suggesting that S for a particular sample and separation should be measured, rather than estimated, but this is not critical to prep-LC method development as described here).

The product peak b^* is estimated to comprise about 4 percent of the total sample, based on relative peak areas in Figure 7.12 (a better value could be obtained by the analysis of the sample, but since b^* is an unknown constituent of the sample, this was not feasible in the present case). Continuing with the application of Equation (7.9) for the purpose of estimating α_0, for the separation of Figure 7.12(c) we have a 70–100 percent MeOH gradient ($\Delta \phi = 0.3$), $t_G = 45$ min, and $(t_{R,b} - t_{R,a}) = (22.10 - 21.36) = 0.74$ min for peaks c and b^*, and $(22.77 - 22.10) = 0.67$ min for peaks b^* and a. Since the three peaks appear to be of comparable size, the lesser separation for peaks b^* and a (and a smaller value of α_0) should be used in Equation (7.9). This then gives log $\alpha_0 = [(0.3)(13.3)/45](0.67) = 0.06$, or $\alpha = 1.15$. Similarly [Equation (2.13)], $k^* = t_G F / 1.15 V_m \Delta \phi S = (45)(2)/[(1.15)(1.5)(0.3)(13.3)] = 13$. From Table 7.1, $w_x = 0.4$ mg for a 150 × 4.6 mm column, or $(0.4)(250/150) = 0.7$ mg for the present 250 × 4.6-mm column (suggesting a *total* sample weight of injected sample equal to $0.7/0.04 = 17$ mg). This suggests that 0.7 mg of b^* can be isolated from a single T-P separation with the conditions of Figure 7.12(c) and a sample weight of 17 mg.

The next step is to verify that 0.7 mg of b^* can be injected for T-P separation, by carrying out a separation with this sample weight (and conditions of Fig. 7.12c). Various possible results of this experiment (as in Fig. 7.7) can be anticipated, leading to a possible revision of the required sample weight for T-P separation and the resulting yield of purified product. Let us assume that a result as in Figure 7.7(b) is obtained, confirming that 17 mg of injected sample (or 0.7 mg of product) corresponds to T-P separation. At this point, a further strategy for developing a final prep-LC method can be laid out. If we accept the separation of Figure 7.12(c), repetitive runs are estimated to each provide about 0.7 mg of purified product in a time of about 50 min (allowing for an inter-run equilibration of 5 min, or about 4 column volumes (Section 5.3). Because the product peak leaves the column at 22 min, however, the gradient can be changed at this point to a step gradient to 100 percent MeOH, with an isocratic hold of 3 min (70–85–100–100 percent B at 0–22–22.1–25 min). This will effectively reduce the run time to about 30 min, again allowing 5 min for between-run column equilibration. So, repetitive sample injection under these conditions is estimated to provide about 1.4 mg/h of purified b^*, and 4 h of repeated injections would generate about 6 mg of product. If this suffices, further method development is not needed.

If the required amount of purified product is $\gg 6$ mg, then various alternatives to the above proposal (which yields 6 mg of pure product in 4 h) should be considered. One option is to attempt an increase in the resolution of b^* by increasing its value of α_0. Further, trial-and-error changes in gradient time and temperature are a first step. Or, if computer simulation is available (Section 3.4), the four runs of Figure 7.10($b-e$) can be used to predict conditions for the maximum resolution of b^*. Figure 7.12(f) shows the resulting optimized separation; for a gradient time of 57 min at 52°C, the resolution of b^* is increased from $R_s = 2.1$ in (c) to 3.4 in (f), corresponding to an approximate increase in $\alpha_0 - 1$ by the same ratio [Equation (3.4)], that is, by a factor of $(3.4)/(2.1) = 1.6$. This yields a value of $\alpha_0 - 1$ for the separation of Figure 7.12(e) equal to $(1.6)(0.15) = 0.24$, or $\alpha_0 = 1.24$ for the separation of (e). From Table 7.1, this suggests an increase in T-P sample weight by a factor of about 2, or a recovered product weight of 1.4 mg per run. Thus, for the conditions of Figure 7.12(f), about 12 mg of purified product can be obtained in a half-day of repetitive injections [the time required per run is about the same for (c) and (f)].

A further reduction in separation time is often possible by beginning the gradient at a higher %B, while maintaining gradient steepness b constant [Equation (2.11)]. In the present case, this led to a loss in resolution of peak b^* which was not adequately compensated for by the shorter gradient time (note that the initial gradient starts at 70 percent B in this example). Further increase in α_0 might also be pursued, by changing the other separation conditions of Table 3.4. Because of the desirability of maximum α_0 in prep-LC, and because gradient time and temperature are less effective for controlling separation selectivity and values of α_0, other more powerful changes in conditions that affect selectivity should not be overlooked (Section 3.6), for example, a change in the organic solvent type and/or mobile phase pH, while avoiding substantial ionization of the product during separation.

If a much larger weight of purified product were required, the next step is to consider scale-up and a larger column. Reference to Table 7.3 suggests possible column sizes that can yield a predictable increase in the weight of purified product from each run. Run time will vary in proportion to column length, but column diameter will have no effect on run time, as long as flow rate can be increased as indicated in Table 7.3 for larger-diameter columns. Note that a few experiments for each sample, plus some simple calculations as summarized above, can be used to design an effective strategy for prep-LC method development. This approach often can save a good deal of time and effort, because the amount of work required is adjusted to the need for a given weight of purified product.

7.3.2.1 Step Gradients

Many prep-LC separations that appear to require gradient elution are more conveniently carried out by means of a step gradient between two (or more) isocratic conditions. Thus, when scaling up a prep-LC separation, these step gradients can be carried out on the same equipment that is used for isocratic separation, by using a switching valve between two solvent reservoirs and the pump. If solvents are recovered for re-use, step gradients also minimize the need for recovery of pure A and B solvent from the solvent mixtures in the two reservoirs [because there should be little mixing of these two solvent mixtures, only the sample needs to be removed from recovered solvent (e.g., by distillation)]. The guiding principle in the substitution of a step-gradient for a continuous gradient is as follows. The initial isocratic step should elute the product peak with $k \approx k^*$ for the corresponding (optimized) gradient separation (Fig. 7.12f in this case). The remainder of the sample will then be washed from the column by a sudden increase in %B (second step). When the step gradient is performed in this way, a very similar separation of the product peak should result. This is illustrated for the sample of Figure 7.12 in the step gradient of Figure 7.12(g). The resulting value of α_0 equals 1.23 (essentially the same as in Figure 7.12f), so the same sample size can be injected in (g) as in (f) for T-P separation. Run time for the final separations of Figure 7.12(f and g) is not much different, assuming a 70–87–100 %B gradient at 0–25–27 min for the separation of Figure 7.12(f).

7.3.3 Sample-Volume Overload

The effect of sample volume on gradient vs isocratic separation can be somewhat different. In isocratic elution, the sample is preferably introduced as a mixture in the mobile phase, in which case Equation (7.6) applies. In gradient elution, the usual practice is to dissolve the sample in mobile phase that corresponds to the start of the gradient. However, due to the strong retention of the sample at the start of the gradient, much larger sample volumes will be allowed, compared with the prediction of Equation (7.6). On the other hand, the solubility of the sample in the (lower %B) sample solvent will often be very much reduced for gradient vs isocratic elution, when the foregoing practice is used. From a practical standpoint, increased amounts of sample in soluble form can be introduced in gradient elution either by using larger volumes of sample solvent [compared with Equation (7.6)], or the sample can be dissolved in a stronger solvent, for example, in a mobile phase of composition ϕ_e that corresponds to the gradient at the time the product leaves the column. Either practice may work for a given separation. Otherwise, the problem of sample solubility and sample-volume overload can be addressed as in Sections 7.2.1.3 and 7.2.2.4 for isocratic prep-LC, for example, use of the "at-column dilution" technique.

7.3.4 Possible Complications of Simple Touching-Peak Theory and Their Practical Impact

Two deviations from simple T-P theory as described above should be noted, as they may occasionally be encountered. First, we have assumed equal column capacities w_s for the two adjacent peaks to be separated. If the value of w_s for the more retained peak (larger k) is much smaller than that of the preceding peak, so-called "*crossing isotherms*" can result [2]. In this case, as sample weight increases, the separation factor α decreases for both isocratic and gradient elution, until the two overloaded peaks merge together with complete loss in resolution. Our second assumption above is that values of $S = d(\log k)/d\phi$ are constant for the two peaks we are trying to separate. When this is not the case in prep gradient elution, the sample size for touching peaks can be either greater or less than estimated from Equation (7.4).

7.3.4.1 Crossing Isotherms An example of the crossing-isotherm phenomenon is illustrated in the isocratic separations of Figure 7.13 [2]. In each case (Fig. 7.13a–d), a mixture of two compounds is injected, and the use of different detection wavelengths allows the monitoring of each peak separately [dashed or solid lines in (c) and (d)]. For the two small-sample separations in (a) and (b), values of N_0, k, and α_0 ($= 1.12$) for each peak or peak pair were quite similar. Consequently, similar column overload and peak resolution would be expected from Equation (7.4) for the two samples in (a) and (b), when equal, larger weights of each sample are injected in (c) and (d). For the example of (c), T-P separation results with minor overlap of the two peaks. For (d), however, almost complete peak overlap results. The latter can be attributed to a larger value of w_s for the first peak (PE) than for the second peak (C); consequently, the second compound overloads more quickly, and then overtakes the first compound during migration through the column. Similar results as in Figure 7.13 can be expected in gradient elution, when isotherms cross or values of w_s vary for two adjacent peaks.

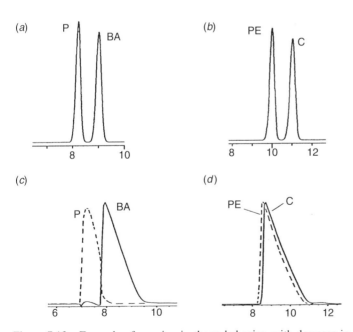

Figure 7.13 Example of crossing-isotherm behavior, with decrease in allowed sample weight for touching peak separation. Conditions: 150 × 4.6 mm, 5 μm C_{18} column; methanol–water mobile phases; 1.0 mL/min. (a) 3 μg phenol (P) and 7 μg benzyl alcohol (BA); (b) 3 μg 2-phenylethanol (PE) and 7 μg p-cresol (C); (c) same as (a), except 1 mg of phenol plus 3 mg of benzyl alcohol; (d) same as (b), except 3 mg 2-phenylethanol and 0.5 mg p-cresol. Adapted from [2]; see text for details.

7.3.4.2 Unequal Values of S

The importance of $S = d(\log k)/d\phi$ in gradient elution has been stressed repeatedly. Because k^* is inversely proportional to S, samples with larger S values (larger molecules) require generally flatter gradients for reasonable peak resolution (Section 6.1.1). Differing S-values for "irregular" samples (Section 1.6.2) mean that large changes in relative retention (α_0) can result when isocratic %B or gradient time is varied (Section 2.2.3). Consequently, optimizing gradient time is an important first step in the separation of such samples, so as to maximize α_0.

When two adjacent peaks have different values of S, this also can affect the sample weight for T-P separation [15, 17, 19, 20], as illustrated conceptually in Figure 7.14. Figure 7.14(a) illustrates a T-P gradient separation for the case of equal S values for two compounds ("*parallel*" case). At the top of the figure is a plot of log k^* vs ϕ^* (or %B) for each peak (see the similar comparison of Fig. 2.8). The dotted lines connect the log $k^* - \phi^*$ plot for compounds A and B to (narrow) peaks in the chromatogram below, corresponding to a small weight of each compound. The values of k^* and ϕ^* for each peak are determined by gradient conditions [i.e., Equation (2.13) for k^* and Equation (2.14) for ϕ^*], with a gradient time of 30 min for each separation in Figure 7.14. Now assume that a large enough sample weight has been injected to allow peak B to cover the space between the two small-sample peaks (T-P separation), giving the wide cross-hatched peaks in the

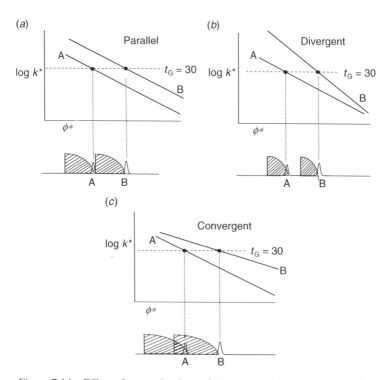

Figure 7.14 Effect of unequal values of S on the overload separation of two peaks by gradient elution. See text for details.

chromatogram of Figure 7.14(a). We can see that the vertical separation of the two log k^* vs ϕ plots is constant and equal to log α for each value of ϕ. Thus, at the beginning of elution of overloaded peak B (at a lower value of ϕ^*, corresponding to the elution of a small sample of A), α is the same ($=\alpha_0$) as at the end of elution, that is, the separation factor is not a function of sample weight. (Note that Equation (7.4), which relates sample weight for T-P separation to values of α_0 assumes approximately equal values of S for the two adjacent peaks.)

Figure 7.14(b) is similar to that of 7.14(a) [same weight of injected sample for T-P separation in (a)], except that now the plots of log k^* vs ϕ^* are no longer parallel, but diverge for lower values of ϕ^* ("*divergent*" case), that is, the value of S for compound B is greater than for compound A. For higher loading of the column (at lower values of ϕ^*), the vertical separation of the two log $k^*-\phi^*$ curves increases, corresponding to an increase in α with increasing sample weight. A larger value of α means a larger sample weight for T-P separation [Equation (7.4)], so the same injected weight of sample as in (a) is no longer sufficient to cause the peaks to touch. That is, the divergent case allows a larger weight of injected sample (other factors equal), compared to the equal S case of Equation (7.4) and Figure 7.14(a).

Figure 7.14(c) illustrates the third possibility: log $k^*-\phi^*$ plots that converge for smaller ϕ^* ("*convergent*" case); that is, S for compound B is less than for compound A. Now α decreases with increasing sample weight, and injection of the same weight of sample as in (a) for T-P separation leads to a more rapid column overload with overlap of the two peaks. Method development that begins with experiments where gradient time is varied (as in Fig. 7.12) allows estimates of the relative values of S for each peak and permits us to anticipate which of the three situations of Figure 7.14 will apply, in turn providing information on whether Equation (7.4) will give reliable (a), low (b), or high (c) estimates of the allowed sample weight for T-P separation. Thus, if two adjacent peaks that are to be separated by prep-LC (product and closest adjacent peak) show decreasing resolution for a longer gradient time, this is a strong indication of "convergent" behavior (with a reduction in allowed sample weight for T-P separation), and vice versa if resolution increases with gradient time. More exact predictions of this kind can be carried out if values of S for the two peaks are calculated from two experiments where only gradient time is varied (Section 9.3.3). When convergent behavior can be anticipated, further changes in separation conditions should be considered, with the goal of reversing the elution order of the two peaks (product and nearest impurity bands). A similar approach can also be used to minimize the problem of crossing isotherms (Section 7.3.4.1). For a further discussion of the consequences of unequal S values in gradient prep-LC, see [15, 17, 19, 20].

7.4 SEVERELY OVERLOADED SEPARATION[†]

Severely overloaded prep-LC can generate grams to kilotons of purified product. Although the initial goal of prep-LC method development is often the recovery of

---

[†]Contributed by G. B. Cox.

smaller amounts of product, there exists the possibility that the separation may eventually be scaled to a much larger size; therefore, the connection between the initial small-scale T-P separation and a later, larger scale, severely overloaded separation ought to be kept in mind. The choice of initial conditions and materials should be considered with the possibility of using similar conditions and materials in a large-scale separation; see Section 7.2.2 for further details.

7.4.1 Is Gradient Elution Necessary?

In prep-LC, usually only a single sample component requires purification and recovery; remaining components are generally impurities or are otherwise of no interest. As the separation is scaled up from a few hundred milligram or gram quantities to tens or hundreds of grams, the separation conditions selected become critical, as these affect the speed of the purification (important at any scale) and the cost of the purification (vital at production scale). The reasons for selecting gradient instead of isocratic elution include (a) a wide retention range for the components of the sample (i.e., large differences in polarity), and (b) samples composed of large molecules (e.g., peptides, proteins, or organic polymers; Chapter 6). For wide-retention-range samples, stepwise elution as in Figure 7.12(g) can be used to isolate the product and strip later-eluting peaks from the column.

An alternative is to carry out an initial crude separation (which may include a gradient for complex samples), followed by recovery of the partially purified product. Subsequently, a final (more efficient) *isocratic* purification can be used for faster replicate injections and the elimination of inter-run column equilibration. The latter approach has a further advantage in allowing the use of quite different separation conditions for the two steps, for example, reversed phase separation to exclude very polar components, followed by normal-phase separation of the less polar fraction. The sample size that can be injected in the first step will normally be many times greater than is possible for separation of the product from adjacent impurity peaks. This procedure has other potential advantages, including improved separation as a result of very different selectivities for the two separations.

Large-scale separations by gradient vs isocratic elution have been compared theoretically [21]. The results of this study were found to be dependent upon the goals, which can be expressed quantitatively by an "objective function" which is to be optimized; the product of production rate and recovery yield was selected in this study. (Note that recovery usually decreases and production rate increases for an increase in the weight of injected sample.) Provided that isocratic separation does not require a reconditioning of the column in order to remove strongly retained contaminants, the latter study found that isocratic elution is usually preferred on economic grounds. When isocratic separation requires column reconditioning between runs, isocratic and gradient elution give similar performance, with a preference for steeper gradients. However, the "best" objective function is not necessarily the same for all preparative separations. Thus, the material to be separated may be quite valuable, having been the result of a multistage chemical synthesis (as in a pharmaceutical setting); a higher recovery may then be paramount, which would then determine whether isocratic or gradient elution is preferred. The analysis of

[21] may also be restricted to separations where $1.2 \leq \alpha \leq 1.5$, the values used in that study; prep-LC separations where $\alpha \leq 1.2$ can be especially challenging, and relevant guidelines are less certain.

In the case of large-molecule samples, sample S values can be large enough so that it is difficult to obtain reproducible isocratic separations (Section 6.1.1). Consequently, for the prep-LC separation of peptides and small proteins, RP-LC gradient elution is the norm; such separations represent an important application area (Section 7.4.4). Prep-LC is also widely used for the purification of enantiomers; however, these separations are almost always carried out by isocratic elution. For information on the prep-LC separation of enantiomers, see [22].

7.4.2 Displacement Effects

Another factor in the choice of separation conditions for prep-LC concerns the relative concentration of the desired product. The product concentration can range from very small, for example in the isolation of an ingredients in a natural product (as in Fig. 7.12), to ~100 percent for the final purification ("polishing") of a product that has been through prior purification steps. An important factor for product "polishing" is the elution order of the main component and the nearest impurity peak. As noted above (Section 7.2.3), peak displacement can occur by the action of a larger later-eluting peak on a smaller, earlier-eluting peak [see Figs 7.8*e* (isocratic) and 7.10*b* (gradient)]. Displacement effects are generally much more pronounced for larger samples (severe overload) and are an important factor in the choice of load and separation conditions. It is usually advantageous (if possible) to select conditions such that the impurity peak elutes first (as in Figs 7.8*e* and 7.10*b*). As the load increases to that required for touching peaks, displacement effects are frequently rather minor and the front of the product peak becomes contaminated by the less-retained impurity. Further increases in sample size increase the displacement effect and also further reduce the retention of the impurity peak. These combined effects favor the increased purity of the major component as sample size increases.

There is unfortunately a down-side to the above benefits of sample displacement. Increasing sample size beyond touching-peak separation often works in the opposite way for a *later*-eluting impurity peak, as seen in Figures 7.8(*f*) and 7.10(*c*). As a result of the "tag-along" effect, the impurity peak is smeared under the product peak, so that all fractions from this peak can be contaminated by the impurity. Likewise, the impurity peak is similarly overlapped by the product peak. To summarize, *it is generally preferable that the impurity peak elute before the product peak.* Often a change in separation conditions can result in a reversal of retention for two peaks, so as to position the impurity peak earlier.

7.4.3 Method Development

The choice of severely overloaded vs touching-peak (T-P) prep-LC separation depends on the amount of purified product that is required. While severely overloaded separations are more efficient and economical, their optimization can require considerably more method-development effort. Because of the complexity of severely overloaded separation (whether carried out by isocratic or gradient elution), and the economic advantage of an optimized separation, computer modeling may be justified for the selection of final separation conditions [21]. Computer modeling requires detailed information about the adsorption isotherms for the solutes in a given phase system; in the case of gradient elution, such information is required for a range in mobile phase compositions, rendering computer modeling somewhat less practical.

Method development for a severely overloaded gradient separation is usually carried out empirically (using an analytical-scale column for both convenience and minimum expense). Typically, two or three gradients will be selected; one gradient is based on the existing analytical separation, while the other gradients will be flatter and steeper, respectively. For each of the three gradients, a series of increasing sample sizes will be injected, starting a little above the sample size for T-P separation and increasing by a maximum of 5- to 10-fold, for example, sample sizes of 1, 2, 4, 7, and 10 (relative to a value of "1" for T-P separation). For each run with a given sample size, fractions are collected throughout the peak envelope (as in Fig. 7.1c) and are then analyzed to determine their composition. The data are subsequently used to calculate the purity and recovery possible for combinations of various fractions. The required product purity and recovery will usually be specified at the start of method development, which then allows the determination of the best sample size for each gradient run. The purity and recovery for each gradient run can then be plotted vs gradient time t_G to obtain the value of t_G which gives the best production rate: maximum value of (purified product weight)/(run time), taking column equilibration between runs into account. The preferred gradient time should then be confirmed experimentally. Scale-up from the analytical column can proceed as for the T-P case (Section 7.2.2.3).

As the scale of a prep-LC separation increases, the cost of the solvent becomes a dominant factor. For this reason, the step-gradient approach of Figure 7.12(g) may be preferred, especially for the separation of small molecules. In the case of large-molecule separations (see following Section 7.4.4), a linear gradient is usually more practical, because of the difficulty in controlling the mobile phase composition within $\pm 0.1 - 0.3$ percent B. Solvent consumption in the latter case can be minimized by using a very short gradient range (e.g., 15–20 percent B), which also minimizes run time when a single product peak is to be recovered.

7.4.4 Separations of Peptides and Small Proteins

As discussed above, a common application of preparative gradient elution is in the purification of peptides, oligonucleotides, and small proteins. Often, a rather flat gradient is selected in order to maximize the retention time differences (or value of α_0) between the desired component and its impurities, although, as discussed in Section 6.2.1.3, an intermediate gradient steepness may result in maximum α_0 and therefore be preferable. Sample displacement is a common (and possibly surprising) feature of peptide and small protein separations based on severely overloaded gradient elution.

The first report of this behavior was by Parker et al. [23], who found extremely sharp boundaries between severely overloaded adjacent peaks. An apparently single peak was found to disguise two highly pure peptide peaks, as seen in the example of Figure 7.15; the overlap between the two peptides is shown as the very narrow shaded area. This group subsequently used sample displacement as the basis for the purification of synthetic peptides [24, 25].

A similar study of the separation of mixtures of small proteins [26] demonstrated that sample displacement was not unique to the peptide mixture of [23], but is rather a general feature of large-molecule gradient separations. In the study of [26], peak overlap could be distinguished by the use of proteins which have differing UV absorption characteristics; in this case cytochrome C absorbs at longer UV wavelengths and can thus be differentiated from lysozyme which does not. Mixtures of these two proteins clearly showed extremely sharp boundaries between the two (severely overloaded) peaks, similar to the example of Figure 7.15. Although computer modeling suggests that such sample displacements are expected, variable protein conformation during gradient elution appears to preclude an easy determination of the adsorption isotherms; consequently, the further study of these separations had to be carried out experimentally.

Subsequent experimental work [26] was carried out with mixtures of two smaller proteins: bovine and porcine insulins, compounds which are structurally very similar, and therefore more representative of most prep-LC separations. The results of this study clearly showed similar behavior (Fig. 7.16). In this investigation, multiple fractions were collected and analyzed, which resulted in the reconstructed chromatograms shown in Figure 7.16. The examples of Figures 7.15 and 7.16 suggest a clear and significant conclusion for the severely overloaded separation of peptides and small proteins by RP-LC gradient elution: remarkable separations

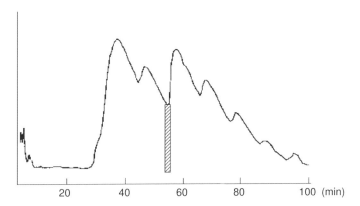

Figure 7.15 Gradient chromatogram of a 1:1 binary peptide mixture showing the overlap zone (shaded) between the two components. A 20 mg load on an Aquapore C_8 column, 220 × 4.6 mm; gradient 0.1 percent/min acetonitrile in 0.5 percent aqueous trifluoroacetic acid at 1 mL/min flow rate. Adapted from [23].

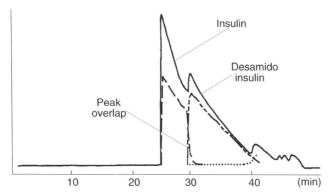

Figure 7.16 Gradient separation of a 1:1 mixture of two insulins. The solid line is the observed chromatogram; (-------) porcine insulin peak; (......) desamidoporcine insulin peak. Conditions: 150 × 4.6 mm C_{18} column; 10–90 percent acetonitrile–buffer gradient in 10 min; 0.5 mL/min; 2.5 mg sample size. See text for details. Adapted from [26].

can be expected between closely similar species, once the sample size is sufficiently high.

7.4.5 Column Efficiency

The value of N_0 required for T-P separation can be surprisingly low, as indicated in Table 7.1 for $\alpha \geq 1.5$. In more highly overloaded separations, however, this is not necessarily the case, because the separation is then governed by sample displacement. During the above study of the separation of bovine and porcine insulins [26], it was noted that the recovery of the products was highly dependent upon the particle size (and thus the efficiency) of the columns used. In the 100 mm long columns employed for the separation, it was found that acceptable recovery of purified product (85–90%) was possible only with particles smaller than 10 μm. Sample displacement was substantially degraded with the use of 20 μm particles and was totally eliminated when 50 μm particles were used. In the case of the cytochrome C–lysozyme separation [20], a similar influence of efficiency was observed, although in this example the separation began to degrade only with particle sizes larger than 15 μm. This means that for heavily overloaded separations, where sample displacement is important, a more efficient column than is needed for touching peaks is generally required.

7.4.6 Production-Scale Separation

Surprisingly few applications of gradient elution at the production scale have been reported in the literature, although such separations are fairly common in practice. The earliest report of production-scale reversed-phase gradient elution described the final purification of recombinant human insulin [27]. The latter report (which likely differs from the actual process-scale separation, because of a desire not to share proprietary information) demonstrates very clearly how a separation can be scaled from the laboratory to the production plant. This procedure, which certainly relied upon the displacement effects noted earlier, was carried out using columns packed with 10 μm particles to achieve the efficiency (40,000 plates/m) required for the high recovery needed for this valuable product. This process (or a variant), which increased the insulin purity from 83 to 98.6 percent with a recovery of 82 percent is still in use. It is likely that similar procedures are used in other human insulin production facilities.

Unfortunately, the pharmaceutical industry is generally reluctant to provide details of production-scale processes. Although there are many products that are purified by RP-LC gradient elution (notably peptides and small proteins), few details of such processes have been published at the time this book went to press.

> whatsoever things are pure ... think on these things.
>
> —Paul, *II Corinthians*

REFERENCES

1. L. R. Snyder, G. B. Cox, and P. E. Antle, *Chromatographia* 24 (1987) 82.
2. G. B. Cox and L. R. Snyder, *J. Chromatogr.* 483 (1989) 95.
3. S. M. C. Buckenmaier, D. V. McCalley, and M. R. Euerby, *Anal. Chem.* 74 (2002) 4672.
4. U. D. Neue, T. E. Wheat, J. R. Mazzeo, C. B. Mazza, J. Y. Cavanaugh, F. Xia, and D. M. Diehl, *J. Chromatogr. A* 1030 (2004) 123.
5. D. V. McCalley, *LCGC* 23 (2005) 162.
6. J. H. Knox and H. M. Pyper, *J. Chromatogr.* 363 (1986) 1.
7. L. R. Snyder, J. J. Kirkland, and J. L. Glajch, *Practical HPLC Method Development*, 2nd edn, Wiley-Interscience, New York, 1997.
8. L. R. Snyder and G. B. Cox, *LCGC* 6 (1988) 894.
9. L. R. Snyder and J. J. Kirkland, *Introduction to Modern Liquid Chromatography*, 2nd edn, Wiley-Interscience, New York, 1979, pp. 634–636.
10. U. D. Neue, C. B. Mazza, J. Y. Cavanaugh, Z. Lu, and T. E. Wheat, *Chromatographia Suppl.* 57 (2003) S-121.
11. J. E. Eble, R. L. Grob, P. E. Antle, and L. R. Snyder, *J. Chromatogr.* 405 (1987) 51.
12. M. Z. El Fallah and G. Guiochon, *Anal. Chem.* 63 (1991) 859.
13. G. Crétier and J. L. Rocca, *J. Chromatogr. A* 658 (1994) 195.
14. Y.-B. Yang, K. Harrison, D. Carr, and G. Guiochon, *J. Chromatogr.* 590 (1992) 35.
15. G. Crétier, M. El Khabchi, and J. L. Rocca, *J. Chromatogr.* 596 (1992) 15.
16. L. R. Snyder, G. B. Cox, and P. E. Antle, *J. Chromatogr.* 444 (1988) 303.
17. L. R. Snyder, J. W. Dolan, and G. B. Cox, *J. Chromatogr.* 540 (1991) 21.
18. J. W. Dolan, L. R. Snyder, N. M. Djordjevic, D. W. Hill, D. L. Saunders, L. Van Heukelem, and T. J. Waeghe, *J. Chromatogr. A* 803 (1998) 1.
19. F. D. Antia and Cs. Horváth, *J Chromatogr.* 484 (1989) 1.
20. L. R. Snyder, J. W. Dolan, and G. B. Cox, *J. Chromatogr.* 484 (1989) 437.
21. A. Felinger and G. Guiochon, *J Chromatogr. A* 796 1998 59.
22. G. B. Cox (ed.), *Preparative Enantioselective Chromatography*, Blackwell, Oxford, 2005.
23. J. M. R. Parker, C. T. Mant, and R. S. Hodges, *Chromatographia* 24 (1987) 832.

24. T. W. L. Burke, C. T. Mant, and R. S. Hodges, *J. Liq. Chromatogr.* 11 (1988) 1229.
25. R. S. Hodges, T. W. Lorne-Burke, and C. T. Mant, *J. Chromatogr.* 444 (1988) 349.
26. G. B. Cox, *J. Chromatogr.* 599 (1992) 195.
27. E. P. Kroeff, R. A. Owens, E. L. Campbell, R. D. Johnson, and H. I. Marks, *J Chromatogr.* 461 (1988) 45.

CHAPTER 8

OTHER APPLICATIONS OF GRADIENT ELUTION

Three, unrelated topics are examined in the present chapter: (a) liquid chromatography with mass spectrometric detection (LC-MS), using RP-LC gradient elution; (b) gradient separations other than RP-LC; and (c) the use of more complex, ternary- or quaternary-solvent gradients. LC-MS is today a very important analytical technique, often based on gradient RP-LC. While the general principles of LC-MS separation are quite similar to those for the use of UV detection (Chapters 1–3, 6 and 7), LC-MS presents some important additional requirements that are the subject of Section 8.1.

Three separation modes other than RP-LC were discussed in Chapter 6 for the separation of large-molecule samples: hydrophobic interaction chromatography (HIC, Section 6.2.2.1), ion-exchange chromatography (IEC, Section 6.2.2.2), and hydrophilic interaction chromatography (HILIC, Section 6.2.2.3). In the case of HIC, values of log k vary linearly with mobile phase composition (%B) – just as for RP-LC; HIC separation is therefore described quantitatively by the same equations as for RP-LC (i.e., the LSS model applies quantitatively for HIC). Consequently, gradient separations by either HIC or RP-LC can be interpreted and developed in the same general way. In this chapter, we will examine the fundamental basis of two separation modes that are *not* described quantitatively by the LSS model: ion-exchange chromatography (Section 8.2) and normal-phase chromatography (Section 8.3, including HILIC). Reasons for the use of IEC and NPC, as well as practical information on their application, can be found in more general HPLC references [1]. In this chapter, we will minimize details of this kind and instead focus on the qualitative similarity of all gradient separations (by RP-LC, IEC, or NPC). In this way, general conclusions which have been developed for RP-LC gradient elution can be extended to other separation modes, even when the LSS model may not be quantitatively applicable. This approximate use of the LSS model can contribute to an easier understanding of separations based on linear-gradient IEC or NPC. Method development for these separations can also be carried out in a similar fashion as for gradient RP-LC. Because the following discussion of IEC and NPC is closely related to corresponding gradient separations by RP-LC, Chapter 3 should be reviewed before reading Sections 8.2 and 8.3.

High-Performance Gradient Elution. By Lloyd R. Snyder and John W. Dolan
Copyright © 2007 John Wiley & Sons, Inc.

Most gradient RP-LC separations use aqueous buffer as the A solvent and either methanol or acetonitrile as the B solvent. For the further control of separation selectivity, it can be worthwhile to use a *mixture* of two or three organic solvents as the B solvent, or to vary the ratios of these organic solvents independently during the gradient. Section 8.4 reviews the use of such ternary or quaternary solvent gradient procedures.

8.1 GRADIENT ELUTION FOR LC-MS

> God in the beginning formed matter in solid, massy, hard, impenetrable, movable particles, of such sizes and figures, and with such other properties, and in such proportion to space, as most conduced to the end for which he formed them.
>
> —Isaac Newton, *Optics*

Since about 1990, the use of the mass spectrometer (MS) as a detector for HPLC has changed from a research tool used by workers highly skilled in mass spectrometry, to a routine detector for many applications in the analytical laboratory. Much of this change has resulted from a reduction of cost, improvement of the LC-MS interface for reversed-phase mobile phases, increased reliability and simplicity of vacuum pumps, and simpler, more intuitive computer interfaces. The major challenge for the LC-MS interface is to convert a mostly aqueous HPLC mobile phase into the gaseous state, add a charge to the analyte molecules, and reduce the pressure from atmospheric to $10^{-5}-10^{-6}$ torr, all within a path length of a few centimeters. Today's LC-MS systems accomplish this with such reliability that the instruments are used by many laboratories for the routine quantitative analysis of 100 or more samples per instrument per day.

For many of the same reasons that gradient elution is chosen for applications using UV detectors, gradients often are the first choice for LC-MS as well. These include rapid development of methods based on generic scouting runs, the easy convertability of gradient scouting runs into either gradient or isocratic separations, the self-cleaning nature of gradient methods, and narrow, sharp peaks that aid quantification in trace analysis.

MS detectors come in two popular configurations. The single-stage detector, sometimes called an MSD (mass selective detector), is used to measure a single ionic species for each analyte, often the protonated molecular ion (M + H). (Within a given run, more than one analyte ion can be monitored by switching back and forth between different m/z values.) Instruments using this type of detection are referred to as LC-MS. A more complex detector design isolates the primary ionic species (parent or precursor ion), fragments it into additional ions (daughter or product ions), and monitors one or more of these product ions. This process, sometimes called multiple reaction monitoring (MRM), gives added selectivity when the transition from parent to product ion is used as a "signature" of a specific analyte. Such systems are referred to as LC-MS/MS. In this chapter, we will refer to LC-MS/MS when this specific technique is used, and LC-MS for the single-stage methodology or when it is not important whether the system is LC-MS or LC-MS/MS.

8.1.1 Application Areas

As more and more laboratories acquire LC-MS instrumentation, the applications of this technique continue to expand. Several of the more popular applications are mentioned here. In general, an application will focus on either qualitative or quantitative analysis; optimization for one of these goals often compromises the other, because the increased mass resolution (higher accuracy of mass data) required for qualitative analysis requires more time. For quantitative applications, speed of analysis (the number of samples to be run) often is a critical factor, as is the minimum detection limit – these are attained at the sacrifice of mass resolution, resulting in decreased qualitative information.

The pharmaceutical industry uses LC-MS for many different purposes. Early in the discovery process, LC-MS provides a quick screening of compound purity from combinatorial synthesis, and/or supporting data for structural identification or confirmation of structure [2]. For such applications, fast (e.g., <5 min) runs are desirable, and generic methods that can be used for different possible reaction products are often selected. In such cases, either standard methods or standardized rapid-method-development procedures [3] are used to minimize the time spent before samples can be analyzed. The study of drug metabolism is facilitated by qualitative and quantitative LC-MS [4]. One of the more widely used applications in the drug development process is the monitoring of drug levels in plasma or other tissues ("bioanalytical" applications), in order to determine pharmacokinetic parameters, dosing levels, and toxicological information (the *Journal of Chromatography B*, Elsevier, is a source of hundreds of articles on such applications). Section 8.1.5 outlines a procedure for the development of bioanalytical methods.

LC-MS also is used widely outside the pharmaceutical industry. The applications of LC-MS include forensics, pesticide manufacture, clinical monitoring, and monitoring chemical residues in the environment [2]. Drugs, pesticides, growth regulators, and other chemicals are monitored in the food supply or the environment by LC-MS [2, 5, 6]. Proteomics and related "omics" fields have been advanced by the use of capillary HPLC columns, miniaturized interfaces (nanospray), and customized software to get the maximum information from LC-MS systems for protein sequencing [7].

8.1.2 Requirements for LC-MS

Because of the high degree of selectivity provided by the MS detector (and even more by MS-MS), the cost of MS instrumentation, as well as the end use of the data, "best" conditions for LC-MS applications differ somewhat from the usual requirements for gradient methods with UV detection. In the early days of LC-MS, it was thought that detector selectivity was sufficient to obviate the need for HPLC separation, and many LC-MS methods were developed with little or no real separation ($R_s \approx 0$) and run times of 1–2 min. Now, however, it is recognized that interferences can compromise LC-MS results in various ways, particularly interferences that suppress ionization of the analyte in the detector interface (Section 8.1.6.3). Consequently, the HPLC component of LC-MS now receives greater attention than in the past. Resolution requirements for LC-MS are still lower than for LC-UV (where $R_s \geq 1.7-2.0$); $R_s \geq 1.0$ generally is desired for quantitative LC-MS (although lower resolution can be adequate, see Section 8.1.6.5).

There is a recognized need for fast LC-MS separations, because of the number of samples to be run, and because of the desire to make efficient use of the expensive MS detector. The need for fast separations, combined with lower resolution requirements, means that shorter columns and faster gradients are usually used. The typical LC-MS gradient method is 4–10 min, including re-equilibration, as compared with 10–30 min for LC-UV methods. Routine bioanalytical methods use LC-MS or LC-MS/MS for trace analysis, where detection sensitivity and assay specificity become major issues. Low sample concentrations (e.g., pg/mL to ng/mL of analyte in plasma) give small peaks that are inherently more noisy, but LC-MS methods with a precision and accuracy of 15–20 percent RSD are acceptable for this application [8]. Finally, because the LC-MS interface is more variable than coupling a UV detector to an HPLC instrument, and because extensive sample cleanup may be involved, internal-standard calibration usually is preferred in order to correct for variations in these processes.

8.1.3 Basic LC-MS Concepts

A detailed description of gradient separations by LC-MS and LC-MS/MS requires a mass spectrometry (MS) background that is beyond the scope of this chapter. However, a brief primer on several important MS concepts will help to orient the reader to different MS and MS/MS detectors. For a more detailed general discussion, see references [2, 9, 10] or other texts on LC-MS. *Readers familiar with LC-MS instrumentation may wish to skip this section.*

8.1.3.1 The Interface MS detectors manipulate and detect ions in the gaseous phase; so for the MS to be useful as an HPLC detector, the mobile phase must be evaporated and sample ions must be generated. This is the function of the MS interface. The mobile phase must be converted from liquid to gas, an expansion in volume of ≈ 1000-fold; at the same time, the pressure must be reduced from atmospheric pressure (760 torr) to $10^{-5}-10^{-6}$ torr in the 10–20 cm flow path of the interface. Pressure is reduced by pumping most of the vaporized sample and mobile phase to waste (no concentration takes place); only a tiny fraction of the sample is drawn into the MS itself.

The two most popular interfaces are electrospray ionization (ESI) and atmospheric pressure chemical ionization (APCI). The ESI interface (Fig. 8.1) adds a charge to analytes in the mobile phase by placing a potential (e.g., 3–5 kV) on the stainless-steel nebulizer-spray-tip ("capillary" in Fig. 8.1). Mobile phase is sprayed into the heated interface, where solvent evaporates, leaving ions in the gaseous state. ESI is the most commonly used interface for bioanalytical applications because it is a "softer" ionization technique and it is less likely to cause undesirable analyte degradation.

The APCI interface (Fig. 8.2) vaporizes the mobile phase first, then uses a corona discharge to add a charge to the analyte in the gas phase. This technique is

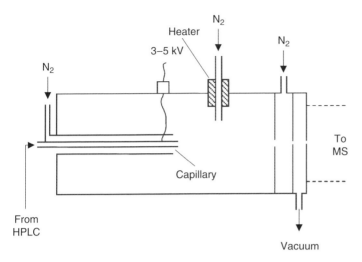

Figure 8.1 Electrospray ionization (ESI) interface.

used for compounds that do not ionize well with ESI (often more stable, smaller-molecular-weight compounds and some nonpolar compounds), but uses harsher conditions, so it is more likely than ESI to cause sample degradation, especially with heat-labile compounds. Also, APCI has been shown to have fewer matrix ionization problems than ESI. APCI and ESI have different ionization mechanisms, so the response and selectivity may vary significantly between the two interfaces. Either interface can be operated in the positive or negative ion mode, resulting in the generation of positively or negatively charged sample ions (most commonly achieved by adding or removing a proton from the analyte molecule).

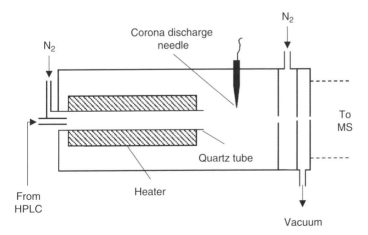

Figure 8.2 Atmospheric pressure chemical ionization (APCI) interface.

8.1.3.2 Column Configurations To minimize the work required by the interface, a smaller column i.d. is selected, so as to reduce the mobile-phase flow rate. Although LC-MS interfaces can operate with a flow rate of 1 mL/min, they are more reliable with lower flow rates. The use of 2.1 mm i.d. columns allows the use of flow rates of 0.2–0.5 mL/min, with linear velocities (and separation) comparable to flow rates used with conventional 4.6 mm i.d. columns (1.0–2.5 mL/min). Short, 30–50 mm long columns packed with 3–5 μm particles provide fast separations of the usual (simple) mixtures encountered in bioanalytical applications, that is, an analyte, an internal standard, and one or two metabolites. For more complex mixtures, longer column lengths (100–150 mm) may be required, in order to obtain larger column plate numbers (with longer run times).

8.1.3.3 Quadrupoles and Ion Traps Two designs are predominant for LC-MS (single stage) applications: quadrupoles and ion traps. Quadrupoles use a set of four rods and a carefully controlled electric field to isolate selected ions from the sample. Ions of a selected mass-to-charge ratio (m/z) are then passed to an electron multiplier for detection, providing a selective response for the desired analyte. Ion traps use a ring electrode in combination with end-cap electrodes to accomplish the same isolation of desired ions, followed by detection. Both quadrupoles and ion traps can be set up to change rapidly from monitoring one mass to another, and thus generate a spectrum (scan) across a range of masses. An alternate mode of operation allows the detector to "simultaneously" detect co-eluting compounds, such as an analyte and internal standard, by switching back and forth between data collection channels for each mass during the elution of the peaks. As discussed below, quadrupole MS detectors are favored for quantitative analysis, whereas ion traps have advantages for qualitative (structural) applications.

Single-stage MS detectors of the above kind are used in less expensive LC-MS units; however, additional structural discrimination is needed for more selective detection. The triple quadrupole (Fig. 8.3*a*), or tandem MS, can provide additional selectivity compared with that obtained with a single quadrupole unit. Sample ions generated in the interface (A^+, B^+, C^+, D^+ in Fig. 8.3*b*) enter the first quadrupole. The ions of a given m/z (A^+) are isolated in the first quadrupole and sent to a second quadrupole (collision cell), which is filled with an inert gas (nitrogen or argon). The ions are fragmented ($A^+ \rightarrow A_a^+$, A_b^+, A_c^+) in the collision cell and passed to a third quadrupole. The third quadrupole then isolates specific ion fragments (e.g., A_b^+) and passes them to the electron multiplier for measurement. The transition from the initial ion (precursor or parent) to the fragment ion (product or daughter ion) provides a unique "signature" ($A^+ > A_b^+$ in Fig. 8.3*b*) for an analyte, and greatly increases the selectivity of the triple-quadrupole (MS/MS) over the single quadrupole detector. (Note that the conventional notation is "$A^+ > A_b^+$" to represent the transition signature of the precursor A^+ to the product ion A_b^+. We will use this shorthand, while using "$A^+ \rightarrow A_a^+$, A_b^+, A_c^+" to represent the fragmentation process itself.)

Ion traps accomplish multiple-stage fragmentation and the isolation of a preferred product ion in the same physical space (vs in different parts of the detector as in the triple quadrupole of Fig. 8.3). First, ions are generated in the interface and

8.1 GRADIENT ELUTION FOR LC-MS   329

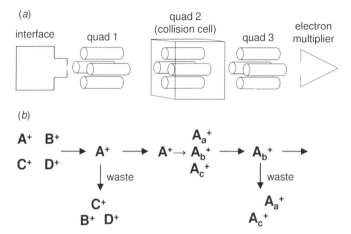

Figure 8.3 (a) Triple quadrupole mass spectrometer, (b) MS/MS experiment for $A^+ > A_b^+$ (precursor > product ion) transition.

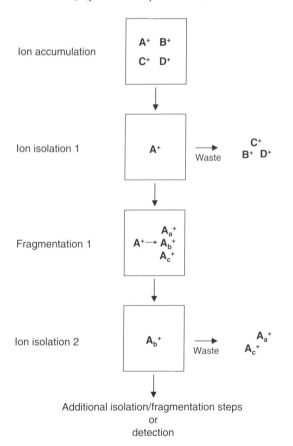

Figure 8.4 Ion trap mass spectrometer. MS/MS experiment for $A^+ > A_b^+$ (precursor > product ion) transition.

passed into the ion trap (ion accumulation, Fig. 8.4). Ions of a desired m/z are held, while the remaining ions are sent to waste (ion isolation 1). The isolated ions (A^+ in Fig. 8.4) are then fragmented ($A^+ \rightarrow A_a^+, A_b^+, A_c^+$) and the desired fragment m/z is isolated (A_b^+, ion isolation 2). The ions can then be sent to the electron multiplier for detection or the process can be continued (further fragmentation of A_b^+, isolation, etc.). The ion trap is capable of performing this operation over and over, isolating and breaking ion fragments into successively smaller fragments (with a corresponding loss of sensitivity for each fragmentation step). This is useful for structural identification, but historically the ion trap has not been as good for quantitative work as the quadrupole, because of space-charge effects (ion interactions within the detector) and variability in the output signal intensity. Thus quadrupoles (single and triple) tend to be more widely used for routine quantitative work, whereas ion traps are preferred when structural identification is needed, such as in metabolite isolation and identification.

8.1.4 LC-UV vs LC-MS Gradient Conditions

Table 8.1 compares the major differences between LC-MS and LC-UV methods as discussed here (the LC-UV requirements listed in Table 8.1 are the same as those in Table 3.2). Mobile phase volatilization is easier with higher-organic mobile phases, so the column and mobile phase conditions are selected to favor higher %B. With LC-UV methods, there are no compelling reasons to choose between a C_8 or C_{18} column, but LC-MS methods somewhat favor a C_{18} column; for comparable retention times, the more retentive C_{18} column will require, on average, about 5 percent v/v more organic solvent. Otherwise there is little reason to favor a C_{18} over a C_8 column for LC-MS.

To reduce the volume of mobile phase that must be evaporated, LC-MS systems typically use 2.1 mm i.d. columns – these result in a 5-fold reduction in mobile phase flow rate for the same linear velocity when compared with 4.6 mm i.d. columns. The added selectivity of the MS detector means that $R_s \geq 1.0$ is often sufficient when compared with LC-UV methods, which typically need $R_s \geq 1.7-2.0$; therefore, 30–50 mm long columns usually provide satisfactory separation, especially when packed with 3.0 μm particles. When isotopically labeled internal standards are used, $R_s > 1.0$ may not be possible, because of the extreme structural similarity between the analyte and internal standard. In such cases, the user must rely upon the selectivity of the MS detector to distinguish between the two compounds (see additional discussion in Section 8.1.6.5). The smaller-volume columns used for LC-MS require a corresponding reduction in system dwell volume (Sections 4.4, 5.2.1, 8.1.6.1) for practical use with gradient elution; otherwise, run times at low flow rate can be excessive.

LC-UV methods often operate at wavelengths $\leq$220 nm, so the added UV transparency of acetonitrile (vs methanol) favors the use of this solvent with UV transparent buffers, such as phosphate or acetate. Volatility, not UV transparency, is critical for LC-MS, however, so either ACN or methanol is useful. (MeOH is the preferred organic modifier when using APCI, because of the tendency of ACN to form deposits on the corona discharge needle.) Tetrahydrofuran is used with LC-UV for different selectivity from ACN or MeOH, but generally is useful only at wavelengths >240 nm because of its high UV absorbance. THF cannot be

Variable		LC-UV[a]	LC-MS	Comment for MS
Column				
	Type	C_8 or C_{18} (type B)	C_{18} (type B)	Stronger retention preferred
	Dimensions	150 × 4.6 mm	50 × 2.1 mm	
	Particle size	5 μm	3 μm	
Resolution		≥1.7–2.0	≥1.0	<1.0 often OK with isotopic standards
Dwell volume		1.0–4.0 mL	0.1–1.0 mL	Smaller is better
Mobile phase				
	Organic	ACN	ACN or MeOH	UV absorbance not important
	Buffer	Phosphate, acetate	Formate, acetate, ammonia	Must be volatile
	Ion pairing (bases)	Sulfonates	TFA or HFBA	Must be volatile
Flow rate		1.5–2.0 mL/min	0.2–0.5 mL/min	Smaller flow rate desired
Temperature		30–35°C	30–35°C	
Gradient		5–100 percent B in 15 min	5–100 percent B in 4 min (0.5 mL/min) or 10 min (0.2 mL/min)	$k^* \approx 5$
Sample				
	Volume	≤50 μL	≤10 μL	Injection solvent percent B ≤ mobile phase percent B
	Weight	≤10 μg	≤1 μg	Proportional to column volume
	k^*	~5	~5	
	Standardization	Internal or external	Usually internal	Stable-label internal standard preferred

[a] See Table 3.2 and associated discussion for LC-UV methods.

used with LC-MS systems that contain PEEK components (valve rotors, tubing, etc.), because it extracts materials which interfere with the MS signal.

Phosphate buffers are not volatile, but buffers containing acetate, formate, ammonia, or other volatile components can be used with LC-MS. Sulfonic acid ion pairing reagents are used to ion-pair basic compounds for LC-UV, but are not sufficiently volatile for MS detection. Instead, trifluoroacetic acid or heptafluorobutyric acid (HFBA) are used for ion pairing with LC-MS (however, these reagents can cause ion suppression [11, 12], so they should be used with care). The column temperature is a useful variable in the development of both LC-UV and LC-MS methods. The gradient conditions with LC-MS methods need to be adjusted for smaller columns and lower flow rates, so as to obtain reasonable k^* values. Similarly, smaller columns may require a reduction in the injection volume and/or sample mass to avoid peak-shape problems and column overload. The MS interface adds variability to the method, so internal standardization (preferably with an isotopically labeled version of the analyte) is favored to obtain the most reliable results from LC-MS methods. LC-UV methods generally require internal standardization only to correct for sample preparation variability, not instrument variability.

8.1.5 Method Development for LC-MS

This section presents a general scheme to develop LC-MS methods for bioanalytical samples based on the steps outlined in Table 8.2. It roughly parallels the steps outlined in Table 3.1 for conventional LC-UV methods, but is modified somewhat based on the laboratory experience of one of the authors (J.W.D.). Although the analysis of bioanalytical samples using LC-MS/MS is emphasized, the same general approach can be taken for any gradient LC-MS or LC-MS/MS method.

This section focuses on the HPLC portion of the LC-MS method; because of the intentionally limited scope of the present discussion, a comprehensive treatment of MS detector tuning and optimization is not included. The assumption is made that the reader is not a novice in mass spectrometry and that initial MS (or MS/MS) experiments already have been performed. These include infusion experiments to find the molecular ion, selection of the appropriate interface (ESI or APCI) and mode (positive or negative ion), MS/MS fragmentation conditions, and so forth. For help with these processes, one should consult an experienced LC-MS user and/or reference materials, such as [2, 10].

8.1.5.1 Define Separation Goals (Step 1, Table 8.2) The general principles for LC-UV methods discussed in Section 3.1.1 also apply to LC-MS separations. The following are typical LC-MS separation objectives:

- preliminary sample assessment;
- development of a routine assay procedure;
- development of a "generic" separation.

The conditions of Table 8.1 (see discussion in Section 8.1.4) can be used for *preliminary sample assessment*. A faster, but less comprehensive, approach to

TABLE 8.2 Steps in the Development of a Routine Bioanalytical LC-MS Method[a]

Step	Comment
1. Define separation goals (Section 8.1.5.1)	
2. Collect sample information (Section 8.1.5.2)	(a) MS conditions (b) Internal standard selection (c) Mobile phase buffering required? (d) Sample pretreatment required?
3. Carry out initial separation (run 1, Section 8.1.5.3)	(a) 4 min gradient (run 1); Table 8.1 (b) Any problems? (Sections 3.2, 8.1.5.3, Fig. 3.5) (c) Isocratic separation possible? (Section 3.2.1, Fig. 3.3)
4. Optimize gradient retention k^* (Section 8.1.5.4)	The conditions of Table 8.1 should yield an acceptable value of $k^* \approx 5$
5. Optimize separation selectivity α^* (Section 8.1.5.5)	(a) Increase gradient time by 3-fold (run 2, 12 min; Sections 3.3.2, 8.1.5.5) (b) Increase temperature by 20 °C (runs 3 and 4; Sections 3.3.2 and 8.1.5.5)
5a. If best resolution from step 5 is $R_s < 1$, or if run times are too short or too long, vary further conditions to optimize peak spacing (for maximum R_s or minimum run time)	(a) Replace acetonitrile by methanol and repeat runs 1–4 (b) Replace column and repeat runs 1–4 (c) Change pH and vary runs 1–4 (d) Consider use of segmented gradients (Section 3.3.4)
6. Adjust gradient range and shape (Section 8.1.5.6)	(a) Select best initial and final values of percent B for minimum run time with acceptable R_s (Sections 3.3.3, 8.1.5.6) (b) Add a steep gradient segment to 100 per cent B for "dirty" samples (e.g., Fig. 3.8b) (c) Add a steep gradient segment to speed up separation of later, widely spaced peaks (Fig. 3.8c)
7. Vary column conditions (Section 3.3.3)	With best separation from step 5 or 6, choose best compromise between resolution and run time (Section 8.1.5.7)
8. Determine necessary column equilibration between successive sample injections (Section 8.1.5.8)	Using the procedure developed above, carry out two successive, identical separations with the equilibration time between runs varied; select the shortest equilibrium time which results in no change in separation between adjacent runs (Section 5.3)

[a]See Table 3.1 for a parallel treatment of gradient methods with UV detection.

develop bioanalytical LC-MS/MS methods is presented in [3]: a standardized three-test process (matrix-effect test, interference test, and standard-curve linearity test) is used with protein-precipitated plasma to shorten the development cycle for bioanalytical methods that support early drug development.

The present discussion emphasizes on the *development of routine assay procedures* for drugs in biological fluids ("bioanalytical" methods). In such cases, short runs are desired (typically $\leq 5-7$ min) and usually five or fewer analytes plus metabolites are to be monitored.

Generic methods can be useful for "high throughput screening," as is used to semiquantitatively screen samples from combinatorial synthesis for purity, or for "walk-up" systems, where a generic method is set up on an LC-MS system in order to support multiple users or applications. In such cases, the generic method is designed for the elution of a broad range of sample polarities and with sufficient resolution for a high probability of adequate separation on the first run.

8.1.5.2 Collect Information on Sample (Step 2, Table 8.2)   As for analysis by LC-UV (Section 3.1.2), any sample information that is available prior to starting LC-MS method development may help to reduce development time. Often this information can be obtained by doing a literature search (*Journal of Chromatography B*, Elsevier, is an excellent source of bioanalytical applications) or consulting methods or workers internal to your company for information about previous assays of the analyte(s) or similar compounds. The conditions of Table 8.1 (possibly with a longer, 100×2.1 mm i.d. column), at low and/or high mobile phase pH, may be satisfactory generic separation conditions for many applications.

MS conditions (separation conditions and/or MS instrument settings) may be the same as, or similar to, conditions used for other LC-MS methods; this information may be helpful in choosing initial experimental conditions. An infusion experiment is used to optimize the MS system. Typically, a dilute solution (e.g., 1 μg/mL of sample in MeOH) is infused into the mobile phase at $300-500$ μL/h through a tee fitting mounted between the HPLC column outlet and the inlet to the MS interface (same plumbing setup as for ion suppression experiments, see Fig. 8.8). The mobile phase should be selected to approximate the composition (%B) in which the analyte is eluted from the column. This setup bleeds a steady-state concentration of analyte, diluted in column effluent, into the MS, simplifying optimization of the interface conditions and other settings in the mass spectrometer. This process is called "tuning," and is necessary to adjust the MS for optimum response.

Selection of the internal standard (IS) should be made before starting method development experiments, because the separation of the IS from the analyte(s) and the recovery of the IS during sample pretreatment are important for a successful LC-MS method. If a literature search is made, check for the identity and source of internal standards used in similar methods. Compounds of similar structure ("analog" internal standards) often are readily available as failed exploratory compounds in drug discovery. If analogs are available, they are usually the best standards to use initially, because of the delay and expense involved in the synthesis of stable-label internal standards. A stable-label internal standard is a compound with the same structure as the analyte, but with multiple ^{13}C or ^{2}D atoms substituted for ^{12}C and ^{1}H, respectively. Such compounds have the same, or similar, chemical and chromatographic properties, so that they mimic the analyte during sample preparation and analysis; that is, have similar recoveries and appropriate retention, respectively.

It is best to have a stable-label standard that is at least 3–5 amu higher mass than the analyte, so as to minimize problems due to the natural abundance of ^{13}C in the analyte. Generally the chromatographic separation between a ^{13}C internal standard and the nonlabeled analyte will be very small, but the separation of a ^{2}D standard from its nonlabeled analyte is often significant. When chromatographic separation is not possible, one must rely on the mass selectivity of the MS detector for discrimination between the internal standard and analyte. Stable-label internal standards have many advantages for LC-MS, but residual unlabeled compound in the labeled standard can confuse data interpretation, and care must be taken that the analyte and internal standard do not suppress each other in the ESI interface [13] – selection of the proper concentration of internal standard usually will overcome these problems. The improved data quality obtained (e.g., closely mimicking extraction and recovery in sample preparation, lower percentage RSD of results) with stable-label standards usually outweighs any problems due to chromatographic co-elution. (See Section 8.1.6.5 for an additional discussion of resolution requirements for internal standards.)

Mobile phase buffering usually is required in order to obtain reproducible retention of sample acids or bases. Buffering also may be important to obtain the desired ionization conditions in the MS interface. One should follow the general guidelines of Section 3.2, with the exception that the buffer must be volatile (e.g., no phosphate allowed). In the absence of other information, *start at low pH with 0.1 percent formic acid for pH control*, for the same reasons that low pH mobile phases are preferred for LC-UV methods. Buffer and additive concentrations in the 5–10 mM range are recommended for LC-MS methods. Table 8.3 lists some common mobile phase additives used for LC-MS and their nominal pH values and buffering ranges.

Sample preparation can represent more work than selecting the chromatographic and MS conditions for a bioanalytical method. A primer on sample preparation is beyond the scope of this book; the reader is encouraged to consult

TABLE 8.3 Buffers and pH Control for LC-MS

Additive	Typical concentration	pH	Buffering range
Formic acid	0.1%	2.7	—
Acetic acid	0.1%	3.3	—
Trifluoroacetic acid	0.1%	2.0	—
Ammonium formate	5–10 mM	—	2.7–4.7
Ammonium acetate	5–10 mM	—	3.7–5.7
Ammonium carbonate[a]	5–10 mM	—	6.6–8.6

[a]Should be formulated daily to avoid pH drift due to evaporation of CO_2.

comprehensive texts on sample preparation [14], technical articles [15], and websites and literature produced by suppliers of sample preparation products [16, 17]. Only a brief overview of sample preparation is presented here; although this discussion will emphasize bioanalytical samples, the usefulness of these various cleanup processes for other LC-MS applications should be obvious. Five standard techniques are used widely for bioanalytical sample preparation:

- dilution;
- protein precipitation;
- solid-phase extraction (SPE);
- liquid–liquid extraction (LLE);
- on-line cleanup.

Dilution may be the only sample preparation necessary if the analyte(s) is present at a sufficient concentration and materials in the sample matrix do not cause other analytical problems (e.g., column blockage, fouling, ion suppression). For bioanalytical samples, dilution generally is limited to protein-free samples, such as urine. For other LC-MS applications, such as process analysis or impurity assays, dilution may be the favored sample preparation technique, because it does not (unintentionally) remove sample components that may be of interest. Filtration (e.g., through a 0.5 μm porosity membrane filter) may be used to provide further cleanup, but this adds cost and raises questions regarding selective loss of analyte(s) on the filter, and/or contamination of the sample by filter components. An alternative chosen by many workers is to use centrifugation (e.g., for 5 min at >1500g), as the final sample preparation step to remove suspended particulate matter so as to protect the column from blockage.

Protein precipitation is a "quick-and-dirty" procedure that is fast and simple, but one that provides a low level of cleanup. Typically, plasma proteins are precipitated with a 2- to 3-fold excess of ACN [18], followed by vortexing and centrifugation. This procedure is acceptable for projects with only a small number of samples, or samples that are not amenable to other cleanup techniques. However, a fairly large concentration of proteins, pigments, lipids, and other contaminants is left suspended in the sample extract, which can coat or block the column (shortening column lifetime) or lead to ion suppression (Section 8.1.6.3).

Solid-phase extraction is a crude chromatographic separation that is carried out on individual SPE cartridges or in sets of cartridges in a 96-well microtiter-plate format. Many stationary phases are available, including ion exchange, reversed-phase, and mixed-mode (reversed-phase plus ion exchange) products. Typically, the cartridges are activated with a methanol wash followed by a water rinse. Sample, in a weak solvent, is loaded onto the cartridge under conditions in which the analyte is retained. This is followed by various wash and elution steps, such that the sample is separated from most of the potential contaminants. The most effective cleanup is accomplished if a SPE phase is selected that uses a different mechanism of retention than the analytical column, as this favors the removal of interfering compounds that might overlap analytes in the RP-LC separation. Thus, an ion-exchange or mixed-mode phase is a good choice when using a reversed-phase analytical column. SPE manufacturers provide generic method development guides [16, 17] that can help new workers learn how to use the technique. SPE is amenable to automation, but can be expensive both in terms of materials and support equipment (robots, automated evaporators, specialized centrifuges, etc.). SPE cleanup is widely applicable to other LC-MS applications, such as forensics or environmental analysis.

Liquid–liquid extraction is a traditional cleanup technique in which the analyte is partitioned between two immiscible liquid phases, so that impurities enter one phase and analytes the other. Solvent polarity and pH can be adjusted to fine-tune the cleanup process so as to increase the discrimination between analytes and interferences. Sometimes multiple extraction and back-extraction steps can be used to obtain cleaner extracts. One popular LLE method [19] uses methyl-*tert*-butyl ether (MTBE) as the organic phase and pH-adjusted plasma as the aqueous phase. The analyte(s) and internal standard are partitioned into the MTBE phase, which is then evaporated to dryness and reconstituted in an aqueous injection solvent. LLE is very flexible, can produce very clean extracts, and often is less expensive than SPE. Liquid–liquid extraction is applicable to any LC-MS application in which desirable and undesirable sample components have significantly different polarity, especially when ionic materials are present. Although LLE works well for plasma, tissues, and many aqueous samples, LLE is susceptible to emulsion problems with some types of samples (e.g., some plant extracts) – in such cases, SPE may be a better choice.

On-line cleanup procedures generally are based on a re-usable cleanup column (a specialized SPE cartridge or a guard column) mounted on an automated valve. Two popular configurations are widely used. In the "stripping" configuration, the sample is loaded onto the cartridge and unwanted material (e.g., protein) is retained (Fig. 8.5*a*), allowing the analyte(s) to elute directly to the analytical column. Then the valve is switched (Fig. 8.5*b*) and the cartridge is backflushed for cleaning while the analysis is completed. In the "enrichment" mode, the unwanted materials are flushed to waste while the desired components are held on the cartridge (Fig. 8.6*a*). Then the mobile phase is changed and the valve is switched to flush the analytes onto the analytical column (Fig. 8.6*b*). In this configuration, a large volume of sample can be loaded onto the enrichment column in a weak solvent, then eluted in a small volume for injection onto the analytical column. For either type of column-switching application, method development involves selection of the appropriate cleanup column, the load, wash, and elution solvents, and determination of the appropriate valve timing. On-line cleanup is simple and reliable for routine operation. Home-built systems can be fashioned from commercially available parts (e.g., valves from Rheodyne, Rohnert Park, CA, USA; or Valco, Houston, TX, USA) or dedicated cleanup systems can be purchased (e.g., Spark Holland, The Netherlands; or Cohesive Technologies, Franklin, MA, USA). On-line sample cleanup is useful in a wide variety of LC-MS applications.

With the possible exceptions of protein precipitation and on-line cleanup, all of these cleanup techniques have less than 100 percent recovery and the recovered volume fraction may vary from one sample to the next. This represents a compelling reason to use internal standard calibration, which can correct for such losses.

338 CHAPTER 8 OTHER APPLICATIONS OF GRADIENT ELUTION

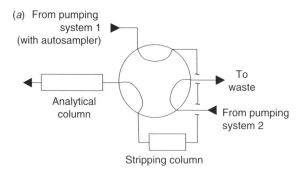

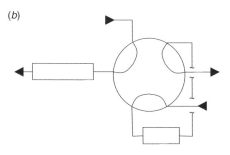

Figure 8.5 Column switching for removal of interferences. (*a*) Stripping-column traps interferences while analyte(s) pass to analytical column; (*b*) analytical-column is eluted while stripping-column is backflushed to waste.

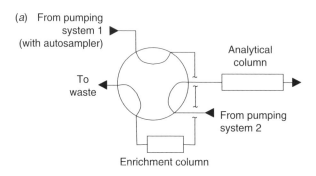

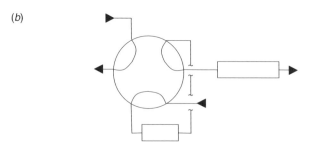

Figure 8.6 Column switching to enrich sample. (*a*) Sample is loaded onto enrichment column while analytical column is equilibrated; (*b*) concentrated sample is transferred from enrichment column onto analytical column.

It may make sense to use a graduated approach to sample preparation. For example, during initial method development or for small numbers of samples, protein precipitation may be preferred, because it is quick and simple. For large sample sets (e.g., clinical trials generating thousands of samples), the investment in the development of an SPE cleanup technique should pay off in terms of increased column life, reduced potential for interferences, less variability in the analytical data, and more reliable method operation.

8.1.5.3 Carry Out Initial Separation (Run 1, Step 3, Table 8.2)

Initial runs for the development of gradient LC-MS methods should generally be performed with reference standards dissolved in either the starting mobile phase, the anticipated injection solvent, or a similar-strength solvent. This avoids complications or problems arising from the sample matrix (e.g., plasma). As soon as conditions are found that give reasonable retention, extracted samples should be used so that matrix-related problems can be addressed. Conditions as in Table 8.1 should be used; for bioanalytical methods, conditions should be targeted to give a retention for the first peak of interest of $\geq 4t_0$ so that sufficient time is allowed for the elution of the ion-suppression region seen near t_0 in many samples (see Section 8.1.6.3 and Fig. 8.9).

Inspect the chromatogram for problems. Early elution, late elution, or no elution (e.g., Fig. 3.5a, b) may require a different mobile phase pH, column, or separation mode (see Section 3.2.1.2).

Evaluate the initial separation using the procedure outlined in Section 3.2.1 and Figure 3.3 to determine if an isocratic separation is possible or whether a gradient is needed. It should be noted that the self-cleaning characteristic of gradients is a compelling reason to use gradient elution for bioanalytical methods, even if isocratic separations are possible.

8.1.5.4 Optimize Gradient Retention k* (Step 4, Table 8.2)

The initial conditions of Table 8.1 (4 min gradient at 0.5 mL/min or 10 min gradient at 0.2 mL/min) should give $k^* \approx 5$ for small molecules (e.g., molecular weight <500 Da), so no further adjustment should be necessary. For complex samples and other LC-MS applications requiring more resolution than typical bioanalytical separations, it may be best to start with a 100–150 × 2.1 mm i.d. column packed with 3 μm particles. In such cases, the remaining gradient conditions should be adjusted to achieve $k^* \approx 5$ (Equation 3.3) and acceptable backpressure. For larger molecules such as peptides or proteins, flatter gradients are usually a better choice; see the discussion of Table 6.1, while noting the need for smaller-i.d. columns and lower flow rates for LC-MS.

8.1.5.5 Optimize Selectivity α* (Step 5, Table 8.2)

In many ways, the optimization of gradients for LC-MS/MS methods is easier than for LC-UV methods. Because of the selectivity of the MS detector, $R_s \geq 1$ is usually sufficient for the analyte and an analog internal standard; $R_s < 1$ may be acceptable when an isotopically labeled internal standard is used (Section 8.1.6.5). Bioanalytical LC-MS/MS samples generally have a small number of analytes of interest — ≤ 5 in most cases – so their separation is not as challenging as for more complex mixtures. The major challenge with bioanalytical methods is often separation of the analytes from potential interferences present in the sample matrix (ion suppression, Section 8.1.6.3). A further advantage of MS detection, with its better detection selectivity, is that peak tracking (Section 3.4.7) is seldom a problem during method development.

As with conventional HPLC separations (Section 3.3.2), selectivity changes for LC-MS separations are best approached by first changing the gradient time 3-fold and then the temperature by 20°C (Table 8.2). These data can be used to optimize gradient time and temperature as described in Section 3.3.2, either manually or with the help of computer simulation (Section 3.4).

If gradient time and temperature in combination do not produce satisfactory results, explore other variables that influence selectivity, as discussed in more detail in Sections 3.3.2 and 3.6. For example, exchange the mobile phase organic solvent, ACN, for MeOH (or vice versa) and repeat the gradient and temperature experiments; or try a different column – in the experience of one of the authors (J.W.D.), an embedded-polar-phase column from the same manufacturer as the initial C_{18} (or C_8) column is a good next choice for a second column. If retention times are too short, an embedded-polar-phase column can be tried, using a mobile phase of near-100 percent water (with 0.1 percent formic acid or other additive) may result in acceptable retention. (The information in Appendix III can be used to help chose a replacement column of different selectivity.) Because mobile phases at or near 100 percent water decrease desolvation efficiency in the MS interface (requiring a further decrease in flow rate), a change to hydrophilic interaction chromatography (HILIC, Section 8.3.3) should permit the use of higher percentage-organic mobile phases, and thus more efficient desolvation. A change in pH often will make a large difference in retention, especially if basic compounds were initially run at low pH. In one of the authors' (J.W.D.) experience, use of a base-stable column (such as Waters Corp.'s XTerra) allows successful separations to be obtained with a mobile phase pH above the pK_a of many basic compounds (e.g., operation at pH 9–10).

8.1.5.6 Adjust Gradient Range and Shape (Step 6, Table 8.2)
If further adjustment of the starting and ending gradient conditions will speedup the separation, it should be done at this stage, using the guidelines of Section 3.3.3. Be cautious when adjusting the value of initial %B (ϕ_0), because early elution (due to large ϕ_0) can compromise method performance as a result of ion suppression. *As a general rule, adjust conditions so that the retention of the first peak is* $>4t_0$, *so as to avoid the ion-suppression region found early in the run* (t_0-4t_0) *for many methods (Section 8.1.6.3).*

Conventional segmented gradients, as described in Section 3.3.4, are of less use with LC-MS than in LC-UV, but many workers use a hold-elute-flush sequence with stepped gradients. An initial gradient delay is used to allow polar materials to be flushed from the column, then an isocratic step gradient or a short gradient is used to elute the compound(s) of interest, followed by a steep gradient or step gradient to high %B to flush the column prior to the next run. Such conditions usually are obtained as the result of empirical optimization; depending on the sample, either ACN or MeOH may be more effective at removing unwanted contaminants from the column [20]. *Because plasma and other biological matrix samples often contain strongly-retained interferences, even after extensive sample cleanup, it is recommended to flush the column with strong solvent between runs; a step flush will accomplish this in the minimum amount of time.*

8.1.5.7 Vary Column Conditions (Step 7, Table 8.2) Because initial gradient development is usually performed on a short, small-particle column in LC-MS, further gains in throughput by reducing column length may be marginal. However, if large numbers of samples are to be run (such as tens of thousands of samples in some Phase III clinical studies), method adjustments that gain only 0.5–1.0 min in run time may pay off handsomely over time. For example, increasing the flow rate during flushing, and loading the injector loop during equilibration can help to trim wasted time from high-throughput methods.

8.1.5.8 Determine Inter-Run Column Equilibration (Step 8, Table 8.2) Column equilibration time can be a significant portion of the total run time for gradient LC-MS methods, and the re-equilibration delay is one argument that some workers use for avoiding gradients at all cost. However, the between-run equilibration time often can be made much shorter than previously recognized (Section 5.3), so this may not represent a valid objection to present-day gradient LC-MS. For bioanalytical methods, the precision and accuracy requirements ($\pm 15-20$ percent [8]) are such that a minor deterioration in precision and accuracy, due to incomplete (or even variable) equilibration, may not be noticed. In any event, it is important to determine the shortest possible inter-run equilibration time, especially when large numbers of samples are to be run.

8.1.6 Special Challenges for LC-MS

In the discussion above, it should be apparent that many aspects of the development of gradient methods for LC-MS are no different than those for LC-UV methods. However, there are some unique challenges with LC-MS that need to be considered. In this section, the following topics are discussed, especially as they apply to LC-MS methods:

- dwell volume;
- gradient distortion;
- ion suppression;
- co-eluting compounds;
- resolution requirements;
- use of computer simulation software;
- isocratic methods;
- throughput enhancement.

8.1.6.1 Dwell Volume Section 4.4 contains a discussion of the practical impact of system dwell volume V_D on the separation. An excessive dwell volume will result in a longer run time, and small changes in dwell volume among different gradient systems can lead to significant changes in separation, especially for the low flow rates and short gradient times that characterize LC-MS. A practical rule of thumb (Section 4.4) is that the dwell volume should be no more than about 10 percent of the gradient volume ($V_G = t_G F$). For the recommended starting conditions of Table 8.1, a 4 min gradient run at 0.5 mL/min would generate $V_G = 2.0$ mL, so $V_D \leq 200$ µL is desired. Larger dwell volumes also mean longer runs and fewer separations per hour.

The dramatic reduction in sample throughput when large dwell volumes are used with small-volume gradients means that conventionally plumbed HPLC systems with $V_D \approx 1.5$–4 mL are unsuitable for gradient LC-MS methods. Either a system specialized for LC-MS applications must be obtained, or an existing system should be modified to reduce the dwell volume. High-pressure mixing systems usually are simple to modify by replacing the conventional mixer with a micromixer purchased from an aftermarket HPLC parts supplier (e.g., Upchurch, Oak Harbor, WA, USA). This technique was used in the laboratory of one of the authors (J.W.D.) to reduce the dwell volume of a conventional high-pressure mixing HPLC system from $\sim$2.3 mL to $\sim$300 µL, including a 100 µL injection loop. Satisfactory sample throughput with LC-MS methods can be obtained with this setup. Low-pressure mixing systems are difficult, if not impossible, to convert to low-dwell-volume applications, so if low-pressure mixing is desired for LC-MS, it is recommended to purchase a low-dwell-volume system.

8.1.6.2 Gradient Distortion The dwell volume V_D includes the mixing volume V_M of the gradient system plus additional plumbing volume (Section 9.2.2). This mixing volume can create distortion of the gradient shape, if the volume is large relative to the gradient volume (Table 9.2). This is illustrated in Figure 8.7 for two different gradients [21]. A low-pressure mixing system ($V_D \approx 1$ mL, $V_M \approx 0.24$ mL)

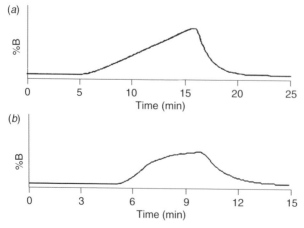

Figure 8.7 Gradient distortion. (*a*) Gradient of 20–20–50–50–20 percent B at times 0.0–0.1–10–11–11.1 min; (*b*) 20–20–40–40–20 percent B at 0.0–0.1–1.5–5–5.1 min. Data from [20].

was used with a 100 × 2.1 mm i.d. column at 0.2 mL/min, so the gradient reached the detector at ~5 min (see gradient trace in Fig. 8.7a and b). A 9.9 min (2 mL, $V_M/V_G \approx 0.12$) gradient generated the expected linear profile with little apparent distortion, as seen in Figure 8.7(a). This gradient was followed by a 1 min isocratic hold and a 0.1 min step gradient back to the initial conditions.

For Figure 8.7(b), a steep, 1.4 min (0.28 mL, $V_M/V_G \approx 0.86$) gradient was run, followed by a 3.5 min isocratic hold and a 0.1 min step back to the initial conditions. Notice the severe distortion of the gradient, with the programmed gradient (time ~6–7 min in Fig. 8.7b) and the isocratic hold (~7–10 min) merged together into a two-phase curve. Although it may be possible to generate reproducible gradients under the conditions of Figure 8.7(b) on one instrument, it is unlikely that a steep-gradient method with large V_M/V_G would transfer to a second instrument without problems.

Further evidence of severe distortion with a short gradient can be seen in Figure 8.7(a and b) during the re-equilibration phase. The 0.1 min step gradient ($V_M/V_G \approx 2.4$ in both cases) takes several minutes to return to baseline. If the mixing volume is reduced to less than 10 percent of the gradient volume (Table 9.2), gradient distortion will be minimized, and total run-time (including column equilibration) will be reduced.

8.1.6.3 Ion Suppression

Under reversed-phase separation conditions, it is often desirable to suppress sample ionization during separation, so that sample acids and bases are more strongly retained and do not leave the column too soon. However, MS detection only functions with ions, so the interface between the column and the MS must convert (neutral) analytes in the liquid mobile phase to ions in the gas phase before they enter the MS. Thus, ionization is a necessary condition for MS detection; suppression of ionization in the MS interface results in a reduction in signal intensity. *When discussing ion suppression in the LC-MS context, it is important to distinguish between ion suppression during separation for purposes of reducing peak tailing (good) and in the MS detector (bad); unfortunately the same terminology has been used for both processes.*

A simple experimental setup [22] can be used to identify ion suppression problems in the MS interface. This is illustrated in Figure 8.8, where a dilute solution of the analyte is infused into the column effluent from the HPLC system. The mass

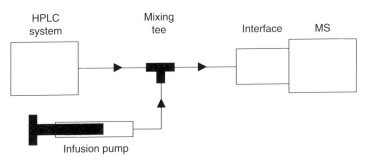

Figure 8.8 Instrument setup to test for ion suppression in the MS.

spectrometer can be operated under normal detection conditions – in the single-ion monitoring (SIM) mode to look for the molecular ion of interest (no fragmentation), or in the multiple-reaction monitoring mode to look for the precursor > product ion transition (operated in the MS/MS configuration). A typical output signal at steady state is illustrated by the 1.2–3.0 min region in the trace of Figure 8.9(a), representing a constant high signal from the detector when the analyte is passed through at a constant concentration. Once a stable (high) baseline is obtained, an injection of a blank sample extract (extracted matrix without analyte) is made. The resulting mass chromatogram may appear as in the example of Figure 8.9(a), where a blank plasma extract is injected while paclitaxel is infused into the column effluent at steady state. Any peaks eluting from the column that suppress ionization of the analyte in the interface will result in a reduction in the steady state signal. This dip in the baseline at $\approx 0.7-1.0$ min in Figure 8.9(a). It is important that separation conditions are selected so that the analyte(s) of interest does not elute during this suppression region; otherwise, a falsely low concentration may be reported. The mass chromatogram of Figure 8.9(b) shows that paclitaxel ($\sim$2 min) and one of its metabolites ($\sim$1.6 min) are eluted well after the ion suppression region, so problems from ion suppression are unlikely in this assay. The ion suppression dip near t_0 is typical of many LC-MS methods, much like the large "garbage" peak at the beginning of LC-UV chromatograms, representing poorly retained, polar materials originating from the sample matrix.

Although a generic ion suppression region early in the chromatogram (as in Fig. 8.9) is common, ion suppression can be generated by any compound that is eluted from the column and reduces the ionization of the desired analyte(s). Lipid-related compounds are a compound class that is of much concern in the determination of pharmaceutical compounds by LC-MS/MS [20]. The impact of ion suppression by lipids is illustrated in Figure 8.10. Figure 8.10(a) shows the steady

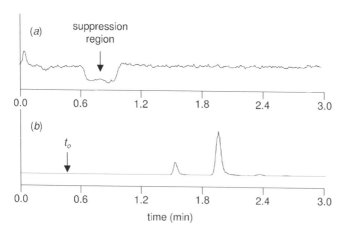

Figure 8.9 A simple case of ion suppression. (a) MS output for the injection of blank extracted plasma into a continuous infusion of paclitaxel into the HPLC mobile phase, using the setup of Figure 8.8; (b) mass chromatogram for paclitaxel ($\sim$2 min) and one of its metabolites ($\sim$1.6 min).

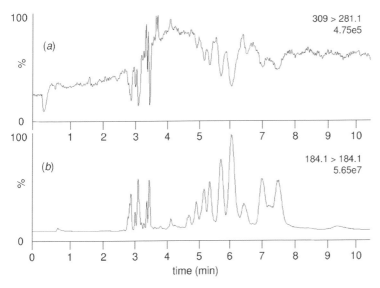

Figure 8.10 Ion suppression from glycophospholipids under gradient conditions (proprietary method). (a) Continuous infusion of Xanax into the HPLC mobile phase, monitoring the 309 > 281.1 m/z MS-MS transition; (b) mass chromatogram for 184.1 m/z signal typical of lipid-related compounds. Data from [22].

state infusion of Xanax monitored for the 309 > 281.1 m/z transition (precursor ion = 309 and measured product ion = 281.1). It can be seen that there are many regions of ion suppression (negative dips in the steady state baseline). The lower trace (Fig. 8.10b) is a mass chromatogram of a blank plasma extract monitored at 184.1 m/z, a fragment mass common to lipid-related compounds. Note that there is a strong correlation between the lipid peaks in Figure 8.10(b) and the suppression regions in Figure 8.10(a), providing strong evidence for ion suppression by these compounds. If the separation conditions were selected so that the compounds of interest were eluted at <8 min, the method for Xanax should be free of ion-suppression due to lipid components of the sample.

Ion suppression can result from the co-elution of an analyte with any compound that suppresses ionization, whether the suppressing compound originates in the sample, mobile phase, or other reagents. Ion-pairing reagents, such as trifluoroacetic acid can generate significant ion suppression under certain conditions [11, 12]. It is wise to check for ion suppression, as a standard part of method development for LC-MS and LC-MS/MS methods.

8.1.6.4 Co-Eluting Compounds
Compounds that co-elute with analytes of interest are of less concern with LC-MS methods than with LC-UV methods, because MS detection has added selectivity – especially if the MS/MS technique is used. As long as a co-eluting substance does not cause ion suppression (Section 8.1.6.3) or have the same molecular ion (MS) or transition (MS/MS), co-elution should not be a problem. One technique that can help determine the impact of

co-eluting substances on the assay is to inject neat standards of suspected interferences, such as co-administered drugs. If the suspect compounds do not co-elute, they can be ignored. If they co-elute, they are checked for ion suppression and the chromatographic conditions can be adjusted, if necessary, to separate the problem compound from the analyte(s) of interest. If a co-eluting stable-label internal standard is used, it is important to check for suppression of the analyte by the IS, and vice versa.

8.1.6.5 Resolution Requirements
There are several factors that contribute to a lower resolution requirement ($R_s \geq 1.0$), or even preference, for LC-MS:

- specificity of the detector;
- data acquisition process;
- compensation of experimental errors by the internal standard;
- potential for ion suppression.

The *specificity* of LC-MS, especially when operated in the MS-MS mode, should allow accurate quantification of co-eluting compounds, because each compound has a specific mass and, for the case of MS/MS, a unique precursor > product ion transition. (There are exceptions to this, such as some positional isomers and other compounds with the same m/z.) However, because of the characteristics of the *data acquisition process*, the detector can acquire data from only one mass at a time. If two peaks co-elute, they can both be detected, but only if the detector rapidly switches back and forth between the ions of interest during the elution of the peaks. Any time *not* spent counting a given mass means a smaller signal will be generated, and thus smaller signal-to-noise values. For example, if two compounds of interest co-elute and half the time is spent counting each compound (e.g., a detector "dwell time" of 100 ms for each analyte), the signal for each will be half as large as could be obtained if twice the time (e.g., 200 ms) were spent counting just one compound.

An internal standard is used for conventional HPLC methods to *compensate for experimental errors*, such as extraction differences, dilution errors, or injection volume variability. With MS there is additional variability in the vaporization and ionization of the sample in the MS interface and routing of sample ions into the MS itself. An internal standard that co-elutes with the analyte should do a better job of correcting for interface variability than one that is chromatographically separated from the analyte.

Finally, co-elution of sample components that cause *ion suppression* (Section 8.1.6.3) is not desirable – $R_s \gg 1$ form such interferences is preferred in such cases. Thus, accuracy in LC-MS is favored by the use of internal standards that overlap the analyte peak ($R_s < 1$), but at the possible expense of decreased precision for low concentrations of the analyte (because of a reduced peak area count). However, this is only true if ion suppression effects can be avoided. The use of an internal standard that is separated from the analyte with $R_s > 1$ will result in increased precision for low concentrations of the analyte, but at the expense of possibly reduced accuracy (bias).

The above factors all contribute in varying degree for different sample types and instrument setups. Our recommendation of $R_s \geq 1.0$ for LC-MS methods is a suitable compromise in most situations, although $R_s < 1.0$ generally is acceptable for stable-label internal standards.

8.1.6.6 Use of Computer Simulation Software   Computer simulation software (Section 3.4) can be used to advantage for the development of both LC-MS and LC-UV methods. Typically, bioanalytical LC-MS samples have less than five components of interest, so the separation challenge may be less (with less need for computer simulation) than with LC-UV methods, especially LC-UV methods used for stability indication or the assay of impurities. However, more complex LC-MS sample applications (protein and peptide digests, impurity profiles, etc.) will provide many of the same separation challenges of conventional LC-UV samples, and thus similar benefits for computer simulation. Computer simulation software also can be useful to adjust the gradients for less complex samples, so as to avoid ion suppression (if the retention times of ion suppressing peaks are entered along with the retention times of the analytes of interest, the latter can be separated from the former). Remember that the short, small volume columns typical of LC-MS methods are more susceptible to extracolumn effects, so additional care must be taken to minimize these effects and to enter the correct extracolumn volume into the software.

8.1.6.7 Isocratic Methods   Many workers prefer isocratic methods over gradients for LC-MS, because isocratic methods are simple, have less stringent equipment requirements (e.g., dwell volume is less of an issue), and provide adequate separation for the typically small number of analytes in bioanalytical applications. However, late-eluting compounds can appear in subsequent runs, causing ion suppression (Section 8.1.6.3) or interference (Section 8.1.6.4), so a step gradient to a strong flush solvent often is necessary with isocratic methods. When step gradients are used for this purpose, allowance needs to be made for the dwell volume (Section 8.1.6.1), and care should be taken to ensure that the analyte(s) of interest is not eluted during the step because of the potential for gradient distortion (Section 8.1.6.2) and related reproducibility problems. For these reasons, even when isocratic conditions are used, it is wise to use a low-dwell-volume system. If a step-gradient is not used to elute highly retained compounds, then a suppression check should be made after multiple injections of a matrix blank, not a single injection, because late-eluting compounds may not exit the column until several runs later.

8.1.6.8 Throughput Enhancement   Because of the high cost of the MS detector, many applications of LC-MS strive for minimum run times, so as to minimize the per-sample cost of analysis. In this context, any time spent not running samples is added expense. One way to increase throughput is to use parallel chromatography [23] (also called multiplexing) so that the effluent from more than one column is fed into the MS. By running the same method on two columns in parallel, one can adjust the injection times such that one column is running a sample while the other is re-equilibrating for the next injection. Thus, the time wasted during column flushing and equilibration does not tie up the mass spectrometer. (With appropriate valve timing it also is possible to vent the t_0 "garbage" and the ion-suppression region to waste, so as to reduce contamination of the MS system.) A diagram for one such setup is shown in

Figure 8.11. This example combines two gradient pumping systems, one autosampler, and two columns. Each column runs a separate set of calibrators and control samples, so small differences in retention or response for the two systems do not have to be compensated. In Figure 8.11(*a*), sample is injected onto column 2, the gradient is run by pumping system 2, and a mass chromatogram is recorded. Meanwhile, column 1 is flushed and re-equilibrated by pumping system 1. As soon as the components of interest are eluted from column 2, the valves (V1 and V2) are switched to the configuration shown in Figure 8.11(*b*). Now the method is run on column 1 (with pumping system 2) while column 2 is flushed and re-equilibrated by pumping system 1. This setup requires just one autosampler and one MS detector. The plumbing diagram of Figure 8.11 uses two six-port valves controlled by the external events outputs of the HPLC or MS system. A similar result can be obtained with a single 10-port switching valve, but lacks the flexibility for other applications that two 6-port valves afford. Other multiplexing scenarios also exist, such as the "MUX" interface (multiplexed ESI, Waters Micromass, Milford, MA, USA), in which multiple columns are sequentially parsed into the MS system, for the simultaneous collection of data from several HPLC columns.

Column switching for sample cleanup or enrichment, as described in Section 8.1.5.2, can increase sample throughput, if the process is designed such that the cleanup step does not slow down the overall analysis. If the analytical column must wait for the cleanup step to be completed, the overall throughput may suffer. Clever system design, such as timing the cleanup to occur during elution of the previous sample or combining on-line cleanup with parallel chromatography, can help to minimize any delays during the cleanup step(s).

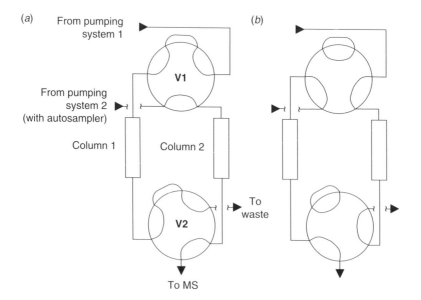

Figure 8.11 Parallel chromatography to improve sample throughput. (*a*) Equilibration of column 1, elution of column 2; (*b*) equilibration of column 2; elution of column 1.

8.2 ION-EXCHANGE CHROMATOGRAPHY (IEC)

IEC is widely used for the separation of large biomolecules [24, 25], inorganic ions [26], and to a lesser extent in combination with other HPLC procedures for two-dimensional or multistep separation [27–29]. As reviewed in Chapter 6, the IEC separation of most proteins, oligonucleotides, and oligosaccharides can be described quantitatively by the LSS model (when the molecular charge $|z| \geq 3$), and therefore needs little further discussion. Mixtures of inorganic ions are usually separated and analyzed by ion chromatography, a specialized form of HPLC that is beyond the scope of the present book. The gradient IEC separation of organic ions with molecular weights <1000 Da is the primary subject of the present section; however, such separations are today less popular, especially when compared with the use of RP-LC.

Positively charged sample ions X^+ can be separated using *cation-exchange* columns, which typically contain porous particles with negatively charged acidic functional groups R^-. These columns can bind cationic species such as protonated bases BH^+ by ionic interaction. Anion-exchange columns carry positively charged, or basic, functional groups R^+ that are capable of binding sample anions X^- such as ionized carboxylic acids $RCOO^-$. The mobile phase will usually contain a buffer to maintain constant pH, plus varying concentrations of a salt (counter-ion) to control the retention k of sample ions. The charge on the counter-ion will have the same sign as that of the sample ion; for example, K^+ can be used to control the retention of protonated bases BH^+ by cation exchange, while Cl^- can be used for the separation of ionized acids $RCOO^-$ by anion exchange. However, it should be noted that high concentrations of Cl^- can attack stainless steel and should therefore be avoided. Nitrate, phosphate, or sulfate counter-ions are less corrosive alternatives.

8.2.1 Theory

The LSS model can be useful for relating gradient elution to "corresponding" isocratic separations (Section 2.2.1.1) that is, where only the concentration of the counter-ion (or B solvent) changes. In the case of RP-LC,

$$\log k = \log k_w - S\phi \qquad (2.9)$$

The isocratic retention factor k of a compound varies linearly with the volume-fraction ϕ of the B solvent in the mobile phase; the solute-parameter S is constant when only ϕ varies. The gradient retention factor k^* is analogous to the isocratic retention factor k. Changes in either k^* or k result in similar changes in selectivity, peak width, and resolution for "corresponding" separations. For RP-LC separation,

$$k^* = 1/1.15b$$
$$= (t_G F)/(V_m \Delta \phi S) \qquad (2.13)$$

Here, b is the intrinsic gradient steepness, t_G is gradient time, F is mobile phase flow rate, V_m is the column dead volume (proportional to column length and internal diameter-squared), and $\Delta\phi$ is the change in ϕ during the gradient. For linear-gradient IEC, a similar equation for k^* can be derived (see below)

$$k^* = 1/1.15b$$
$$= (t_G F)/1.15[V_m m \log(C_f/C_0)] \tag{8.1}$$

Here, m is the absolute value of the charge z on the sample compound ($|z|$; e.g., equal 1 for either protonated aniline or ionized acetic acid), and C_0 and C_f are the molar concentrations of counter-ion (salt) in the mobile phase at the beginning and end of the gradient, respectively. A monovalent counter-ion is assumed (charge of ± 1) in Equation (8.1).

A number of examples have been provided in earlier chapters for the effects of various changes in conditions on RP-LC separation. Both RP-LC and IEC separation will vary with gradient conditions in similar fashion, for comparable changes in k^*. Some representative examples for RP-LC gradient elution are summarized in Figure 8.12, for the "regular" sample of Table 1.3. Assuming the initial separation of Figure 8.12(a), a decrease in gradient time (Fig. 8.12b) results in a decrease in resolution and an increase in peak heights, due to the resulting decrease in k^*. Similarly, an increase in gradient time (Fig. 8.12c) leads to increased k^* and resolution, and decreased peak heights. Beginning the gradient at a higher %B (Fig. 8.12d) while maintaining gradient steepness ($\Delta\phi/t_G$) constant, can result in a lower value of k^* for early peaks, which then means increased peak heights for these peaks with a loss of resolution; there is less effect on k^*, resolution, and peak height for later peaks when initial %B is increased. Similar examples for the "irregular" sample of Table 1.3 are provided in Figure 3.9. In the case of IEC [or NPC, Section 8.3), qualitatively similar changes in resolution and peak heights can be expected for analogous changes in conditions (gradient time, initial and final counter-ion concentrations (C_0 and C_f), column size, and flow rate] that result in changed values of k^* [Equation (8.1)].

When column length, flow rate, or gradient range $\Delta\phi$ is changed for an RP-LC separation, it has been recommended to maintain k^* constant (Section 3.3), by varying gradient time so as to keep $(t_G F)/(V_m \Delta\phi)$ constant. This procedure simplifies method development by avoiding changes in relative retention or selectivity, when changing conditions in order to either increase resolution or decrease run time. A similar recommendation can be made for IEC; when changing column length, flow rate, or gradient range (C_f/C_0); keep k^* constant by varying gradient time [Equation (8.1)].

The following quantitative treatment of IEC retention and separation is somewhat detailed and has limited practical application; the reader may wish to skip to the following section, starting again at Section 8.2.2.

Isocratic retention in IEC is governed by a competition between sample ions and mobile phase counter-ions for interaction with stationary phase ionic groups of opposite charge. IEC retention can be illustrated by the cation-exchange retention of a protonated basic compound BH^+ using K^+ as the counter-ion:

$$BH^+ + R^-K^+ \iff K^+ + R^-BH^+ \tag{8.2}$$

8.2 ION-EXCHANGE CHROMATOGRAPHY (IEC)   351

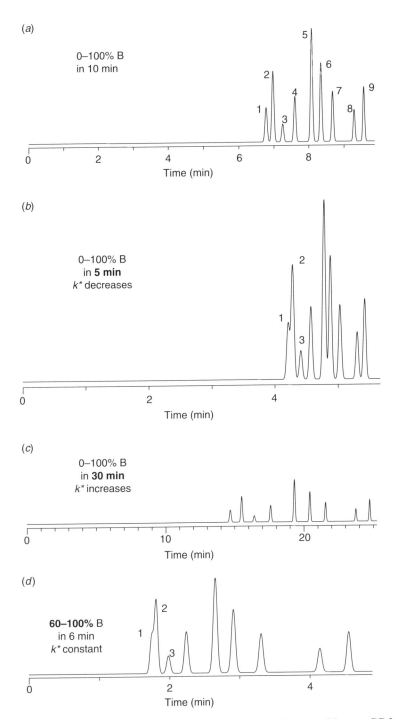

Figure 8.12 Examples of the effect of a change in gradient conditions on RP-LC separation for the "regular" sample of Table 1.3. Conditions for separation of (*a*): 150 × 4.6 mm C_{18} column (5 μm particles); 0–100 percent methanol/water gradient in 10 min; 2 mL/min; ambient temperature. For (*b–d*), change in conditions from (*a*) are indicated in bold.

Here, R^- refers to an anionic group attached to the column packing, which can bind either the sample ion BH^+ or the mobile-phase counter-ion K^+ by electrostatic interaction. Equation (8.1) can be generalized for sample ions which have a charge $|z| \equiv m$, as well as counter-ions with charge $|p| \equiv n$, where m and p can assume any positive, integral values:

$$nX^{+m} + mR_n^- Y^{+n} \iff nX^{+m}R_m^- + mY^{+n} \qquad (8.3)$$

If the counter-ion Y carries a charge of $+1$ (i.e., Y^-, $n = 1$), we can write Equation (8.3) as

$$X^{+m} + mR^- Y^+ \iff X^{+m}R_m^- + mY^+ \qquad (8.3a)$$

In the following discussion, we will assume a monovalent counter-ion (Y^+) and Equation (8.3a), which is often the case in practice. Equations identical to Equations (8.3) and (8.3a) apply equally for retention in anion-exchange chromatography of a sample ion X^{-m} and a counter-ion Y^{-n}.

Values of the retention factor k in IEC for a univalent counter-ion $Y^\pm$ and *either* cation or anion exchange can be derived from the equilibrium of Equation (8.3a):

$$\log k = \log K' - m \log C \qquad (8.4)$$

where C is the molar concentration of the counter-ion $Y^\pm$ in the mobile phase, $\log K'$ equals $\log k$ for $C = 1$ M, and m is the absolute value of the charge z on the solute molecule X. K' is a constant (for a given sample compound, column, salt, buffer, mobile phase pH, and temperature); $\log K$ and m can vary with mobile phase pH. An illustration of Equation (8.4) is shown in Figure 8.13 for the anion-exchange separation of four polyphosphates with z equal -3, -4, -6, and -8 (tri-, tetra-, hexa-, and octa-phosphates, respectively). Numerous examples of the validity of Equation (8.4) for isocratic IEC have been reported [30, 31].

A linear salt gradient in IEC as a function of time t during the gradient can be described by

$$C = C_0 + (\Delta C/t_\text{G})t \qquad (8.5)$$

where C_0 is the concentration of the counter-ion $X^\pm$ at the start of the gradient, and $\Delta C = (C_\text{f} - C_0)$ is the change in C during the gradient. Retention time t_R in linear-gradient IEC can be derived from the fundamental equation for gradient retention [Equation (9.2)], using values of C from Equation (8.5) as a function of time t, and derived values of k from Equation (8.4) (assumes that values of K' and z are known or can be measured) [32–34]:

$$\begin{aligned} t_\text{R} = &\{C_0^{m+1} + [V_\text{m}K(m+1)\Delta C]/t_\text{G}\}^{1/(m+1)} \\ &- (C_0 t_\text{G}/\Delta C) + t_0 + t_\text{D} \end{aligned} \qquad (8.6)$$

Peak width W in IEC will be the same as in RP-LC:

$$W = (4N^{*-1/2})t_0(1 + k_\text{e}) \qquad (2.17)$$

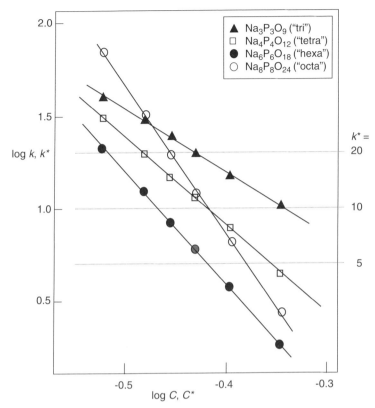

Figure 8.13 Illustration of the dependence of log k on log C in isocratic IEC. Sample: four polyphosphates as shown in figure; conditions: 500 × 4.0 mm TSKgel SAX anion exchange column; aqueous KCl salt solutions as mobile phase; 30°C. Adapted from [30].

where k_e is the value of k when the band elutes from the column. The value of k_e can be determined from Equation (8.4), with C determined from Equations (8.5) and (8.6) ($t = t_R - t_0 - t_D$).

Because of differences in the dependence of retention k on mobile phase composition C in IEC [log–log, Equation (8.4)] vs RP-LC [log–linear, Equation (2.9)], the LSS model used for RP-LC is not quantitatively applicable for IEC. This can be seen in Figure 8.14, where values of log k from Equation (8.4) for IEC (with $m = 1$) are plotted vs C over a useful range in k ($1 \leq k \leq 10$), rather than vs log C as suggested by Equation (8.4). In the case of RP-LC, an approximately linear fit of values of log k vs mobile phase composition (with %B ≡ C in Fig. 8.14) is expected, as in Figs 6.1 and 9.1. However, the deviation from linearity (shown by the dashed curve) of the IEC plot in Figure 8.14 suggests that the LSS model is a relatively poor approximation for this example; the standard deviation (SD = 0.10) of the plot of log k vs C in Figure 8.14 (for $1 \leq k \leq 10$) represents an uncertainty in values of k of ±26 percent. The latter error decreases for more highly charged solutes (larger values of m or $|z|$), in approximate proportion to the value of m, for example,

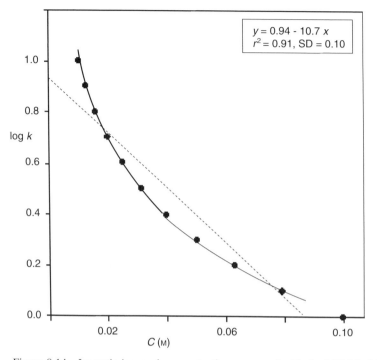

Figure 8.14 Isocratic ion-exchange retention compared with the LSS Model. Plot of log k vs counter-ion (salt) concentration based on Equation (8.2) with $z = 1$. The dashed curve is the best linear fit to data for $1 \leq k \leq 10$. See text for details.

± 9 percent for $|z| = 3$. Consequently, the linear-gradient IEC separation of molecules with $|z| \geq 3$ (e.g., proteins, oligonucleotides, polysaccharides) can be described semiquantitatively by the LSS model [35–37]. However, linear-gradient IEC with small-molecule samples often results in $|z| < 3$; for such samples, the LSS model can still be useful, but it is only qualitatively applicable.

Consider next the derivation of an approximate expression for k^* in linear-gradient IEC, starting with Equation (2.13) for RP-LC. In RP-LC, the term $\Delta \phi S$ is equal to log (k_f/k_0), where k_f is the final value of k at the end of the gradient, and k_0 is the initial value of k at the start of the gradient (corresponding to mobile phase compositions in RP-LC of ϕ_f and ϕ_0). From Equation (8.4) for IEC, we have a corresponding relationship:

$$\log (k_f/k_0) = m \log (C_f/C_0) \tag{8.7}$$

so for IEC we can replace $\Delta \phi S$ in Equation (2.13) with $m \log (k_f/k_0)$, to obtain a value of k^* for IEC [Equation (8.1)]. A comparison of Equations (8.1) and (2.13) shows that the quantities S and $\Delta \phi$ in RP-LC are replaced in IEC by m and $\log(C_f/C_0)$, respectively. Whereas values of S can be predicted only approximately for RP-LC separation (Section 9.4), values of m (or z) in IEC are often known from the structure of the sample molecule and mobile phase pH. For example, $m = 1$ for partly or fully ionized monofunctional acids or bases. Resolution R_s in linear-gradient NPC can be described by the same equation as for RP-LC:

$$R_s = (1/4)[k^*/(1+k^*)](\alpha^* - 1)N^{*1/2} \quad (2.21)$$

As in the case of RP-LC, plots of either log k vs C or log k^* vs C^* for "corresponding" separations will fall on the same curve, as illustrated in Figure 9.1(d) and the examples of [37, 38]. That is, both isocratic and gradient retention as a function of counter-ion concentration C are equivalent in "corresponding" IEC separations. Any change in conditions which affects values of k and α in isocratic IEC will have a similar effect on values of k^* and α^* in gradient IEC. Thus an understanding of how separation varies with conditions in isocratic IEC can be applied directly to gradient IEC.

The usual need for a buffer in IEC means that the concentration of the buffer counter-ion should be added to the concentration of the mobile-phase counter-ion, when calculating values of C_0 and C_f. However, because the buffer counter-ion may be retained more or less strongly than the mobile phase counter-ion, the *effective* values of C_0 and C_f in Equation (8.1) are only approximately the sum of salt and buffer counter-ion concentrations at the start and end of the gradient, respectively. A corresponding uncertainty therefore exists in values of k^* calculated by means of Equation (8.1). However, this does not detract from the *qualitative* value of the LSS model for the interpretation of IEC separation as a function of gradient conditions. At the same time, a rigorous quantitative treatment of gradient IEC is complicated to the point of being impractical for practical application.

In RP-LC, values of k^* are relatively constant for both early- and late-eluting peaks in a linear-gradient chromatogram. This is less true for linear-gradient IEC, as summarized in Table 8.4. Relative to an *average* value of k^* calculated from Equation (8.1) (k^*_{avg}), and assuming constant m, values of k^* are smaller for peaks

TABLE 8.4 Variation of Gradient Retention k^* in Linear-Gradient Ion-Exchange Chromatography as a Function of (k_f/k_0) and Relative Retention Time [38]

$(t_R - t_0 - t_D)/t_G$	Value of k^*/k^*_{avg} for indicated value of C_f/C_0			
	$C_f/C_0 = 100$	30	10	3
0^a	0.05	0.1	0.3	0.6
0.05	0.3	0.3	0.4	0.6
0.1	0.5	0.5	0.5	0.7
0.15	0.7	0.6	0.6	0.7
0.2	1.0	0.8	0.7	0.8
0.3	1.4	1.1	0.9	0.9
0.4	2.0	1.4	1.1	1.0
0.6	2.5	2.0	1.7	1.3
0.8	3.3	2.5	2.0	1.4
1^b	5.0	3.3	2.5	1.7

[a]Corresponds to elution at start of gradient.
[b]Corresponds to elution at completion of gradient.

that elute early in the chromatogram, and larger for later peaks. This difference in values of k^* at the start and end of the gradient increases for larger values (C_f/C_0) (Table 8.4). As a practical consequence of this difference in values of k^* for different peaks in the chromatogram, resolution will be relatively poorer at the beginning of the chromatogram, and relatively better at the end. Similarly, peaks will be somewhat narrower (and taller) for early peaks, and wider (and shorter) for later peaks. An example is provided by the IEC separations of Figure 1.3 (and to a lesser extent, Fig. 8.15a), which can be compared with similar (linear-gradient) RP-LC separations in previous chapters (e.g., Fig. 1.1d).

8.2.2 Dependence of Separation on Gradient Conditions

Both RP-LC and IEC separation will vary with change in gradient conditions in a similar fashion. The various changes in conditions illustrated in Figure 8.12 do not cause changes in relative retention for this "regular" RP-LC sample; that is, peak spacing and retention order remain the same in each example. As discussed in Section 2.3, and illustrated in Figure 3.9, however, changes in conditions which result in changes in k^* for the RP-LC separation of "irregular" samples often lead to peak reversals or other changes in relative retention. Such changes in selectivity for RP-LC arise when two adjacent peaks have different values of the solute parameter S (Section 2.2.3.1). Similar changes in relative retention are expected in IEC for two adjacent peaks with different ionic charges z (e.g., X^+ and Y^{++}). This is the case for the isocratic example of Figure 8.13 (where $z = -3, -4, -6, -8$ for tri-, tetra-, hexa-, and octa-polyphosphate); the retention sequence varies with k^* as follows:

$$k^* = 20 \quad \text{hexa} < \text{tetra} < \text{octa} < \text{tri}$$
$$k^* = 10 \quad \text{hexa} < \text{tetra} = \text{octa} < \text{tri}$$
$$k^* = 5 \quad \text{hexa} < \text{octa} < \text{tetra} < \text{tri}$$

Further changes in retention order can be visualized for $k^* \gg 20$ or $\ll 5$. Since k^* increases for larger gradient time t_G, later elution of the "octa" peak should occur as gradient time increases (relative to the other three peaks of Fig. 8.13). This predicted trend in separation with increasing gradient time has been verified for this sample [30].

Since retention in IEC is usually strongly affected by the value of m for each sample compound; compounds with $m = 1$ often elute first, followed by compounds with $m = 2$, $m = 3$, and so on. As a result, there is less likelihood that two compounds with different values of m will elute with similar retention times. Consequently, the occurence of "irregular" samples whose relative retention changes markedly with changes in gradient conditions may be somewhat less likely in IEC.

8.2.3 Method Development for Gradient IEC

8.2.3.1 Choice of Initial Conditions
Many different columns are available for IEC, for example, packings with particle sizes of 2–10 μm and pore diameters

of 8–100 nm, packed into columns of varying length and diameter. The choice of cation- or anion-exchange columns depends on the nature of the sample (acids or bases), and within each column category different suppliers offer further options [1]. For example, IEC columns are available in "strong" and "weak" forms; weak IEC columns exhibit a charge or column capacity that varies with mobile phase pH [1], whereas strong IEC columns maintain their charge independent of mobile phase pH. Fast separation is favored by small-particle, nonporous columns (Section 3.3.7), as illustrated by the separation of Figure 8.15(a) (nonporous, $d_p = 2.1$ μm) vs that of Figure 8.15(b) (porous, $d_p = 10$ μm).

In most cases, an aqueous mobile phase will be used for IEC, with the pH of the A and B solvents determined by the nature of the sample. For separations of basic compounds by cation exchange, the pH should be low enough to at least partially ionize (and retain) the various sample components, for example, pH < 6 for the separation of weak bases such as anilines or pyridines, and pH < 11 for the separation of strong bases such as aliphatic amines. The retention of acidic compounds such as carboxylic acids requires a pH > 4.

For RP-LC separation, we have seen that preferred gradients will have $1 \leq k^* \leq 10$, which allows an informed choice of IEC gradient conditions based on Equation (8.1). For this purpose, it is convenient to rearrange Equation (8.1) to

$$t_G = 1.15 k^* [V_m m \log(C_f/C_0)]/F \qquad (8.7)$$

For example, assume the separation of a mixture of monofunctional acids and/or bases ($m = 1$). Further assume a typical gradient from 0 to 0.5 M of the counter-ion and a buffer concentration of 0.005 M, so that $\log(C_f/C_0) = \log(0.505/0.005) \approx 2.0$. If we further assume a 150 × 4.6 mm column [for which $V_m \approx 1.5$ mL; Equation (3.3b)] and a flow rate of 2 mL/min, and we desire $k^* \approx 5$, then the required gradient time will be $t_G = (1.15 \times 5 \times 1.5 \times 1 \times 2)/2 \approx 8$ min. However, recognizing the uncertainty in calculated values of C_0 and C_f (because of the presence of a buffer; see Section 8.2.1 above), as well as smaller values of k^* for initially eluting peaks (Table 8.4), the latter estimate of a preferred gradient time should be taken as merely a first guess, one that can be modified after observing the initial separation.

8.2.3.2 Improving the Separation The variation of gradient time and k^* leads to predictable changes in separation, as illustrated by the RP-LC examples of Figure 8.12(a–c). Further improvements in separation can be obtained by a change in selectivity (values of α^*). As noted above, relative retention or selectivity in IEC may be less affected by changes in gradient time or k^*, because compounds with different z-values tend not to overlap. The primary variables used for varying IEC selectivity are mobile phase pH and the choice of counter-ion or buffer, for example, replacing acetate by phosphate. If the pK_a values of adjacent peaks can be estimated, predictable changes in relative retention will result for a given change in pH [1]; that is, retention increases for increasing ionization of a sample compound, and the change in ionization of a sample compound with change in pH is greatest when pH $\approx pK_a$. Following the optimization of retention and selectivity using a linear gradient, additional improvement in separation may be possible

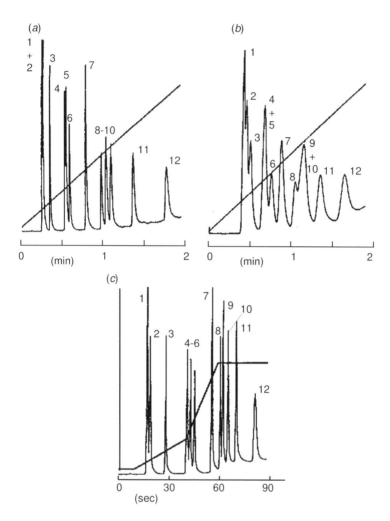

Figure 8.15 Separation of 5′-mono-, di-, and tri-nucleotides. Sample: 1, UMP; 2, CMP; 3, AMP; 4, GMP; 5, UDP; 6, CDP; 7, ADP; 8, UTP; 9, CTP; 10, GDP; 11, ATP; 12, GTP. Conditions: 33 × 8.0 mm columns coated with polyethyleneimine; aqueous salt gradients with pH 6.3 KH_2PO_4; 5 mL/min; 25°C. (*a*) Nonporous, 2.1 µm particles; 0.01–0.50 M gradient in 2 min; (*b*) LiChrospher Si-4000 (10 µm particles); 0.01–0.50 M gradient in 2 min; (*c*) same as (*a*), except for gradient of 0.01–0.01–0.14–0.50 M at 0–0.2–0.7–1.0 min. Reprinted with permission from [37].

by means of segmented gradients (Fig. 8.15*c* vs *a*); see the related discussion of Section 3.3.4 for RP-LC. The use of convex gradients to improve the resolution of homo-oligonucleotides has also been suggested [39]. Finally, further improvements in separation can be realized using changes in column length and flow rate (while varying gradient time in order to maintain k^* constant), so as to either increase resolution or decrease run time; see the related RP-LC examples of Figure 3.11.

8.3 NORMAL-PHASE CHROMATOGRAPHY (NPC)

Compared with RP-LC, normal-phase chromatography is typically carried out with a more polar column packing and a less polar mobile phase. NPC is normally used for the separation of nonionized samples, but ionized samples are also potential candidates for NPC. Nonaqueous mobile phases are the rule, except for hydrophilic interaction chromatography (HILIC, Section 8.3.3). With the exception of HILIC, NPC is not often used for routine sample analysis. However, it is often preferred for preparative separations (Chapter 7), with silica as the most common column packing. Gradient elution with silica columns is subject to certain practical difficulties (slow column equilibration, Section 9.2.1.1; solvent demixing, Section 9.2.1.2), and for this reason should be avoided if possible. If isocratic elution with NPC is not possible, step gradients may be a preferred alternative. For this and other reasons, preparative separations that require gradient elution are carried out with RP-LC, if possible. Alternatively, gradient NPC is possible with few problems, by using polar-bonded columns (cyano, diol, amino) and nonaqueous mobile phases [1]. In a change with past practice, more and more gradient NPC separations of small molecules are being carried out with polar-bonded columns and aqueous/organic mobile phases (HILIC; Section 8.3.3), in which case linear gradient elution presents no more practical difficulty than analogous separations by RP-LC.

8.3.1 Theory

NPC retention can usually be described by the Soczewinski equation:

$$\log k = \log c - m \log \phi \quad (8.8)$$

Here, for a given solute and only the concentration of the B solvent C_B varying, c and m are constants. Equation (8.8) is seen to be of the same (log–log) form as Equation (8.4) for IEC, so for linear gradients resulting expressions for NPC retention take a similar form as Equations (8.1)–(8.6) for IEC [40, 41], for example,

$$t_R = \{\phi_0^{m+1} + [V_m c(m+1)\Delta\phi]/t_G\}^{1/(m+1)}$$
$$- (\phi_0 t_G/\Delta\phi) + t_0 + t_D \quad (8.9)$$

$$k^* = (t_G F)/1.15[V_m m \log(\phi_f/\phi_0)] \quad (8.10)$$

Values of ϕ_0 and ϕ_f in Equations (8.9) and (8.10) can be equated with the volume-fraction ϕ of the polar B solvent at the beginning and end, respectively, of the gradient; $\Delta\phi = \phi_f - \phi_0$. Note that ϕ_0 cannot equal zero in Equations (8.9) and (8.10), because Equation (8.8) becomes invalid for small values of ϕ.

For separation on silica and polar-bonded stationary phases, the value of m is often ~ 1, but it can increase for larger solute molecules and/or an increased number of polar substituents [42–44]. For an initial separation by linear-gradient NPC, gradient time is given by

$$t_G = 1.15 k^* [V_m m \log(\phi_f/\phi_0)]/F \quad (8.11)$$

where $k^* \approx 5$ is a good initial choice, and it is assumed that the gradient does not begin with $\phi_0 = 0$. For a gradient from 5–100 percent B, a 150 × 4.6 mm column, a flow rate of 2 mL/min, and assuming a value of $m = 1$, the initial value of $t_G \approx 6$ min. For larger values of m, the preferred gradient time will increase proportionately, so that a gradient time of 10–15 min may represent a good first choice. For a more detailed discussion of the theory of isocratic NPC (which forms the basis for gradient NPC), see [42, 45].

8.3.2 Method Development for Gradient NPC

A systematic approach to method development for gradient NPC is not as well developed as for RP-LC, partly because of the much less frequent application of gradient NPC. However, the general approach outlined in Figure 3.1 for RP-LC is appropriate for NPC as well. For isocratic NPC, columns packed with unbonded silica are often used. For gradient separations, however, polar-bonded columns, such as cyano or diol, are preferred. Retention with silica columns can be strongly affected by small changes in water concentration (which are difficult to avoid [1]), and this problem is further exacerbated in gradient elution. The selection of the mobile phase depends on the wavelength to be used for UV detection, since many NPC solvents are not transparent below 250 nm. For detection at wavelengths as low as 215 nm, hexane as A solvent can be used with methyl-t-butyl ether (MTBE) as B solvent [46]. Samples with a wide retention range may require a more polar B solvent, in which case both solvent demixing (Section 9.2) and solvent immiscibility are potential problems; for example, the nonpolar A solvent hexane and the very polar B solvent acetonitrile cover a wide range in elution strength, but are immiscible. The addition of small amounts of a co-solvent such as MTBE can result in miscible mixtures of hexane (A) and acetonitrile (B) for a wide range of %B. See also the method development procedure of Jandera for gradient NPC [47].

Relay gradient elution (RGE) with a diol column has been proposed as a way of dealing with samples that contain both nonpolar and very polar solutes [48], that is, a very wide retention range for the sample. For such samples, the use of a binary mobile phase can be impractical, because the combination of a nonpolar A solvent (e.g., hexane) with a very polar B solvent (e.g., water) would necessarily involve both solvent demixing and solvent immiscibility. RGE begins with a nonpolar to moderately polar gradient step (e.g., hexane to ethylacetate), followed by a second step of increasing polarity (ethylacetate to acetonitrile), followed by a final step to a maximally polar B solvent (acetonitrile to water). As a result, each gradient step avoids problems which would occur if a gradient from hexane to water were used. RGE is seldom needed, however, because most samples do not exhibit the very wide range in polarity and NPC retention that requires the use of RGE.

Following an initial experiment, as described above (e.g., a hexane to MTBE gradient), further improvements in resolution can be sought by first varying selectivity. As in RP-LC (Section 3.3.2), changes in both gradient time and temperature are convenient for initial exploration. Further changes in selectivity can be achieved by varying the B solvent; for example, methylene chloride or ethyl acetate. Different polar solvents can be classified according to their selectivity as *nonlocalizing, basic localizing*, and *nonbasic localizing* [1, 49, 50]. Nonlocalizing (less polar) B solvents include halogenated alkanes, such as methylene chloride or chloroform; basic localizing solvents include aliphatic ethers, such as MTBE; nonbasic localizing solvents include aliphatic esters, ketones, and nitriles. Varying the proportions of methylene chloride, MTBE, and acetonitrile in the B solvent can provide a considerable control over selectivity [1, 50], in turn leading to maximum sample resolution. Following the adjustment of selectivity for adequate resolution, changes in column length and flow rate can be used for further improvements in either resolution or run time.

8.3.3 Hydrophilic Interaction Chromatography

HILIC, which can be regarded as normal-phase chromatography with an aqueous–organic mobile phase [1, 51–53], can be used to address certain limitations of RP-LC:

- samples that contain very polar compounds that are not retained adequately in RP-LC;
- samples with limited solubility in water or mobile phases rich in water;
- a need for a change in separation selectivity.

Very polar compounds can elute near t_0 in RP-LC, which may prevent their resolution and reliable quantitation; their strong retention in HILIC overcomes this problem. If a sample is not adequately soluble in mobile phases that are predominantly water, this may prevent the use of gradient RP-LC for its separation. In the case of HILIC, the gradient usually begins with only 3–10 percent water, so that dissolution of the sample in nearly-pure organic solvent becomes feasible. Finally, relative retention often changes dramatically in HILIC vs RP-LC, allowing the separation of peaks which might overlap in RP-LC; see the example of Figure 8.16.

HILIC can be a very useful complement to RP-LC, an observation which needs to be emphasized. The present section will therefore address this technique in somewhat greater detail. The advantages of HILIC (compared with NPC, Section 8.3) include (a) its relative freedom from slow-equilibration and solvent-demixing problems, (b) its avoidance of commonly used NPC solvents which can limit detection or sample solubility, and (c) its operational similarity to RP-LC. HILIC separations can be carried out with hydrophilic columns such as unbonded-silica (used less often), diol-silica, or amide-bonded silica. Poly-2-hydroxyethyl aspartamide (polyhydroxyethyl A; PolyLC, Columbia, MD, USA) is one of the more widely used HILIC columns at present, comparable to C_{18} for RP-LC.

It should be noted that the essence of HILIC separation is the use of a relatively polar bonded phase with water-containing mobile phase. Sample retention decreases as mobile phase polarity (or percentage water) increases. Consequently, water–acetonitrile gradients are carried out in "reverse" fashion, with percentage water *increasing* during the gradient. Polar molecules tend to be retained more strongly in HILIC, whereas the reverse is true in RP-LC. Moreover, changes in relative retention or selectivity can be pronounced for HILIC vs RP-LC, as

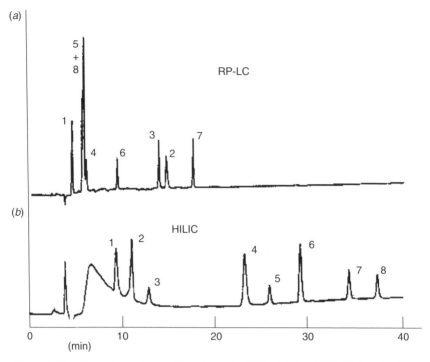

Figure 8.16 Separation of a peptide mixture by RP-LC (*a*) and HILIC (*b*). Conditions: 250 × 4.6 mm columns; 1.0 mL/min; 0.1 percent TFA added to water; (*a*) C_{18} column; 5–55 percent acetonitrile–water in 83 min; (*b*) HILIC column (TSK gel Amide-80); 3–45 percent water–acetonitrile in 70 min. Reprinted with permission from [55].

seen in the example of Figure 8.16 for the separation of a peptide mixture by each procedure. HILIC has been used for the separation of carbohydrates [51, 52], peptides [53–57], proteins [58], oligonucleotides [51], and miscellaneous small molecules [59, 60]. Gradient elution with a more polar column and typical RP-LC conditions (e.g., a gradient from buffer to organic) is sometimes (incorrectly) referred to as HILIC; no further discussion of the latter pseudo-HILIC separations will be offered.

HILIC retention can be described by an equation of the same form (log–log) as Equation 8.8 for NPC:

$$\log k = \log k_{ACN} - m_{HILIC} \log \phi_{H_2O} \qquad (8.12)$$

Here, k_{ACN} is the value of k for ACN as mobile phase, ϕ_{H_2O} is the volume-fraction of water in the mobile phase, and m_{HILIC} is the slope of plots of $\log k$ vs ϕ_{H_2O}. An example of the validity of Equation (8.12) is shown in Figure 8.17(*a*) for the separation of several peptides. A linear-gradient HILIC separation of the peptide sample of Figure 8.17(*a*) is illustrated in Figure 8.17(*b*). Equations for HILIC retention comparable to Equations (8.9) and (8.10) for NPC are applicable, where k_{ACN}

replaces c and m_{HILIC} replaces m:

$$t_R = \{\phi_{\text{H}_2\text{O},o}{}^{m_{\text{HILIC}}+1} + [V_m k_{\text{ACN}}(m_{\text{HILIC}} + 1) \times \Delta\phi_{\text{H}_2\text{O}}]/t_G\}^{1/(m_{\text{HILIC}}+1)} \quad (8.13)$$
$$- (\{\phi_{\text{H}_2\text{O},o} t_G/\Delta\phi_{\text{H}_2\text{O}}) + t_o + t_D$$
$$k^* = (t_G F)/1.15[V_m m_{\text{HILIC}} \log(\phi_{\text{H}_2\text{O},f}/\phi_{\text{H}_2\text{O},o})] \quad (8.14)$$

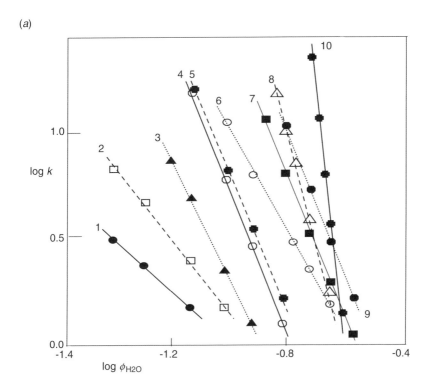

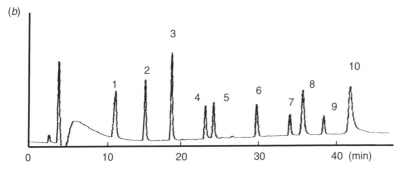

Figure 8.17 HILIC separation of a mixture of peptides. Conditions: 250 × 4.6 mm TSK gel Amide-80 column; 1.0 mL/min; 40°C. (*a*) Isocratic retention as a function of $\phi_{\text{H}_2\text{O}}$; (*b*) gradient separation for 3–45 percent water in 70 min. Adapted from [56].

Here, $\phi_{H_2O,o}$ and $\phi_{H_2O,f}$ refer to the values of ϕ_{H_2O} at the beginning or end of the gradient; $\Delta\phi_{H_2O}$ refers to the difference in values of ϕ_{H_2O} at the end and beginning of the gradient. The relative effect of a change in gradient conditions on linear-gradient HILIC separations can be predicted from corresponding changes in k^* [Equation (8.14)], as summarized in Figure 8.12. While *relative* changes in k^* can be estimated for HILIC from Equation (8.14), it is at present not possible to predict reliable values of m_{HILIC} for different sample compounds. The above discussion of NPC retention suggests that m is approximately equal to the number of polar groups n within a solute molecule, whereas the discussion of Section 6.2.2.3 suggests that m_{HILIC} is related to the number n of polar substituent groups in the sample molecule by

$$m_{HILIC} \approx n^{0.6} \qquad (8.15)$$

which for small-molecule samples (molecular weight <1000) suggests a value of m_{HILIC} of $1-3$, similar to values of m for other NPC separations. Equation (8.15) may be a consequence of variable *localization* of polar groups in the sample molecule [42, 45].

8.3.3.1 Method Development for Gradient HILIC

Method development for the HILIC separation of a small-molecule sample can be carried out in similar fashion to that for RP-LC or NPC. A $150-250 \times 4.6$ mm HILIC column is first selected, for example, Polyhydroxyethyl A (PolyLC, Columbia, MD). An initial gradient is carried out from 5 to 50 percent B, where the A solvent is acetonitrile and the B solvent is water; 0.1 percent of an acidic buffer (trifluoroacetic, acetic, or formic acid) is added to each solvent. An initial gradient time can be estimated from Equation (8.11), assuming $k^* = 5$, $m \equiv m_{HILIC} \approx 2$, and a flow rate of 2.0 mL/min (for the separation of samples with molecular weights >1000, see the discussion in Section 6.2.2.3).

Following the completion of the latter linear-gradient experiment, separation can be improved by changes in selectivity. Little has been reported concerning the

TABLE 8.5 Conditions Which Can be Varied in Order to Change Selectivity in HILIC, Arranged in Approximately Decreasing Promise

Condition	Comment	Reference
Gradient time (k^*)	Requires different values of m_{HILIC} for adjacent peaks; several samples have shown such differences	51, 56
pH	Changes in sample ionization will have a profound effect on HILIC retention; greater sample ionization means stronger retention; ammonium acetate is a useful buffer for the control of pH in HILIC	
Buffer type	Trifluoroacetic, acetic, and formic acids can each provide a differing selectivity	55
Column type	Minor differences seen for amide, diol, and silica columns	57
B solvent	Water can be replaced by either methanol or ethanol	54

best way to change retention order and selectivity in HILIC, but some options are summarized in Table 8.5. The usual changes in column length and flow rate are also available as means for increasing resolution or decreasing run time, while holding k^* constant [Equation (8.14)].

For a further discussion of gradient elution based on IEC or NPC, see [1, 32, 61, 62].

8.4 TERNARY- OR QUATERNARY-SOLVENT GRADIENTS

RP-LC gradients usually involve water or buffer as A solvent, and acetonitrile, methanol, or occasionally another organic solvent as B solvent, that is, so-called binary gradients that are based on a single organic solvent. It is also possible to use a mixture of two or three organic solvents as the B solvent, primarily as a means of achieving additional control over selectivity. Such ternary- or quaternary-solvent gradients were first reported in the early 1980s as a means of maximizing relative retention and resolution [61, 63]. Since that time, other (less complicated) means have been reported for the control of selectivity in gradient RP-LC, so ternary- and quaternary-solvent gradients are used infrequently today.

> Any intelligent fool can make things ... more complex ... It takes a touch of genius – and a lot of courage – to move in the opposite direction.
>
> —E.F. Schumacher

Two kinds of multisolvent gradients have been described, designated [62] as "elution strength" gradients, and "selectivity" gradients. In "elution strength" gradients, the B solvent is a *constant* mixture of two or three organic solvents, usually chosen from acetonitrile, methanol, and/or tetrahydrofuran. By varying the proportions of these three solvents in the B solvent, changes in "solvent selectivity" are achieved which are analogous to the use of quaternary-solvent mobile phases in isocratic elution [64]. An example of an "elution strength" gradient is shown in Figure 8.18(*c*). Figure 8.18(*a* and *b*) shows the separation of a 14-component mixture by means of an acetonitrile/water gradient in (*a*) and a tetrahydrofuran/water gradient in (*b*). Neither of the latter separations provides acceptable resolution; $R_s = 0.6$ in (*a*) and 0.4 in (*b*) (arrows mark critical peak pairs). However, mixtures of acetonitrile and tetrahydrofuran result in separations of intermediate selectivity (relative retention), with a better chance of obtaining a satisfactory resolution. As seen in the separation of Figure 8.18(*c*), which uses a mixture of 60 percent acetonitrile plus 40 percent tetrahydrofuran as B solvent, baseline resolution is achieved ($R_s = 1.6$).

"Selectivity" gradients refer to a change in the composition of the B solvent *during* the gradient, in order to achieve a desired peak spacing. An example of such a gradient is shown in Figure 8.19(*c*). In Figure 8.19(*a*), for an acetonitrile–water gradient, peaks 8 and 9 are seen to overlap. Similarly, for a methanol–water gradient (Fig. 8.19*b*), peaks 2 and 3 are poorly resolved. With the "selectivity" gradient of Figure 8.19(*c*), the B solvent is initially rich in ACN, which favors the

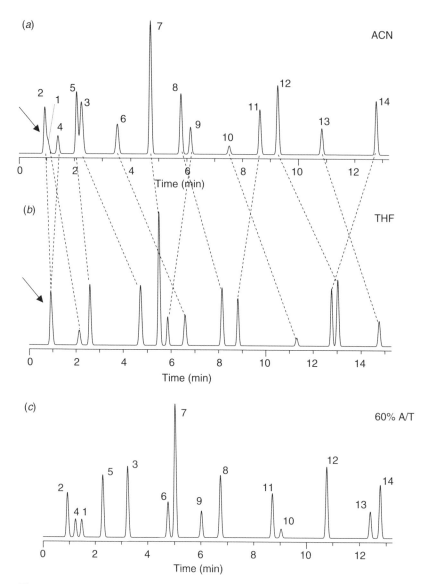

Figure 8.18 Example of an "elution strength" gradient. (*a*) Separation with a 17–84 percent acetonitrile–water gradient in 20 min; (*b*) separation with a 12–59 percent tetrahydrofuran–water gradient in 20 min; (*c*) separation with a 14–69 percent B "elution strength" gradient in 20 min, where the B solvent is a mixture of 60 percent acetonitrile plus 40 percent tetrahydrofuran. The sample is a mixture of diverse organic compounds (see [63] for peak numbering). Other conditions: 15 × 0.46 C_{18} column; 3 mL/min; 35°C. Recreated from data in [63].

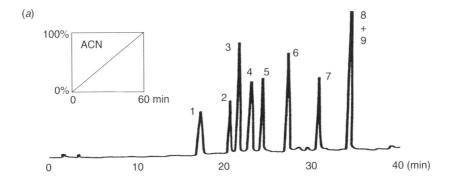

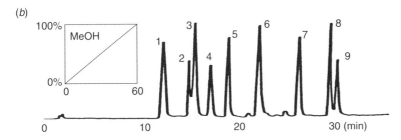

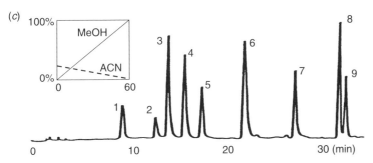

Figure 8.19 Example of a "selectivity" gradient. (*a*) Separation with 0–100 percent acetonitrile–water in 60 min; (*b*) 20–100 percent methanol–water gradient in 60 min; (*c*) "selectivity" gradient of 0–100 percent B in 60 min; B solvent varies as shown in (*c*). The sample is a mixture of phenols. Other conditions: 300 × 4.2 mm C_{18} column; 1.0 mL/min; ambient temperature. Reprinted with permission from [61].

separation of peaks 2 and 3. As the gradient progresses, however, the B solvent becomes largely methanol, which favors the separation of peaks 8 and 9 near the end of the gradient. The result of the "selectivity" gradient of (*c*) is to provide an overall better separation than either of the binary-solvent gradients of (*a*) or (*b*).

Whether using either "elution strength" or "selectivity" gradients, the effect of further changes in gradient conditions on separation will be similar to the examples of Figure 8.12.

REFERENCES

1. L. R. Snyder, J. J. Kirkland, and J. L. Glajch, *Practical HPLC Method Development*, 2nd edn, Wiley-Interscience, New York, 1997.
2. M. C. McMaster, *LC/MS: a Practical User's Guide*, Wiley, New Jersey, 2005.
3. X. Xu, J. Lan, and W. A. Korfmacher, *Anal. Chem.* 77 (2005) 389A.
4. K. Levsen, H.-M. Schiebel, B. Behnke, R. Dötzer, W. Dreher, M. Elend, and H. Thiele, *J. Chromatogr. A* 1067 (2005) 55.
5. A. A. M. Stolker, U. A. Th. Brinkman, *J. Chromatogr. A* 1067 (2005) 15.
6. M. Petrović, M. D. Hernando, M. S. Díaz-Cruz, and D. Barceló, *J. Chromatogr. A* 1067 (2005) 1.
7. Y. Ishihama, *J. Chromatogr. A* 1067 (2005) 73.
8. *Guidance for Industry: Bioanalytical Method Validation*, US Food and Drug Administration (May 2001); available at www.fda.gov/cder/guidance/index.htm.
9. R. Willoughby, E. Sheehan, and S. Mitrovich, *A Global View of LC/MS*, Global View Publishing, Pittsburgh, PA, 1998.
10. R. B. Cole, ed., *Electrospray Ionization Mass Spectrometry: Fundamentals, Instrumentation, and Applications*, Wiley, New York, 1997.
11. A. Apffel, S. Fishcher, G. Goldberg, P. C. Goodley, and F. E. Kuhlman, *J. Chromatogr. A* 712 (1995) 177.
12. F. E. Kuhlmann, A. Apffel, S. M. Fischer, G. Goldberg, and P. C. Goodley, *J. Am. Soc. Mass Spectrom.* 6 (1995) 1121.
13. H. R. Liang, R. L. Foltz, M. Meng, and P. Bennett, *Rapid Commun. Mass Spectrom.* 17 (2003) 2815.
14. D. A. Wells, *High Throughput Bioanalytical Sample Preparation: Methods and Automation Strategies*, Elsevier, Amsterdam, 2003.
15. R. E. Majors, *Sample Prep. Perspectives*, a periodic column in *LCGC*.
16. Waters Corporation; www.waters.com, sample preparation section.
17. Varian Inc.; www.varianinc.com, sample preparation section.
18. C. Polson, P. Sarkar, B. Incledon, V. Raguvaran, and R. Grant, *J. Chromatogr. B* 785 (2003) 263.
19. M. S. Alexander, M. M. Harrington, T. Culley, J. R. Kern, J. W. Dolan, J. D. McChesney, J. Zygmunt, and S. J. Bannister, *J. Chromatogr. B* 785 (2003) 253.
20. J. L. Little, M. F. Wempe, and C. M. Buchanan, *J. Chromatogr. B* 833 (2006) 219.
21. G. Hendriks, J. P. Franke, and D. R. A. Uges, *J. Chromatogr. A* 1089 (2005) 193.
22. R. Bonfiglio, R. C. King, T. V. Olah, and K. Merkle, *Rapid Commun. Mass Spectrom.* 13 (1999) 1175.
23. M. D. Nelson and J. W. Dolan, *LCGC* 22 (2004) 338–343.
24. C. T. Mant and R. S. Hodges, eds, *High-performance Liquid Chromatography of Peptides and Proteins*, CRC Press, Boca Raton, FL, 1991.
25. K. M. Gooding and F. E. Regnier, eds, *HPLC of Biological Macromolecules*, 2nd edn, Marcel Dekker, New York, 2002.
26. P. R. Haddad and P. E. Jackson, *Ion Chromatography – Principles and Applications*, Elsevier, Amsterdam, 1990.
27. N. Tanaka, H. Kimura, D. Tokuda, K. Hosoya, T. Ikegami, N. Ishizuka, H. Minakuchi, K. Nakanishi, Y. Shintani, M. Furuno, and K. Cabrera, *Anal. Chem.* 76 (2004) 1273.
28. M. Vollmer, R. Horth, and E. Nagele, *Anal. Chem.* 76 (2004) 5180.
29. D. R. Stoll and P. W. Carr, *J. Amer. Chem. Soc.* 127 (2005) 5034.
30. Y. Baba and G. Kura, *J. Chromatogr. A* 550 (1991) 5.
31. J. E. Madden and P. R. Haddad, *J. Chromatogr. A* 850 (1999) 29.
32. P. Jandera and J. Churáček, *Gradient Elution in Column Liquid Chromatography*, Elsevier, Amsterdam, 1985.
33. E. S. Parente and D. B. Wetlaufer, *J. Chromatogr.* 355 (1986) 29.
34. S. R. Gallant, S. Vunnum, and S. M. Cramer, *J. Chromatogr. A* 725 (1996) 295.
35. M. A. Quarry, R. L. Grob, and L. R. Snyder, *Anal. Chem.* 58 (1986) 907.
36. T. Sasagawa, Y. Sakamoto, T. Hirose, T. Yoshida, Y. Kobayashi, and Y. Sato, *J. Chromatogr.* 485 (1989) 533.
37. G. Jilge, K. Unger, U. Esser, H.-J. Schafer, G. Rathgeber, and W. Muller, *J. Chromatogr.* 476 (1989) 37.

38. R. W. Stout, S. I. Sivakoff, R. D. Ricker, and L. R. Snyder, *J. Chromatogr.* 353 (1986) 439.
39. Y. Baba and M. K. Ito, *J. Chromatogr.* 485 (1989) 647.
40. P. Jandera and M. Kučerová, J. Chromatogr. A, 759 (1997) 13.
41. P. Jandera, *J. Chromatogr. A* 965 (2002) 239.
42. L. R. Snyder, *Principles of Adsorption Chromatography*, Marcel Dekker, New York, 1968.
43. L. R. Snyder, *Anal. Chem.* 46 (1974) 1384.
44. L. R. Snyder and T. C. Schunk, *Anal. Chem.* 54 (1982) 1764.
45. L. R. Snyder, in *High-performance Liquid Chromatography: Advances and Perspectives*, Vol. 3, Cs. Horváth, ed., Academic Press, New York, 1983, p. 157.
46. V. R. Meyer, *J. Chromatogr.* 768 (1997) 315.
47. P. Jandera, *J. Chromatogr. A* 797 (1998) 11.
48. L. R. Treiber, *J. Chromatogr. A* 696 (1995) 193.
49. L. R. Snyder, J. L. Glajch, and J. J. Kirkland, *J. Chromatogr.* 218 (1981) 299 .
50. J. L. Glajch, J. J. Kirkland, and L. R. Snyder, *J. Chromatogr.* 239 (1982) 268.
51. A. Alpert, *J. Chromatogr.* 499 (1990) 177.
52. S. C. Churms, *J. Chromatogr. A* 720 (1996) 75.
53. A. J. Alpert, M. Shukla, A. K. Shukla, L. R. Zieske et al., *J. Chromatogr. A* 676 (1994) 191.
54. T. Yoshida, *J. Biochem. Biophys. Methods.* 60 (2004) 265.
55. T. Yoshida, *Anal. Chem.* 69 (1997) 3038.
56. T. Yoshida, *J. Chromatogr. A* 811 (1998) 61.
57. T. Yoshida and T. Okada, *J. Chromatogr. A*, 840 (1999) 1.
58. S. W. Taylor, J. H. Waite, M. M. Ross, J. Shabanowitz, and D. F. Hunt, *J. Am. Chem. Soc.* 116 (1994) 10803.
59. M. A. Strege, *Anal. Chem.* 70 (1998) 2439.
60. B. A. Olsen, *J. Chromatogr. A* 913 (2001) 113.
61. P. Jandera, J. Churáček, and H. Colin, *J. Chromatogr.* 214 (1981) 35.
62. P. Jandera, *Adv. Chromatogr.* 43 (2004) 1.
63. J. J. Kirkland and J. L. Glajch, *J. Chromatogr.* 255 (1983) 27.
64. J. L. Glajch, J. J. Kirkland, K. M. Squire, and J. M. Minor, *J. Chromatogr.* 199 (1980) 57.

CHAPTER 9

THEORY AND DERIVATIONS

> Is man an ape or an angel? I, my lord, I am on the side of the angels. I repudiate with indignation and abhorrence those newfangled theories.
>
> —Benjamin Disraeli, Speech at Oxford Diocesan Conference (1864)

9.1 THE LINEAR SOLVENT STRENGTH MODEL

The present chapter provides a more detailed account of the theory of gradient elution, with emphasis on the linear solvent strength (LSS) model. Several relationships presented in earlier chapters will be derived, more complete and/or detailed forms of these relationships will be described, and a number of related topics will be examined. Unless noted otherwise, reversed-phase linear-gradient elution will be assumed. See the treatments of [1–5] for additional background and details.

The development of a theory of separation by gradient elution began in the mid-1950s, with the derivation of a fundamental equation for gradient retention [6–8]. The basis of this derivation is a model which assumes the passage of differential volume elements dV of mobile phase (of fixed composition) *through* the band center as the band migrates through the column (equivalent to a multistep gradient with an infinite number of isocratic steps). For isocratic elution, the retention volume V_R is given as $V_m(1+k)$, where V_m is the column dead-volume and k is the retention factor. However, the total volume of mobile phase passing through the band center is the corrected retention volume $V'_R = (V_R - V_m) = V_m k$. Therefore, after passage of an infinitesimal volume element dV through the band center, the band will have moved a fractional distance $dx = dV/(V_m k)$ through the column. We can sum these fractional migrations dx, with $\Sigma dx = 1$ when $\Sigma dV = V'_R$; or,

$$\int_0^{V'_R} (1/V_m)(dV/k_i) = 1 \tag{9.1}$$

V is the cumulative volume of mobile phase that has entered the column at any time t after the start of the gradient (and passed through the band center), and k_i is the instantaneous value of k for the solute band (for each volume element dV) during its migration through the column. A qualitative illustration of Equation (9.1) is

---

High-Performance Gradient Elution. By Lloyd R. Snyder and John W. Dolan
Copyright © 2007 John Wiley & Sons, Inc.

provided by Figure 1.5 and the accompanying text (Section 1.4.1). Equation (9.1) can be restated in terms of time (rather than volume) units:

$$\int_0^{t'_R} (1/t_0)(dt/k_i) = 1 \qquad (9.2)$$

Here, $t_R' = t_R - t_0 = t_0 k$ is the corrected retention time, t_R is the retention time, and t_0 is the column dead time. Equation (9.2) recognizes that the fractional migration of a band (dx) during a time increment dt is simply dt divided by the corrected retention time t_R'. The derivation of Equation (9.2) is analogous to that of Equation (9.1).

Unless noted otherwise, we will initially assume:

- Linear gradients where the concentration ϕ of the B solvent leaving the gradient mixer and entering the column is given by

$$\phi = \phi_0 + (\Delta\phi/t_G)t \qquad (9.3)$$

The quantity ϕ_0 is the value of ϕ (volume-fraction of B solvent, equal to 0.01 percent B) at the start of the gradient, $\Delta\phi$ is the change in ϕ during the gradient, t_G is gradient time, and t refers to time after initiation of the gradient (Fig. 1.4).

- Isocratic retention as a function of ϕ can be approximated by

$$\log k = \log k_w - S\phi \qquad (9.4)$$

Equation (9.4) has been described previously [Equation (2.9)] and is examined further in Sections 9.2.1 and 9.4 below. For normal-phase and ion-exchange chromatography, the dependence of k on mobile phase composition is described by a different general relationship (Sections 8.2 and 8.3), leading to non-LSS retention for a linear gradient.

Equations (9.3) and (9.4) combine to give the value of k at the column inlet as a function of time $(k_{0,t})$ (regardless of the actual position of the band in the column)

$$\log k_{0,t} = \log k_w - S(\phi_0 + \Delta\phi t/t_G)$$
$$= (\log k_w - S\phi_0) - (S\Delta\phi/t_G)t \qquad (9.5)$$

Because k_w, S, ϕ_0, $\Delta\phi$, and t_G are constant for a given solute and defined linear-gradient conditions,

$$\log k_{0,t} = (\text{constant}) - (\text{constant})t \qquad (9.6)$$

By definition, gradient separations which can be described (even approximately) by Equation (9.6a) are said to exhibit LSS behavior. Note that values of $k_{0,t}$ and t from Equation (9.6) apply for each volume element dV in Equation (9.1).

It is important to distinguish LSS *behavior* from the LSS *model*, and also to distinguish qualitative from quantitative relationships that describe gradient separation. Thus, although ion-exchange and normal-phase chromatography with linear gradients are not described quantitatively by Equation (9.6), *nonlinear* gradients of appropriate shape can be visualized which would provide a close fit to

Equation (9.6) (and therefore LSS behavior) for either IEC or NPC. Alternatively, while Equation (9.6) is a relatively poor approximation for some linear-gradient separations that involve normal-phase or ion-exchange chromatography, the LSS model can still provide a useful (if qualitative) basis for understanding and predicting changes in these separations as a function of gradient conditions (Sections 8.2 and 8.3). Similarly, separations which obey Equation (9.4) (i.e., reversed-phase) may involve nonlinear gradients of various kinds, yet these separations can also be understood qualitatively by means of the LSS model. A number of such examples are provided in Chapter 2. Thus, the LSS model can be applied to all of gradient elution, even if only approximately for some separations.

9.1.1 Retention

We begin by rearranging Equation (9.5), which expresses isocratic retention as a function of mobile phase composition ϕ during the gradient:

$$\log k_{0,t} = (\log k_w - S\phi_0) - (t_0 S \Delta\phi / t_G)(t/t_0)$$
$$= \log k_0 - b(t/t_0) \qquad (9.7)$$

where $\log k_0 = \log k_w - S\phi_0$, and k_0 is the value of k at the start of the gradient (at time zero, for $\phi = \phi_0$); b is a fundamental measure of gradient steepness (so-called "intrinsic gradient steepness"; see the discussion preceding Equation (1.5) for a "conceptual" appreciation of b)

$$b = t_0 S \Delta\phi / t_G \qquad (9.8)$$
$$= V_m \Delta\phi S / (t_G F) \qquad (9.8a)$$

Note that some workers [2, 5, 9–11] use different symbols for Equation (9.8), as well as natural logarithms in Equation (9.4),

$$\ln k = \ln k_w - S\phi \qquad (9.9)$$

The use of Equation (9.9) leads to values of k_w, S and b that are 2.3 times greater than in Equation (9.4). See the Glossary of Symbols at the front of this book for symbols used in alternative versions of Equation (9.8), as well as their relation to corresponding symbols used in this book.

The insertion of Equation (9.7) into Equation (9.2) (with $k_i \equiv k_{0,t}$) then yields the gradient retention time t_R as a function of gradient condition [12]:

$$t_R = (t_0/b) \log(2.3 k_0 b + 1) + t_0 \qquad (9.10)$$

Equation (9.10) does not take into account the dwell volume V_D of the gradient equipment (Chapter 4 and Section 9.2.3). If it is assumed that k_0 is large (often a reasonable approximation in gradient elution), and a significant dwell volume exists, Equation (9.10) becomes

$$t_R = (t_0/b) \log(2.3 k_0 b) + t_0 + t_D \qquad (9.11)$$

where t_D is the dwell ("gradient delay") time, equal to V_D/F (Section 2.3.6, Chapter 4). When k_0 is large for all solutes, the effect of an increase in gradient dwell volume is simply to increase the retention time of each peak by t_D, with no other change in the separation; that is, the entire chromatogram appears to be shifted to the right by t_D min. Note also that if the gradient volume $V_G \equiv t_G F$ is maintained constant for a given column (for changes in t_G or F), values of t_R/t_G and relative retention remain unchanged [13].

When k_0 is small and $V_D \neq 0$, Equation (9.11) must be modified as described in Section 9.1.1.2 [Equation (9.24)]. Regardless of the value of k_0, however, Equation (9.11) is usually an acceptable approximation for practical use (as in Chapters 2 and 3). Given values of t_R for two gradient runs where only b is varied (usually by varying gradient time t_G), it is possible to determine values of $\log k_w$ and S for each solute (Section 9.3.3).

Because values of k (or k_i) for a band in gradient elution vary as the band migrates through the column, it will prove useful to define a *median* or "equivalent" value of k for the entire separation. This median value of k (k^*) can be defined as the value of k when the band has migrated half way through the column For migration of the band some fraction r of the column length, the integral of Equation (9.2) can be set equal to r, rather than 1, which then yields a fractional retention time

$$t_R = (t_0/b) \log(2.3 k_0 br + 1) + rt_0 + t_D \tag{9.12}$$

which for large k_0 gives

$$t_R \approx (t_0/b) \log(2.3 k_0 br) + rt_0 + t_D \tag{9.12a}$$

(see the further discussion in Section 9.1.1.2 of k^* when k_0 is small). A value of k as a function of migration distance r can be obtained from Equations (9.7) and (9.12a); ($t_R - rt_0 - t_D$) corresponds to the time t in Equation (9.7), so k after elution of the band a fractional distance r through the column is given as

$$k = 1/2.3br \tag{9.13}$$

For $r = 0.5$ (at the column midpoint),

$$k \equiv k^* = 1/(1.15b) \tag{9.14}$$

Similarly, the value of k when the peak elutes from the column (i.e., for $r = 1$) is

$$k_e = 1/2.3b \equiv k^*/2 \tag{9.14a}$$

Note that prior to 1996, gradient retention k^* was represented by the symbol $\bar{k}$.

Resolution is usually the primary aim of method development and the improvement of separation; as will be seen, Equation (9.14) provides an important link between separation in "corresponding" isocratic and gradient separations. By "corresponding" separations, it is understood that all separation conditions are the same (sample, column, flow rate, temperature, the B-solvent, pH, etc.); however, ϕ is constant for isocratic elution, while varying in gradient elution. As we will see (Section 9.1.3), the resolution of two adjacent peaks in "corresponding" isocratic and gradient separations will be similar, if k^* (gradient) $\approx k$ (isocratic). This equivalence of isocratic and gradient separation enables the chromatographer to use the same strategy for the optimization of resolution and the development of both isocratic and gradient methods. For most people, isocratic separation is easier to understand, so the use of Equation (9.14) renders gradient separation more readily comprehensible, and gradient methods easier to develop. Because k^* is approximately constant for different peaks in a gradient separation, but k varies in isocratic elution, k and k^* can only be equal for two closely adjacent peaks. However, method development, for either isocratic or gradient elution, usually emphasizes just one pair of adjacent peaks (the "critical" pair). See the further discussion of gradient method development in Section 3.3.

9.1.1.1 Gradient and Isocratic Retention Compared
Given that there exists a gradient retention factor k^* which is chromatographically equivalent to k for a "corresponding" isocratic separation (in terms of resolution), this implies that Equation (9.4) also applies for gradient elution:

$$\log k^* = \log k_w - S\phi^* \quad (9.15)$$

where $\log k_w$ and S have the same values for both isocratic and gradient elution, and ϕ^* is the value of ϕ corresponding to k^* when the band is at the column midpoint. It is possible to determine values of $\log k_w$ and S from two gradient runs (Section 9.3.3), which enables the calculation of k^* for any gradient run as a function of separation conditions [Equations (9.8a) and (9.14)]:

$$k^* = (t_G F)/(1.15 V_m \Delta \phi S) \quad (9.16)$$

Values of ϕ^* for associated values of k^* can be obtained as follows. First, the value of ϕ at elution (ϕ_e) is known from Equation (9.3), noting that t at this time is equal to $t_R - t_0 - t_D$. From Equations (9.14a) and (9.15) we can determine the quantity $\phi^* - \phi_e = (\log 2)/S \equiv 0.3/S$. From Equation (9.3) and the latter relationship, we then have

$$\phi^* = \phi_0 + (\Delta\phi/t_G)[t_R - t_0 - t_D - 0.3(t_0/b)] \quad (9.16a)$$

The proposal that isocratic and gradient retention [as expressed by Equations (9.4) and (9.15)] are equivalent for corresponding separations can be tested by comparing values of $\log k$ vs ϕ (isocratic) with values of $\log k^*$ vs ϕ^* (gradient), using overlapping plots of the two data sets. If these two relationships [Equations (9.4) and (9.15)] are valid with the same values of $\log k_w$ and S for each solute, all data points (both isocratic and gradient) should overlap the same, approximately straight, best-fit line. Examples of such a test are shown in Figure 9.1 for several different solutes and separation conditions. Figure 9.1(a) shows data for a small molecule (dimethoxybenzophenone), Figure 9.1(b) plots data for the small protein insulin (molecular weight 9 kDa), and Figure 9.1(c) compares retention data for a 50 kDa polystyrene fraction. The deviation of the isocratic (solid curve) and gradient (dashed curve) data in (c) corresponds to a shift in ϕ of only 0.0015 units (0.15 percent by volume B), which might reasonably be attributed to small errors in the delivery of the gradient pumping system, solvent demixing, or other second-order effects discussed in Section 9.2.

Figure 9.1(d) differs from the other three plots, as these data are for the separation of two proteins by *ion-exchange chromatography*; a linear salt-gradient was used for the gradient data. These data are plotted as $\log k$ vs $\log C$ (or $\log k^*$

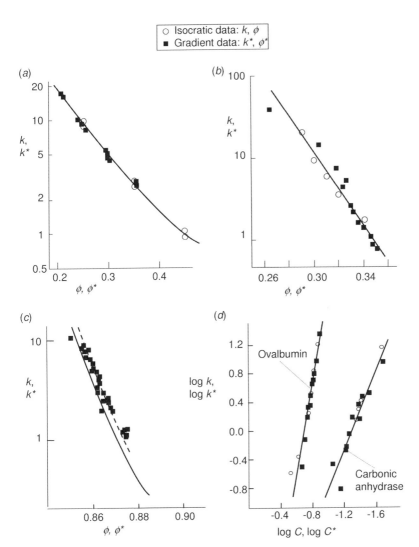

Figure 9.1 Comparison of retention vs mobile phase composition for gradient (k^* vs ϕ^*) and isocratic (k vs ϕ) separation. (a) 4,4'-Dimethoxybenzophenone, acetonitrile/water mobile phases [14]; (b) insulin, acetonitrile–pH 2 buffer mobile phases [15]; (c) 50 Da polystyrene, tetrahydrofuran–water mobile phases [16] (isocratic data points not shown; solid curve is best fit to data); (d) ovalbumin and carbonic anhydrase, sodium bromide in water mobile phase [17] (ion-exchange separation). Adapted from original figures.

vs log C^*), where C or C^* is the (varying) salt concentration of an aqueous mobile phase, rather than log k vs C – due to a differing form of the relationship in ion-exchange chromatography for k or k^* vs mobile phase composition [Equation (8.4)]. Again, data for isocratic retention k and gradient retention k^* fall on the same curve for these corresponding IEC separations. See Section 8.2 for a further discussion of IEC retention.

The ion-exchange retention of multiply charged solutes [as in Fig. 9.1(d)] can also be approximated (Section 8.2, [16]) by

$$\log k \approx a + cC \quad (9.17)$$

where a and c are constants for a given solute and salt (c is the absolute value of the charge on the solute ion divided by the charge on the salt counter-ion). When linear gradients are used, ion exchange separations which can be approximated by Equation (9.17) are also described quantitatively by LSS equations similar to those derived for reversed-phase separation.

Additional examples of the equivalence of Equations (9.2) and (9.15) (as in Fig. 9.1) have been reported [18–20]; a similar behavior of isocratic and gradient elution can be expected in most cases. It should be kept in mind, however, that very nonlinear plots of $\log k$ vs ϕ for reversed-phase isocratic elution have been observed occasionally [21, 22], and attributed to mixed-retention processes or changes in solute molecular conformation as a function of ϕ. Separation under these conditions cannot be interpreted readily in terms of the LSS model.

9.1.1.2 Small Values of k_0

Prior equations in this book for retention and separation in gradient elution have usually assumed strong sample retention at the start of the gradient, that is, large values of k_0. When k_0 is small, some of these prior equations become more approximate. In the absence of equipment dwell volume ($V_D = 0$), retention time is given by Equation (9.10), regardless of the value of k_0. The value of k^* for small k_0 can be derived [1, 4], by noting that t in Equation (9.7) corresponds to $t_R - rt_0 - t_D$ in Equation (9.12):

$$\text{(small } k_0\text{)} \quad k^* = 1/[1.15b + (1/k_0)] \quad (9.18)$$

For the same gradient steepness b, Equation (9.18) states that smaller values of k_0 correspond to smaller values of k^* (for the same value of b), and – assuming equal values of N and α – reduced resolution R_s [Equation (9.38)]. The reduction in R_s is relatively modest, however. For example, a 25 percent decrease in R_s for $k_0 = 2$, compared with $k_0 = 100$. Note for very flat gradients (b small) that $k^* \approx k_0$; that is, early peaks elute isocratically.

For $V_D > 0$ and k_0 small, retention time is no longer given exactly by Equation (9.11) because of the significant migration of early peaks during passage of the dwell volume through the column. The precise retention time can be derived using the representation of Figure 9.2, which shows a band at the column inlet at the time of sample injection (a), and its subsequent migration (b) a fractional distance Δx through the column until the dwell volume (with $\phi = \phi_0$) has passed through the band center. The total retention time t_R, taking the dwell volume into account, will be the time for the band to travel the distance Δx through the column ($t_{R,1}$), plus the time ($t_{R,2}$) for the band to travel the remaining distance ($1 - \Delta x$) through the column. We can regard the above example as equivalent to two connected columns (1 and 2) of fractional lengths Δx and $(1 - \Delta x)$, respectively [23], with $t_R = t_{R,1} + t_{R,2}$ (Fig. 9.2b).

The dead times for columns 1 and 2 in Figure 9.2 are $t_{01} = \Delta x t_0$ and $t_{02} = (1 - \Delta x)t_0$, respectively, where t_0 is the value for the total column. The retention

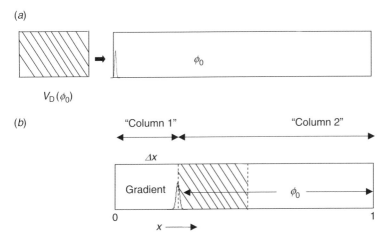

Figure 9.2 Modeling the effect of dwell volume on solute retention, when k_0 for the solute is small (the "two-column" analogy). (*a*) After sample injection; (*b*) after passage of a volume V_D of initial mobile phase ($\phi = \phi_0$) through the center of the solute band. See text for details.

time for column 1 can be written either as

$$t_{R,1} = t_{01} + t_D = \Delta x\, t_0 + t_D \tag{9.19}$$

or

$$t_{R,1} = t_{01}(1 + k_0) = \Delta x\, t_0(1 + k_0) \tag{9.20}$$

Solving for Δx from Equations (9.19) and (9.20),

$$\Delta x = t_D/(t_0 k_0) \tag{9.21}$$

The retention time for column 1 is then

$$t_{R,1} = (t_D/k_0) + t_D \tag{9.22}$$

The retention time for the second column can be calculated as follows. First, the value of b for column 2 will be $b_2 = t_{02}\Delta\phi S/t_G = (1 - \Delta x)b$ (where b refers to the value for the total column). Inserting b_2 and t_{02} into Equation (9.10), we have

$$\begin{aligned}
t_{R,2} &= (t_0/b)\log(2.3 k_0[1 - \Delta x]b + 1) + (1 - \Delta x)t_0 \\
&= (t_0/b)\log\{2.3 k_0 b[1 - (t_D/t_0 k_0)] + 1\} + [1 - (t_D/t_0 k_0)]t_0
\end{aligned} \tag{9.23}$$

The total retention time t_R is then the sum of $t_{R,1}$ [Equation (9.22)] plus $t_{R,2}$ [Equation (9.23)]:

$$\begin{aligned}
t_R &= (t_D/k_0) + t_D + (t_0/b)\log\{2.3 k_0 b[1 - (t_D/t_0 k_0)] + 1\} + [1 - (t_D/t_0 k_0)]t_0 \\
&= (t_0/b)\log\{2.3 k_0 b[1 - (t_D/t_0 k_0)] + 1\} + t_0 + t_D
\end{aligned} \tag{9.23a}$$

An equivalent derivation has been given by Schoenmakers [9]. Note that if $t_D \geq t_0 k_0$, the band elutes isocratically, with $t_R = t_0(1 + k_0)$.

Pre-elution (movement of the band before the gradient reaches the center of the migrating band) as discussed above can be contrasted with late elution (isocratically) after the gradient is complete; for late elution, the value of k_0 is necessarily large, so that Equation (9.11) applies. During a time $t_G + rt_0 + t_D$ for the end of the gradient to arrive at the center of the migrating band, the band will have migrated some fraction r of the column length [Equation (9.12)]. The peak will then be eluted isocratically from the column in a time $(1 - r) t_0(1 + k_f)$, where k_f is the value of k at the end of the gradient; the total retention time t_R is then given as $t_G + t_0 + t_D + (1 - r) k_f$. A value of r can be obtained from Equation (9.12), by replacing t_R by $(t_G - rt_0 - t_D)$ and solving for r. For further details, see [5, 9, 24, 25].

A value of k^* for the case where k_0 is small and V_D is significant has also been derived [25]:

$$\text{(small } k_0, V_D > 0) \quad k^* = k_0/\{2.3b[(k_0/2) - (V_D/V_m)] + 1\} \quad (9.24)$$

For k_0 large and $V_D = 0$, Equation (9.24) reduces to Equation (9.13). For k_0 small, and $V_D = 0$, Equation (9.24) reduces to Equation (9.18). As V_D increases from zero, the value of k^* increases, as discussed further in Section 3.3.3.1.

Retention for segmented gradients (Fig. 1.4d) can be calculated in similar fashion as in Figure 9.2. After the first gradient segment has passed through the band center, the band will have moved a fractional distance Δx through the column, following which the band will be eluted from the remainder of the column by the second gradient segment. This approach can be extended to gradients with additional segments.

9.1.2 Peak Width

In the following discussion, we distinguish "band" from "peak", corresponding respectively to before and after the solute leaves the column. Just prior to the isocratic elution of a band, the band will have a baseline width on the column that is equal to some fraction x of the column length. Assuming that the band front is exactly at the end of the column (Fig. 9.3a), the further time required to elute point i and the band from the column (the peak width W) will be given as $xt_0(1 + k)$, since $t_0(1 + k)$ is the peak retention time t_R, corresponding to $x = 1$ (this assumes no further broadening of the band during its elution from the column, nor any significant change in k during band elution; both assumptions are reasonable, because the band will be relatively narrow for reasonable values of N). Then, since $W = x\, t_0(1 + k)$, $t_R = t_0(1 + k)$, and $N \equiv 16(t_R/W)^2$,

$$N = 16/x^2$$

or

$$x = 4/N^{1/2} \quad (9.25)$$

The movement through the column of a band in either isocratic or gradient elution should result in the same bandwidth (x) on the column just prior to elution (Fig. 9.3a), assuming that N in isocratic elution can be approximated by N^*, the

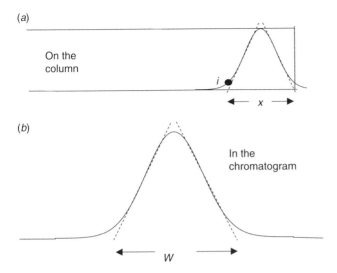

Figure 9.3 Peak width in gradient elution. Illustration of elution of a band from the column during gradient elution. (a) The band at the outlet end of the column; (b) the eluted band (after leaving the column).

value in gradient elution when the band has migrated half-way through the column; Equation (9.25) also assumes that gradient compression (Section 9.1.2.1) is ignored. Since the width W of the eluted peak (Fig. 9.3b) is the time required for point i in Figure 9.3(a) to leave the column, W in gradient elution is

$$\text{(gradient)} \quad W = (4 N^{*-1/2}) t_0 (1 + k_e) \tag{9.26}$$

where k_e is the value of k at elution. Equation (9.26) is exactly analogous to the equation for peak width in isocratic elution (derivable from the equation for N):

$$\text{(isocratic)} \quad W = (4 N^{-1/2}) t_0 (1 + k) \tag{9.27}$$

The value of k_e in Equation (9.26) is given by Equation (9.14a); inserting $k_e = 1/(2.3\,b)$: into Equation (9.26) gives another version of the peak width W in gradient elution

$$W = (4N^{*-1/2}) t_0 (1 + [1/2.3b]) \tag{9.28}$$

$$\equiv (4N^{*-1/2}) t_0 (1 + [k^*/2]) \tag{9.29}$$

Assuming that values of S do not vary much for the different solutes in a sample (a reasonable approximation for many samples), values of b will be approximately the same for all peaks in the sample [Equation (9.8)], so peak widths W in gradient elution will also be approximately constant – in contrast to isocratic elution, where values of W increase in proportion to retention time t_R (compare Fig. 1.3a and b). The preceding conclusions assume that N is approximately constant for all peaks in the chromatogram, which is not necessarily the case (Section 9.5). Equation (9.28) also assumes that k_0 is large; when this is not the case, Equation (9.29) is applicable, with k^* given by Equation (9.18) ($V_D \approx 0$) or Equation (9.24) ($V_D > 0$).

9.1.2.1 Gradient Compression

The above derivation of peak width W in gradient elution ignores "gradient compression" [26], which theoretically should result in narrower peaks than are predicted by Equations (9.21) or (9.22). Gradient compression *within the column* is illustrated in Figure 9.4 for a step-gradient (weaker mobile phase 1, followed by stronger mobile phase 2). In (*a*), a band is shown just prior to being overtaken by mobile phase 2 (cross-hatched). As mobile phase 2 overtakes the band, the trailing end of the band (labeled "*i*") begins to move more quickly in mobile phase 2, while the front of the band (labeled "*ii*") continues to move more slowly in the weaker mobile phase 1. As a result, by the time mobile phase 2 overtakes the *front* of the band (point *ii*, Fig. 9.4*b*), the width of the band has narrowed (compare W_2 with W_1) relative to band width in the absence of gradient compression. The relative widths of the bands in (*b*) vs (*a*) can be defined by the ratio $G_{12} = W_2/W_1$, and the value of G_{12} is determined by the values of k for the band in mobile phases 1 and 2 (k_1 and k_2, respectively). Note that band migration has been exaggerated in Figure 9.4 to better visualize the effects of gradient compression.

A value of G_{12} can be derived as follows. First, define the fractional migration R of a band along the column, where $R = 1/(1+k)$ (as in thin-layer chromatography, where $R \equiv R_F$). For movement of the mobile phase some distance Δx along the column, the band (*or any point on the band*) will move a distance $\Delta x R$. In Figure 9.4(*a*), mobile phase 2 has just touched the trailing edge of the band (point *i*). By the time mobile phase 2 has reached the leading edge of the band (point *ii*, Fig. 9.4*b*), the band is compressed by the ratio $W_2 : W_1$ (the latter quantities refer to band widths *within* the column). During the passage of mobile phase 2 through the band (over a distance Δx along the column), point *i* moves a distance equal to $\Delta x R_2$, where $R_2 = 1/(1+k_2)$. The width of the compressed band in (*b*) is then

$$W_2 = \Delta x(1 - R_2) \tag{9.30}$$

Consider next the migration of point *ii* during the time that mobile phase 2 is moving through the band, illustrated in (*c*) with "start" and "finish" bands. While mobile phase 2 is overtaking point *ii* of the band (measured by the distance Δx in Fig. 9.4*c*), point *ii* will be moving through the column in mobile phase 1. The distance moved by point *ii* before it is overtaken by mobile phase 2 is $\Delta x R_1$, so that

$$\Delta x = W_1 + \Delta x R_1 \tag{9.31}$$

Eliminating Δx between Equations (9.30) and (9.31) then gives

$$W_2/W_1 = G_{12} = (1 - R_2)/(1 - R_1)$$

or

$$G_{12} = \{1 - [1/(1+k_2)]\}/\{1 - [1/(1+k_1)]\} \tag{9.32}$$

In gradient elution, a continuous, linear gradient can be visualized as composed of a series of small segments [as in the derivation of Equation (9.1)], each of which (as it passes through a band on the column) leads to a (very small)

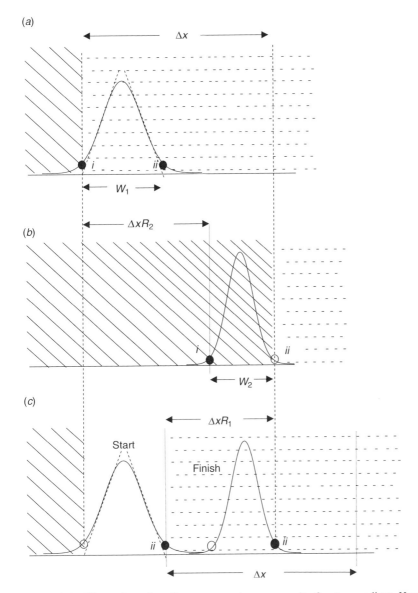

Figure 9.4 Illustration of gradient compression as a result of a step-gradient. Hypothetical visualization of a band that has migrated part-way down a column, (a) before, and (b) after being overtaken by the step gradient. W_1 and W_2 refer to baseline band width values. (a) The band at the time when the step-gradient (mobile phase 2, cross-hatched) has reached the trailing edge of the band (point i); (b) migration of point i until the step gradient overtakes the front of the band (point ii); (c) illustration of the migration of point ii in the initial mobile phase 1 (dashed lines) between times (a) and (b). The relative band migration in this example is intentionally exaggerated. See text for details.

compression of the band by the factor G_{12} of Equation (9.32). By the time the band has passed through the column and is about to elute, the band will have significantly compressed by a cumulative factor G, so that Equations (9.26), (9.28), and (9.29) must be modified to give corresponding expressions for the observed peak width W:

$$W = (4N^{*-1/2})G t_0(1 + [1/2.3b]) \quad (9.33)$$

$$\equiv (4N^{*-1/2})G t_0(1 + [k^*/2]) \quad (9.34)$$

$$\equiv (4N^{*-1/2})G t_0(1 + k_e) \quad (9.34a)$$

The gradient compression factor G can be related to gradient steepness b (see [27] for the full derivation). First define the quantity p

$$\begin{aligned} p &= 2.3k_0b/(k_0 + 1) \\ &\approx 2.3b \end{aligned} \quad (9.35)$$

for large k_0. G is then given in terms of p as

$$G = \{(1 + p + [p^2/3])/(1 + p)^2\}^{1/2} \quad (9.36)$$

Values of G vary with gradient steepness b as follows: for $0.05 < b < 2$ (corresponding to $17 > k^* > 0.4$), $1 < G < 0.6$; that is, large b or small k^* corresponds to smaller G. However, in practice it has been reported that $G \approx 1$ for many RP-LC gradient separations [28]; that is, experimental peak widths in gradient elution are often closer to values predicted by Equation (9.28) ($G = 1$) than by Equation (9.33) ($G < 1$). This failure to observe gradient compression experimentally was previously attributed either to (a) equipment extra-column volume, or (b) a variation of values of N with k [4]. Subsequently, a careful study of peak width in gradient elution was carried out in an effort to resolve this anomaly [29]. Five possible causes for the observed increase in experimental peak widths were considered:

1. Extra-column peak broadening.
2. Variation in N with mobile phase composition (%B).
3. Viscous fingering [30].
4. Stationary-phase diffusion [31–33].
5. Error in predictions of W from Equation (2) due to curvature of plots of log k vs ϕ.

The study of [29] reported experimental peak widths that were closer to theory [Equation (9.33)] than previously observed, but still averaging about 7 percent higher than values calculated from Equation (9.34a) with experimental values of k_e. This much closer match of experiment and theory was attributed to the use of Equation (9.34a) in place of previously used Equation (9.33). Equation (9.33) is susceptible to the slight concave curvature of experimental plots of log k vs ϕ which is generally observed. That is, error in peak-width calculations (#5 above) appears to be the main reason for the larger errors that have been found in previous comparisons of experimental and calculated peak widths.

Extra-column peak broadening (1) and variation in N (2) were ruled out as significant contributions to the residual 10 percent error in calculated values of W observed in the study of [29]. Viscous fingering (3) arises when a less viscous solvent phase follows a more viscous solvent in a chromatographic column; the less viscous solvent tends to penetrate the boundary between the two solvents, resulting in mixing of the two solvents with increased dispersion of any band present at the solvent boundary. However, the study of [29] suggests that viscous fingering is likely insignificant for linear gradient elution, although it can be quite significant for step-gradient elution. The latter observations appear to leave stationary-phase diffusion (4) as the most likely cause of the observed 10 percent increase in peak widths compared with theory in [29]. Stationary-phase diffusion in RP-LC is about one-third as fast as diffusion in the mobile phase, but a band that is initially strongly retained in gradient elution can still broaden as a function of time before the band begins to move appreciably through the column. This additional band-broadening increases the value of W predicted by Equations (9.33)–(9.34a).

9.1.3 Selectivity and Resolution

For the case of isocratic elution [34], resolution can be expressed as

$$R_s = (1/4)(k/[1+k])(\alpha - 1)N^{1/2} \quad \text{(isocratic)} \tag{9.37}$$

{for a derivation of Equation (9.37), see [35]}. If isocratic and gradient retention are related for "corresponding" systems as implied by Equations (9.4) and (9.15) (and illustrated in Fig. 9.1), and if $\phi \approx \phi^*$ for each of "critical" adjacent peaks 1 and 2, it follows that values of $\alpha = k_2/k_1$ (isocratic) and $\alpha^* = k_2^*/k_1^*$ (gradient) are also equal. Given that the term $(k/[1+k]) \approx (k^*/[1+k^*])$ [1, 4], we then have a fundamental relationship for gradient elution:

$$R_s = (1/4)(k^*/[1+k^*])(\alpha^* - 1)N^{*1/2} \quad \text{(gradient)} \tag{9.38}$$

Thus, for $\phi = \phi^*$ and therefore $k = k^*$, the same resolution of two adjacent peaks is expected for both isocratic and gradient elution. Likewise, the same means by which we select isocratic separation conditions for maximum selectivity (α) and resolution for the whole sample can be duplicated for the selection of optimal conditions in gradient elution, that is, the successive optimization of k^*, α^*, and N^* as for isocratic separation (Section 2.1 and [34]). Because k^* (and α^*) depends on gradient time t_G and flow rate F, changes in either t_G or F can be used to vary separation selectivity in gradient elution [36–38]; similar changes in selectivity for isocratic separation can be achieved by varying %B.

Consider next a more rigorous and detailed derivation of Equation (9.38) [4], beginning with the definition of resolution R_s for adjacent peaks 1 and 2

$$R_s = 2[t_{R2} - t_{R1}]/(W_1 + W_2) \tag{2.6}$$

where subscripts 1 and 2 refer to values for each peak. For peaks 1 and 2, Equation (9.11) becomes

$$t_{R1} = (t_0/b)\log(2.3k_{0,1}b) + t_0 + t_D \tag{9.39}$$

and

$$t_{R2} = (t_0/b)\log(2.3k_{0,2}b) + t_0 + t_D \tag{9.39a}$$

assuming equal values of S (and therefore b) for each peak. Again, subscripts 1 and 2 refer to peaks 1 and 2, respectively. Similarly [Equation (9.33)], the widths W of each peak will also be equal, so

$$R_s = (t_{R2} - t_{R1})/W \tag{9.40}$$

Equations (9.33), (9.39), and (9.39a) can be substituted into Equation (9.40) to give

$$R_s = \{(t_0/b)\log(k_{0,2}/k_{0,1})\}/\{(4N^{*-1/2})Gt_0(1+[1/2.3b])\} \tag{9.41}$$

For small values of x, $2.3 \log x \approx (x-1)$, and for equal values of b for each peak, $(k_{0,2}/k_{0,1}) = \alpha^*$ is constant for all values of ϕ. For closely adjacent peaks ($\alpha < 1.1$), Equation (9.41) can then be restated as

$$R_s = (1/4)(\alpha^* - 1)N^{*1/2}\{1/[G(2.3b+1)]\} \tag{9.42}$$

A comparison of Equations (9.38) and (9.42) shows that $(k^*/[1+k^*])$ must equal $\{1/[G(2.3b+1)]\}$, if Equation (9.38) is valid. This is very closely the case for $1 \leq k^* \leq 20$ (the useful range in k^*), as shown by a comparison of these two quantities in Table 9.1. For most values of k^* the difference between Equations (9.38) and (9.42) is <1 percent, and the difference is never greater than 4 percent when $k^* = k > 1$. The accuracy of Equation (9.38) decreases somewhat for two peaks whose values of S are unequal, as discussed in [4].

When comparing detection sensitivity in isocratic vs gradient elution, peak width W is the primary consideration. W in gradient elution can be approximated by Equation (9.34), and compared with Equation (9.27) for isocratic elution. Assuming that $k = k^*$ for equal values of ϕ and ϕ^* (i.e., "corresponding" separations), and noting that k at elution in gradient elution (i.e., $k_e = k^*/2$), the ratio

TABLE 9.1 Demonstration of the Equivalence of "Corresponding" Isocratic and Gradient Separations in Terms of Resolution [The Validity of Equation (9.38)]: $R_s = (1/4)(\alpha^* - 1)N^{*1/2}(k^*[1+k^*])$. See text for details

k^*	$k^*/(1+k^*)$	$\{1/[G(2.3b+1)]\}$	Difference (%)
20	0.952	0.952	0.0
15	0.938	0.937	0.1
10	0.909	0.908	0.1
8	0.889	0.887	0.2
6	0.857	0.854	0.3
4	0.800	0.795	0.7
2	0.667	0.655	1.8
1	0.500	0.480	4.1

of peak widths in gradient vs isocratic elution should equal $G[1 + (k^*/2)]/(1 + k^*)$, or a value of about $1/2$ for $1 \leq k^* \leq 10$. This should mean narrower peaks in gradient elution by a factor of 2, other factors being equal, and correspondingly increased detection sensitivity. However, this conclusion ignores the typically greater baseline noise in gradient vs isocratic elution. A significant net increase in detection sensitivity for gradient elution is therefore unlikely to be observed for individual peaks in corresponding separations.

9.1.4 Advantages of LSS Behavior

Gradient elution carried out under conditions such that Equation (9.6) applies (LSS behavior) has some potential advantages of varying importance:

- easy calculation or estimation of retention and resolution as a function of experimental conditions;
- near-equivalent isocratic and gradient separation when conditions are selected for $k^* = k$ ("corresponding" conditions assumed);
- more nearly equal resolution for both early- and late-eluting peaks (on average);
- an absence of peak-splitting as a result of the gradient (a *very* rare event with highly efficient columns).

The preceding and following equations in this chapter (and their experimental verification) should convince the reader that *LSS gradient separation can be readily predicted* when separation conditions are known, and the fundamental sample parameters k_w and S can be measured experimentally (Section 9.3.3). In the mid-1980s, this initiated the development and use of computer simulation for gradient elution (Section 3.4, [38, 39]), which in turn has greatly facilitated the development of gradient elution methods.

The *near equivalence of gradient and isocratic separation* when $k^* = k$ was first proposed on the basis of theoretical considerations [1] and subsequently confirmed experimentally (Fig. 9.1 and related discussion). This in turn has enabled chromatographers to use essentially the same method-development strategy for both isocratic and gradient elution. Because isocratic separation is relatively simple, easily understood and the subject of a vast literature, whereas gradient elution *appears* much more complicated, the LSS model provides a simple approach to gradient method development (Section 3.3), one that is closely patterned after isocratic method development (with which many chromatographers are more experienced).

When LSS gradients are employed, and the retention of early-eluting compounds is fairly strong ($k_0 > 10$), values of k^* will usually be similar for all the peaks in the chromatogram. This in turn leads to *comparable resolution of both early- and late-eluting peaks* which have similar values of α^*, as well as similar peak widths and detection sensitivity. While this tends to be so for all gradient separations, it is even more true for LSS separation [40], as can be appreciated from the discussion of Table 8.4 for IEC gradient elution (i.e., non-LSS separation). Minor exceptions to this observation occur for homologous or oligomeric series, where values of S increase continuously for larger, later-eluting sample molecules.

In these cases, convex or segmented gradients can provide a more similar resolution of early and late peaks in the chromatogram, as seen in Figure 2.26.

Peak-splitting as a result of gradient elution is essentially unheard of today, because peaks are usually quite narrow, and peak-tailing is much less common than in the past. However, when this is not the case, it can be shown that LSS gradients are less likely to result in peak-splitting [41].

9.2 SECOND-ORDER EFFECTS

Everything should be made as simple as possible, but not simpler.

—Albert Einstein

By second order ("nonideal") effects, we refer to the causes of (usually minor) deviations between experimental data and predictions from Equations (9.1)–(9.42). These second-order effects, which often can be ignored in the practical application of gradient elution, may arise from either imprecise assumptions about values of ϕ (and k) during gradient elution (Section 9.2.1), or nonideal gradient equipment (Section 9.2.2). In addition, differences between experimental and predicted gradient separations can arise from errors in (or poor control of) mobile phase composition and temperature, as well as variable column performance. However, the latter contributions to separation variability have similar effects on both isocratic and gradient elution, and will not be discussed further.

9.2.1 Assumptions About ϕ and k

The LSS model assumes that values of k and ϕ are related to time t as described by Equation (9.7). Significant deviations of actual values of ϕ and/or k as a function of t can arise from the following "nonideal" contributions to gradient retention:

- incomplete column equilibration;
- solvent demixing;
- nonlinear plots of log k vs ϕ;
- variation of V_m with ϕ.

9.2.1.1 Incomplete Column Equilibration

This has been discussed in Section 5.3 from the standpoint of how to minimize any deleterious consequences; Figure 9.5 provides additional insight into its effect on retention in gradient elution. For purposes of illustration, $\Delta\phi$ is assumed equal to 1.00 (a gradient from 0 to 100 percent B), and there is no dwell volume. It is further assumed that at the end of the gradient (which could include a gradient hold at 100 percent B) the entire column is now equilibrated with the B solvent (which is close to the case in practice). Thus, in Figure 9.5(*a*), the condition of the column at the end of the gradient is represented by curve *i*, which shows that the complete column is in

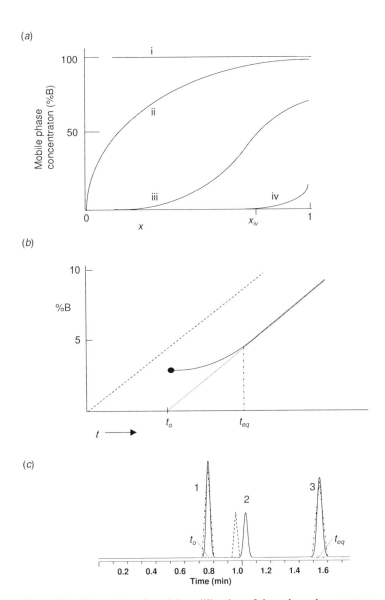

Figure 9.5 Illustration of partial equilibration of the column between two gradient runs. (a) Plot of relative saturation of stationary phase by the B solvent as a function of fractional distance x between column inlet (0) and outlet (1); (b) distortion of the gradient leaving the column for case iv of (a). (- - - -) Gradient entering the column; (·····) gradient leaving the column, assuming a fully equilibrated column; (——), actual gradient leaving the column; (c) chromatogram assuming partial equilibration as in (b) (- - - -) or full equilibration (——); (d) experimental data of [42] for effect of column equilibration on retention time for peak 6; (e) effect of 1.5 min equilibration period on retention ($-\delta t_R$) of different sample peaks as a function of t_R; (f) effect of 1.5 min equilibration period on retention difference ($-\delta t_R$) of different sample peaks as a function of t_R. Conditions for (d–f): 150 × 4.6 mm C_{18} column; 10–80 percent acetonitrile–buffer gradient in 10 min; 2 mL/min. See text for details.

388 CHAPTER 9 THEORY AND DERIVATIONS

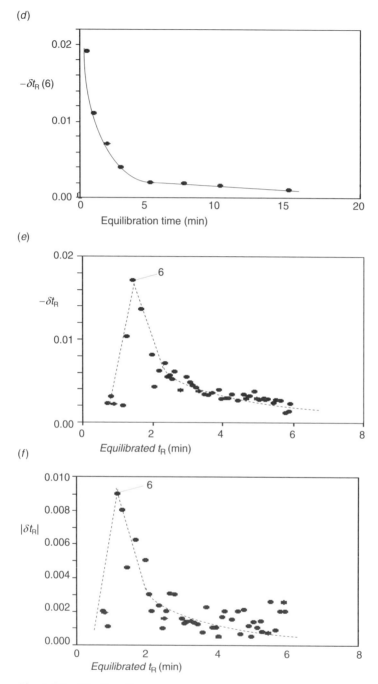

Figure 9.5 (*Continued*)

equilibrium with 100 percent B [from the column inlet ($x = 0$) to the outlet ($x = 1$)]. If the column is subsequently flushed with the initial mobile phase used in the gradient ($\phi = 0$; column re-equilibration step), the condition of the column can be represented successively by curves *ii*, *iii*, and *iv* at later times t_2, t_3, and t_4 ($t_2 < t_3 < t_4$). The concentration of B in the stationary phase does not return immediately to 0 percent B for the entire column after the passage of a single column volume of initial mobile phase. The patterns (*ii*)–(*iv*) illustrated in Figure 9.5(*a*) can arise for various reasons; including (*a*) a significant capacity of the stationary phase for retention of the B solvent, (*b*) a slow release or diffusion of B from the stationary phase into the mobile phase, and (*c*) mixing of the two mobile phases (0 and 100 percent B) by dispersion within the equipment or column during column equilibration.

Curve *iv* of Figure 9.5(*a*) represents the condition of the column at some time t_4 after the end of the gradient (i.e., after significant, but incomplete equilibration of the column with the A solvent). The inlet region of the column will have become equilibrated with the A solvent (0 percent B in the stationary phase), but not the region near the column outlet. Assuming only partial equilibration of the column between gradient runs, the corresponding effect on the mobile phase leaving the column at the start of the next gradient separation is visualized in Figure 9.5(*b*), where %B in the mobile phase is shown as a function of time *t* during this run. The dashed curve corresponds to the gradient selected for this separation, that is, the value of %B entering the column (assumes no equipment hold-up or "dwell" volume). The dotted curve represents the expected (assuming complete column equilibration) value of %B vs time for mobile phase leaving the column (shifted to the right by the column dead time t_0). The actual curve for the composition of exiting mobile phase (partial equilibration of the column) is represented by the solid curve of Figure 9.5(*b*). Values of %B in the latter curve are initially higher than predicted by the dotted curve, because the column is incompletely equilibrated at the start of the gradient, as in curve *iv* of Figure 9.5(*a*). At some time t_{eq} after the start of the gradient, mobile phase leaving the column will have a value of %B that matches the expected (dotted line) curve of Figure 9.5(*b*); that is, the column and mobile phase will be (approximately) in equilibrium with each other at this time (ignoring any solvent demixing; see below).

The effect of non-equilibration of the column as in Figure 9.5(*b*) is illustrated for the hypothetical chromatogram of Figure 9.5(*c*). The solid peaks correspond to a fully equilibrated column prior to the start of the gradient, while the dashed peaks represent the chromatogram for a column that is partially equilibrated – as represented by Figure 9.5(*b*) and curve *iv* of Figure 9.5(*a*). Peak 1 of Figure 9.5(*c*) leaves the column at time $t = t_0$; because peak 1 is unretained ($k_0 = 0$), the lack of complete column equilibration (or a change in %B) has no effect on its retention; the solid line and dashed peaks therefore coincide (same retention times). Peak 3 leaves the column at a time $t > t_{eq}$, meaning that the peak has migrated through a column that is near-equilibrated. The retention time of peak 3 is therefore similar for both the equilibrated (solid-line peak) and partially equilibrated column (dashed peak). Only for peak 2, which is retained during gradient elution, but which migrates through the column under partially equilibrated conditions, are the retention times for the two cases significantly different. The dashed peak 2 for partial equilibration is less retained, because %B in the mobile and stationary phases is slightly greater during its migration than for a fully equilibrated column.

An experimental example of the effects of incomplete column equilibration is illustrated in Figures 9.5(d–f) (data from [42]). A mixture of 46 neutral and ionizable compounds was separated in a 10–80 percent acetonitrile/low-pH buffer gradient in 10 min. Retention times were first measured for conditions of near-complete inter-run equilibration, then retention times were compared for various shorter equilibration times. In Figure 9.5(d), the *change* in retention time ($-\delta t_R$) of peak 6 (equilibrated $t_R = 1.4$ min) is plotted as a function of equilibration time (incomplete equilibration reduces t_R). There is a rapid increase in t_R (decrease in $-\delta t_R$, by 0.017 min) as the equilibration time is increased to 5 min, followed by a slower change over the following 10 min; that is, complete column equilibration requires a fairly long time. In Figure 9.5(e), $-\delta t_R$ is plotted for each sample compound as a function of retention time, for a short equilibration period (1.5 min). As anticipated from Figure 9.5(c), maximum changes in retention occur for compounds with intermediate retention [$1 < t_R < 3$ min in (e)]. In Figure 9.5(f), changes in the *difference* in retention times (absolute values, $\delta\delta t_R$) for adjacent peaks (proportional to resolution) are plotted vs retention time. The maximum value of $|\delta\delta t_R|$ is 0.009 min, which corresponds to a change in resolution of only $\delta R_s \approx 0.1$ for this representative separation, that is, an insignificant change in resolution due to column nonequilibration. We also see that the maximum retention shift for peak 6 of 0.017 min corresponds to a change in ϕ_e (and presumably ϕ^*) of only $0.017/10 = 0.002$. The effect of such a small change in ϕ during elution on α^* would be negligible, even for a significant difference in values of S for two adjacent peaks (Section 2.2.3.1). It thus appears that relatively short column equilibration times can be compatible with negligible changes in resolution. For some practical rules on the selection of the equilibration time, see Section 5.3.3.

The above, simplified picture for gradient elution with a nonequilibrated column, combined with the assumption of a Langmuir isotherm for various mobile phase components, logically leads to the following conclusions:

- the more strongly retained and/or the slower the release of the B solvent from the stationary phase, the longer the time required for complete column equilibration;
- equilibration should be slowest for small concentrations of strongly retained mobile phase additives, such as certain ion-pair reagents;
- the effects of column nonequilibration on peak retention and separation should be reduced for larger ϕ_0 and smaller $\Delta\phi$;
- column equilibration will be faster for columns having a reduced affinity and/or capacity for solvent B, for example, wide pore (low-surface-area) columns, C_3 vs C_{18} columns, more hydrophilic columns such as cyano or phenyl, and so on;

- differences in peak retention for equilibrated vs nonequilibrated columns will be greatest for peaks that elute in an intermediate region between $(t_0 + t_D)$ and $(t_{eq} + t_0 + t_D)$.

Strongly retained, retention-affecting compounds may have been added to the mobile phase, such as hydrophobic ion-pair reagents or silanol-suppressing compounds (e.g., triethylamine). When the concentrations of these mobile-phase components are relatively low (e.g., <10 meq/L), it is sometimes observed that column equilibration can be quite slow [43]. (Note that adding equal concentrations of the ion-pair reagent, IPR, to both the A and B solvents of the gradient does *not* ensure that the IPR is in equilibrium during the gradient, because of changes in retention of the IPR with change in ϕ.) One study [44] of column equilibration for gradient elution with added ion-pair reagent has proposed means for achieving relatively fast column equilibration. Ion-pair reagents that are more strongly retained (e.g., C_8- or C_{10}-alkylsulfonate) should be avoided, especially for IPR concentrations of <10 mM. See some further recommendations concerning column equilibration with ion-pair reagents in [34, 44]. Column equilibration is considerably slower with tetrahydrofuran as B solvent, than for methanol or acetonitrile [45]. This can be attributed to the lesser polarity and stronger retention of THF in RP-LC.

Slow column equilibration can be much more pronounced for normal-phase chromatography, especially separations on bare silica [5]. The presence of trace amounts of water in the usually employed nonaqueous mobile phases can lead to column equilibration times measured in hours. However, the use of dried solvents with %B values >6 percent speeds up equilibration time and improves retention reproducibility [46]. Other studies claim that water in the mobile phase is unimportant, and that reproducible normal-phase separations are possible with an equilibration time of only 2 min [47]. However, the latter study was accompanied by several qualifications, and results were reported only for a few consecutive gradient runs.

9.2.1.2 Solvent Demixing

This refers to the preferential uptake by the stationary phase of organic solvent (or other strongly retained components present in the B solvent) from the mobile phase. As a result, the mobile phase composition ϕ in the early stages of gradient elution is reduced, leading to an opposite effect on %B vs time than is observed in Figure 9.5(b) for a nonequilibrated column. An exaggerated example is illustrated in Figure 9.6(a), which can be contrasted with Figure 9.5(b) for a nonequilibrated column. The solid curve in Figure 9.6(a) is a representation of %B vs time for mobile phase leaving the column, while the dotted curve is the expected result in the absence of solvent demixing. Initially, the stationary phase retains some of the organic solvent (B solvent) that enters the column, but eventually the column becomes saturated with B solvent, so that the actual and expected plots of %B vs time coincide.

The effect of solvent demixing on sample retention is illustrated in Figure 9.6(b), which can be compared with Figure 9.5(c) for partial equilibration of the column. The retention of unretained peaks (peak 1) will be unaffected

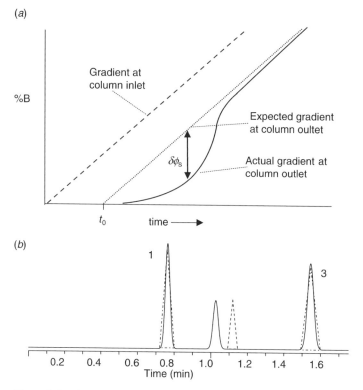

Figure 9.6 Illustration of solvent demixing (exaggerated for easier visualization). (*a*) (- - - -) Gradient entering the column; (·····) gradient leaving the column, assuming no solvent demixing; (——) actual gradient leaving the column, due to solvent demixing; (*b*) chromatogram assuming solvent demixing as in (*a*) (- - - -). See text for details.

by solvent demixing, early-eluting, retained peaks (e.g., peak 2, $k > 0$) in the chromatogram will be *more* retained, due to reduced %B in the mobile phase during their movement through the column. The retention of later-eluting peaks (peak 3) will be unaffected by solvent-demixing, because these bands migrate through the column after solvent demixing occurs and the gradient has returned to its programmed composition (as in Fig. 9.6*a*). Incomplete equilibration of the column between sequential gradient runs can offset the effects of solvent demixing, and in some circumstances these two opposing effects might mutually cancel.

Solvent demixing can be quite important in normal-phase chromatography (e.g., thin-layer chromatography [48]), because small changes in the concentration of very polar B solvents can lead to large changes in k [49], and such solvents are strongly retained by the column. For RP-LC separation, however, solvent demixing is seldom noticeable. An approximate theoretical treatment of solvent demixing in RP-LC gradient elution has been reported [50], with the following conclusions. First, the maximum change in %B due to solvent demixing ($\delta\phi_s$, see Fig. 9.6*a*) is usually small, seldom more than 1 percent by volume. Second, the value of $\delta\phi_s$ is proportional to V_m/V_G (or to $b/\Delta\phi$) and the surface area of the packing. Third, for each 20 percent by volume increase in ϕ_0, the maximum value of $\delta\phi_s$ decreases by a factor of roughly 2.

9.2.1.3 Nonlinear Plots of log k vs φ

A slight nonlinearity in plots of $\log k$ vs ϕ is typical for RP-LC separation (Section 9.4). This modest failure of Equation (9.4) for gradient RP-LC is of interest in connection with (a) the verification of the theory of gradient elution and the LSS model (Section 9.3.1.1), and (b) potential errors in the quantitative prediction of separation and the use of computer simulation (Section 9.3.4). For the most reliable comparisons of experimental vs theoretical values of retention, curve-fitting can be used to obtain more accurate values of k as a function of ϕ, followed by the numerical integration of Equation (9.6). An alternative is to approximate the true $\log k$ vs ϕ relationship by a series of linear segments [51]. For the effect of nonlinear plots of $\log k$ vs ϕ on peak width, see [29].

In the use of computer simulation for gradient method development (Section 3.4), accurate predictions of resolution are more important than accurate values of k. Fortunately, errors in k due to nonlinear plots of $\log k$ vs ϕ tend to cancel in the calculation of resolution, so that predictions of resolution are correspondingly more reliable [52] than are predictions of retention time. A similar result is seen in Figure 9.5(*e*, *f*) for incomplete column equilibration, where the retention of peak 6 is decreased by 0.17 min, but the *separation* of peaks 6 and 7 changes by only 0.009 min.

9.2.1.4 Dependence of V_m on φ

There is still disagreement over the significance of the dead volume V_m and the best procedure for its measurement [53], but this uncertainty is of limited practical importance. Values of V_m are commonly assumed to be constant for a separation where only mobile phase composition ϕ is varied. However, values of V_m and t_0 can vary with ϕ by $\pm 10-15$ percent [50], with minimum t_0 occurring for $\phi \approx 0.5$. If an average (constant) value of t_0 is assumed for the calculation of values of k or k^*, resulting predictions of retention (and especially resolution) should not be much affected by actual variations in t_0 (note the related discussion of Section 9.3.1.1).

9.2.2 Nonideal Equipment

The equipment used for gradient elution is affected by four factors that can affect the accuracy of the above equations for predicting retention and peak width:

- dwell volume;
- gradient distortion;
- flow-rate errors;
- extra-column peak broadening.

Dwell volume has been discussed previously, mainly its origin (Chapter 4) and its affect on retention [Equation (9.24)]. A change in dwell volume V_D can affect resolution (Section 2.3.6), thereby complicating method transfer (Section 5.2.1).

The value of V_D is equal to the volume of the gradient mixer plus the additional volume of the flow path between the mixer and the column inlet (Figs 4.1 and 4.2); V_D can be measured in various ways, as described in Section 4.3.1.2. A value of V_D can also be determined by means of computer simulation; given experimental runs 1, 2, and 3 with gradient times $t_{G1} < t_{G2} < t_{G3}$ (other conditions being the same), data from runs 1 and 2 can be used to predict retention times for run 3. A correct value of V_D (obtained by trial-and-error) will provide the closest average agreement between experimental and predicted retention times for run 3.

Gradient distortion can arise from mobile phase compressibility (Section 4.1.2.4), error in the design of the gradient former, and/or gradient "rounding" (Section 5.2.2.2). Modern equipment for gradient elution attempts to correct for the compression of the A and or B solvents during gradient elution [54] and is usually free of design errors that can lead to significant gradient distortion. The main remaining source of gradient distortion is gradient "rounding," as illustrated in Figure 9.7(*a*) (arrows). As the gradient is formed and transferred to the column

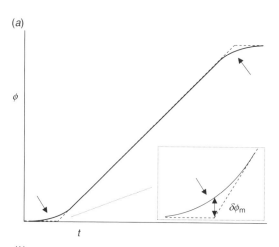

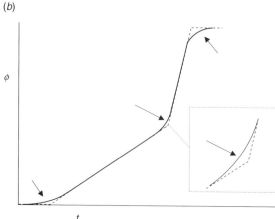

Figure 9.7 Effect of "gradient rounding" by the equipment on the gradient entering the column. (———), Actual gradient; (- - - -) gradient in the absence of rounding. (*a*) Gradient rounding at start and finish of a linear gradient; (*b*) gradient rounding for a segmented gradient. Arrows mark gradient rounding. Area of rounding is expanded within dotted boxes, for better visualization.

inlet, it undergoes dispersion or mixing during its movement through connecting lines, sample valve, and other parts of the gradient-former, so there is then no longer a sharp boundary that separates the start and finish of the gradient from preceding or following mobile phase. The distortion of the gradient $\delta\phi_m$ as a result of gradient rounding (see Fig. 9.7a) is proportional to V_M/V_G, where V_M is the "mixing volume" of the gradient system, and V_G is the gradient volume ($=t_G F$). The smaller V_M/V_G is, the less distortion there is of the gradient. The gradient of Figure 9.7(a) assumes a short gradient hold (100 percent B) at the end of the gradient.

A semitheoretical relationship for gradient distortion as a function V_M, V_G, and time t during the gradient can be derived [54]; resulting values of $\delta\phi_m$ as a function of V_M/V_G and t are summarized in Table 9.2. Older gradient systems can have values of V_M on the order of ~2 mL or larger (see [54] for the estimation of V_M from the geometry of the gradient system), and $V_M \approx 2$ can be compared with representative values of $V_G = 12$–50 for a 150 × 4.6 mm column and $k^* \approx 2$–10 ($\Delta\phi = 1$). For such equipment and these gradient conditions, values of $V_M/V_G \approx$ 0.05–0.2 would be observed. From Table 9.2, we see a resulting gradient distortion $\delta\phi_m \approx 2$–7 percent B at the beginning ($t = t_D$) and end of the gradient ($t = t_D + t_G$). Newer, better plumbed equipment is characterized by much smaller values of V_M and correspondingly reduced gradient rounding. However, when such equipment is used with narrow-diameter columns (with proportionally smaller values of V_m and therefore V_G), resulting values of V_M/V_G may still be as large as 0.1 or

TABLE 9.2 Gradient Distortion $\delta\phi_m$ as a Result of the "Mixing Volume" V_M of the Gradient Equipment (for Gradient as it Enters the Column, Based on 0–100 Percent B Gradient[a]) [54]

	$\delta\phi_m$ for indicated values of V_m/V_G				
$(t - t_D)/t_G^a$	$V_m/V_G = 0.2$	0.1	0.05	0.02	0.01
−0.2	0.006	0.000	0.000	0.000	0.000
−0.1	0.021	0.000	0.000	0.000	0.000
0.0	0.074	0.037	0.018	0.007	0.004
0.1	0.045	0.014	0.003	0.000	0.000
0.2	0.027	0.005	0.000	0.000	0.000
0.3	0.016	0.002	0.000	0.000	0.000
0.4	0.010	0.001	0.000	0.000	0.000
0.5	0.006	0.000	0.000	0.000	0.000
0.6	0.004	0.000	0.000	0.000	0.000
0.7	0.000	0.000	0.000	0.000	0.000
0.8	−0.004	0.000	0.000	0.000	0.000
0.9	−0.021	0.000	0.000	0.000	0.000
1.0	−0.074	−0.037	−0.018	−0.007	−0.004
1.1	−0.045	−0.014	−0.002	0.000	0.000
1.2	−0.027	−0.005	0.000	0.000	0.000

[a] Time during the gradient, corrected for dwell volume; expressed as a fraction of gradient time t_G.

larger – with significant gradient rounding. A detailed discussion of the origin and estimation of values of V_M as a function of equipment design is given in [54]. Values of V_M are usually of similar magnitude as values of the dwell volume V_D, but somewhat smaller.

Gradient rounding does not normally have much effect on the separation, except when initial peaks in the chromatogram have values of $k_0 < 10$. However, for $V_M/V_G > 0.1$, the gradient can become sufficiently distorted (as in Fig. 8.7b) to preclude accurate predictions of retention as a function of gradient condition or the use of computer simulation. A "two chamber model" correction for such gradient distortion has been described [55], which improves the prediction of severely distorted gradients. Predictions of retention for such gradients can then be made via numerical integration [Equation (9.2)].

Gradient rounding can also occur in the *middle* of the gradient, when segmented gradients are used (Fig. 9.7b). Resulting changes in separation can be more serious than for gradient rounding at the start and finish of the gradient (as in Fig. 9.7a), because rounding is now more likely to affect peak retention. For this reason, gradient separations that use segmented gradients may not transfer as well as linear-gradient methods. The effect of gradient rounding in the middle of the gradient mainly affects peaks that immediately follow the break between the two segments [56]. The extent of gradient rounding as in Figure 9.7(b) will be a function of the relative steepness of the two gradient segments, being less for gradient segments of more similar slope (and with no rounding at all when the slopes are equal, i.e., a nonsegmented gradient). As in the related example of slow column equilibration (Fig. 9.5), the effect of gradient rounding on resolution will usually be much less than the effect on retention times.

Flow-rate errors are of two kinds: (a) errors arising from mobile phase compression under pressure; and (b) change in mobile phase volume upon mixing the A and B solvents (Section 4.1.2.4). Modern gradient equipment is designed to correct for errors due to mobile phase compression (a), while errors due to solvent mixing (b) should occur only for high-pressure mixing systems (Section 4.1.2.3). Flow-rate errors usually have little effect on either the final separation or quantitative predictions based on computer simulation (Section 3.4), when experimental gradient runs are used to predict gradient separation. However, larger errors can result when gradient runs are used to predict isocratic separation. For a further discussion of flow-rate errors and their effect on gradient separation, see [54, 57].

Extra-column peak broadening can occur for both isocratic and gradient elution; it is the result of a broadening of sample peaks (a) between the sample injector and column inlet, and (b) between the column outlet and the detector flow-cell outlet. Because the sample is usually strongly retained at the start of the gradient, peaks that have been broadened before the column (step a) will be compressed at the column inlet, thereby minimizing any effect of pre-column peak broadening on the separation. However, post-column peak broadening (step b) should be similar for both isocratic and gradient elution. Thus, extra-column peak broadening is expected to be less in gradient vs isocratic elution. For a theoretical discussion of post-column peak broadening as a function of the dimensions of connecting tubing and the volume of the flow cell, see [58]; note that this theory is oversimplified and therefore somewhat approximate [59].

9.3 ACCURACY OF GRADIENT ELUTION PREDICTIONS

Equations presented in this chapter allow the prediction of gradient separations: retention times, peak widths, and resolution. Qualitative predictions can be used to guide and interpret experiments used for method development, as well as help diagnose problems encountered in routine analysis. Quantitative predictions have found increasing use in computer simulation (Section 3.4). The accuracy of these predictions is of interest with respect to (a) the validity and completeness of our understanding of gradient elution based on Equation (9.2), and (b) the reliability of computer simulation based on two initial experiments with different gradient times (Sections 9.3.3 and 9.3.4).

9.3.1 Gradient Retention Time

9.3.1.1 Confirmation of Equation (9.2) Numerous studies [60–73] and additional references cited in [4] have examined the accuracy of Equation (9.2) (retention times) and Equation (9.29) (peak width, generally ignoring gradient compression) for reversed-phase, normal-phase, and ion-exchange chromatography. For the most part, these studies have ignored the second-order effects summarized in Section 9.2; the agreement of experimental and predicted retention times varies from ± 2 to 10 percent, with an average value of about ± 5 percent. An error $\delta\phi$ can be related to δt_R as

$$\delta\phi = \delta t_R \Delta\phi / t_G \tag{9.43}$$

For RP-LC gradient elution, we can assume an average value of $t_R \approx t_G/2$, and $\Delta\phi = 1$, so that

$$\delta t_R / t_G \approx 0.5 \text{ (average percentage error)}/100 \tag{9.44}$$

or

$$\delta\phi \approx 0.005 \text{ (average percentage error)} \tag{9.45}$$

Thus, an average error in predicted values of t_R of 5 percent corresponds to an error in ϕ ($d\phi \equiv d\phi_e$) of about ± 0.025 units (or ± 2.5 percent B).

For a more critical test of the fundamental equation for gradient elution [Equation (9.2)], one study [50] reported results for 30 RP-LC gradient runs (widely different gradient times and flow rates; two columns and two temperatures) for a five-component sample. These 150 values of t_R were compared with predicted retention times from Equation (9.11) (using values of log k_0 and S determined from isocratic measurements). In the latter study, (a) $k_0 \geq 98$ for each of the five solutes, (b) precise and accurate gradient equipment was used (errors in delivered values of $\phi < 0.1$ percent B), and (c) most of the second-order effects of Section 9.2 were either measured or estimated, so that values of t_R could be corrected for these latter contributions to retention. The resulting overall agreement of experimental vs predicted values of t_R [from Equation (9.11)] was equivalent to an average overall value of $\delta\phi^* = 0.010$, corresponding to an error of ± 2 percent in t_R.

Random errors of various kinds that could not be corrected for were equivalent to about $\delta\phi = \pm 0.003$, and the day-to-day variability of retention time measurements was $\delta\phi = 0.002$, for a total experimental variability of $(0.003^2 + 0.002^2)^{0.5} = 0.004$. Thus, the observed error in predicted retention times ($\delta\phi = 0.010$) was slightly greater than the experimental uncertainty ($\delta\phi = 0.004$).

It was also observed (Table XI of [50]) that *average* errors (or bias) in predicted gradient retention times (values of $\delta\phi$) correlated with ϕ^*, varying from $\delta\phi = -0.001$ for $\phi^* = 0.1$ to $+0.010$ for $\phi^* = 0.5-0.6$, and -0.010 for $\phi^* = 1.0$. Values of t_0 correlate with ϕ in similar fashion, resulting in a strong correlation of the bias (average values of $\delta\phi$) with values of t_0:

$$\delta\phi = 0.085 - 0.070 t_0$$

$$r^2 = 0.83, \text{SD} = 0.004$$

Note that the uncertainty in this relationship (SD = 0.004) is the same as the experimental uncertainty in measured retention times. This suggests that the manner in which variable t_0 was corrected for in the study of [50] may have been responsible (in whole or in part) for the observed bias and any excess error in predicted values of t_R. For this reason, and in view of the approximate nature of some of the corrections that were made for second-order effects, *the study of [50] appears to represent a quite satisfactory agreement of experimental data with Equation (9.11)* (for large values of k_0).

The experimental variability of gradient retention in the study of [50] was also compared with the variability of isocratic retention in "corresponding" separations. Errors δt_R in predicted retention times can be related to errors ($\delta\phi_e$) in ϕ at elution, or to equivalent errors (δk) in k at elution (k_e). An error $\delta\phi$ can be related to δt_R by Equation (9.43), which corresponds to an error in log k [$\delta \log k$; Equation (9.4)] of

$$\log(1 + [\delta k/k]) = -S\ \delta\phi_e \qquad (9.46)$$

or for small errors $\delta k/k$,

$$\delta k/k \approx -2.3 S\ \delta\phi_e \qquad (9.47)$$

Equation (9.46) can also be obtained from Equation (9.11). Equation (9.47) can be applied to either isocratic (values of k) or gradient elution (values of k_e), allowing a comparison of retention variability for "corresponding" separations. Day-to-day retention variability can arise from variable conditions (temperature, mobile phase pH, etc.), changes in column retention over time, and other factors that affect retention. Many of these factors should affect isocratic values of k and gradient values of k_0 similarly, leading to similar average errors $\delta\phi$ or $\delta k/k$ for "corresponding" isocratic and gradient separations. As an example, replicate isocratic and gradient retention data were obtained for "corresponding" separations of the same sample over a period of 2 months in the study of [50]. The average coefficient of variation for isocratic values of k was ± 1.2 percent, and the average variation of gradient retention times was $\delta t_R = \pm 0.02$ min. From Equation (9.43) for the latter gradient data, $\delta\phi = 0.02(0.9/10) = \pm 0.0018$, and Equation (9.47) then gives (with $S \approx 3$

for this sample) $\delta k/k = (\pm 2.3)(3)(0.0018) = 0.012$, or ± 1.2 percent – the identical variability to that for isocratic retention (values of k).

9.3.1.2 Computer Simulation

Computer simulation (Section 3.4) makes use of two initial experiments to calculate values of log k_w and S for each solute in the sample (Section 9.3.3), following which it is possible to carry out predictions of both isocratic or gradient separation as a function of mobile phase %B or gradient conditions. For the case of linear gradients, retention times can be calculated from Equation (9.24), while peak widths can be estimated as described in Section 9.3.2. Alternatively, experimental peak widths from the two runs used to initiate computer simulation can be used to estimate (more reliable) peak widths for simulated separations by interpolation of derived values of N^*. For the case of segmented gradients, retention can be calculated by summing the retention times for each gradient segment [Equation (9.12)], similar to the derivation of Equation (9.24).

Numerous comparisons have been carried out between experimental and computer-simulated values for gradient retention time and resolution [39, 56, 74–79]. Retention time predictions are usually accurate to $\leq \pm 1$ percent, while the accuracy of resolution predictions (i.e., predictions of *differences* in retention time for adjacent peaks) is usually ± 10 percent or better. Predictions as reliable as this are in most cases adequate for purposes of method development. Additional studies have been reported which relate the accuracy of computer simulation to the experimental conditions used, as well as examining the accuracy of isocratic predictions from experimental gradient data [52, 80–84]. The precision of retention time predictions can be improved in various ways [84–87], but improved predictive accuracy is rarely needed. As an aside, these comparisons of experimental retention times with predictions by computer simulation do not represent a direct test of Equation (9.2), which would require the calculation of gradient retention times from *isocratic* data (a much more stringent challenge, as illustrated by the study of [50]).

9.3.2 Peak Width Predictions

Peak width W in gradient elution is given by Equations (9.33) and (9.34a), each of which requires a value of N^* for the peak of interest. In the past, values of N^* have been assumed equal to values of N measured in corresponding isocratic separations. Earlier studies of peak broadening in gradient elution [2, 70, 78] often showed somewhat higher experimental values of W than predicted, by as much as a factor of two for larger values of b (or lower k^*). A later study [29] concluded that this discrepancy was most likely due to curvature in plots of log k vs ϕ, plus the use of Equation (9.33), rather than Equation (9.34a), for the prediction of W. When Equation (9.34a) was used instead in one study [29], experimental values of W were only 7 percent higher than predicted, and this discrepancy was attributed mainly to the neglect of stationary-phase diffusion (Section 9.1.2.1). Plots of log k vs ϕ tend to be more linear for larger solute molecules (with larger values of S). For these compounds, values of W predicted by Equation (9.33) should be more reliable, as observed for several peptides [88].

An accurate estimate of peak width is required for computer simulation, in order to predict reliable values of resolution. In this case, a value of N^* for each peak can be calculated from a value of W and Equation (9.33), since the two initial computer-simulation experiments allow the calculation of log k_w and S (and therefore b) for each peak. One study [81] examined the accuracy of peak width predictions carried out in this way, finding an average accuracy of ± 13 percent for separations that involve interpolation, but a larger error for extrapolation.

9.3.3 Measurement of Values of S and log k_0

For large values of k_0, Equation (9.11) accurately describes linear-gradient retention in RP-LC. For this case, it is possible to calculate values of log k_w and S for each compound in any sample, based on two experimental gradient runs where only gradient time is varied. Thus, assume gradient times for the two experiments of t_{G1} and t_{G2} ($t_{G1} > t_{G2}$), with a ratio $\beta = t_{G1}/t_{G2}$. Given values of t_R for a given solute in run 1 (t_{R1}) and run 2 (t_{R2}), a value of b_1 can be calculated:

$$b_1 = (t_0 \log \beta)/[t_{R1} - (t_{R2}/\beta) - (t_0 + t_D)(\beta - 1)/\beta] \qquad (9.48)$$

Similarly,

$$\log k_0 = [b_1(t_{R1} - t_0 - t_D)/t_0] - \log(2.3 b_1) \qquad (9.49)$$

Insertion of b_1 into Equation (9.8) allows the calculation of a value of S, while log k_w is then calculable as log $k_0 + S\phi_0$. Predictions of retention time by computer simulation are based on Equations (9.48) and (9.49); the accuracy of these predictions provides a further confirmation of these two equations.

When the value of k_0 is $\ll 100$, Equations (9.48) and (9.49) yield less accurate values of log k_0 and S. In this latter case, once approximate values of b_1 and log k_0 have been determined from Equations (9.48) and (9.49) (which assume large k_0), iterative trial-and-error variation of b_1 and log k can be used with Equation (9.24) to give a best fit of retention times t_{R1} and t_{R2} for each peak in the chromatogram, in this way providing final values of log k_0 and S that are corrected for small values of k_0 {computer software (e.g., Excel "solver") is also available for this purpose [89]}. For the case of high-molecular-weight solutes (>20,000 Da), with corresponding large values of S ($S > 30$), the application of Equations (9.48) and (9.49) for the determination of values of log k_w and S is subject to greater potential error (Section 6.1.5.3), depending upon how the experiments are carried out [16, 90–92]. A proper choice of experimental conditions with an appropriate treatment of the data can reduce, but not eliminate, these errors in measured values of S for large-molecule solutes. More accurate values of log k_w and S for large molecules are favored by values of $\beta \geq 3$, as well as by higher values of ϕ_0 (as long as $k_0 > 100$, which is easily achieved for large molecules, because of very large values of k_w).

9.4 VALUES OF S

For a review of Equation (9.4) ($\log k = \log k_w - S\phi$) and values of S as a function of solute structure and separation conditions, see [93]. Any discussion of S as a function of the solute and separation conditions must recognize that Equation (9.4) is a purely empirical relationship. While attempts have been made to derive Equation (9.4) in terms of solubility parameter theory [5, 9] or the competition between solute and mobile phase molecules for a place in the stationary phase [94], neither of these models is at all adequate. Solubility parameter theory is only reliable for nonpolar solutions, which excludes the water–organic mobile phases used in RP-LC. Similarly, a simple competition model ignores interactions between solute and solvent molecules in the mobile phase. Solute interactions with both mobile and stationary phases, *as well as* competition, must each play a role in determining k as a function of %B (therefore k as a function of ϕ must be a very complex relationship).

Plots of $\log k$ vs %B are seldom exactly linear over a wide range in values of %B [16, 40, 92, 94], as illustrated by the examples of Figure 9.8 [compare actual (—) vs linear (- - -) curves for each solute]. Purely empirical fitting functions have occasionally been used to improve the fit to $\log k$–ϕ data as in Figure 9.8, for example,

$$\log k = A\phi + B\phi^2 + C \quad [40] \tag{9.50}$$

or

$$\log k = a + bE_T(30) \quad [95] \tag{9.51}$$

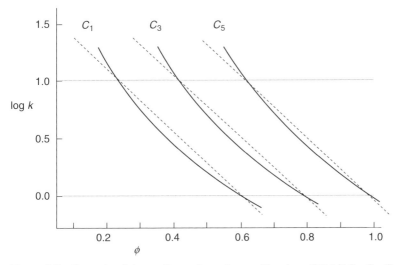

Figure 9.8 Example of the nonlinear dependence of $\log k$ on %B (ϕ) for C_1, C_3, and C_5 dialkylphthalates. C_{18} column, acetonitrile–water mobile phase. Data from [50]; see text for details.

Here, A, B, a, and b are constants for a given solute and only ϕ varying, while $E_T(30)$ is a measure of mobile phase polarity derived from spectroscopic measurements.

Depending on which two experimental values of ϕ are chosen to calculate values of log k_w and S [using Equation (9.4)], somewhat different values of these two quantities will often result. Measured values of log k_w and S are therefore usually *not* fundamental properties of the solute. Severe curvature of log k–ϕ plots for values of $1 \geq k \geq 10$ can also compromise the accuracy of computer simulation. Equation (9.4) is usually a better approximation for methanol as B solvent than for acetonitrile or tetrahydrofuran.

9.4.1 Estimating Values of *S* from Solute Properties and Experimental Conditions

A rough estimate of S for a given sample can facilitate its separation by gradient elution. Thus, for a reasonable compromise between run time, peak detection, and resolution, values of $1 \leq k^* \leq 10$ are recommended (Section 3.3.1, i.e., similar to the case of isocratic elution). Values of k^* are determined by values of b [Equation (9.14)], which is a function of S [Equation (9.8)]. The primary factor that determines values of S is solute molecular weight M, as discussed in Section 6.1.1. For a wide range in values of M, S can be approximated (Fig. 6.2) by

$$S \approx 0.25 M^{1/2} \tag{6.4}$$

or

$$\log S \approx \log(0.25) + 0.5 \log M \tag{9.52}$$

A plot of values of log S vs log M is shown in Figure 9.9(*a*) for 67 compounds with molecular weights between 50 and 400. The solid curve through the data of Figure 9.9(*a*) [Equation (9.52)] is seen to fit these data with an average error of ± 0.7 units in S (about ± 20 percent). For a limited range in values of M, plots of S vs M will typically be somewhat scattered; as the range in M increases, the correlation improves (e.g., Fig. 6.2).

Apart from a general increase in S with increasing solute molecular size [Equation (9.52)], few general rules for estimating values of S for individual solutes have been reported. One oft-cited generalization [40, 93] is a correlation of values of S and log k_w:

$$S = p + q \log k_w \tag{9.53}$$

Equation (9.53) is consistent with the observation that in RP-LC both retention and values of S increase with solute molecular size, in turn suggesting that values of S for solutes which overlap in a gradient separation (have similar retention times) will be similar. This in turn implies that a change in gradient steepness will be ineffective for the separation of previously overlapping peaks [see Equation (9.54) below]. Fortunately, this is often not the case, and indeed Equation (9.53) is unreliable except for solutes of very similar molecular structure (it is exact for some homologous series [96]). The inexactness of Equation (9.53) is illustrated in Figure 9.9(*b*), where values of S are plotted vs log k_w for 67 small molecules with widely varying

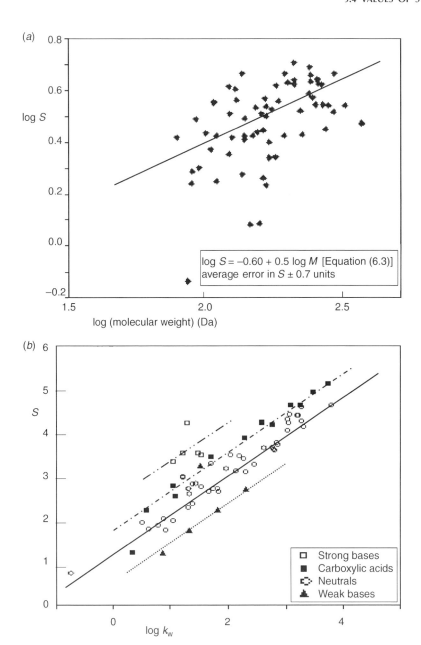

Figure 9.9 Dependence of S on (a) Solute molecular weight M [Equation (9.52)] and (b) log k_w [Equation (9.53)] for 67 compounds with $50 < M < 400$ Da. Conditions: 50% acetonitrile–water, 35°C. Data from [96]; see text for details.

molecular structures (some of the same compounds as in Fig. 9.9a; data of [97]). There is a reasonable correlation for the 45 *neutral* solutes of Figure 9.9(b) (open circles, $S = 1.31 + 0.90 \log k_w$, $r^2 = 0.93$, SD = 0.25), but significant average deviations (bias) from the latter correlation exists for other solute types: fully protonated, strong bases, +1.1 units in S; carboxylic acids, +0.4 units; partly protonated weak bases, −0.8 units. This dependence of S on solute ionization (and therefore mobile phase pH) has been noted in other studies [98], especially for basic solutes.

Values of S can also vary with molecular conformation, as noted for separations of both peptides [99, 100] and proteins [14] by RP-LC. Usually S (and sample retention) increases for less compact conformations that expose a greater hydrophobic area to the stationary phase. Thus, helical peptides have larger values of S than nonhelical peptides, as do denatured vs native proteins.

Values of S do not vary significantly with temperature for most solutes [101]. For more than a hundred compounds of widely different molecular structure, and a wide range in mobile phase conditions, an increase in temperature of 20–40°C resulted in an average change in S of only -2 ± 6 percent, that is, essentially no change within experimental variability. Values of S tend to be lower for less-hydrophobic columns such as cyano [93], and less polar B solvents such as THF [102].

The difference in values of S for two adjacent bands i and j also determines how α^* and resolution will vary as gradient steepness b (and ϕ^*) is varied (Section 2.2.3.1):

$$\log \alpha^* = (\log k_{wj} - \log k_{wi}) - (S_j - S_i)\phi \qquad (9.54)$$

Consequently, the ability to anticipate sizable differences in S for two overlapping peaks in a given gradient separation is potentially useful. At present, however, it is seldom possible to predict values of S (from solute molecular structure) with sufficient accuracy for the useful application of Equation (9.54); an accuracy of ± 0.1–0.2 units in S would be required. However, the convenient experimental determination of values of S as in Section 9.3.3 *does* allow the prediction of α^* vs ϕ^* by means of Equation (9.54).

9.5 VALUES OF N IN GRADIENT ELUTION

The effect on resolution of the column plate number N (or N^*) was discussed in Section 3.3.5. N varies with column length L, flow rate F, and particle size d_p, and these so-called *column conditions* can be used either to increase resolution or decrease run time (but not usually both simultaneously!). Because k^* and selectivity also vary with column volume V_m (proportional to column length) and flow rate F [Equations (9.8a) and (9.14)], changes in L or F should usually be accompanied by a change in gradient time t_G, so as to maintain k^* constant. In this way, N^* can be changed without accompanying changes in selectivity (after selectivity has been optimized). N also varies with temperature, mobile phase composition, the nature of the sample (mainly its molecular weight), and how well the column is packed. For the moment we will assume that $N \approx N^*$, so the following discussion can be in terms of N. The so-called "Knox equation" has been widely used to predict N

as a function of experimental conditions and the sample being separated:

$$h = Av^{0.33} + B/v + Cv \tag{9.55}$$

Here, h is the reduced plate height, v is the reduced mobile phase velocity and the various constants in Equation (9.55) can be approximated (roughly) by $A = 1$, $B = 2$ and $C = 0.05$ (for temperatures near ambient and samples with molecular weights <500 Da). The reduced parameters h and v are in turn defined by

$$h = H/d_p \tag{9.56}$$

and

$$v = ud_p/D_m \tag{9.57}$$

The plate height H (cm) is

$$H = L/N \tag{9.58}$$

allowing the calculation of values of N from Equation (9.55). L is column length (cm), d_p is particle diameter (cm), u is mobile phase velocity (cm/s), and D_m is the solute diffusion coefficient in the mobile phase (cm^2/s). Because D_m decreases for larger molecules, the effect of an increase in solute molecular weight is to increase v proportionately [Equation (9.57)]. Thus, the effect on N of an increase in sample molecular weight M is similar to an increase in flow rate, and therefore N will decrease for increasing M in most cases.

Equation (9.55) inversely mirrors the dependence of column plate number on flow rate, as shown in the plots of Figure 9.10 for different particle sizes d_p and

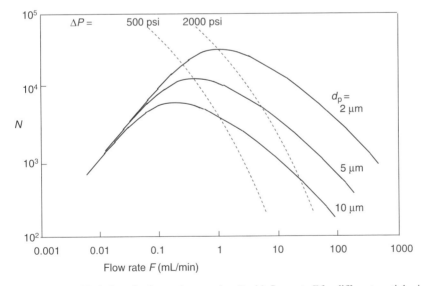

Figure 9.10 Variation of column plate number N with flow rate F for different particle sizes d_p. Assumes 150 × 4.6 mm column, acetonitrile–water mobile phase, near-ambient temperature, and a sample molecular weight < 500 Da; (- - -) connects points on curves of N vs F for $P = 500$ or 2000 psi. Based on Equation (9.51) with $A = 1$, $B = 2$, and $C = 0.05$; see text for details.

150 × 4.6 mm columns. For a column packed with 10 μm particles ($d_p = 10$ μm), a maximum plate number of $N \approx 6000$ is seen for $F \approx 0.2$ mL/min. As particle size decreases to 5 and 2 μm, respectively, the maximum value of N increases to 13,000 and 32,000, while the corresponding flow rate increases to 0.4 and 1 mL/min, respectively. Since run time decreases in inverse proportion to flow rate (for the same column size: 150 × 4.6 mm in Fig. 9.10), smaller-diameter particles are seen to result in higher plate numbers in shorter run times. However, the column pressure drop ΔP also increases as the particles become smaller (note dashed curves for $\Delta P = 500$ and 2000 psi in Fig. 9.10). A particular HPLC system is usually limited to some maximum pressure drop, so either the flow rate or column length must be decreased for smaller d_p. A practical goal is maximum N for some defined run time, which in turn depends on particle size and the maximum allowable column pressure, as well as on the sample, column length, flow rate, and other conditions. The choice of "best" conditions for maximum N and resolution in minimum time will be addressed next.

Equation (9.55) predicts a minimum value of $h \approx 2$ for an intermediate (optimum) value of v (≈ 3) or mobile phase flow rate. It can be shown that the maximum possible column plate number N for some run time t will always correspond to conditions such that h is a minimum ($h \approx 2$). Furthermore, maximum values of N (assuming minimum h) increase with t_0 (or retention time) and with column pressure, *provided that* column length and flow rate are selected such that minimum h and maximum allowable column pressure result. That is, the maximum attainable value of N increases with run time and pressure, because longer columns can be used. The column pressure-drop P can be approximated by

$$P = 3000 L \eta / (t_0 d_p^2) \tag{9.59}$$

where η is the viscosity of the mobile phase (Appendix IV). In gradient elution, η will vary during the gradient because of the changing mobile phase composition, as will P. For acetonitrile–buffer gradients, pressure usually drops toward the end of the gradient, while for methanol–buffer gradients, pressure first rises, then drops.

Given the above relationships [Equations (9.55)–(9.59)], it is possible to predict maximum possible values of N for some maximum column pressure as a function of d_p and separation time t, where t can be equated to retention time $t_R = t_0(1 + k)$ in isocratic elution. The goal of maximum N for a given separation time and some maximum pressure can only be achieved by a particular choice of particle size d_p, column length L, and flow rate F, as well as allowed maximum run time t. For a more detailed discussion of the theory of column efficiency N and the practical application of Equation (9.55), see [58, 103–105].

If we next assume that N is not dependent on the value of k (approximately true, but see [32]), it is possible to derive an optimum value of k for maximum resolution in minimum time. For isocratic separation, and a given value of t_0, the variation of resolution with k is proportional to $[k/(1+k)]$ [Equation (9.37)], and run time t_R varies with k as $t_0(1+k)$. Therefore, resolution per unit time varies as the ratio of these two quantities: $[k/(1+k)]/t_0(1+k)$. The latter function yields a maximum value of $k = 1$ (for a given value of t_0). We can assume a similar relationship for gradient elution (k^* replacing k), so that for maximum resolution per unit time a value of $k^* = 1$ will be optimum. From Equation (3.3a) we then have $k^* = 1 = t_G/(1.15 t_0 \Delta\phi S)$, or

$$t_G/t_0 = 1.15 \Delta\phi S \qquad (9.60)$$

For a full-range gradient (5–100 percent B) and $S \approx 4$, we then have $t_G/t_0 \approx 4$ for maximum resolution per unit time, and (gradient run time)$/t_0 \approx (t_G + t_0 + t_D)/t_0 \approx t_G/t_0$. Note that comparing run time in isocratic and gradient separation corresponds to $k = 3$ for isocratic elution, or (isocratic run time)$/t_0 = 4$. It should also be kept in mind that we are presently ignoring possible changes in α^* with k^*, which can be of critical importance in the separation of a given sample; that is, values of α^* are assumed constant.

A useful summary of the above dependence of maximum resolution or N on run time, particle size, and column pressure has been provided by Guiochon [58], as summarized in Figure 9.11. Each point in this figure represents the largest

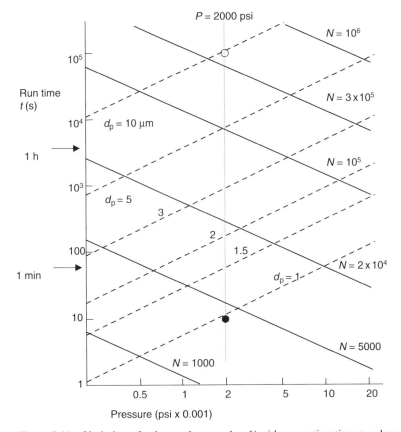

Figure 9.11 Variation of column plate number N with separation time t, column pressure P, and particle diameter d_p. (———) Lines representing different values of N; (- - - -) lines representing different particle sizes; (·····) line representing a column pressure of 2000 psi. All values in figure are for minimum plate height, $h = 2$ (so that $v = 3$) and $k^* = 1$. Adapted from [58] for a viscosity of 0.6 cP and $D_m = 10^{-5}$ cm^2/s (corresponds to a sample molecular weight of 300; ●, separation time of 10 s, pressure of 2000 psi). See text for details.

possible value of N for a given particle size, pressure, and run time (reduced plate height $h \approx 2$ with reduced velocity $v \approx 3$). Thus, if the pressure is specified, a given run time in Figure 9.11 defines the maximum possible value of N. At the same time, this value of N requires a specific particle size, which in turn defines a specific column length and flow rate. The particle size also determines a mobile phase velocity u from Equation (9.57) (since $v \approx 3$). The original calculations of [58] are for isocratic separation with $k = 3$, with $t_R/t_0 = (1 + k) = 4$; however, roughly the same conclusions result for gradient elution with an optimum value of $t_G/t_0 \approx 4$; that is, run times for both isocratic and gradient elution are equivalent, when $k = 3$ and $k^* = 1$ [see discussion of Equation (9.60) above]. The results of Figure 9.11 assume that column length and flow rate have been selected for minimum h (and maximum N), for specified values of d_p, run time t, and column pressure P. In Figure 9.11, the vertical dotted line indicates a column pressure $P = 2000$ psi (a reasonable maximum pressure for the routine operation of many HPLC systems). As run time increases, it is seen that both N^* and the optimum particle size d_p increase for $P = 2000$ psi; for example, for a run time of 10 s, $N^* \approx 4000$ and $d_p \approx 1$ μm (marked by ●). For a run time of 100,000 s (28 h), $N^* = 400,000$ with $d_p = 10$ μm (marked by ○).

Figure 9.12 provides a practical illustration of the application of Figure 9.11: the separation of two samples [the "irregular" sample in (b) and compounds 1, 3–5, 9, 11 of the irregular sample in (a)] for conditions taken from Figure 9.11 ($k^* = 1$, 40°C, 0–100 percent acetonitrile–buffer gradient). In each separation shown in Figure 9.12, the combination of column length and flow rate have been chosen for $P = 2000$ psi and minimum h. The sample of (a) is relatively easy to separate; a resolution of $R_s = 2$ can be obtained with $N^* = 6700$. The sample in (b) is much more difficult to resolve, requiring $N^* = 220,000$ for $R_s = 2$. The separation of sample (a) with $R_s = 1.6$ can be achieved with 1 μm particles in a time of 0.2 min, while the separation of sample (b) with $R_s = 1.4$ requires 100 min and the use of 5 μm particles. A summary of the separations of Figure 9.12 is provided in Table 9.3. The maximum value of N (for the conditions of Fig. 9.12) is plotted vs run time in Figure 9.13, based on the data of Table 9.3. If a required plate number is specified for a given gradient separation with $k^* = 1$, Figure 9.13 provides the particle size and minimum run time required. Note that a million theoretical plates can be achieved with a particle size of about 15 μm, but a run time of about 100 h is required.

If the optimization of selectivity for a given sample results in $k^* \neq 1$, the use of Figure 9.11 proceeds similarly. The only change is that the time required for the separation will be approximately proportional to k^*. The optimum choice of column length, flow rate, and particle size for a required value of N will remain the same.

Still faster separations for a given value of N are possible by increasing the column pressure (>2000 psi). Equipment and columns for routine use at pressures of 10,000 psi and higher are now available [106], and as seen in Figure 9.11 there is a steady increase in maximum N for a given run time t as pressure increases, for example, from $N = 3700$ for $t = 10$ s and $P = 2000$ psi, to $N = 8200$ for $P = 10,000$ psi and $t = 10$ s. Values of N in Figure 9.13 increase as

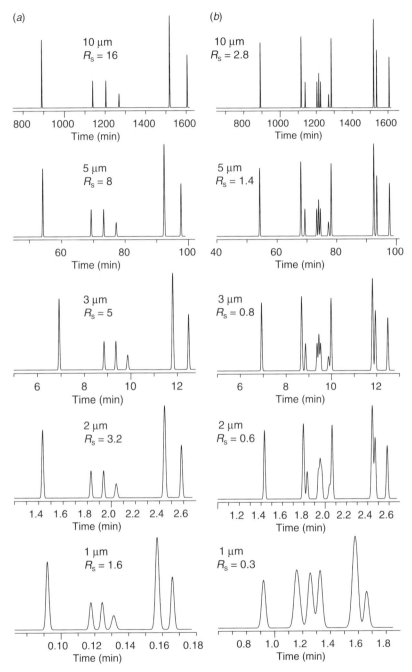

Figure 9.12 Separations optimized for maximum plates within a given time, for different particle sizes d_p. Conditions: 0–100 percent acetonitrile–buffer gradients, 40°C, and other conditions of Table 9.3. (*a*) Sample is compounds 1, 3–5, 9, 11 of irregular sample of Table 1.3; (*b*) sample is compounds 1–11 of irregular sample. See text for details.

TABLE 9.3 Separation Conditions for Achieving Maximum Values of N^* for a Given Gradient Separation Time. Based on Data of Figure 9.11; Assumes $t_G = 4t_0$, $k^* = 1$, a Column Pressure of 2000 psi, a Temperature of 40°C and 0–100 Percent Acetonitrile–Buffer Gradients. See Corresponding Chromatograms of Figure 9.12 and Text for Details [Values Below are Taken from Computer Simulations Based on DryLab Software; Small Differences Between Table 9.3 and Figure 9.11 Exist, Due to the Use of an Expanded Version of Equation (9.55) in DryLab]

d_p (μm)	t_G (min)[a]	N	F (mL/min)	L (mm)
1	0.3	4,400	2.5	9
2	2.8	17,000	1.25	0
3	14	38,000	0.85	230
5	110	110,000	0.51	1080
10	1700	430,000	0.25	8700

[a] For values of k^* other than 1, change these values of t_G in proportion to k^*.

$P^{1/2}$, so that an increase in the allowed pressure from 2000 to 8000 psi will increase N in Figure 9.13 by a factor of 2. Likewise, an increase in temperature T reduces mobile phase viscosity and increases solute diffusion D_m, which allows higher values of N for the same run time and maximum column pressure [107]. While the use of a higher column pressure P can result in potential improvements in separation (due to the larger values of N for a given run time), a value

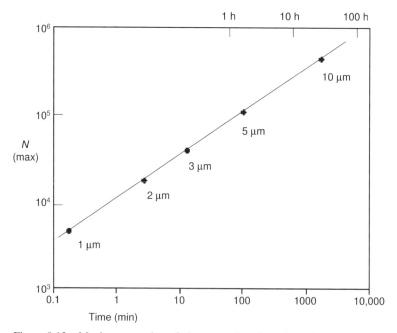

Figure 9.13 Maximum number of plates as a function of gradient time t_G. Conditions of Figure 9.12 and Table 9.3.

of $P > 3000$ psi can by itself result in changes in retention, selectivity and N, possibly making method development somewhat more complicated and difficult [108].

Values of N depend on solute diffusion [Equation (9.57)], and the diffusion coefficient D_m decreases with increasing sample molecular weight M. However this dependence of N on M can be compensated by reducing the mobile phase flow rate in proportion to D_m. An increase in M also means an increase in solute molecular size, and this can introduce an additional contribution to N apart from Equation (9.55). Thus, if the hydrodynamic radius of the solute molecule is greater than about one-quarter that of pores within the particle, the pore walls will induce a frictional resistance to diffusion within the pore, leading to an effective decrease in D_m. See [3] for further details.

> Scientists should have nothing to do with reality. Their business is to construct models that account for their observations.
>
> —Neils Bohr, Quoted in *The Child in Time* by Ian McEwan

REFERENCES

1. L. R. Snyder, in *High-performance Liquid Chromatography. Advances and Perspectives*, Vol. 1, Cs. Horváth, ed., Academic Press, New York, 1980, p. 207.
2. P. Jandera and J. Churáček, *Gradient Elution in Column Liquid Chromatography*, Elsevier, Amsterdam, 1985.
3. L. R. Snyder and M. A. Stadalius, in *High-performance Liquid Chromatography. Advances and Perspectives*, Vol. 4, Cs. Horváth, ed., Academic Press, New York, 1986, p. 195.
4. L. R. Snyder and J. W. Dolan, *Adv. Chromatogr.* 38 (1998) p. 115.
5. P. Jandera, *Adv. Chromatogr.* 43 (2004) 1.
6. B. Drake, *Arkiv Kemi* 8 (1955) 1.
7. E. C. Freiling, *J. Am. Chem. Soc.* 77 (1955) 2067.
8. E. C. Freiling, *J. Phys. Chem.* 61 (1957) 543.
9. P. J. Schoenmakers, H. A. Billiet, R. Tijssen, and L. De Galan, *J. Chromatogr.* 149 (1978) 519.
10. G. Crétier and J. L. Rocca, *J. Chromatogr. A* 658 (1994) 295.
11. A. Felinger and G. Guiochon, *J. Chromatogr. A* 724 (1996) 27.
12. P. Jandera and J. Churáček, *J. Chromatogr.* 91 (1974) 207, 223.
13. H. Engelhardt and H. Elgass, *J. Chromatogr.* 158 (1978) 249.
14. M. Kunitani, D. Johnson, and L. Snyder, *J. Chromatogr.* 371 (1986) 313.
15. M. A. Stadalius, H. S. Gold, and L. R. Snyder, *J. Chromatogr.* 296 (1984) 31.
16. M. A. Quarry, R. L. Grob, and L. R. Snyder, *Anal. Chem.* 58 (1986) 907.
17. A. N. Hodder, M. I. Aguilar, and M. T. W. Hearn, *J. Chromatogr.* 476 (1989) 391.
18. R. W. Stout, S. I. Sivakoff, R. D. Ricker, and L. R. Snyder, *J. Chromatogr.* 353 (1986) 439.
19. C. H. Lochmüller and M. B. McGranaghan, *Anal. Chem.* 61 (1989) 2449.
20. G. Jilge, K. Unger, U. Esser, H.-J. Schäfer, G. Rathgeber, and W. Müller, *J. Chromatogr.* 476 (1989) 37.
21. K. E. Bij, Cs. Horváth, W. R. Melander, and A. Nahum, *J. Chromatogr.* 203 (1981) 65.
22. M. T. W. Hearn, in *High-performance Liquid Chromatography of Peptides and Proteins*, C. T. Mant and R. S. Hodges, eds, CRC Press, Boca Raton, FL, 1991, p. 105.
23. P. Jandera and J. Churáček, *Adv. Chromatogr.* 19 (1981) 79.
24. P. Schoenmakers, F. Fitzpatrick, and R. Grothey, *J. Chromatogr. A* 965 (2002) 93.
25. A. P. Schellinger and P. W. Carr, *J. Chromatogr. A* 1077 (2005) 110.
26. L. R. Snyder and D. L. Saunders, *J. Chromatogr. Sci.* 7 (1969) 195.
27. H. Poppe, J. Paanakker, and M. Bronkhorst, *J. Chromatogr.* 204 (1981) 77.
28. J. D. Stuart, D. D. Lisi, and L. R. Snyder, *J. Chromatogr.* 485 (1989) 657.

29. U. D. Neue, D. H. Marchand, and L. R. Snyder, *J. Chromatogr. A* 1111 (2006) 32.
30. M. Czok, A. M. Katti, and G . Guiochon, *J. Chromatogr.* 550 (1991) 705.
31. J. C. Giddings, *Dynamics of Chromatography, Part 1, Principles and Theory*, Marcel Dekker, New York, 1965.
32. R. W. Stout, J. J. DeStefano, and L. R. Snyder, *J. Chromatogr.* 282 (1983) 263–286.
33. J. H. Knox and J. P. Scott, *J. Chromatogr.* 282 (1983) 297–313.
34. L. R. Snyder, J. J. Kirkland, and J. L. Glajch, *Practical HPLC Method Development*, 2nd edn, Wiley-Interscience, New York, 1997.
35. L. R. Snyder and J. J. Kirkland, *Introduction to Modern Liquid Chromatography*, 2nd edn, Wiley-Interscience, New York, 1979, pp. 34–36.
36. K. A. Cohen, J. W. Dolan, and S. A. Grillo, *J. Chromatogr.* 316 (1984) 359.
37. J. L. Glajch, M. A. Quarry, J. F. Vasta, and L. R. Snyder, *Anal. Chem.* 58 (1986) 280.
38. J. W. Dolan, L. R. Snyder, and M. A. Quarry, *Chromatographia* 24 (1987) 261.
39. J. W. Dolan, D. C. Lommen, and L. R. Snyder, *J. Chromatogr.* 485 (1989) 91.
40. P. J. Schoenmakers, H. A. H. Billiet, and L. De Galan, *J. Chromatogr.* 185 (1979) 179.
41. L. R. Snyder, *J. Chromatogr.* 13 (1964) 415.
42. A. P. Schellinger, D. R. Stoll, and P. W. Carr, *J. Chromatogr. A*, (2006) in press.
43. L. R. Snyder, J. J. Kirkland, and J. L. Glajch, *Practical HPLC Method Development*, 2nd edn, Wiley-Interscience, New York, 1997, p. 338
44. M. Patthy, *J. Chromatogr.* 592 (1992) 143.
45. L. R. Snyder and J. L. Glajch, unpublished studies.
46. P. Jandera, private communication.
47. V. R. Meyer, *J. Chromatogr.* 768 (1997) 315.
48. E. Geiss, *Fundamentals of Thin Layer Chromatography*, Huethig, Heidelberg, 1987.
49. L. R. Snyder, *Principles of Adsorption Chromatography*, Marcel Dekker, New York, 1968, pp. 213–216.
50. M. A. Quarry, R. L. Grob, and L. R. Snyder, *J. Chromatogr.* 285 (1984) 19.
51. P. Nikitas and A. Pappa-Louisi, *J. Chromatogr. A* 1068 (2005) 279.
52. J. W. Dolan, L. R. Snyder, L. C. Sander, P. Haber, T. Baczek, and R. Kaliszan, *J. Chromatogr. A* 857 (1999) 41.
53. C. A. Rimmer, C. R. Simmons, and J. G. Dorsey, *J. Chromatogr. A* 965 (2002) 219.
54. M. A. Quarry, R. L. Grob, and L. R. Snyder, *J. Chromatogr.* 285 (1984) 1.
55. G. Hendriks, J. P. Franke, and D. R. A. Uges, *J. Chromatogr. A* 1089 (2005) 193.
56. D. D. Lisi, J. D. Stuart, and L. R. Snyder, *J. Chromatogr.* 555 (1991) 1.
57. J. P. Foley, J. A. Crow, B. A. Thomas, and M. Zamora, *J. Chromatogr.* 478 (1989) 287.
58. G. Guiochon, in *High-performance Liquid Chromatography. Advances and Perspectives*, Vol. 2, Cs. Horváth, ed., Academic Press, New York, 1980, p. 1.
59. J. G. Atwood and M. J. E. Golay, *J. Chromatogr.* 218 (1981) 97.
60. L. R. Snyder and H. D. Warren, *J. Chromatogr.* 15 (1964) 344.
61. M. Popl, V. Dolansky, and J. Mostecky, *Anal. Chem.* 44 (1972) 2082.
62. P. Jandera and J. Churáček, *J. Chromatogr.* 93 (1974) 17.
63. S. R. Abbott, J. R. Berg, P. Achener, and R. L. Stevenson, *J. Chromatogr.* 126 (1975) 421.
64. P. Jandera and J. Churáček, *J. Chromatogr.* 104 (1975) 9.
65. P. Jandera and J. Churáček, *J. Chromatogr.* 115 (1975) 9.
66. H. Engelhardt and H. Elgass, *J. Chromatogr.* 158 (1978) 249.
67. P. Jandera and J. Churáček, *J. Chromatogr.* 170 (1979) 1.
68. P. Jandera and J. Churáček, *J. Chromatogr.* 174 (1979) 35.
69. R. A. Hartwick, C. M. Grill, and P. R. Brown, *Anal. Chem.* 51 (1979) 34.
70. J. W. Dolan, J. R. Gant, and L. R. Snyder, *J. Chromatogr.* 165 (1979) 31.
71. P. Jandera, J. Churáček, and H. Colin, *J. Chromatogr.* 214 (1981) 35.
72. M. A. Stadalius, H. S. Gold, and L. R. Snyder, *J. Chromatogr.* 296 (1984) 31.
73. J. Schmidt, *J. Chromatogr.* 485 (1989) 421.
74. T. Sasagawa, Y. Sakamoto, T. Hirose, T. Yoshida, Y. Kobayashi, and Y. Sato, *J. Chromatogr.* 485 (1989) 533.
75. R. G. Lehmann and J. R. Miller, *J. Chromatogr.* 485 (1989) 581.

76. D. J. Thompson and W. D. Ellenson, *J. Chromatogr.* 485 (1989) 607.
77. J. W. Dolan and L. R. Snyder, *LCGC* 5 (1987) 970.
78. J. D. Stuart, D. D. Lisi, and L. R. Snyder, *J. Chromatogr.* 485 (1989) 657.
79. R. Chloupek, W. S. Hancock, and L. R. Snyder, *J. Chromatogr.* 594 (1992) 65.
80. B. F. D. Ghrist, B. S. Cooperman, and L. R. Snyder, *J. Chromatogr.* 459 (1989) 1.
81. N. Lundell, *J. Chromatogr.* 639 (1993) 97.
82. P. Chaminde, A. Baillet, and D. Ferrier, *J. Chromatogr.* 672 (1994) 67.
83. G. Vivó-Truyols, J. R. Torres-Lapasió, and M. C. Garcia-Alvarez-Coque, *J. Chromatogr. A* 1018 (2003) 169.
84. G. Vivó-Truyols, J. R. Torres-Lapasió, and M. C. Garcia-Alvarez-Coque, *J. Chromatogr. A* 1057 (2004) 31.
85. P. Chaminade, A. Baillet, and D. Ferrier, *J. Chromatogr.* 672 (1994) 67.
86. F. Fitzpatrick, H. Boelens, and P. Schoenmakers, *J. Chromatogr. A* 1041 (2004) 43.
87. R. G. Wolcott, J. W. Dolan, and L. R. Snyder, *J. Chromatogr. A* 869 (2000) 3.
88. M. A. Stadalius, H. S. Gold, and L. R. Snyder, *J. Chromatogr.* 327 (1985) 27.
89. F. Fitzpatrick, B. Staal, and P. Schoenmakers, *J. Chromatogr. A* 1065 (2005) 219.
90. C. H. Lochmüller and M. B. McGranaghan, *Anal. Chem.* 61 (1989) 2449.
91. C. H. Lochmüller, C. Jiang, Q. Liu, and V. Antonucci, *Crit. Revs. Anal. Chem.* 26 (1996) 29.
92. J. C. Ford and J. Ko, *J. Chromatogr. A* 727 (1996) 1.
93. K. Valkó, L. R. Snyder, and J. L. Glajch, *J. Chromatogr.* 656 (1993) 501.
94. X. Geng and F. E. Regnier, *J. Chromatogr.* 296 (1984) 15.
95. B. P. Johnson, M. G. Khaledi, and J. G. Dorsey, *Anal. Chem.* 58 (1986) 2354.
96. P. Jandera, *J. Chromatogr.* 314 (1984) 13.
97. N. S. Wilson, M. D. Nelson, J. W. Dolan, L. R. Snyder, and P. W. Carr, *J. Chromatogr. A* 961 (2002) 195.
98. J. A. Lewis, J. W. Dolan, L. R. Snyder, and I. Molnar, *J. Chromatogr.* 592 (1992) 197.
99. T. J. Sereda, C. T. Mant, and R. S. Hodges, *J. Chromatogr. A* 695 (1995) 205.
100. A. W. Purcell, G. L. Zhao, M. I. Aguilar, and M. T. W. Hearn, *J. Chromatogr. A* 852 (1999) 43.
101. P. L. Zhu, J. W. Dolan, and L. R. Snyder, *J. Chromatogr. A.* 756 (1996) 41.
102. L. R. Snyder, M. A. Quarry, and J. L. Glajch, *Chromatographia* 24 (1987) 33.
103. L. R. Snyder and J. J. Kirkland, *Introduction to Modern Liquid Chromatography*, 2nd edn, Wiley-Interscience, New York, 1979, Chap. 5.
104. S. G. Weber and P. W. Carr, in *High Performance Liquid Chromatography*, P. R. Brown and R. A. Hartwick, eds, Wiley-Interscience, New York, p. 1.
105. L. R. Snyder and M. A. Stadalius, in *High-performance Liquid Chromatography. Advances and Perspectives*, Vol. 4, Cs. Horváth, ed., Academic Press, New York, 1986, p. 195.
106. K. D. Patel, A. D. Jerkovich, J. C. Link, and J. W. Jorgenson, *Anal. Chem.* 76 (2004) 5777.
107. H. Chen and Cs. Horváth, *J. Chromatogr. A* 705 (1995) 3.
108. M. Martin and G. Guiochon, *J. Chromatogr. A* 1090 (2005) 16.

APPENDIX I

THE CONSTANT-S APPROXIMATION IN GRADIENT ELUTION

The examples of Section 2.2.1.1 for a "regular" sample assume that values of S are approximately constant for peaks that elute either early or late in the chromatogram. As a result, plots of log k^* vs ϕ^* can be used to visualize the effect of a change in gradient time on separation. This is illustrated in Figure I.1, for compounds 1, 5, and 9 of the "regular" sample of Table 1.3. If values of S were the same for all three compounds, a horizontal (dotted) line through the data-point for compound 1

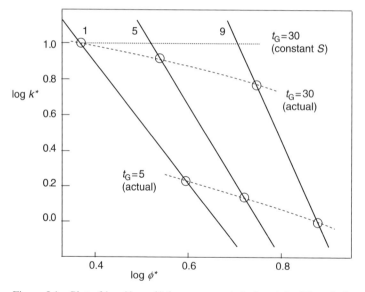

Figure I.1 Plot of log k^* vs ϕ^* for compounds 1, 3 and 5 of "regular" sample of Table 1.3. Conditions: 150 × 0.46 mm C_{18} column; 0–100 percent B gradient; 2.0 mL/min; see Table 1.3 for other conditions and text for details.

High-Performance Gradient Elution. By Lloyd R. Snyder and John W. Dolan
Copyright © 2007 John Wiley & Sons, Inc.

(with a gradient time of $t_G = 30$ min) should intersect the retention plots for compounds 5 and 9 to give the same value of k^*. The horizontal spacing along this dotted line between adjacent peaks should also be proportional to differences in retention time and resolution for each peak pair.

However, it was noted in Section 1.6.1 and Figure 1.11(a) that values of S often increase for later peaks in the chromatogram, as is seen to be the case for the compounds in Figure I.1: $S_{(1)} < S_{(5)} < S_{(9)}$. As a result [Equation (2.13)], values of k^* decrease for later-eluting peaks 5 and 9 in the same gradient separation (where $t_G = 30$ min). The dashed line connecting values of k^* for compounds 1, 5, and 9 corresponds to actual values of k^* for $t_G = 30$, and is different from the dotted line, which assumes no difference in values of S for these three compounds. The consequences of the (true) dashed line vs the (assumed) dotted line is a slight increase in the actual peak spacing (values of ϕ^*) and related resolution values. However, this does not detract significantly from conclusions that can be drawn from plots such as that of Figure 2.10, which assume no significant change in S when approximating the effect of a change in t_G by the use of horizontal lines.

APPENDIX II

ESTIMATION OF CONDITIONS FOR ISOCRATIC ELUTION, BASED ON AN INITIAL GRADIENT RUN

It is assumed that an initial gradient run has been carried out, for example, as described in Table 3.2. From the resulting chromatogram, identify the first and last peaks of interest, A and Z, respectively. Let their retention times be $t_{R,A}$ and $t_{R,Z}$, respectively. From values of $t_{R,A}$ and $t_{R,Z}$, it is possible to determine whether isocratic separation is possible (Section 2.2.3.2). If isocratic elution is feasible, we need an estimate of the value of %B for an isocratic separation that will exhibit acceptable retention: either $1 \leq k \leq 10$ or $0.5 \leq k \leq 20$. Gradient retention time t_R is given by

$$t_R \approx (t_0/b) \log(2.3 k_0 b) + t_0 + t_D \quad (2.12)$$
$$= (t_0/b)[\log k_0 + \log(2.3) + \log b] + t_0 + t_D \quad (II.1)$$

Define an average retention time $(t_R)_{avg} \equiv (t_{R,A} + t_{R,Z})/2$, and assume values of S and therefore b are equal for compounds A and Z. The isocratic values of k for the two compounds are

$$\log k_A = \log k_{0,A} - S\phi_0 \quad (II.2)$$

and

$$\log k_B = \log k_{0,B} - S\phi_0 \quad (II.3)$$

From Equation (II.1), the value of $(t_R)_{avg}$ is then

$$(t_R)_{avg} = 0.5(t_0/b)[\log(k_A \, k_B)] + (t_0/b)[\log(2.3) + \log b] + t_0 + t_D \quad (II.4)$$

From Equation (II.4) it is seen that $(t_R)_{avg}$ is related to isocratic retention for bands A and B as $0.5 [\log (k_A \, k_B)]$, so for either $1 \leq k \leq 10$ or $0.5 \leq k \leq 20$, $0.5 [\log (k_A \, k_B)]$ equals 0.5, corresponding to $(k_A \, k_B)^{1/2} = 10^{1/2} = 3.2$. The latter corresponds to k at elution (k_e) equal to 3.2. The value of ϕ at elution (ϕ_e) corresponding to $(t_R)_{avg}$ is also given by Equation (2.15) with $(t_R)_{avg} = t_R$:

$$\phi_e = \phi_0 + (\Delta\phi/t_G)[(t_R)_{avg} - t_0 - t_D] \quad (II.5)$$

High-Performance Gradient Elution. By Lloyd R. Snyder and John W. Dolan
Copyright © 2007 John Wiley & Sons, Inc.

The value of k at elution (k_e) is equal to $k^*/2$ (Section 9.1.1), which will in general differ from the desired average value of k (k_{iso}) for the isocratic separation, with $k_{iso} \approx 3.2$ and $\phi \equiv \phi_{iso}$. Therefore [Equation (2.9)],

$$\phi_e = [\log(k_w/k_e)]/S \tag{II.6}$$

and

$$\phi_{iso} = [\log(k_w/k_{iso})]/S \tag{II.7}$$

where ϕ_{iso} refers to the value of ϕ corresponding to k_{iso}. From Equations (II.6) and (II.7), with $k_{iso} = 3.2$,

$$\phi_{iso} - \phi_e = [\log(k_e/3.2)]/S \tag{II.8}$$

Combining Equations (II.5) and (II.8), with $k_e = k^*/2$ [Equation (9.14a)], we have

$$\phi_{iso} = (\Delta\phi/t_G)[[t_R]_{avg} - t_0 - t_D] + \phi_0 + [\log(k^*/6.4)]/S \tag{II.9}$$

Given $S \approx 4$ for typical "small-molecule" samples, as well as experimental conditions for the initial gradient run (which determines a value of k^*), a value of ϕ_{iso} for isocratic separation of the sample can be estimated from Equation (II.9).

In Chapter 3 we recommend experimental conditions for an initial gradient run (Table 3.2), with $\Delta\phi/t_G = 0.95/15 = 0.0633$, $t_0 = V_m/F \approx 1.5/2 = 0.75$, $\phi_0 = 0.05$ and $k^* \approx 5$. Inserting these values into Equation (2.26), we have

$$\phi_{iso} = 0.0633[[t_R]_{avg} - t_D] - 0.02 \tag{II.10}$$

Or if we express ϕ_{iso} as %B,

$$(\text{isocratic \%B}) = 6.33[(t_R)_{avg} - t_D] - 2 \tag{II.11}$$

Equation (II.11) becomes somewhat less reliable, for small values of $(t_R)_{avg}$, due to corresponding smaller values of k_0. That is, we have assumed that Equation (2.12) is valid, but this equation assumes $k_0 \gg 10$. A correction can be made for small values of k_0, by replacing Equation (2.12) by Equation (2.10). However, the latter assumes a small value of t_D; see the related discussion of Section 9.1.1.2. When both k_0 is small, and t_D is significant, a rigorous derivation of ϕ_{iso} becomes sufficiently complex to be of little practical value. In any case, Equation (II.11) is only intended as an initial *estimate* of a value of isocratic %B that will provide an acceptable retention range for the sample.

If the gradient used for the initial separation stops short of 100 percent B at completion [due to buffer solubility limitations (Section 3.2)], the above procedures for assessing isocratic elution can still be used. However, it is necessary to maintain the same gradient slope: $(\Delta\phi/t_G) = 6.3$ percent B/min. Predictions of isocratic separation as above [using values of Δt_R and $(t_R)_{avg}$] can then be carried out in identical fashion. While Equation (3.1) only applies for the gradient recommended in Table 3.2 ($t_G = 15$ min; $\Delta\phi = 0.95$), it can be adapted for other values of t_G, $\Delta\phi$, and initial %B;

$$\text{isocratic \%B} \approx (89/[t_G\Delta\phi])[(t_R)_{avg} - t_D] - 7 + (\text{initial \%B}) \tag{II.12}$$

APPENDIX III

CHARACTERIZATION OF REVERSED-PHASE COLUMNS FOR SELECTIVITY AND PEAK TAILING

The choice of column for a RP-LC separation should consider several aspects of the column [1]:

- efficiency or plate number N;
- reproducibility;
- stability;
- peak shape (symmetry);
- selectivity.

Assuming a well-packed column, its length and particle size determine its value of N (Section 9.5). The use of columns supplied by reputable manufacturers usually ensures that *column efficiency* and *reproducibility* will be acceptable. *Column stability* is a special concern when the mobile phase pH is <3 or >7, and/or for temperatures >50°C [especially for the *combination* of extreme pH ($2.5 \geq$ pH ≥ 7.5) and a higher temperature]. The column manufacturer's literature should be consulted in order to assure that a column will be stable for the conditions likely to be used for a given separation. Peak shape and selectivity are of primary concern here. Each of these column properties can be related to certain column properties that will be described next.

Symmetrical peaks are highly desirable in HPLC separation. Peak symmetry can be characterized either by the asymmetry function As or by the tailing factor TF [1] (Fig. III.1). Assuming a well-packed column that has not been degraded by use, and a sample size which is not too large, the usual cause of tailing peaks is an interaction of protonated bases with ionized silanols or contaminating metals in the silica used for the column packing (especially Al[III] and Fe[III]). Columns with low metal concentrations are referred to as "type B" ("basic"), in contrast to older, metal-contaminated columns ("type A," "acidic"). Consequently, type B RP-LC columns are preferred for the separation of basic compounds, in order to

---

High-performance Gradient Elution. By Lloyd R. Snyder and John W. Dolan
Copyright © 2007 John Wiley & Sons, Inc.

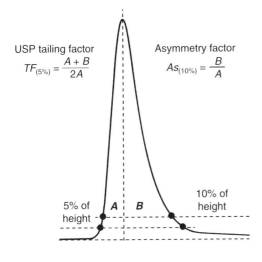

Figure III.1 Definitions of peak asymmetry factor As (measured at 10 percent of peak height) and USP tailing factor TF (at 5 percent height).

minimize peak tailing. The classification of alkylsilica columns as either type A or B is provided in Table III.1 for several hundred alkylsilica columns.

Column selectivity is also an important column characteristic. The selectivity of columns of a given type from a manufacturer can change over time, or the column may no longer be available. In either case, a column of equivalent selectivity from a different source will need to be identified, for the continued use of a method based on the original column. This requires the characterization of different commercial RP-LC columns in terms of their selectivity. Such a characterization procedure has been described [2], by which five different column selectivity parameters can be measured: hydrophobicity, **H**; steric resistance to penetration, **S***; hydrogen-bond acidity, **A**, and basicity, **B**; cation-exchange capacity, **C**. If two columns have sufficiently similar values of these five column parameters, the two columns can be regarded as equivalent in terms of selectivity, and equivalent for use in a given RP-LC method. Table III.1 summarizes values of **H**, **S***, and so on, for more than 300 columns. Note that the column parameter **C** varies with mobile phase pH; a value of **C** for a given mobile phase pH can be obtained by interpolation of values of **C** between pH 2.8 (**C**[2.8]) and pH 7.0 (**C**[7.0]). Column similarity or equivalency can be measured by the function F_s:

$$F_s = \{[12.5(\mathbf{H}_2 - \mathbf{H}_1)]^2 + [100(\mathbf{S}_2^* - \mathbf{S}_1^*)]^2 + [30(\mathbf{A}_2 - \mathbf{A}_1)]^2 \\ + [143(\mathbf{B}_2 - \mathbf{B}_1)]^2 + [83(\mathbf{C}_2 - \mathbf{C}_1)]^2\}^{1/2} \tag{III.1}$$

Here, $\mathbf{H}_1$ and $\mathbf{H}_2$ refer to values of **H** for columns 1 and 2, respectively (and similarly for values of $\mathbf{S}_1^*$ and $\mathbf{S}_2^*$, etc.). F_s can be regarded as the distance between two columns whose values of **H**, **S***, and so on, are plotted in five-dimensional space, with the weighting factors (12.5, 100, etc.) determined for a 67-component sample of "average" composition. It was found [2] that if $F_s \leq 3$ for two columns 1 and 2, average variations in α should be ≤ 3 percent, so that the two columns are likely to provide equivalent selectivity and separation for different samples and conditions

TABLE III.1 Column Selectivity Parameters for Commercial RP-LC Columns. See Text and [2] for Details. Values Below Differ Slightly from Values Reported in [2], as a Result of Added Columns, Repeat Measurements for Original Columns, and the Elimination of Several Transcription Errors

Name	H	S*	A	B	C (2.8)	C (7.0)	Retention[a]	Type[b]	Source
Zorbax C_{18}	1.089	0.055	0.474	0.060	1.489	1.566	10.7	A	Agilent
Zorbax C_8	0.974	−0.041	0.216	0.176	0.974	1.051	8.3	A	Agilent
Zorbax Eclipse XDB-C_{18}	1.077	0.024	−0.063	−0.033	0.055	0.088	9.1	B	Agilent
Zorbax Eclipse XDB-C_8	0.919	0.025	−0.219	−0.008	0.003	0.012	6.6	B	Agilent
Zorbax Extend C_{18}	1.098	0.050	0.012	−0.041	0.030	0.016	8.4	B	Agilent
Zorbax Rx-C_{18}	1.077	0.037	0.309	−0.038	0.096	0.415	7.7	B	Agilent
Zorbax Rx-C_8	0.792	−0.076	0.116	0.018	0.012	0.948	5.0	B	Agilent
Zorbax StableBond 300 Å C_{18}	0.905	−0.050	0.045	0.043	0.254	0.701	2.2	B	Agilent
Zorbax StableBond 300 Å C_3	0.526	−0.122	−0.195	0.047	0.057	0.357	0.7	B	Agilent
Zorbax StableBond 300 Å C_8	0.701	−0.085	0.002	0.047	0.146	0.820	1.3	B	Agilent
Zorbax StableBond 80 Å C_{18}	0.996	−0.032	0.264	−0.001	0.136	1.041	7.6	B	Agilent
Zorbax StableBond 80 Å C_3	0.601	−0.124	−0.081	0.038	−0.084	0.810	2.8	B	Agilent
Zorbax StableBond 80 Å C_8	0.795	−0.079	0.137	0.018	0.014	1.020	5.1	B	Agilent
Bonus RP	0.654	0.107	−1.046	0.373	−2.971	−1.103	4.5	EP	Agilent
Zorbax SB-AQ	0.593	−0.120	−0.083	0.038	−0.136	0.736	2.5	EP	Agilent
Zorbax SB-Phenyl	0.623	−0.161	0.065	0.038	0.033	1.089	2.7	Phenyl	Agilent
Zorbax XDB-Phenyl	0.665	−0.127	−0.242	0.019	0.063	0.584	3.2	B	Agilent
Zorbax SB-CN	0.502	−0.108	−0.224	0.042	−0.146	1.047	1.7	CN	Agilent
Zorbax Eclipse XDB-CN	0.456	−0.068	−0.312	0.003	0.074	0.994	1.3	CN	Agilent
Kromasil 100-5C_{18}	1.051	0.036	−0.071	−0.023	0.039	−0.057	12.5	B	Akzo Nobel
Kromasil 100-5C_4	0.733	0.003	−0.335	0.015	0.009	−0.004	5.0	B	Akzo Nobel
Kromasil 100-5C_8	0.864	0.013	−0.213	0.019	0.054	−0.001	7.6	B	Akzo Nobel
Kromasil KR60-5CN	0.440	−0.135	−0.578	−0.014	0.216	1.036	2.0	CN	Akzo Nobel
Advantage 300	0.867	−0.001	0.123	0.020	0.597	1.110	1.7	B	Analytical Sales & Service
Advantage Armor C_{18} 120 Å	0.962	−0.014	−0.076	−0.004	0.077	0.261	8.6	B	Analytical Sales & Service
Armor C_{18} 3 μm	0.964	−0.016	−0.079	−0.002	0.122	0.296	8.5	B	Analytical Sales & Service

Ultrasphere Octyl	0.896	0.016	0.003	0.086	0.157	0.547	5.7	B	Beckman
Ultrasphere ODS	1.085	−0.014	0.173	0.068	0.279	0.382	8.7	B	Beckman
BAS MF-8954	0.979	−0.069	0.181	0.022	1.081	1.397	6.9	A	BioAnalytical Systems
EU Reference Column	1.004	0.001	0.264	0.006	0.178	0.449	9.3	A	Bischoff
ProntoSIL 120-5 C_{18} SH	1.031	0.018	−0.109	−0.024	0.113	0.402	8.7	B	Bischoff
ProntoSIL 120-5 C_8 SH	0.739	−0.062	−0.081	0.013	0.076	0.526	4.9	B	Bischoff
Prontosil 120-5-C_1	0.413	−0.079	−0.085	0.020	0.042	0.656	1.2	B	Bischoff
ProntoSIL 120-5-C_{18} H	1.005	0.008	−0.106	−0.004	0.125	0.156	9.7	B	Bischoff
ProntoSIL 120-5-C_{18}-AQ	0.974	0.007	−0.083	0.003	0.137	0.223	8.1	B	Bischoff
Prontosil 120-3-C_{30}	0.919	−0.130	0.571	−0.003	0.507	1.788	6.9	B	Bischoff
Prontosil 300-5-C_{30} EC	0.925	−0.047	−0.018	0.012	0.303	0.458	2.8	B	Bischoff
Prontosil 200-5-C_{18} AQ	0.974	−0.007	−0.084	0.003	0.137	0.223	8.1	B	Bischoff
Prontosil 200-5 C_8 SH	0.761	−0.026	−0.195	0.024	0.125	0.238	2.7	B	Bischoff
ProntoSIL 200-5-C_{18} H	0.955	−0.001	−0.121	0.016	0.163	0.218	4.8	B	Bischoff
Prontosil 200-5-C_{30}	0.909	−0.099	0.347	0.007	0.305	1.171	4.4	B	Bischoff
Prontosil 200-5-C_4	0.549	−0.063	−0.221	0.038	0.086	0.511	1.3	B	Bischoff
ProntoSIL 300-5 C_8 SH	0.739	−0.041	−0.131	0.028	0.156	0.405	1.8	B	Bischoff
ProntoSIL 300-5-C_{18} H	0.956	−0.012	−0.090	0.015	0.238	0.249	3.2	B	Bischoff
Prontosil 300-5-C_{30}	0.893	−0.107	0.322	0.030	0.401	1.547	3.0	B	Bischoff
Prontosil 300-5-C_4	0.471	−0.093	−0.074	0.055	0.115	0.786	0.6	B	Bischoff
ProntoSIL 60-5 C_8 SH	0.929	−0.015	0.161	−0.017	−0.313	1.005	8.4	B	Bischoff
ProntoSIL 60-5-C_{18} H	1.158	0.041	0.066	−0.078	0.102	0.263	12.2	B	Bischoff
Prontosil 60-5-C_4	0.686	−0.072	0.108	0.001	−0.056	1.201	4.1	B	Bischoff
ProntoSil CN	0.370	−0.114	−0.414	−0.028	0.168	0.668	0.9	CN	Bischoff
Prontosil 120-5-CN EC	0.427	−0.053	−0.320	0.015	0.019	0.768	1.2	CN	Bischoff
ProntoSIL 120-5-C_{18} ace-EPS	0.772	0.042	−0.590	0.228	−0.304	0.041	8.3	EP	Bischoff
ProntoSIL 120-5-C_{18} AQplus	0.947	−0.017	0.214	0.041	−0.133	0.605	9.1	EP	Bischoff

(*continued*)

TABLE III.1 Column Selectivity Parameters for Commercial RP-LC Columns. See Text and [2] for Details. Values Below Differ Slightly from Values Reported in [2], as a Result of Added Columns, Repeat Measurements for Original Columns, and the Elimination of Several Transcription Errors

Name	H	S*	A	B	C (2.8)	C (7.0)	Retention[a]	Type[b]	Source
Prontosil 120-5-C_8 ace-EPS	0.532	−0.007	−0.852	0.213	−0.282	0.094	3.7	EP	Bischoff
ProntoSIL 200-5-C_{18} ace-EPS	0.765	0.021	−0.566	0.214	0.026	0.143	4.7	EP	Bischoff
ProntoSIL 300-55-C_{18} ace-EPS	0.762	0.025	−0.579	0.211	−0.054	0.136	2.9	EP	Bischoff
ProntoSIL 120-5-Phenyl	0.568	−0.158	−0.201	0.022	0.176	0.712	2.5	Phenyl	Bischoff
ProntoSIL 60-5-Phenyl	0.705	−0.194	−0.003	−0.010	0.411	1.510	4.5	Phenyl	Bischoff
ProntoSIL HyperSORB 120-5-ODS	0.951	−0.065	0.039	−0.021	0.795	1.315	5.2	B	Bischoff
ProntoSIL SpheriBOND 80-5-ODS1	0.700	−0.190	0.367	0.010	1.453	2.400	3.8	A	Bischoff
ProntoSIL SpheriBOND 80-5-ODS2	1.010	−0.026	0.153	−0.037	0.731	1.008	7.7	A	Bischoff
Acclaim C_{18}	1.032	0.018	−0.143	−0.027	0.086	−0.002	10.1	B	Dionex
Acclaim C_8	0.857	0.004	−0.274	0.011	0.086	0.016	6.0	B	Dionex
Acclaim300 C_{18}	0.957	0.018	−0.170	0.019	0.261	0.222	2.9	B	Dionex
Acclaim PA C_{16}	0.855	−0.068	−0.116	0.023	−0.270	0.357	6.8	B	Dionex
Acclaim Polar Advantage II	0.429	0.089	−0.992	0.050	−0.726	0.262	6.3	special	Dionex
Acclaim Organic Acid	0.761	−0.068	−0.345	−0.039	−0.543	0.213	7.0	special	Dionex
SynChropak RP8	0.639	−0.099	0.109	0.029	0.223	0.940	1.2	A	Eprogen
SynChropak RPP	0.746	−0.115	0.230	0.033	0.259	1.286	1.8	A	Eprogen
SynChropak RPP 100	0.918	−0.059	−0.072	0.123	0.225	0.317	5.4	A	Eprogen
Chromegabond WR C_{18}	0.979	0.026	−0.159	−0.003	0.320	0.283	5.4	B	ES Industries
Chromegabond WR C_8	0.855	0.025	−0.279	0.024	0.200	0.144	3.6	B	ES Industries
Inertsil Ph-3	0.526	−0.179	−0.133	0.040	0.121	0.735	2.6	B	GL Science
Inertsil C_8-3	0.830	−0.004	−0.268	−0.017	−0.334	−0.362	7.1	B	GL Science
Inertsil ODS-2	1.007	0.045	−0.079	−0.014	−0.139	0.446	8.2	B	GL Science
Inertsil ODS-3	0.990	0.022	−0.146	−0.023	−0.474	−0.334	10.9	B	GL Science
Inertsil ODS-P	0.978	−0.028	0.611	−0.039	0.234	1.479	11.2	B	GL Science
Inertsil WP300 C_{18}	0.938	−0.015	−0.117	0.001	0.202	0.163	3.8	B	GL Science
Inertsil WP300 C_8	0.793	−0.015	−0.212	0.013	0.122	0.069	2.2	B	GL Science

							CN	
Inertsil CN-3	0.369	0.049	−0.808	0.083	−2.607	−1.297	1.1	B	GL Science
Inertsil ODS-EP	0.807	0.064	−1.525	0.050	−0.626	−0.075	7.2	B	GL Science
Adsorbosphere (C$_{18}$)	0.989	−0.073	0.070	−0.044	1.496	1.683	7.7	A	Grace/Alltech
Adsorpbosphere UHS C$_{18}$	1.103	0.004	0.402	−0.046	−0.125	−0.125	18.2	A	Grace/Alltech
Allsphere ODS1	0.733	−0.160	0.387	0.002	0.846	1.142	4.7	A	Grace/Alltech
Allsphere ODS2	1.004	−0.040	0.243	−0.028	0.960	1.281	8.0	A	Grace/Alltech
Alphabond (C$_{18}$)	0.845	−0.094	0.061	0.001	0.579	1.760	5.3	A	Grace/Alltech
Econosil (C$_{18}$)	0.966	−0.066	0.376	−0.032	1.026	1.339	8.2	A	Grace/Alltech
Econosphere C$_{18}$	0.818	−0.128	0.036	−0.017	1.046	1.522	5.1	A	Grace/Alltech
Prosphere C$_{18}$ 300 Å	0.903	−0.012	0.176	0.013	0.577	1.266	2.2	A	Grace/Alltech
Alltima AQ	0.882	−0.070	0.301	0.016	0.158	1.157	8.8	B	Grace/Alltech
Alltima C$_{18}$	0.993	−0.014	0.035	−0.013	0.092	0.391	11.5	B	Grace/Alltech
Alltima C$_{18}$-LL	0.780	0.085	−0.165	0.041	−0.056	0.367	5.9	B	Grace/Alltech
Alltima C$_{18}$-WP	0.938	−0.062	0.027	0.002	−0.079	−0.081	4.9	B	Grace/Alltech
Alltima C$_8$	0.756	0.015	−0.279	0.009	−0.062	0.288	5.5	B	Grace/Alltech
Alltima HP C$_{18}$	0.985	−0.020	−0.040	0.006	0.177	0.199	4.9	B	Grace/Alltech
Alltima HP C$_{18}$ High Load	1.080	−0.066	0.066	−0.040	−0.322	−0.244	11.8	B	Grace/Alltech
Alltima HP C$_8$	0.834	0.010	−0.116	0.035	0.122	−0.418	3.1	B	Grace/Alltech
Brava BDS C$_{18}$	0.935	−0.033	0.033	0.012	0.281	0.768	4.7	B	Grace/Alltech
Platinum C$_{18}$	0.615	−0.168	0.335	0.026	0.719	1.729	2.6	B	Grace/Alltech
Platinum C$_8$	0.378	−0.138	0.076	0.031	0.274	1.133	1.1	A	Grace/Alltech
Platinum EPS C$_{18}$	0.615	−0.168	0.335	0.026	0.719	1.729	2.6	B	Grace/Alltech
Platinum EPS C$_8$	0.420	−0.152	0.151	0.026	0.509	1.369	1.1	B	Grace/Alltech
Prevail C$_{18}$	0.888	−0.070	0.315	0.022	0.107	1.206	9.4	B	Grace/Alltech
Prevail C$_8$	0.617	−0.089	0.039	0.041	0.081	1.072	3.4	B	Grace/Alltech
Prevail Select C$_{18}$	0.822	0.029	−0.368	0.141	−1.057	0.455	7.5	B	Grace/Alltech
Alltima HP C$_{18}$ Amide	0.497	−0.026	0.357	0.124	−0.019	0.926	4.8	EP	Grace/Alltech

(*continued*)

TABLE III.1 Column Selectivity Parameters for Commercial RP-LC Columns. See Text and [2] for Details. Values Below Differ Slightly from Values Reported in [2], as a Result of Added Columns, Repeat Measurements for Original Columns, and the Elimination of Several Transcription Errors

Name	H	S*	A	B	C (2.8)	C (7.0)	Retention[a]	Type[b]	Source
Alltima HP C$_{18}$ EPS	0.655	−0.104	0.401	0.036	0.459	0.955	1.2	EP	Grace/Alltech
Platinum EPS C$_{18}$ 300	0.450	−0.058	0.379	0.016	0.247	1.291	3.6	EP	Grace/Alltech
Platinum EPS C$_8$ 300	0.584	−0.113	−0.136	0.089	0.481	0.961	0.8	EP	Grace/Alltech
Prevail Amide	0.862	−0.063	0.251	0.033	0.058	1.209	9.6	EP	Grace/Alltech
Prosphere 300 C$_4$	0.689	−0.015	−0.059	0.027	0.312	0.684	1.0	B	Grace/Alltech
Prosphere 100 C$_{18}$	0.883	−0.073	0.305	0.017	0.181	1.517	6.4	B	Grace/Alltech
Vydac 218MS	0.770	0.182	0.111	−0.373	0.659	1.234	1.7	A	Grace/Vydac
Vydac Everest	0.993	0.049	0.121	0.004	0.065	0.341	2.2	B	Grace/Vydac
Denali (120A C18)	1.052	0.042	0.125	−0.014	0.143	0.222	9.371191	B	Grace/Vydac
Vydac 218TP	0.909	0.009	0.345	−0.005	0.279	0.670	2.0	A	Grace/Vydac
Vydac 201TP	0.901	−0.022	0.409	−0.004	0.394	1.026	2.1	A	Grace/Vydac
Vydac Protein and Peptide C$_{18}$	0.909	0.009	0.345	−0.005	0.279	0.670	2.0	A	Grace/Vydac
Vydac Monomeric C$_{18}$	0.993	0.049	0.121	0.004	0.065	0.341	2.2	B	Grace/Vydac
GROM-SIL 120 ODS-5 ST	1.035	−0.001	0.134	−0.005	0.135	0.121	10.5	B	Grom/Alltech
GROM-SIL 120 Octyl-6 MB	0.872	0.001	−0.007	0.029	−0.017	0.135	5.8	B	Grom/Alltech
GROM Saphir 110 C$_{18}$	1.055	−0.002	0.085	0.000	−0.030	0.115	12.1	B	Grom/Alltech
GROM-SIL 120 ODS-3 CP	1.029	0.019	0.093	−0.005	0.099	0.123	10.2	B	Grom/Alltech
GROM SAPHIR 110 C$_8$	0.835	−0.032	−0.103	0.031	−0.093	0.255	7.1	B	Grom/Alltech
HxSil C$_{18}$	0.848	−0.077	0.303	0.017	0.230	1.054	7.0	B	Hamilton
HxSil C$_8$	0.684	−0.075	0.089	0.030	0.066	0.856	4.4	B	Hamilton
Hichrom RPB	0.964	0.027	0.106	0.003	0.153	0.143	6.4	B	HiChrom
Hichrom 300 5 RPB	0.944	0.028	0.044	0.015	0.226	0.216	2.6	B	HiChrom
Targa C$_{18}$	0.977	−0.019	−0.070	0.000	0.013	0.175	8.6	B	Higgens
Unison UK-C$_{18}$	0.981	−0.019	0.015	−0.011	0.110	0.070	8.9	B	Imtakt
Apex C$_{18}$	0.985	−0.035	0.013	0.042	1.246	2.311	6.1	A	Jones/Alltech
Apex C$_8$	0.869	−0.071	0.235	0.177	1.364	1.373	4.4	A	Jones/Alltech

Column								
Apex II C$_{18}$ (ODS??)	1.008	−0.074	0.235	0.123	2.039	2.690	6.7	A	Jones/Alltech
Genesis AQ 120 Å	0.960	−0.036	−0.157	0.007	0.060	0.233	9.6	B	Jones/Alltech
Genesis C$_{18}$ 120 Å	1.005	0.004	−0.069	−0.007	0.139	0.124	9.8	B	Jones/Alltech
Genesis C$_{18}$ 300 Å	0.974	0.005	−0.086	0.013	0.266	0.270	3.5	B	Jones/Alltech
Genesis C$_4$ 300 Å	0.615	−0.057	−0.397	0.036	0.143	0.249	1.1	B	Jones/Alltech
Genesis C$_4$ EC 120 Å	0.646	−0.058	−0.331	0.027	0.063	0.400	3.4	B	Jones/Alltech
Genesis C$_8$ 120Å	0.829	−0.016	−0.082	0.018	0.055	0.300	6.2	B	Jones/Alltech
Genesis EC C$_8$ 120 Å	0.863	0.005	−0.174	0.023	0.064	0.141	6.9	B	Jones/Alltech
Genesis CN 120 Å	0.424	−0.114	−0.681	−0.013	−0.001	0.573	1.4	CN	Jones/Alltech
Genesis CN 300 Å	0.397	−0.108	−0.645	−0.009	0.025	0.397	0.5	CN	Jones/Alltech
Genesis Phenyl	0.609	−0.140	−0.368	0.031	0.133	0.588	2.9	Phenyl	Jones/Alltech
Nucleosil 100-5-C$_8$ HD	0.865	−0.008	−0.174	0.029	0.045	0.188	6.3	A	Machery Nagel
Nucleosil 100-5-C$_{18}$ HD	0.961	−0.021	−0.126	0.009	0.089	0.150	8.8	A	Machery Nagel
Nucleosil 100-5-C$_{18}$ Nautilus	0.702	0.003	−0.483	0.268	−0.441	0.486	5.4	EP	Machery Nagel
Nucleosil C$_8$	0.575	−0.134	0.038	0.017	0.282	1.122	2.7	A	Machery Nagel
Nucleosil C$_{18}$	0.906	−0.052	0.012	−0.030	0.321	0.730	7.3	A	Machery Nagel
Nucleosil ODS	0.860	−0.081	−0.008	0.014	0.453	0.984	2.7	A	Machery Nagel
Nucleodur 100-C$_{18}$ Gravity	0.868	0.032	−0.240	0.000	−0.158	0.631	6.6	B	Machery Nagel
Nucleodur C$_{18}$ Gravity	1.056	0.041	−0.097	−0.025	−0.080	0.316	11.0	B	Machery Nagel
EC Nucleosil 100-5 Protect 1	0.544	0.048	−0.411	0.309	−3.213	−0.573	2.7	EP	Machery Nagel
ACE 300 C$_8$	0.786	−0.003	−0.112	0.032	0.145	1.456	1.8	B	MacMod/ACT
ACE C$_4$	0.674	−0.018	−0.178	0.026	0.090	0.316	2.5	B	MacMod/ACT
ACE 5 C$_4$-300	0.710	−0.014	−0.183	0.039	0.166	0.356	1.3	B	MacMod/ACT
ACE 5 C$_{18}$	1.000	0.027	−0.096	−0.007	0.143	0.096	7.9	B	MacMod/ACT
ACE 5 C$_{18}$-300	0.983	0.025	0.046	0.012	0.262	0.237	3.0	B	MacMod/ACT
ACE 5 C$_8$	0.830	−0.004	−0.268	−0.017	−0.334	−0.298	4.9	B	MacMod/ACT
ACE 5CN	0.409	−0.107	−0.729	−0.008	−0.086	0.441	1.0	CN	MacMod/ACT
ACE 5 CN-300	0.460	−0.074	−0.165	0.030	0.151	0.856	0.4	CN	MacMod/ACT

(*continued*)

TABLE III.1 Column Selectivity Parameters for Commercial RP-LC Columns. See Text and [2] for Details. Values Below Differ Slightly from Values Reported in [2], as a Result of Added Columns, Repeat Measurements for Original Columns, and the Elimination of Several Transcription Errors

Name	H	S*	A	B	C (2.8)	C (7.0)	Retention[a]	Type[b]	Source
ACE AQ	0.804	−0.051	−0.129	0.034	0.009	0.167	4.7	EP	MacMod/ACT
ACE Phenyl	0.647	−0.138	−0.296	0.027	0.132	0.466	2.8	Phenyl	MacMod/ACT
ACE Phenyl-300	0.599	−0.105	−0.234	0.032	0.164	0.548	1.1	Phenyl	MacMod/ACT
Precision CN	0.431	−0.114	−0.485	0.019	−0.041	0.606	1.3	CN	MacMod/Higgins
PRECISION C$_{18}$	1.002	0.003	−0.042	−0.010	0.079	0.341	9.5	B	MacMod/Higgins
PRECISION C$_8$	0.821	−0.014	−0.180	0.022	0.095	0.241	4.9	B	MacMod/Higgins
Precision C$_{18}$-PE	0.976	−0.018	−0.085	−0.001	0.005	0.168	8.6	EP	MacMod/Higgins
Precision C$_8$-PE	0.814	−0.021	−0.159	0.017	0.051	0.279	5.0	B	MacMod/Higgins
Precision Phenyl	0.595	−0.136	−0.296	0.027	0.099	0.508	2.6	Phenyl	MacMod/Higgins
Purospher RP-18	0.841	0.235	0.155	0.300	−0.964	0.901	4.9	B	Merck
LiChrosorb RP$_{18}$	0.909	−0.070	0.151	−0.080	0.714	14.404	7.1	A	Merck
LiChrospher 100 RP$_{-18}$	1.006	−0.021	0.183	−0.036	0.646	5.039	9.5	A	Merck
Chromolith RP18e	1.003	0.029	0.008	−0.014	0.103	0.187	3.1	B	Merck
LiChrospher 60 RP-Select B	0.747	−0.060	−0.042	0.006	0.108	4.538	5.1	B	Merck
Purospher STAR RP$_{18}$e	1.003	0.013	−0.071	−0.037	0.018	0.044	10.6	B	Merck
Cogent UDC Cholesterol	0.625	0.227	0.528	0.069	0.745	1.212	3.4	Special	MicroSolv
Cogent HPS C18	1.021	−0.011	−0.071	−0.014	0.106	0.089	10.2	B	MicroSolv
Cogent HQ C18	0.9080	−0.0659	0.3766	0.0046	0.1898	2.1800	8.1	C	MicroSolv
Cogent Bidentate C18	0.950	0.059	−0.130	0.004	0.785	2.266	5.6	B	MicroSolv
Purospher STAR RP$_{18}$e	1.003	0.013	−0.071	−0.037	0.018	0.044	10.6	B	Merck
Superspher 100 RP-18e	1.030	0.025	−0.028	−0.011	0.352	1.621	9.3	B	Merck
COSMOSIL AR-II	1.017	0.011	0.126	−0.029	0.116	0.494	8.1	B	Nacalai Tesque
COSMOSIL MS-II	1.031	0.042	−0.132	−0.014	−0.118	−0.027	8.1	B	Nacalai Tesque
Cosmosil 5-C$_{18}$-PAQ	0.822	−0.027	−0.342	0.053	−0.353	0.047	5.5	EP	Nacalai Tesque
Develosil C$_{30}$-UG-5	0.976	−0.036	−0.196	0.011	0.158	0.176	7.8	B	Nomura
Develosil ODS-HG-5	0.980	0.015	−0.172	−0.008	0.187	0.221	8.2	B	Nomura

Column								
Develosil ODS-MG-5	0.963	−0.036	−0.165	−0.003	−0.012	0.051	11.2	B	Nomura
Develosil ODS-UG-5	0.996	0.025	−0.146	−0.004	0.150	0.155	8.4	B	Nomura
Bondclone C$_{18}$	0.824	−0.056	−0.125	0.044	0.078	0.347	4.5	A	Phenomenex
Partisil C$_8$	0.749	−0.071	−0.099	0.074	0.035	0.546	4.5	A	Phenomenex
Partisil ODS(3)	0.810	−0.079	−0.007	0.002	0.317	0.902	5.4	A	Phenomenex
Sphereclone ODS(2)	0.975	−0.045	0.278	−0.051	0.866	1.326	7.6	A	Phenomenex
Jupiter300 C$_{18}$	0.945	0.031	−0.225	0.008	0.234	0.218	2.9	B	Phenomenex
Jupiter300 C4	0.698	0.008	−0.426	0.019	0.152	0.141	1.3	B	Phenomenex
Jupiter300 C5	0.729	0.021	−0.382	0.016	0.129	0.330	1.5	B	Phenomenex
Luna C$_{18}$	1.018	0.025	0.072	0.008	−0.361	−0.036	10.9	B	Phenomenex
Luna C$_{18}$(2)	1.002	0.024	−0.124	−0.007	−0.269	−0.174	9.6	B	Phenomenex
Luna C$_5$	0.800	0.035	−0.252	0.003	−0.278	0.114	5.9	B	Phenomenex
Luna C$_8$	0.875	0.037	−0.015	0.024	−0.400	0.133	7.0	B	Phenomenex
Luna C$_8$(2)	0.889	0.041	−0.222	−0.001	−0.300	−0.170	7.2	B	Phenomenex
Prodigy ODS(2)	0.995	0.030	−0.114	−0.001	−0.091	0.237	7.7	B	Phenomenex
Prodigy ODS (3)	1.023	0.025	−0.131	−0.012	−0.195	−0.134	10.1	B	Phenomenex
Selectosil C$_{18}$	0.911	−0.054	0.034	−0.009	0.296	0.743	7.0	A	Phenomenex
Synergi Max-RP	0.989	0.028	−0.008	−0.013	−0.133	−0.034	9.5	B	Phenomenex
Ultracarb ODS (30)	1.114	0.016	0.377	−0.050	−0.311	0.731	18.2	B	Phenomenex
Aqua C$_{18}$	0.966	−0.030	0.033	0.009	0.068	0.276	8.8	B	Phenomenex
Luna CN	0.452	−0.112	−0.323	−0.024	0.439	1.321	1.3	CN	Phenomenex
Polaris C$_{18}$-Ether	0.943	−0.013	−0.122	0.027	0.164	0.553	5.5	EP	Phenomenex
Polaris C$_8$-Ether	0.705	−0.023	−0.312	0.040	0.095	0.269	1.8	EP	Phenomenex
Synergi Hydro-RP	1.022	−0.006	0.169	−0.042	−0.077	0.260	11.3	EP	Phenomenex
Synergi Polar-RP	0.654	−0.148	−0.257	−0.007	0.057	0.778	3.9	EP	Phenomenex
Luna Phenyl-Hexyl	0.782	−0.118	−0.277	−0.004	0.004	0.387	5.2	Phenyl	Phenomenex
Prodigy Phenyl-3	0.529	−0.195	0.055	0.022	0.230	1.467	2.3	Phenyl	Phenomenex

(*continued*)

TABLE III.1 Column Selectivity Parameters for Commercial RP-LC Columns. See Text and [2] for Details. Values Below Differ Slightly from Values Reported in [2], as a Result of Added Columns, Repeat Measurements for Original Columns, and the Elimination of Several Transcription Errors

Name	H	S*	A	B	C (2.8)	C (7.0)	Retention[a]	Type[b]	Source
Curosil-PFP	0.695	−0.079	−0.267	−0.004	0.119	0.379	4.0	F	Phenomenex
Onyx Monolithic C$_8$	0.824	0.003	−0.006	0.004	−0.020	0.441	1.9675	B	Phenomenex
Onyx Monolithic C$_{18}$	1.012	0.021	0.227	−0.018	0.120	0.430	3.197279	B	Phenomenex
Gemini C18 110 Å	0.967	−0.008	0.027	0.013	−0.091	0.195	7.998733	B	Phenomenex
Synergi Fusion-RP	0.879	−0.030	−0.014	0.008	−0.238	0.362	7.494937	B	Phenomenex
Ultra AQ C$_{18}$	0.857	−0.115	0.431	0.001	0.122	1.239	8.7	A	Restek
Allure C$_{18}$	1.116	0.043	0.113	−0.044	−0.047	0.067	15.7	B	Restek
Restek Ultra C$_{18}$	1.055	0.030	−0.069	−0.022	0.009	−0.066	12.6	B	Restek
Restek Ultra C$_8$	0.876	0.031	−0.229	0.018	0.043	0.012	7.6	B	Restek
Ultra IBD	0.657	−0.031	−0.022	0.233	−0.512	0.915	3.8	EP	Restek
SepaxBioC18	0.915	0.028	−0.157	0.019	0.227	0.240	2.6	B	Sepax Technologies
SepaxGP-C18	1.014	−0.014	−0.112	−0.019	0.103	0.096	9.0	B	Sepax Technologies
SepaxHP-C18	0.951	0.026	−0.102	0.001	0.070	0.221	8.5	B	Sepax Technologies
Allure PFP Propyl	0.732	−0.157	−0.179	−0.037	0.710	1.485	6.8	F	Restek
Ultra PFP	0.501	−0.089	−0.228	−0.003	−0.033	0.588	1.9	F	Restek
UltraSep ES AMID H RP18P	0.751	−0.013	−0.101	0.259	−0.527	0.855	4.9	EP	SepServe
UltraSep ES PHARM RP18	0.953	−0.061	0.435	−0.057	0.593	1.674	8.5	A C18	Sepserve
Exsil C$_8$	0.756	−0.076	−0.044	−0.014	0.472	0.974	4.7	A	SGE
Exsil ODS	0.992	−0.036	0.292	−0.040	0.836	1.229	7.6	A	SGE
Wakosil 5C$_8$RS	0.802	−0.008	−0.272	0.001	−0.117	0.097	5.6	B	SGE
Wakosil II 5C$_{18}$AR	0.998	0.075	−0.055	−0.034	0.070	0.010	6.2	B	SGE
Wakosil II 5C$_{18}$HG	1.039	0.036	0.015	−0.023	0.009	0.210	7.1	B	SGE
Wakosil II 5C$_{18}$RS	0.964	−0.008	−0.160	−0.009	−0.070	0.046	9.2	B	SGE
CAPCELL C$_{18}$ UG120	1.007	0.036	0.037	−0.012	0.016	0.001	6.9	B	Shiseido
CAPCELL C$_{18}$ AG120	1.030	0.060	0.122	−0.065	0.543	0.628	7.2	B	Shiseido
CAPCELL C$_{18}$ MG	1.005	0.010	0.042	−0.007	0.079	0.007	10.2	B	Shiseido

CAPCELL C$_{18}$ SG120	0.987	0.031	0.093	−0.023	0.121	0.197	6.6	B	Shiseido
CAPCELL C$_{18}$ A Q	0.867	−0.046	−0.068	0.014	−0.093	0.402	7.0	EP	Shiseido
CAPCELL C$_{18}$ ACR	1.025	0.045	0.073	−0.015	0.037	0.111	8.5	B	Shiseido
CAPCELL PAK C$_8$ DD	0.836	0.020	−0.154	0.015	−0.111	−0.075	5.4	B	Shiseido
CAPCELL PAK C$_{18}$ MGII	1.011	0.011	0.047	−0.006	0.007	−0.009	9.9	B	Shiseido
CAPCELL PAK C$_8$ UG120	0.854	0.037	−0.097	−0.013	−0.046	−0.010	4.3	B	Shiseido
Cadenza CD-C$_{18}$	1.057	0.031	0.083	−0.028	0.113	0.042	9.9	B	Imtakt/Silvertone Sciences
Discovery BIO Wide pore C$_{18}$	0.836	0.014	−0.254	0.028	0.121	0.119	3.4	B	Supelco
Discovery BIO Wide pore C$_5$	0.654	−0.019	−0.305	0.029	0.091	0.219	1.1	B	Supelco
Discovery BIO Wide pore C$_8$	0.839	0.018	−0.224	0.034	0.206	0.194	2.2	B	Supelco
Discovery C$_{18}$	0.984	0.027	−0.128	0.004	0.176	0.153	4.8	B	Supelco
Discovery C$_8$	0.832	0.011	−0.238	0.029	0.119	0.143	3.3	B	Supelco
Discovery CN	0.397	−0.110	−0.615	−0.002	−0.035	0.513	0.6	CN	Supelco
Discovery Amide C$_{16}$	0.720	0.013	−0.625	0.218	−0.092	−0.025	4.0	EP	Supelco
Discovery HS PEG	0.318	0.027	−0.713	0.128	−0.531	0.387	0.7	EP	Supelco
Supelcosil LC$_{18}$	1.018	−0.047	0.181	0.162	1.595	1.752	5.9	A	Supelco
Supelcosil LC$_{18}$-DB	0.979	−0.026	0.047	0.114	0.481	0.531	5.7	A	Supelco
Supelcosil LC$_8$	0.834	−0.048	−0.027	0.086	1.117	1.094	3.6	A	Supelco
Supelcosil LC$_8$-DB	0.819	−0.036	−0.072	0.143	0.446	0.554	3.4	A	Supelco
Supelcosil LC-PAH	0.851	−0.025	0.104	−0.030	0.642	0.830	4.4	A	Supelco
Discovery HS F5	0.631	−0.166	−0.325	0.023	0.709	0.940	4.0	F	Supelco
Thermo CN	0.404	−0.111	−0.709	−0.009	−0.029	0.491	0.8	CN	Thermo/Hypersil
Aquasil C$_{18}$	0.805	0.114	0.265	0.011	0.230		7.4	EP	Thermo/Hypersil
Hypersil 100 C$_{18}$	1.048	0.022	0.118	0.031	0.405	0.348	7.9	A	Thermo/Hypersil
Hypersil BDS C$_{18}$	0.993	0.016	−0.095	−0.009	0.337	0.281	5.6	A	Thermo/Hypersil
Hypersil Elite	0.958	0.031	0.151	−0.010	0.314	0.739	6.5	A	Thermo/Hypersil
Hypersil ODS	0.974	−0.026	−0.122	0.020	0.913	0.974	5.5	A	Thermo/Hypersil

(*continued*)

TABLE III.1 Column Selectivity Parameters for Commercial RP-LC Columns. See Text and [2] for Details. Values Below Differ Slightly from Values Reported in [2], as a Result of Added Columns, Repeat Measurements for Original Columns, and the Elimination of Several Transcription Errors

Name	H	S*	A	B	C (2.8)	C (7.0)	Retention[a]	Type[b]	Source
Hypersil ODS-2	0.985	0.016	0.139	−0.011	0.254	0.370	5.6	A	Thermo/Hypersil
Hypersil PAH	0.949	−0.057	0.234	−0.017	1.439	1.724	5.3	A	Thermo/Hypersil
Hypersil Beta Basic-18	0.993	0.033	−0.099	0.001	0.163	0.126	6.4	B	Thermo/Hypersil
Hypersil Beta Basic-8	0.834	0.016	−0.248	0.029	0.110	0.115	4.2	B	Thermo/Hypersil
Hypersil BetamaxNeutral	1.098	0.036	0.067	−0.031	−0.038	0.012	17.0	B	Thermo/Hypersil
BetaBasic CN	0.426	−0.043	−0.453	0.014	0.014	0.904	0.8	CN	Thermo/Hypersil
BetaMax Acid	0.635	0.057	−0.597	0.376	−2.064	−0.510	5.8	EP	Thermo/Hypersil
BetaMax Base	0.470	−0.060	−0.391	0.010	0.014	1.146	2.2	CN	Thermo/Hypersil
Hypersil Bio Basic-18	0.974	0.025	−0.100	0.007	0.253	0.217	3.2	B	Thermo/Hypersil
Hypersil Bio Basic-8	0.821	0.012	−0.233	0.029	0.231	0.210	1.8	B	Thermo/Hypersil
Hypersil GOLD	0.881	0.002	−0.017	0.036	0.162	0.479	3.9	B	Thermo/Hypersil
Hypurity C$_{18}$	0.980	0.025	−0.091	0.003	0.192	0.167	5.6	B	Thermo/Hypersil
HyPurity C$_4$	0.713	0.000	−0.291	0.028	0.121	0.252	1.9	B	Thermo/Hypersil
Hypurity C$_8$	0.833	0.011	−0.201	0.035	0.157	0.161	3.5	B	Thermo/Hypersil
Hypurity Cyano	0.451	−0.049	−0.492	0.021	−0.016	0.839	0.7	CN	Thermo/Hypersil
Hypersil Prism C$_{18}$ RP	0.645	0.089	−0.459	0.301	−2.817[a]	−0.716[a]	4.8	EP	Thermo/Hypersil
Hypersil Prism C$_{18}$ RPN	0.678	−0.001	−0.068	0.230	−0.544	0.625	3.4	EP	Thermo/Hypersil
Hypurity Advance	0.412	−0.056	−0.095	0.249	−1.332[a]	0.785[a]	1.6	EP	Thermo/Hypersil
Fluophase PFP	0.675	−0.129	−0.311	0.065	0.817	1.375	4.5	F	Thermo/Hypersil
Fluophase RP	0.698	0.028	0.103	0.039	1.034	1.417	3.4	F	Thermo/Hypersil
BetaBasic Phenyl	0.582	−0.159	−0.411	0.049	0.104	0.758	1.7	Phenyl	Thermo/Hypersil
Betasil Phenyl-Hexyl	0.707	−0.053	−0.294	0.028	0.054	0.357	4.3	Phenyl	Thermo/Hypersil
TSKgel ODS-100Z	1.032	0.018	−0.135	−0.031	−0.064	−0.161	11.6	B	Tosoh
TSKgel ODS-100V	0.901	−0.043	−0.226	−0.009	−0.060	−0.020	9.0	B	Tosoh

OmniSpher 5 C$_{18}$	1.055	0.051	−0.033	−0.029	0.122	0.058	10.9	B	Varian
Polaris C$_{18}$-Å	0.928	0.007	−0.227	0.061	0.149	0.160	5.2	EP	Varian
Polaris C$_{8}$-Å	0.601	−0.007	−0.609	0.104	−0.074	0.208	2.2	EP	Varian
Pursuit C$_{18}$	1.001	0.004	−0.166	0.012	0.245	0.226	6.2	B	Varian
MicroBondapak C$_{18}$	0.798	−0.077	−0.030	0.016	0.285	0.854	4.6	A	Waters
Nova-Pak C$_{18}$	1.049	0.004	0.098	−0.027	0.546	0.563	6.5	A	Waters
Nova-Pak C$_{8}$	0.899	−0.028	−0.094	0.006	0.611	0.621	3.9	A	Waters
Resolve C$_{18}$	0.968	−0.127	0.335	−0.046	1.921	2.144	7.9	A	Waters
Spherisorb C$_{8}$	0.763	−0.091	−0.032	0.053	0.737	1.142	4.9	A	Waters
Spherisorb ODS-1	0.682	−0.186	0.323	0.018	0.843	1.297	4.6	A	Waters
Spherisorb ODS-2	0.962	−0.076	0.070	0.034	0.908	1.263	8.3	A	Waters
Spherisorb S5 ODSB	0.975	0.027	0.240	0.384	−0.642	1.680	6.9	A	Waters
Atlantis dC$_{18}$ b	0.917	−0.031	−0.193	0.001	0.036	0.087	8.1	B	Waters
DeltaPak C$_{18}$ 100 Å	1.028	0.019	−0.018	−0.011	−0.051	0.024	9.0	B	Waters
DeltaPak C$_{18}$ 300 Å	0.955	−0.013	−0.105	0.016	0.235	0.286	3.0	B	Waters
J'Sphere H80	1.132	−0.059	−0.023	−0.068	−0.242	−0.161	13.3	B	Waters
J'Sphere L80	0.762	0.036	−0.216	−0.001	−0.400	0.345	5.8	B	Waters
J'Sphere M80	0.926	0.026	−0.123	−0.004	−0.294	0.139	9.1	B	Waters
Sunfire C$_{8}$	0.856	0.036	−0.122	0.006	−0.278	0.006	5.8	B	Waters
Sunfire C$_{18}$	1.031	0.034	0.044	−0.014	−0.186	−0.099	9.9	B	Waters
Symmetry 300 C$_{18}$	0.984	0.031	−0.051	0.003	0.228	0.202	3.5	B	Waters
Symmetry 300 C$_{4}$	0.659	−0.016	−0.428	0.014	0.101	0.184	1.4	B	Waters
Symmetry C$_{18}$	1.052	0.063	0.018	−0.021	−0.302	0.123	9.8	B	Waters
Symmetry C$_{8}$	0.893	0.049	−0.205	0.021	−0.509	0.283	7.0	B	Waters
Xterra MS C$_{18}$	0.984	0.012	−0.143	−0.015	0.133	0.051	6.3	B	Waters
Xterra MS C$_{8}$	0.803	0.005	−0.293	−0.005	0.058	−0.009	3.7	B	Waters

(continued)

TABLE III.1 Column Selectivity Parameters for Commercial RP-LC Columns. See Text and [2] for Details. Values Below Differ Slightly from Values Reported in [2], as a Result of Added Columns, Repeat Measurements for Original Columns, and the Elimination of Several Transcription Errors

Name	H	S*	A	B	C (2.8)	C (7.0)	Retention[a]	Type[b]	Source
YMC Basic	0.821	−0.006	−0.235	0.028	0.070	0.093	3.3	B	Waters
YMC Hydrosphere C_{18}	0.937	−0.022	−0.129	0.006	−0.139	0.157	6.8	B	Waters
YMC ODS-AQ	0.965	−0.036	−0.135	0.004	−0.068	0.100	8.6	B	Waters
YMC Pack Pro C18 RS	1.114	0.057	−0.061	−0.056	−0.176	−0.224	12.7	B	Waters
YMC Pro C_{18}	1.015	0.014	−0.120	−0.007	−0.155	−0.006	8.7	B	Waters
YMC Pro C_8	0.890	0.014	−0.215	0.007	−0.323	0.019	6.5	B	Waters
Nova-Pak CN HP 60 Å	0.362	−0.165	0.100	0.000	0.691	1.175	0.4	CN	Waters
Symmetry Shield C_{18}	0.850	0.027	−0.411	0.093	−0.728	0.136	7.3	EP	Waters
Symmetry Shield C_8	0.730	−0.006	−0.550	0.103	−0.623	0.138	5.7	EP	Waters
Xterra C_{18} RP	0.757	−0.043	−0.483	0.097	−0.170	−0.173	4.3	EP	Waters
Xterra C_8 RP	0.657	−0.049	−0.604	0.099	−0.187	−0.198	3.1	EP	Waters
Xterra Phenyl	0.683	−0.079	−0.363	−0.003	0.119	0.029	2.6	Phenyl	Waters
MicroBondapak Phenyl	0.585	−0.152	−0.247	0.021	0.359	0.976	2.1	Phenyl	Waters
Nova-Pak Phenyl	0.704	−0.159	−0.300	0.015	0.767	0.812	2.5	Phenyl	Waters
Acquity UPLC BEH phenyl	0.764	0.077	−0.051	0.062	0.292	0.586	2.0	Phenyl	Waters
Acquity UPLC BEH Shield RP_{18}	0.907	0.016	−0.031	0.133	−0.055	0.416	3.5	EP	Waters
Acquity UPLC BEH C_8	0.855	−0.008	0.095	0.056	0.220	0.777	2.7	B	Waters
ZirChrom-EZ	1.040	0.117	−0.999	−0.001	2.089	2.089	1.1	Other	ZirChrom
ZirChrom-PBD	1.284	0.158	−0.384	−0.072	2.188	2.188	1.0	Other	ZirChrom
ZirChrom-PS	0.589	−0.232	−0.477	0.062	1.750	1.750	0.3	Other	ZirChrom

[a] Retention of ethylbenzene for 50 percent acetonitrile–buffer; 35 °C (see [2] for details).
[b] Column type: "A," type A alkylsilica; "B," type B alkylsilica; "EP," embedded or end-capped phenyl group; "CN," cyano; "phenyl," phenyl; "F," fluoroalkyl or fluorophenyl.

(assuming the same mobile phase and temperature are used for the two columns being compared). For further details on the selection of columns of equivalent selectivity, see [2, 3].

Columns of quite different selectivity may be required for a change in relative retention during method development (Sections 3.32 and 3.6). Such columns are also needed for the development of "orthogonal" separations (Section 3.7), as examined in detail in [4]. The data of Table III.1 also can be used for each of these two applications.

The data in Table III.1 are of further value in the selection of columns which will minimize the tailing of protonated basic compounds. Columns with low values of C have been found to give reduced peak tailing for basic samples [2]. Type B columns generally have values of $C(2.8) < 0.5$. Separations of peptides and proteins are especially sensitive to the presence of ionized silanols in the stationary phase, or higher values of the column ion-exchange capacity C. Since such separations are usually carried out at low pH, values of $C(2.8) < 0.00$ are probably preferable.

For additional information on column specificity for cyano and phenyl columns and various aromatic and/or polar solutes, see [5].

REFERENCES

1. L. R. Snyder, J. J. Kirkland, and J. L. Glajch, *Practical HPLC Method Development*, 2nd edn, Wiley-Interscience, New York, 1997, Chap. 5.
2. L. R. Snyder, J. W. Dolan, and P. W. Carr, *J. Chromatogr. A* 1060 (2004) 77.
3. J. W. Dolan, A. Maule, L. Wrisley, C. C. Chan, M. Angod, C. Lunte, R. Krisko, J. Winston, B. Homeierand, D. M. McCalley, and L. R. Snyder, *J. Chromatogr. A* 1057 (2004) 59.
4. J. Pellett, P. Lukulay, Y. Mao, W. Bowen, R. Reed, M. Ma, R. C. Munger, J. W. Dolan, L. Wrisley, K. Medwid, N. P. Toltl, C. C. Chan, M. Skibic, K. Biswas, K. A. Wells, and L. R. Snyder, *J. Chromatogr. A* 1101 (2006) 122.
5. K. Croes, A. Steffens, D. H. Marchand, and L. R. Snyder, *J. Chromatogr. A* 1098 (2005) 123.

APPENDIX IV

SOLVENT PROPERTIES RELEVANT TO THE USE OF GRADIENT ELUTION

Note that Table 3.3 contains data on UV absorbance of reversed-phase mobile-phase components as a function of wavelength.

TABLE IV.1 Viscosity of RPC Mobile Phases as a Function of Composition and Temperature

(a) Mobile-phase viscosity at 25°C (η_{25}) for reversed-phase systems

| Mobile phase | η_{25} (cP)[a] | | |
(%v organic/water)	MeOH	ACN	THF
0	0.89	0.89	0.89
10	1.18	1.01	1.06
20	1.40	0.98	1.22
30	1.56	0.98	1.34
40	1.62	0.89	1.38
50	1.62	0.82	1.43
60	1.54	0.72	1.21
70	1.36	0.59	1.04
80	1.12	0.52	0.85
90	0.84	0.46	0.75
100	0.56	0.35	0.46

(b) Variation of the viscosity (cP) of methanol–water and acetonitrile–water mixtures with temperature[b]

| Temperature (°C) | Water content (%, v/v) | | | | | | | | | | |
	0	10	20	30	40	50	60	70	80	90	100
15	0.63	1.05	1.40	1.69	1.91	2.02	2.00	1.92	1.72	1.43	1.10
	0.40	0.54	0.70	0.81	0.89	0.98	1.09	1.30	1.23	1.18	1.10
20	0.60	0.93	1.25	1.52	1.72	1.83	1.83	1.75	1.57	1.32	1.00
	0.37	0.50	0.56	0.69	0.81	0.90	0.99	1.13	1.10	1.14	1.00
25	0.56	0.84	1.12	1.36	1.54	1.62	1.62	1.56	1.40	1.18	0.89
	0.35	0.46	0.52	0.59	0.72	0.82	0.89	0.98	0.98	1.01	0.89
30	0.51	0.76	1.01	1.21	1.36	1.43	1.43	1.36	1.23	1.04	0.79
	0.32	0.43	0.45	0.52	0.65	0.74	0.80	0.86	0.87	0.90	0.79
35	0.46	0.69	0.91	1.09	1.21	1.26	1.24	1.19	1.07	0.92	0.70
	0.30	0.39	0.43	0.47	0.59	0.68	0.72	0.76	0.78	0.73	0.70
40	0.42	0.64	0.83	0.98	1.08	1.12	1.11	1.05	0.96	0.82	0.64
	0.27	0.36	0.41	0.44	0.54	0.62	0.65	0.68	0.70	0.72	0.64
45	0.39	0.58	0.76	0.89	0.98	1.02	1.00	0.96	0.87	0.75	0.58
	0.25	0.33	0.38	0.43	0.50	0.58	0.59	0.61	0.64	0.61	0.58
50	0.37	0.54	0.70	0.82	0.90	0.94	0.93	0.89	0.82	0.71	0.54
	0.24	0.31	0.36	0.41	0.46	0.53	0.55	0.57	0.60	0.60	0.54
55	0.36	0.50	0.65	0.76	0.84	0.88	0.88	0.84	0.77	0.67	0.51
	0.23	0.29	0.34	0.38	0.43	0.49	0.51	0.53	0.56	0.53	0.51
60	0.33	0.47	0.61	0.72	0.79	0.81	0.81	0.77	0.70	0.61	0.47
	0.22	0.27	0.31	0.35	0.41	0.46	0.49	0.50	0.53	0.52	0.47
65	0.28	0.45	0.59	0.68	0.72	0.72	0.69	0.64	0.58	0.51	0.40
	—	—	—	—	—	—	—	—	—	—	—

[a] MeOH, methanol; ACN, acetonitrile; THF, tetrahydrofuran (THF values approximate).
[b] The composition is given in % (v/v) of water at 20.5°C. Upper figures, methanol–water mixture; lower figures, acetonitrile–water mixture.

Source: L. R. Snyder and P. F. Andle, Liq. Chromatogr. 3 (1985) 99.
H. Colin, J. D. Diez-Masa, G. Guiochon, T. Czajkowska, and I. Miedziak, J. Chromatogr. 167 (1978) 41.
M. A. Quarry, R. L. Grob, and L. R. Snyder, J. Chromatogr. 285 (1984) 1.

APPENDIX V

THEORY OF PREPARATIVE SEPARATION

V.1 ISOCRATIC SEPARATION

In the present section, we will examine a simplified theory of isocratic separation as a function of sample weight. The effect of sample volume on preparative isocratic separation will also be discussed. Later, we will extend these conclusions to the case of preparative gradient elution. Figure 7.2(a) illustrates the result of injecting successively larger weights of sample, with the development of wider peaks having a right-triangle shape, and ending approximately at the retention time t_R of the peak. We will first discuss how the width W of the peak is related to sample weight and separation conditions. Then we will examine sample size and "best" conditions for a (T-P) separation (as in Fig. 7.1b or 7.2b).

Our analysis here will assume that the final preparative separation of two peaks is a result of the independent migration of each band through the column. That is, the final separation will be equivalent to overlapping the chromatograms for the separate migration of each peak. This is a reasonable assumption for sample weights no larger than those that correspond to T-P separation, as illustrated in the simulated isocratic separations of Figure 7.8. For the small-sample separations of (a) and (b), the two peaks are separated with $\alpha_0 = 1.5$. As sample size is increased in Figure 7.8a and b, the two peaks are separated (c) or (d) to the point where the two peaks touch, there is little overlap of the two peaks in either case, and *the retention of the minor peak is unchanged*. However, as sample size is increased further in (e) and (f), the two peaks increasingly overlap, with distortion of the minor peak and a shift in its retention to lower values; that is, the two peaks no longer migrate through the column independently of each other.

The latter distortion of the minor peak is the result of the competition of the two solutes for a place in the stationary phase. Especially when the early-eluting peak A is present in smaller amount, the more strongly retained peak B will displace A from the stationary phase and push it ahead of B as B moves more rapidly through the column due to column overload. This results in a narrowing of the peak for A (Fig. 7.8e). When B is present in a smaller amount, the presence of a large excess of A will tend to "pull" B through the column, because the weaker retention of A is more than compensated for by its higher concentration at the beginning of

---

High-Performance Gradient Elution. By Lloyd R. Snyder and John W. Dolan
Copyright © 2007 John Wiley & Sons, Inc.

separation – leading to a more effective competition of A with B for a place in the stationary phase (especially during the initial migration of the sample through the column). This competition of A with B for a place in the stationary phase results in the broadened peak for B in Figure 7.8(f).

V.1.1 Peak Width as a Function of Sample Weight and Separation Conditions

Now let us consider peak broadening as a function of separation conditions and the injected weight w_x of a single solute X. Knox and Pyper [1] have derived the following relationships, assuming that uptake of sample by the stationary phase can be described by a Langmuir isotherm (a good general approximation; see also the related discussion of [2, 3]:

$$H = H_0 + H_{th} \tag{V.1}$$

and

$$H_{th} = (L/4)[k/(1+k)]^2(w_x/w_s) \tag{V.2}$$

Here, H is the column plate height for the observed, overloaded separation, H_0 is the plate height for the corresponding separation with a small sample (for which H does not change for sample sizes less than some upper value), H_{th} is the "thermodynamic contribution to plate height," or the increase in H due to a sample weight above a certain value (resulting in "overloaded" peaks). Similarly, L is column length, k is the value of the retention factor for a small sample, and w_s is the "column saturation capacity" (the total weight of X held by the column stationary phase when the stationary phase is completely saturated by X). A typical value of w_s for a 10 μm, 250 × 4.6 mm RP-LC column is 200–300 mg, but this varies with both the solute and the column. A value of w_x can be measured from experimental values of W for a small sample and a sample large enough to increase the value of W [Equation (7.2a)]. The same units must be used for w_x and w_s (μg, mg, g, etc.)

It is convenient to define the "loading function" w_{xn}

$$w_{xn} \equiv N_0[k/(1+k)]^2(w_x/w_s) \tag{V.3}$$

We also have

$$H_0 = L/N_0 \tag{V.4}$$

and

$$H/LN \tag{V.5}$$

where N refers to the value of N for an overloaded peak; for example, referring to Figure 7.2(a), the width W of an overloaded peak (W_3 in this example) can be used to define N (according to [1]) as

$$N = 16(t_R/W)^2 \tag{V.6}$$

Here, t_R refers to the retention time of the non-overloaded peak, that is, t_R for a sufficiently small sample weight.

From Equations (V.1), (V.4), and (V.5),

$$N/N_0 = H_0/(H_0 + H_{th}) \quad (V.7)$$

and from Equations (V.2) and (V.3),

$$H_{th} = (L/4)w_{xn}/N_0 \quad (V.8)$$

Now, from Equations (V.4), (V.7), and (V.8),

$$N/N_0 = (L/N_0)/[(L/N_0) + (L/4)w_{wn}/N_0]$$
$$= 1/[1 + (1/4)w_{xn}] \quad (V.9)$$

or

$$N = N_0/[1 + (1/4)w_{xn}] \quad (V.9a)$$

Rearranging Equation (V.6) (with $t_R = t_0[1 + k]$)

$$W^2 = (16/N)t_0^2(1 + k)^2 \quad (V.10)$$

and combining Equations (V.9a) and (V.10)

$$W^2 = (16/N_0)t_0^2(1 + k)^2 + 4t_0^2 k^2(w_x/w_s)$$
$$\equiv W_0^2 + W_{th}^2 \quad (V.11)$$

Equation (V.11) describes peak width as a function of relevant experimental conditions (N_0, t_0, k) and the ratio of sample weight w_x to column capacity w_s.

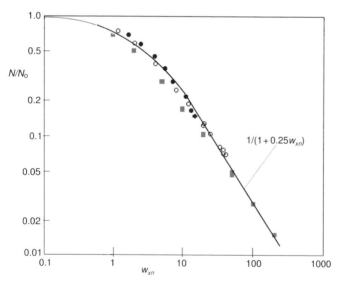

Figure V.1 Variation of column plate number with sample size. Experimental values of N/N_0 (data points) compared with values calculated from Equation (V.9). Adapted from [4].

When the experimental conditions for a small-sample separation (N_0, t_0, k) and column capacity w_s are known, peak width W can be estimated.

The accuracy of the above relationships has been verified in various studies. For example, Figure V.1 provides a test of Equation (V.9) for two different solutes and three different mobile phase-column combinations [4]. The predicted solid curve is seen to agree closely with the experimental data. The values of w_s assumed for each sample and experimental conditions is a best fit to Equation (V.9).

An accurate value of the column capacity w_s can be obtained from Equation (V.11) and experimental values of W for (a) a small sample and (b) a sample weight large enough to significantly broaden the peak. Values of w_s can vary with the nature of the sample, separation conditions and the column (Section 7.2.1.2).

V.1.2 Column Capacity w_s

The surface area SA (m^2) of the stationary phase within the column will be the product of the volume of the column V_c and the surface area per milliliter of the column packing σ_{mL}:

$$SA = (\pi/4) L d_c^2 \sigma_{mL} \qquad (V.12)$$

The density of a porous, silica-base packing within the column will be about 0.7 g/mL, so σ_{mL} can be related to surface area in m^2/g (σ_g) as $\sigma_{mL} = 0.7\,\sigma_g$. Equation (V.12) can then be expressed in more commonly used units (L and d_c in mm) as

$$SA \approx 5.5 \times 10^{-4} L d_c^2 \sigma_g \qquad (V.13)$$

Molecules that lie flat on the stationary phase surface (as in Fig. 7.3a) should yield an approximately constant value of w_s per m^2 of stationary phase surface, and for neutral molecules this is found to be roughly the case [5]:

$$w_s(\text{mg}) \approx 0.4\,SA \qquad (V.14)$$

Combining Equations (V.13) and (V.14) gives

$$w_s \approx 2.2 \times 10^{-4} L d_c^2 \sigma_g \qquad (V.15)$$

For example, for a 250 × 4.6 mm RP-LC column, with a surface area of 200 m^2/g, w_s can be estimated at $(2.2 \times 10^{-4})(250)(4.6^2)(200) = 233$ mg; a 150 × 4.6 cm column as in Table 7.2 would have $w_s = 140$ mg (the value of w_s assumed in Table 7.2).

V.1.3 "Best" Values of w_x/w_s and N_0 for Touching Peak Separation

The largest possible value of α_0 is preferred for touching peak (T-P) separation (Section 7.2.2). Once separation conditions have been selected for maximum α_0, we would like to know the required value of N_0 and w_x/w_s for the separation. The

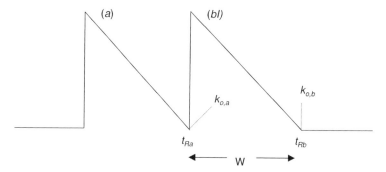

Figure V.2 Hypothetical representation of touching peak separation for gradient elution; see text for details.

value of W for T-P separation (Fig. V.2) is seen to be given by

$$W = t_{R,b} - t_{R,a} = t_0(k_b - k_a) \tag{V.16}$$

or

$$W_b = t_0 k_b[(\alpha_0 - 1)/\alpha_0] \tag{V.16a}$$

Knox [1] has argued that maximum productivity in T-P prep-LC is favored when

$$N_0/N \approx 3 \tag{V.16b}$$

That is, the grams per hour of purified product increases with w_x, but decreases with larger N_0 (because a larger plate number means a longer run time (see discussion of Fig. 9.11). From Equation (V.16b), $(W/W_0)^2 = 3$. Since $W^2 = W_0^2 + W_{th}^2$ [Equation (V.11)], this means that

$$W_b^2 = (3/2)W_{th}^2 \tag{V.17}$$

Also, since $W_{th} = 4t_0^2 k^2 (w_x/w_s)$ [Equation (V.11)] then

$$W_b^2 = 6t_0^2 k^2 (w_x/w_s) \tag{V.18}$$

From Equations (V.16a) and (V.17) we then have

$$(w_x/w_s) = (1/6)[(\alpha_0 - 1)/\alpha_0]^2 \tag{V.19}$$

A value of N_0 for "best" T-P conditions can be derived as follows. The resolution for touching peaks is $R_s = 1$, or since

$$R_s = (1/4)[k_b/(1 + k_b)](\alpha_0 - 1)N^{1/2} \tag{2.7}$$

N is given by

$$N = 16\{[(1 + k_b)/k_b]/(\alpha_0 - 1)\}^2 \tag{V.20}$$

The value of N_0 is then $3N$ [Equation (V.16b)]. Equation (2.7) is adequately reliable (± 10 percent), when $\alpha_0 \leq 1.2$. For larger values of α_0, Equation (2.7) overestimates values of R_s by as much as a factor of 2 for $\alpha_0 = 3$. Equation (V.20) with a correction for the error in Equation (2.7) was used to determine the values of N_0 listed in Table 7.1.

V.1.4 Effect of Sample Volume

Knox and Pyper [1] have argued that sample volumes V_s as large as one-half of the peak volume can be tolerated in preparative separations without adversely affecting separation. Thus, the condition for maximum sample volume is $V_s \leq 0.5\,WF$ (F is flow rate), or

$$V_s \leq 0.5F(t_{R,b} - t_{R,a}) \tag{V.21}$$

Computer simulations [4] suggest that sample volumes as large as $0.5\,WF$ may be slightly too large, but volumes as large as $0.3\,F\,(t_{R,b} - t_{R,a})$ had no effect on the separation. Equation (V.21) assumes that the sample is dissolved in the mobile phase. In this connection, see Section 7.2.2.4.

V.2 GRADIENT SEPARATION

The relative complexity of the above treatment of peak broadening and separation in preparative ("overloaded") chromatography is further increased when it is extended to overloaded gradient elution. However, the use of the LSS model results in an almost complete parallelism between isocratic and gradient prep-LC when values of k^* for nonoverloaded gradient elution are substituted for values of k in a "corresponding" isocratic separation. That is, an overloaded gradient separation with k^* equal to k for a "corresponding" overloaded isocratic separation will respond in very much the same way when sample weight is similar for the two separations (see example of Fig. 7.11 and the related discussion in Chapter 7). Sections V.2.1–V.2.3 will explore this parallelism between gradient and isocratic elution.

V.2.1 Peak Width as a Function of Sample Weight and Separation Conditions

Figure V.2 represents a gradient version of Figure 7.2(b) for isocratic separation, one that we will use to (intuitively) relate peak width in overloaded isocratic and gradient separation. W in Figure V.2 (for isocratic separation) is given by Equation (V.11):

$$W^2 = (16/N_0)t_0^2(1+k)^2 + 4t_0^2k^2(w_x/w_s)$$

$$\equiv W_0^2 + W_{th}^2 \tag{V.11}$$

We will assume that $W^2 = W_0^2 + W_{th}^2$ applies for both isocratic and gradient elution, and we know the value of W_0 for gradient elution [Equation (2.19)]:

$$W_0 \approx 4N^{-1/2}t_0(1 + [1/2.3b]) \tag{V.22}$$

$$= 4N^{1/2}t_0(1 + k_e) \tag{V.23}$$

Here, k_e is the value of k at elution, equal to $(1/2.3b)$ [Equation (9.14a)]. If W_{th} in gradient elution is given by Equation (V.11) (for isocratic elution), but with k_e replacing k, then

$$\text{(gradient)} \quad W_{th} = 2t_0 k_e (w_x/w_s)^{1/2} \tag{V.24}$$

A comparison of Equation (V.24) for gradient elution with Equation (V.11) for isocratic elution

$$\text{(isocratic)} \quad W_{th} = 2t_0 k (w_x/w_s)^{1/2} \tag{V.25}$$

shows an exact parallelism of the effect of sample weight on both isocratic and gradient elution, since $k \equiv k_e$ for isocratic elution. A similar parallelism exists for values of W_0 in isocratic and gradient elution [(cf. Equations (2.16) and (2.17)]. A possibly confusing aspect of comparing values of W in isocratic vs gradient elution is that "corresponding" separations have small-sample values of k^* (gradient) $= k$ (isocratic), but $k^* = (1/2) k_e$. That is, resolution in gradient separation is determined by k^*, but the width of the peak at elution is determined by k_e, which means that values of both W_0 and W_{th} are smaller in "corresponding" gradient elution than in isocratic elution, when $k^* = k$ and resolution is the same. An example of these similarities and differences in "corresponding" isocratic and gradient separations is provided by Figure 7.11.

Several experimental studies have examined the accuracy of Equation (V.24) for overloaded gradient separation, which can also be expressed as

$$\text{(gradient)} \quad W_{th} = k^* t_0 (w_x/w_s)^{1/2}$$
$$= (1/1.15) t_0 / b (w_x/w_s)^{1/2} \tag{V.25a}$$

Figure V.3 shows two such comparisons from [6]. In (a) values of $W_{th}(b/t_0)$ are plotted vs (w_x/w_s) on a log–log scale for benzyl alcohol as sample. For lower values of (w_x/w_s), the data points fall on a straight line corresponding to Equation (V.25a). A similar plot is shown in Figure V.3(b) for caffeine as sample, for two gradient times (and two different values of b). Again, a reasonable agreement with Equation (V.24) is observed. In each case (Fig. V.3a and b), a value of w_s can be calculated from the application of Equation (V.24) to these data. Given a value of w_s, values of W as a function of sample size and gradient conditions can be calculated with good accuracy.

Similar tests of Equation (V.25a) have been reported for four different proteins ranging in molecular weight from 14 to 66 kDa, with variation of gradient time as in Figure V.3b [7]. In each case, a straight line fits the data as in Figure V.3, with an average slope of 0.54 ± 0.08, that is, close agreement with Equation (V.25a).

As in the case of isocratic prep-LC, the value of w_x for T-P separation is given by Equation (V.16).

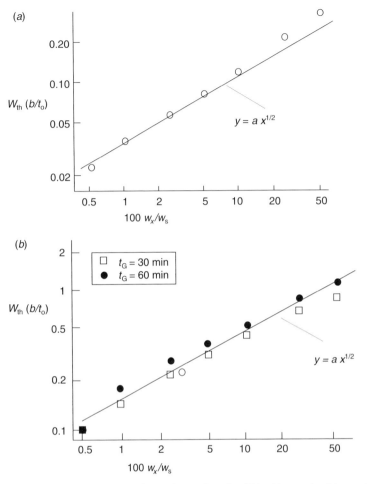

Figure V.3 Comparison of experimental peak widths (data points) in gradient elution with theoretical value (straight lines). (*a*) Benzyl alcohol solute; (*b*) caffeine solute. Adapted from [6].

V.2.2 Determining a Value of α

The use of Table 7.1 for estimating values of w_x and N_0 for T-P gradient elution requires a value of α_0. This can be obtained as follows. Gradient retention times for small samples of solutes *a* and *b* in Figure V.2 are given by [Equation (2.12)]:

$$t_R \approx (t_0/b) \log(2.3k_0 b) + t_0 + t_D \qquad (V.26)$$

k_0 refers to the value of k at the start of the gradient. Adjacent, overloaded peaks are shown in Figure V.2, and we will assume that values of S for each peak are equal; values of $b = V_m \, \Delta\phi \, S/(t_G F)$ are therefore also equal for each peak.

We then have

$$t_{R,b} - t_{R,a} = (t_0/b)\log(k_b/k_a) \tag{V.27}$$

or

$$\begin{aligned}\log\alpha &= (b/t_0)(t_{R,b} - t_{R,a}) \\ &= (\Delta\phi S/t_G)(t_{R,b} - t_{R,a})\end{aligned} \tag{V.28}$$

V.2.3 Sample Size for Touching Peaks

The above analysis (Section V.2.1) suggests that peak width in isocratic and gradient elution will be similar when isocratic values k are equal to gradient values k_e, that is, equal values of k at elution in isocratic and gradient elution. Similarly, touching peak separation will occur for both isocratic and gradient elution for similar sample weights and values of w_x/w_s [Equation (V.24)]. Thus, injecting the same sample weight in these isocratic and gradient separations will give the same sample resolution. *Consequently, Table 7.1 can be used for either isocratic or gradient separation.*

REFERENCES

1. J. H. Knox and H. M. Pyper, *J. Chromatogr.* 363 (1986) 1.
2. H. Poppe and J. C. Kraak, *J. Chromatogr.* 255 (1983) 395.
3. J. E. Eble, R. L. Grob, P. E. Antle, and L. R. Snyder, *J. Chromatogr.* 384 (1987) 25.
4. L. R. Snyder, G. B. Cox, and P. E. Antle, *Chromatographia* 24 (1987) 82.
5. G. B. Cox and L. R. Snyder, *J. Chromatogr.* 483 (1989) 95.
6. G. B. Cox, L. R. Snyder, and J. W. Dolan, *J. Chromatogr.* 484 (1989) 409.
7. G. B. Cox, P. E. Antle, and L. R. Snyder, *J. Chromatogr.* 444 (1988) 325.

APPENDIX VI

FURTHER INFORMATION ON VIRUS CHROMATOGRAPHY[†]

VI.1 ADENOVIRUS STRUCTURE

Adenoviruses are a family of double-stranded DNA viruses which lack the phospholipid-containing membrane that surrounds all cells and some viruses. The virus particle contains 87 percent protein, 13 percent DNA, and no lipid. Its MW is 167×10^6 and its diameter is 98 nm (as measured by light scattering [1]), thus it is vastly larger and more complex than almost all other molecules that can be separated by chromatography. An electron micrograph and structural model of adenovirus can be seen in Figure VI.1 [2]. The virus particle consists of an icosahedral shell of protein surrounding a DNA-containing core and protein fibers at each of the 12 vertices. The outer shell, or capsid, contains 252 subunits, 12 of which have five neighbors (pentons) and 240 of which have six neighbors (hexons) with fibers of the capsid extending at each of the vertices. Electron microscopy and X-ray diffraction studies have elucidated the adenovirus particle structure and that of some of its protein components in remarkable detail [2, 3].

Adenoviruses behave as anions at neutral pH and undergo isoelectric precipitation in the range of pH 5–7. The isoelectric point of hexons is near pH 6 and they dominate the charge behavior of the adenovirus particle because of their abundance and exposure in the outer shell. At pH 7 the hexon protein contributes -23.8 charge units per chain or $-17,136$ charge units per virion (values calculated from Ad5 sequence [4]; and published abundance [2]; i.e., $m \approx 17,000!$). Adenovirus binds more tightly to anion exchange columns, compared with most proteins including hexon, in part because of the large number of binding sites. For example, adenovirus elutes from Resource Q® anion exchange resin at 0.45 M NaCl compared with 0.33 M NaCl for hexon protein [5]. Purified penton and fiber proteins do not bind to this resin at salt concentrations above 0.3 M NaCl. DNA, on the other hand, binds to anion exchangers more tightly than virus, although small nucleic acid fragments can elute throughout the gradient. These properties make anion exchange resins attractive media for chromatography of viruses but also limit the allowable range of conditions.

[†]With Carl Scandella, Paul Shabram, and Gary Vellekamp.

High-Performance Gradient Elution. By Lloyd R. Snyder and John W. Dolan
Copyright © 2007 John Wiley & Sons, Inc.

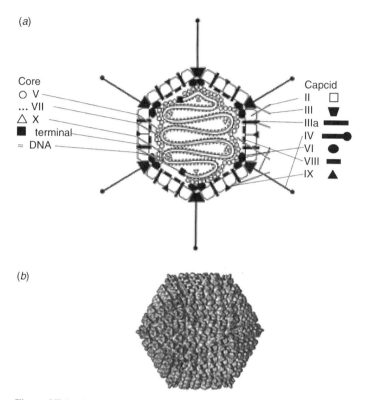

Figure VI.1 (a) A stylized section of the adenovirus virion, indicating the relative positions of the various proteins and DNA. Proteins are grouped according to their location in the core or capsid. (b) An eletron microscope reconstruction at 15 Å resolution of the surface of the virion. Adapted from [2].

More than 50 serotypes of human adenoviruses have been identified on the basis of neutralization assays [6]. These can be divided into six subgroups, all of which share a similar structure and genomic organization. Of these, Ad2 and Ad5 from subgroup C have been particularly well studied. Noncovalent forces hold the adenovirus particle together; thus the proteins can be dissociated and resolved in SDS–polyacrylamide (PAGE) gels. Eleven major structural proteins have been resolved in this manner, seven belonging to the icosahedral capsid (II, III, IIIa, IV, VI, VIII, and IX) and four packaged with DNA in the virus core (V, VII, X, and the terminal protein). Many of these are derived from precursor proteins by the adenovirus 23 kDa protease. Therefore numerous proteins and peptides are also present, sometimes in high copy number. Molecular weights and copy numbers for each protein determined from SDS-PAGE measurements and the molecular weighs have been confirmed by mass spectrometry [2].

The human Ad genome consists of a linear, double-stranded DNA molecule of about 35–36 kilo base pairs. Viral DNA associates with capsid proteins V and VII to form a core structure within the virus particle. As is the case with other viruses, the

Ad genome is efficiently organized to permit the production of virus proteins with minimum replicative apparatus [6]. The first region expressed after infection of the target cell, termed E1A, activates transcription of other genes, so it is a regulatory element. Other early-expressed viral genes included E1B, E2, E3, and E4. The first-generation Ad vectors lacked functional E1 and E3 genes, carrying other genes such p53 in their place. These recombinant viruses were thus unable to replicate in most cells, but were able to deliver the inserted gene. These viruses can replicate in certain cell lines, such as human embryonic kidney cell line 293 or the PER.C6 cell line [7], which are capable of supplying the missing functions. Recombinant adenovirus stocks are normally produced in those cell lines. Unfortunately the first-generation Ad vectors elicited a host immune response when administered to humans and thereby prevented prolonged therapy. It was thought that the immune system responded to a low level of expression of Ad proteins directed by first-generation vectors. Later-generations of Ad vectors addressed these problems by additional genetic manipulations designed to reduce viral protein expression [6].

Adenovirus particles are considered to be rigid and well hydrated with about 21×10^6 water molecules per virion (corresponding to 2.3 g of water per every gram of anhydrous virus) and a buoyant density of 1.34 g/mL while empty capsids band at a lower density of 1.31 g/mL. Hydrodynamic measurements suggest that the particle contains a hard core excluding water of 76 nm diameter. The high degree of hydration suggests that there may be some flexibility in the outer region of the virus.

VI.2 SAMPLE PREPARATION

The ease of virus purification by chromatography or other means depends on such factors as the concentration of virus in the starting material and the nature and concentration of contaminating substances. Fortunately these factors can be optimized and controlled to a large extent, using modern methods of cell culture and genetic engineering. Following the cell growth and virus production phases, the infected cells are harvested and lysed to release the virus. Cell harvesting may be done by centrifugation or filtration. Cells loaded with virus are fragile and easily disrupted, so handling these cells requires care. Adenovirus may be released from the infected cells by one of several lysis methods. On a small scale, freeze–thaw is favored because it is easy to do and requires no special equipment; however, this procedure becomes increasingly difficult at larger scales, because the freezing and thawing of large samples is harder to control. Three cycles of freeze–thaw usually release 90 percent or more of the virus. Nonionic detergents such as Tween, Triton, and Brij also can be used to lyse cells by weakening or dissolving the cell membrane. High-pressure homogenizers such as a French press or Gaulin homogenizer disrupt cells by the sudden drop of pressure caused by passage through an orifice under pressures up to 20,000 p.s.i. With adenovirus the operating pressure is usually limited to around 1000 p.s.i. in order to protect the virus from damage [4, 5]. Tangential flow filtration through microfilters provides for both cell lysis and removal of cell debris.

Cell lysis releases a complex mixture of cellular debris, organelles, nucleic acids, and proteins. Application of such a mixture directly to a chromatography column would soon result in a clogged column. Therefore, the cell debris must be removed, viscosity reduced, and nucleic acids and other interfering substances reduced or eliminated prior to chromatography. Centrifugation, filtration, and flocculation can remove the bulk of the cell debris, while filtration through a filter of 0.2 μm pore size can remove residual fine particulate material. Centrifugation, if used, must be controlled in order to avoid a large losses of product [5, 8]. Nucleic acids can be removed using anion exchange resins, digested by nucleases, or precipitated by specific agents [4, 9]. Low-molecular-weight contaminants can be removed by dialysis, diafiltration, or precipitation of the high-molecular-weight components by ammonium sulfate or other precipitants. The specific method or combination of methods selected for precolumn treatment depends on the product, the scale and what else may be present [4]. On an analytical scale these operations are often done using a microfuge or by solid-phase extraction.

VI.3 FURTHER DETAILS ON VIRUS PURIFICATION BY CHROMATOGRAPHY

The purity of virus purified by column chromatography was compared in [8] to the use of density gradient centrifugation and found to be equal or better by six criteria: SDS PAGE, western blots, $A_{260} : A_{280}$ ratio in SDS, ratio of total virus particles to infectious virus particles, expression of p53 gene product, growth suppression by the gene product, and the presence of host cell protein as measured by immunoassay. Other workers have confirmed and extended these studies, showing the general utility of chromatography for virus purification [4, 5]. At least eight companies have developed and published procedures for purifying adenovirus by chromatography [4]. Anion exchange chromatography is used for the first step, in each case.

VI.4 FURTHER DETAILS ON VIRUS ANALYSIS BY CHROMATOGRAPHY

An early study [10] showed that the concentration of purified virus particles treated with SDS could be measured at 260 nm. One AU at 260 nm corresponds to a virus concentration of 1.2×10^{12} particles per mL. The latter method has been accepted because of its convenience, despite several drawbacks: a large amount of highly purified virus is required, the virus sample is destroyed by the measurement, and intact virus cannot be distinguished from aggregated or disrupted virus. Particle concentration determination by anion exchange chromatography eliminates these disadvantages.

It soon became apparent that a reference standard was needed in order to standardize the assays. A consortium formed consisting of the FDA, other regulatory agencies, academic groups, and industry representatives to produce an Ad5 wild type reference standard. The reference standard facilitates the calibration of chromatographic methods as well as has other uses. The consortium, known as the Adenovirus Reference Material Working Group (ARMWG), produced several lots of virus together with certificates of analysis, characterized the lots by a variety of physical and biological tests including anion exchange and RP-HPLC chromatography, and carried out stability testing. Reference lots were prepared from virus grown in 293 cells, purified by anion exchange chromatography as described [8] and made available through the American Type Culture Collection (ATCC catalog no. VR-1516). Production and test data for the ARM lots can be seen at the web site for the Williamsburg BioProcessing Foundation, www.wilbio.com.

REFERENCES

1. C. J. Oliver, K. F. Shortridge, and G. Belyavin, *Biochim. Biophys. Acta* 437 (1976) 589.
2. J. J. Rux and R. M. Burnett, *Hum. Gene Ther.* 15 (2004) 1167.
3. P. L. Stewart and R. M. Burnett, *Curr. Top. Microbiol. Immunol.* 199 (1995) 25 13.
4. N. E. Altaras, J. G. Aunin, R. K. Evans, A. Kamen, J. O. Konz, and J. J. Wolf, *Adv. Biochem. Engng/Biotechnol.* 99 (2005) 193.
5. P. Shabram, G. Vellekamp, and C. Scandella, in *Adenoviral Vectors for Gene Therapy*, D. T. Curiel and J. T. Douglas, eds, Academic Press, New York, 2002.
6. J. D. Evans and P. Hearing, in *Adenoviral Vectors for Gene Therapy*, D. T. Curiel and. J. T. Douglas, eds, Academic Press, New York 2002.
7. W. W. Nichols, R. Lardenoije, B. J. Ledwith, K. Brouwer, S. Manam, R. Vogels, D. Kaslow, D. Zuidgeest, A. J. Bett, L. Chen, M. van der Kaaden, S. M. Galloway, R. B. Hill, S.V. Machotka, C. A. Anderson, J. Lewis, D. Martinez, J. Lebron, C. Russo, D. Valerio, and A. Bout, in *Adenoviral Vectors for Gene Therapy*, D. T. Curiel and J. T. Douglas, eds, Academic Press, New York 2002.
8. B. G. Huyghe, X. L., S. Sujipto, B. J. Sugarman, M. T. Horn, H. M. Shepard, C. J. Scandella, and P. Shabram, *Hum. Gene Ther.* 6 (1995) 1403.
9. A. Goerke, B. C. S. To, A. L. Lee, S. L. Sagar, and J. O. Konz, *Biotechnol. Bioengng* 91 (2005) 12.
10. J. V. J. Maizel, D. O. White, and M. D. Scharff, *Virology* 36 (1968) 115.

INDEX

Absorbance (UV), mobile phase, 82–83
ACD/LC Simulator, 109
Acetone test. *See* System performance tests
Adsorptive carryover, 204
Air
  bubbles. *See* Bubbles, Degassing
  leaks, 211
  peaks, 186–187, 225
Anion exchange chromatography. *See also* Ion-exchange chromatography
  columns, 34
Anthraquinone separation, 6
APCI. *See* Atmospheric pressure chemical ionization
Area reproducibility problems, 223
Artifacts. *See* Separation artifacts
Artifact peaks. *See* Ghost peaks
Assay procedure, routine, 77
"At-column dilution", 300–301
Atmospheric pressure chemical ionization, interface, 326–327
Autosamplers, 140
  pressure bypass, 207
  problems, 207, 212, 223–224
  reproducibility, 151
  test failure, 212
  wear, 190

Background peaks. *See* ghost peaks
Back-pressure, 136, 211. *See also* Pressure restrictor
Bacterial growth, 184
Ballistic gradients, 75. *See also* Gradient separation, Fast
Band. *See also* Peak
Band migration, gradient elution, 11–13
Band migration, isocratic elution, 10–11

Baseline drift. *See* Drift (baseline)
Baseline noise. *See* Noise (baseline)
Baseline resolution. *See* Resolution, baseline
Beat frequency, 180
Bioanalytical LC-MS, 332
Biomolecules, gradient separation of, 248–271
Blank gradient, 121, 158
  drift, 158
  peaks, 158, 182–185, 225
  test, 217–221
Blockage
  problems, 223–224
  solvent inlet-frit, 207, 210
"Break through" of macromolecules, 273–274
Broadening. *See* Peak shape or Peak width
Bubbles
  autosampler, 212
  injection, 186–187
  problems, 206–207, 209–211, 223–225
  removing from pump, 197
Buffers, 84
  constant buffer-strength gradients, 139
  contamination, 218
  flushing, 159
  good practice, 161
  peptide and protein separations, 252
  phosphate problems, 161
  precipitation, 139, 161
  solubility test, 161
  sources compared, 217–218
Calibrators, 160
Carbohydrates. *See* Biomolecules
Carboxylic acids, IEC separation of, 6–7

High-Performance Gradient Elutions. By Lloyd R. Snyder and John W. Dolan
Copyright © 2007 John Wiley & Sons, Inc.

INDEX **451**

Carryover, 187, 225
  gradient, 174
  isolation, 203–204
Case studies, 180–182, 213–222
Cation exchange chromatography
  columns, 349. *See also* Ion-exchange
    chromatography
Centrifugation, 190
Cereal storage protein, 256–260
Change one thing at a time, 205
Check-valves
  cleaning, 199
  failure, 212, 221–223
  problems, 209, 211, 224
  replacement, 216
  sonication, 214
Chemical composition
    distribution, 275–278
Chromsword, 109
Cleaning glassware, 200
Cleanliness, 159
Coefficient of variation. *See* CV
Column
  anion-exchange, 349
  capacity. *See* Column, saturation
    capacity
  cation-exchange, 349
  characterization of, 416–433
  chemistry change, 225
  cleaning, 161, 187, 204
  comparing two, 419
  dead-volume V_m, 53, 90, 393
  dedicated, 159
  effect of change in size, 169
  efficiency. *See* Plate number N
  equilibration. *See* Equilibration
  equivalent, 157, 419–433
  frictional heating, 170
  LC-MS, 328
  ovens, 170
  overload, 225, 283
  pH stability, 161
  plate number N. *See* Plate number N
  pre-heating, 170
  pressure drop. *See* Pressure
  saturation capacity, 289–292, 439
  selecting reasonable mobile
      phase, 161–162
  selection, 160–161, 416–433
  selectivity, 419–433
  slow equilibration, 159–160. *See also*
    Equilibration
  surface area, 289–290, 439
  temperature, 169–170. *See also*
    Temperature
  temperature problems, 224
  variability, 122, 157
  void, 226
  volume, effect on separation,
    123–124
Column conditions, 29
  effect of diameter on gradient
    separation, 51–55. *See also*
    Column, Volume
  effect of length on gradient
    separation, 51–55
  effect on gradient separation, 102–106
  effect on isocratic separation, 28–31
Column equilibration. *See* Equilibration
Column length. *See* Column conditions
Column-mobile phase equilibration. *See*
    Equilibration
Column switching, 3–4, 347–348. *See*
    also Sample preparation
Complex samples, 89–90
Component failures. *See* Specific
    components
Compressibility-compensation
    errors, 211–212. *See also* Flow rate,
    errors
Compression, gradient. *See* Gradient
    compression
Computer simulation, 18, 80, 108–120.
    See also Resolution maps
  accuracy of, 119, 399. *See also*
    Linear-solvent-strength (LSS)
    model
  column conditions, 112–114
  designated peaks, 117–118, 259–260
  for peptides and proteins, 254
  gradient optimization, 111–112
  isocratic predictions, 115–117
  LC-MS, 347
  options, 116
  resolution maps, 109–111, 110, 113,
    116, 118, 258, 260
  segmented gradients, 117–119
  "two-run" procedures, 119
Conditions, effect on gradient
    separation, 49–72

Conformation of macromolecules, 236–238, 272–273
Convergent case, in preparative separation, 315
"Corresponding" separations, 34–37, 285, 301–302, 373, 374–376
Critical elution behavior, 245–247
Critical mobile phase composition, 278
Critical pair. See Resolution, critical
Critical resolution. See Resolution, critical
Crossing isotherms, 313
CV vs signal-to-noise, 213

Data systems, 141–142
　problems, 225
　sampling rate, 141, 190
Dead-volume. See Column dead-volume
Decomposition, sample, 193–194, 225, 236–238
Degassing, 136, 159, 187, 214, 221
　air peaks, 186–187
　and baseline noise, 182
　helium sparging, 136
　membrane degasser, 136
　problems, 211, 219
　sample, 187
Degradation. See Decomposition, sample
Delay, gradient. See Gradient delay
Denaturation. See Conformation of macromolecules
Designated peaks. See Peaks, designated
Detection, UV absorbance of mobile phase, 82–83
Detectors, 141
　cell, 141, 142
　noise filter, 141
　time constant, 141, 190
"Displacement" effect, 301–302, 304, 318–320
Dissolved air, 186–187. See also Degassing
Divergent case, in preparative separation, 315
Divide-and-conquer strategy, 196, 204–205, 218, 223
Double peaks, 226
Drift (baseline), 176–179, 225
　acetonitrile, 176
　ammonium acetate, 177–178
　ammonium carbonate, 178
　and wavelength, 176
　compensating for, 178
　equimolar buffers, 178
　methanol, 176
　negative, 177
　phosphate, 176–177
　tetrahydrofuran, 176–177
　trifluoroacetic acid and ACN, 179
DryLab® software, 18, 109, 112. See also Computer simulation
Dwell time. See Dwell volume
Dwell volume, 33, 151, 158, 393–394
　adjustment of initial %B, 165–168
　and equilibration, 171–174
　and gradient volume, 151, 169, 343
　compensating for differences, 163–168
　differences, 151, 155, 225
　during method development, 122
　effect of small k_0, 376–378
　effect on separation, 66–67
　high- vs low-pressure mixing, 136–137
　injection delay, 163–164
　isocratic hold, 164–165
　LC-MS, 342
　maximum-dwell-volume methods, 165
　measurement of, 147
　method transfer, 163–168
　typical values, 136

Early elution, 88
Easy vs powerful troubleshooting technique, 205
Eigen peaks. See Ghost peaks
Electrospray ionization (ESI) interface, 326–327
Elution strength gradients, 365–366
Epimer sample, 193–194
Equilibration, 80, 106, 122, 159, 162, 170–175, 386–391
　addition of propanol, 174
　and dwell volume, 174
　effect on blank gradient, 217–218
　incomplete, 169, 172–174, 225
　inter-run, 169
　ion-pair chromatography, 174, 391
　normal-phase chromatography, 174, 391
　practical considerations, 174–175
　primary effects, 171–173

reducing, 174
routine analysis, 174
slow, 173
time, 172
volume, 171
Equipment. *See also* System
 bias, 156
 checkout, 157–163
 comparison of, 135
 design, 133–142
 manufacturers, 143
 "non-ideal", 393–396
 preparative separations, 285–286
 repair service, 144
 selection, 142–145
 special applications, 144
Errors, in linear-solvent-strength
  model, 397–400
ESI interface. *See* Electrospray ionization
Extra peaks, 225. *See also* Ghost peaks,
  Late peaks, t_0 peaks
 in samples, 185–187. *See also*
  Decomposition, sample
Extra-column effects, 189, 225
 peak broadening, 396
Extra-column volume, 142

Fatty acid esters, 114–116
Filter, in-line, 190
Filtration
 mobile phase, 190
 problems, 218–219
 sample, 190
Fingerprint procedure, 77
Flow programming, 3–4
Flow rate. *See also* Column conditions
 effect on gradient separation, 55–58
 errors, 137–139, 396
 measurement, 149–150
 problems, 210–212
Flow test failure, 224
Frit blockage, 190, 197, 207, 210, 226
Fronting peaks, 225

Garbage peak. *See* t_0 peaks
General elution problem, 1–3
Generic separations, 5, 77, 334
 macromolecules, 248
Ghost peaks, 225
 and equilibration time, 183

blank gradient, 182–185
 isolating, 182–185
 organic solvents, 184
 sources, 185
 water, 182
Glassware
 cleaning, 200
 contamination, 218–219
GLP. *See* Good Laboratory Practice
Glycophospholipids, 345
Goals of separation, 75–78
Good Laboratory Practice (GLP),
  145, 157
GPV. *See* Gradient-proportioning-valve
Gradient
 blank. *See* Blank gradient
 compression, 38, 380–383
 linearity failure, 206, 209
 peak width. *See* Peak width, gradient
 performance. *See* System
  performance
 performance test failure, 223–224
 program, 9–10
 rounding, 147, 206, 394–396
 test failures, 205–213
 testing. *See* System performance
 "trimming". *See* Gradient range
Gradient carryover, 174
Gradient conditions, 49
Gradient conditions, effect on
  separation, 49–72
Gradient delay, 7–8
 effect on separation, 63–66. *See also*
  Dwell volume
Gradient distortion, 172, 174, 394–396
 LC-MS, 342–343
Gradient elution. *See also* Gradient
  separation
 basics. *See* Gradient elution, theory
 compared to isocratic elution, 2,
  10–13, 34–37, 39–42,
  304–306, 316–317. *See also*
  "corresponding" separations
 compared to stepwise elution, 2
 history, 3
 reasons for, 4–7
 theory, 13–18, 31–72,
  370–411
 theory for macromolecules,
  242–248

Gradient equilibration. *See* Equilibration
Gradient problems, 88–90, 225–226, 394–396
  causes, 225–226
  solutions, 225–226
  symptoms, 225–226
Gradient range, 8
  adjusting, 87–88
  effect on separation, 58–63
  optimization of, 95–96
Gradient retention, 32–34, 372–378
  compared with isocratic, 374–376
  optimization of, 92
Gradient retention factor k^*, 13, 33–34, 90–91, 370–374
  effect on selectivity, 96–100
Gradient selectivity
  effect of k^* on, 96–100
  optimization of, 92–95
Gradient separation
  effect of final percent-B on, 60–63
  effect of gradient time, 33–34
  effect of initial percent-B, 376–378
  effect of initial percent-B, 58–60
  fast, 106–108, 274
  initial experiment, 76, 79, 80–87, 249–253
  method development, 74–130
  prediction. *See* Computer simulation
  second-order effects, 386–396
Gradient shape, 7–10
  concave, 8
  convex, 8
  curved, 7, 240–241
  effect on separation, 67–71
  linear, 7–9
  nonlinear, 67–71
  segmented, 7–9, 69–71, 100–102, 114, 117–119, 259–260
Gradient steepness
  effect of conditions on, 50–58
  intrinsic (b), 17, 32–33
Gradient step-test. *See* Step-test
Gradient time t_G, 7
  effect on separation, 50–51
Gradient volume, and dwell volume, 151, 343
Gradient-proportioning-valve (GPV)
  test, 148–149
  failure, 208–210

Gradients
  elution strength, 365–366
  selectivity, 365–367
  ternary-solvent, 365–368
  quaternary-solvent, 365–368
Guidelines, avoiding problems, 154–157

Headache = non-linear gradients
Heating. *See also* Column, oven, temperature
High-molecular-weight samples. *See also* Macromolecules
  problem, 221
High-pressure mixing, 133–134
History of gradient elution, 3
Homo-oligomers, 238–242
Hydrophilic interaction chromatography (HILIC), 266–267, 361–365
  method development, 364–365
Hydrophobic interaction chromatography (HIC), 262–264

Impurities at t_0, 225
Incomplete elution, 204
Induced peaks. *See* Ghost peaks
Initial gradient run. *See* Gradient separation, initial experiment
Injection
  delay, 163–164
  disturbance. *See* t_0 peaks
  duplicate, 161–162
  effect of equilibration, 173
  effect on peak shape, 190–193
  effect on sample retention, 190–193
  of air, 186–187
  priming, 159–160
  problems, 225
  sample volume, 190–193, 292, 312
  solvent strength, 190–193
Injection peak. *See* t_0 peaks
Injectors. *See* Autosamplers
In-line filter. *See* Filter, inline
Installation Qualification (IQ), 157–158
Integration, 141
Interfering peaks. *See* Peaks, designated
Intrinsic gradient steepness b, 17, 32–33
Ion exchange chromatography (IEC), 264–266, 349–358
  effect of gradient conditions, 356
  method development, 356–358

INDEX   455

mobile phases, 349
retention process, 350–353
Ion suppression, LC-MS, 343–345
Ion trap MS, 328–330
Ion-pair
  equilibration, 174
  gradient elution of nucleic acid
    fragments, 261
"Irregular" sample. See Sample, "irregular"
IQ. See Installation Qualification
Isocratic elution, 1, 10–11, 23–31
  compared to gradient elution. See
    Gradient elution, compared
    to isocratic elution
Isocratic hold. See Gradient delay
Isocratic retention, 23–24, 27–28
  prediction from a gradient run,
    45–47, 416–417

k. See Isocratic retention
k^*. See Gradient retention factor
Knox equation, 404–405

"Large" molecules. See Macromolecules
Late elution, 88–89, 187, 204, 225
  column washing, 161
LC-MS, 324–348
  applications, 325
  buffers, 332, 335
  challenges, 341–348
  co-eluting compounds, 345–346
  column conditions, 341
  column selection, 328, 330, 340
  column switching, 347–348
  computer simulation, 347
  dwell volume requirements, 342
  equilibration requirements, 341
  generic methods, 334
  gradient distortion, 342–343
  infusion experiments, 343–344
  initial runs, 339
  interface, 326–327
  internal standards, 334–335, 346
  ion suppression, 343–346
  ion trap, 328–330
  isocratic methods, 347
  isotopic standards, 330, 335
  lipid problems, 345
  matrix problems, 341
  method development, 332–341

minimum retention, 339, 340
mobile phase selection, 330, 335
multiplexing, 347–348
MUX interface, 348
parallel columns, 347–348
plasma problems, 341
precision and accuracy, 326, 341
principles, 326–330
removing contaminants, 340–341
requirements, 325–326
resolution, 326, 330, 346
sample preparation, 335–339. See also
  Sample preparation
sample throughput, 347–348
scouting runs, 339
segmented gradients, 340–341
separation goals, 332–334
single ion monitoring, 344
single stage, 324
solvent selection, 340
specificity, 346
stable-label standards, 335
tandem, 328–330
temperature, 340
trifluoroacetic acid, 345
vs LC-MS/MS, 324
vs LC-UV, 330–332
Leaks, 200, 211–212, 223
Linearity test problems, 206, 224
Linear-solvent-strength (LSS)
  gradient, 15
Linear-solvent-strength (LSS)
  model, 13–18, 370–386
  accuracy of, 397–400
  failure of, 393
  measurement of parameters, 400
Linear-solvent-strength behavior,
  advantages of, 385–386
Lipids, and LC-MS, 345
Liquid chromatography under critical
  conditions, 245
  (See also "pseudo-critical conditions")
log k vs %B plot, 36, 40, 44
log k^* vs t_G plot, 37, 41, 44
Low-pressure mixing, 133–134
Lysozyme variants, 263

Macromolecules
  conformation of. See Conformation
    of macromolecules

Macromolecules (*Continued*)
  quaternary structure, 236
  separation of, 228–278
  separation problems of, 271–274
  tertiary structure, 236
  theory of gradient separation, 242–248
  values of N, 235–236
  values of S, 229–235
Mass selective detector, 324
Mass spectrometric detection. *See* LC-MS
Method development, 74–130. *See also* Gradient separation
  avoiding problems, 160–163
  gradient preparative separations, 306–315
  guidelines, 160–163
  hydrophilic interaction chromatography, 364–365
  ion-exchange chromatography, 356–358
  isocratic preparative separations, 292–302
  normal-phase chromatography, 360–361
  preparative separation, 292–302, 306–315, 317–318
  summary, 75, 78
Method instructions, 170
Method transfer
  gradient rounding, 169
  gradient shape, 169
  problems, 163–170
  segmented gradient, 169
Microbial growth, 184
Micromixer, 136, 152
Mixing
  accuracy, 137
  designs, 133–140, 144, 152
  high-pressure. *See* High-pressure mixing
  low-pressure. *See* Low-pressure mixing
  premixing mobile phase, 181
  problems, 224
Mixing volume. *See* Dwell volume
Mobile phase. *See also* Solvent
  absorbance (UV), 82–83
  buffers, 84
  composition change, 225

contamination, 225
filtration, 190
pH selection, 161
premixing, 181, 197–199, 216
selecting reasonable, 161–162
viscosity, 435
Module substitution troubleshooting strategy, 196–197, 205
Molecular weight
  distribution, 275–278
  sample, 229–235
MS. *See* LC-MS
MS/MS. *See* Tandem MS
Myoglobin digest, 56–57

Native proteins, 236
New column test, 223
Noise (baseline), 225
  absorbance matching, 178, 181
  and mixing, 179–180
  beat frequency, 180
  case study, 180–182
  degassing mobile phase, 182
  premixing mobile phase, 181
Non-linear gradients, 206. *See also* Gradient shape, segmented
Non-overloaded separation, 283
Normal-phase chromatography (NPC), 359–365
  dried solvents, 174
  equilibration, 174
  polar solvent addition, 174
  method development, 360–361
  relay gradient elution, 360–361
Nucleic acids. *See also* Biomolecules

Oligomers. *See* Homo-oligomers
Oligonucleotides. *See* Biomolecules
"On-off" elution, 234–235, 244
  of viruses, 268
Operational Qualification (OQ), 146, 157–158
Optimized separation, 75
OQ. *See* Operational qualification
Orthogonal separation. *See* Separation, orthogonal
Osiris, 109
Outgassing, 211. *See also* Degassing
Ovens, column, 170
Overload, 225

Paclitaxel, 344
Parallel case, in preparative
    separation, 314
Parallel chromatography, 347–348
Particle size. *See* Column conditions
Particulate matter, 190
Peak, distortion, 226
Peak area reproducibility problems,
    212–213, 224
Peak asymmetry factor, 188, 418–419
Peak broadening, 142
Peak capacity, 47–49
    required (PC_{req}), 49
    sample (PC^{**}), 49
Peak matching, 94–95
Peak shape
    and injection conditions, 190–193
    broadening, 188–190
    fronting, 188
    in preparative separation, 286–287,
        303–304
    measuring, 188, 418–419
    problems, 188–194, 225
    sample decomposition, 193–194
    sample overload, 188
    split peaks, 190
    tailing, 188, 225
    temperature effects, 170
Peak splitting, 190, 226, 386
Peak tailing, 88, 418–419
    isocratic vs gradient, 6–7
Peak tailing factor, 418–419
Peak tracking, 119–120
Peak width
    and plate number, 189
    broad peaks, 225
    data sampling rate, 190
    extra-column effects, 189
    gradient, 38–39, 378–383, 399–400
    isocratic, 24–25
    preparative separations, 287–288,
        441–443
    synthetic polymers, 241
Peak width change
    and plate number, 189
    and resolution, 189
Peaks
    designated, 117–118, 259–260
    extra, 225
PEEK fittings, 200

Peptide sample, 220–222
Peptides
    gradient separation of, 248–260
    initial experiment, 249–253
    "irregularity" of, 249–250
    method development, 253–256
    preparative separation, 318–320
    reversed-phase columns for,
        252–253, 271
    rhGH digest, 249, 253–256
    separation problems, 215–217, 271
    values of S, 249–250
Perchlorate, for peptide and protein
    separations, 252
Performance Qualification (PQ),157–158
Performance. *See* System performance
Performance test failures, 205–213
pH
    adjustment problems, 218–219
    mobile phase, 125–127
    probe contamination, 219
    Phenylurea sample, 62
Plate number N, 404–411
    for macromolecules, 235–236
    in gradient elution, 38
    isocratic, 25
    maximum achievable values,
        407, 410
    optimization of, 102–106,
        404–411
    preparative separation, 289, 320,
        437–439
    vs change in peak width, 189
Polymers, synthetic, 241, 275–278
    chemical composition
        distribution, 277–278
    molecular weight distribution, 277
Polystyrenes, 68, 198, 231, 239,
    275–277
PQ. *See* Performance Qualification
Precipitation chromatography,
    243–244
Precision vs peak size, 213
Prediction of isocratic separation,
    85–87, 115–117
Pre-elution, theory, 376–378
Premixing mobile phase, 197–199,
    216, 220, 225
    for improved precision, 216–217
Preopt-W, 109

458 INDEX

Preparative separation, 283–321,
    436–444
  column overload, 283
  column saturation capacity,
    289–292, 439
  convergent case, 315
  "corresponding" separations, 285,
    304–306, 316–317
  crossing isotherms, 313
  "displacement" effect, 301–302, 304,
    318–320
  divergent case, 315
  effect of α, 289
  effect of k, 289
  equipment, 285–286
  gradient, 302–321, 441–444
  gradient method
      development, 306–315
  initial isocratic conditions,
    292–295
  isocratic, 286–302, 436–441
  isocratic method
      development, 292–302
  isocratic vs gradient, 285, 304–306,
    316–317
  method development, 292–302,
    306–315, 317–318
  non-overloaded separation, 283
  parallel case, 314
  peak shape, 286–287, 303–304
  peak width, 287–288, 441–443
  peptides, 318–320
  plate number N, 289, 320
  problems, 300–301, 312–315
  production scale, 320–321
  proteins, 318–320
  resolution goal, 296–297
  sample displacement, 318–320
  sample solubility, 300–301
  sample volume, 292, 312
  sample weight, 286–287, 297–300
  scale-up, 298–300
  severe overload, 284–285, 301–302,
    315–321
  step gradients, 311–312
  "tag-along" effect, 301–302,
    304, 317
  touching-peak separation, 283,
    287–289, 306–315, 444
  unequal S values, 314–315

Pressure, 76
  as a function of conditions, 406
  bleed-down test failure, 212, 224
  decay test, 150
  effect of column conditions, 103, 105
  effect on plate number N, 47,
    404–411
  problems, 223
  restrictor, 211
Pressure drop. *See* Pressure
Priming injections, 159–160
"Primitive grid search", 254
Problem isolation
  flowchart, 195
  strategy, 195–197
Problem solving. *See* Troubleshooting
Problems
  guidelines to avoid problems,
    154–157
  use of diagnostic chromatograms, 156
  macromolecule separations,
    271–274
  preparative separation, 300–301,
    312–315
Production scale separation, 320–321
Proline residues, effect on
    conformation, 238
Proportioning problems, 224
Proportioning valve
  calibration, 214
  failure, 207, 214
  problems, 224
  timing, 180
Proteins. *See also* Biomolecules
  cereal storage, 256–260
  denaturation of, 248–249, 252
  gradient separation of, 248–260
  hydrophobic, 272–273
  initial experiment, 249–253
  "irregularity" of, 249–250
  lysozyme variants, 263
  method development for, 256–260
  native, 236
  preparative separation, 318–320
  reversed-phase columns for, 252–253,
    271
  ribosomal, 249, 256–258
  sample loss during separation, 272
  separation problems of, 271–274
  values of S, 249–250

Pseudo peaks. *See* Ghost peaks
"Pseudo-critical" conditions, 278
Pump. *See also* High-pressure or
    low-pressure mixing
  flushing, 139
  maintenance, 216
  piston cleaning, inspection and
    replacement, 200
  purging, 139
  quaternary, 139
  removing air, 197
  seal problems, 209, 211–212,
    223–224,
  seal replacement, 199–200, 216
  seal wear, 190
  ternary, 139
Put-it-back troubleshooting rule, 205

Quadrupole MS, 328–330
Quaternary structure of
    macromolecules, 236
Quaternary-solvent gradients, 365–368

Reagent
  blank, 185
  contamination, 217–220
  quality, 159
Reduced parameters, 405
Reference conditions, 162
Refractive index effect, 185
"Regular" sample. *See also* Sample,
    "regular"
  computer simulation of, 114–115
  homo-oligomers, 238–242
Regulatory recommendations, isocratic
    separation, 27
Relay gradient elution, 360–361
Repeatable separation. *See* Reproducible
    separation
Reproducibility, 205
Reproducibility test, retention,
    150–151
Reproducible separation, 79–80,
    109–110, 120–124
  duplicate runs, 121
  during method development,
    121–122
  during routine analysis, 122–123
  method robustness, 121
Reservoir, cleanliness, 183–184

Resolution
  baseline, 25
  critical, 25–26, 109
  effect of temperature, 28, 109–111
  equations, 25, 26, 91
  gradient, 39–47
  isocratic, 25–27
  LC-MS, 346
Resolution maps, 109–111. *See also*
    Computer simulation,
    Resolution maps
  robust separation, 109–110
Restrictor, back-pressure, 136
Retention. *See also* Gradient retention
  gradient, 397–399
  isocratic, 23–24, 27–28. *See also*
    Isocratic retention
  isocratic, effect of percent-B, 28
  isocratic, prediction from gradient
    run, 85–87, 115–117
  reproducibility, 150–151, 162
  reproducibility problems, 206, 212, 223
  reproducibility test, 150–151
  reproducibility test failure, 224
  test conditions, 150–151
  variation, 213–217, 220, 225
Retention factor, gradient. *See* Gradient
    retention factor
Reversed-phase chromatography,
    assumed unless noted otherwise
rhGH peptides, 253–256
Ribosomal proteins, 256–260
Rounding, gradient. *See* Gradient rounding
Routine applications, suggestions,
    158–160
Rule of One (troubleshooting), 205
Rule of Two (troubleshooting), 205
Run time, shortening, 30, 103–108

S (d([log k]/dϕ), 28, 401–404
  as a function of sample molecular
    weight, 230
  effect on gradient separation,
    42–45, 96–100, 414–415
  macromolecules, 229–235, 247–248
  measurement of from gradient
    data, 400
  peptides and proteins, 249–250
  unequal values in preparative
    separation, 314–315

Sample
   anilines plus carboxylic acids
      ("irregular"), 15
   assessment, 76
   "break through", 273–274
   classification of, 19–21
   cleanliness, 162
   complex, 47–49, 119
   decomposition, 193–194, 225
   displacement, 318–320
   effect of ionization on saturation
      capacity, 290–292
   effect on separation, 78–79
   herbicides ("regular"), 1, 15
   "irregular", 19–21
   "irregular", effect of gradient
      conditions, 92–100
   molecular weight effects,
      229–235
   "regular", 19–21
   solubility, 300–301
Sample preparation, 6, 80, 336–339
   column switching, 337–338, 348
   dilution, 336
   liquid-liquid extraction, 337
   on-line cleanup, 337
   protein precipitation, 336
   recovery, 337
   solid phase extraction, 336–337
   virus, 447–448
Sample pretreatment blank, 185
Sample volume overload, preparative
   separations, 292, 312
Sample weight, effect on
   separation, 286–287
Sampling rate, data, 190
Saturation capacity. See Column,
   saturation capacity
Scale-up, 298–300
Schlieren effect, 185
Second-order effects in gradient
   elution, 386–396
Segmented gradients. See Gradient shape,
   segmented
Selectivity, 124–127. See also Gradient
   selectivity
   effect of different conditions
      compared, 126
   isocratic, 28
   gradients, 365–367

Separation
   isocratic. See Isocratic separation
   orthogonal, 127–130
   two-dimensional, 128–130
Separation artifacts, 175–194
Separation conditions. See Conditions
Separation factor α, 26–27. See also
   Selectivity, Gradient selectivity
Separation goals, 75–78
Severe overload, 284–285, 301–302,
   315–321
Shallow gradients
   premixing, 197–199
   problems, 197–199
Shape. See Gradient shape, Peak shape
Signal-to-noise vs CV, 213
Silanol interactions, 225
Siphon test, 197
Size-exclusion chromatography, 244
Solvent. See also Mobile phase
   compressibility, 137–139
   contamination, 225
   demixing, 391–393
   demixing for normal-phase
      chromatography, 392
   premixing, 216
   purity, 182–185
   siphon test, 197
   uses of several, 139
   viscosity, 137–139. See also Mobile
      phase, viscosity
   volume changes, 137–139
Solvent composition change, 225
Solvent front. See t_0 peaks
Solvent inlet-frit blockage, 207, 210
Solvent proportioning, 135–136. See also
   Mixing
   errors, 136
Sonication, check-valve, 199
Split peaks, 190, 226
Spurious peaks. See Ghost peaks
Standards, 160, 334–335, 346
Stationary phase diffusion, 383
Step gradients, for preparative
   separation, 311–312
Step-test, 247–248
   failure, 206–210, 214, 221–222, 224
Stepwise elution, 1–2, 8
Substitution, module
   (troubleshooting), 205

Switching valves, 337–338
Synthetic polymers. *See* Polymers, Synthetic
System. *See also* Equipment
System cleanliness, 159
System peaks. *See* Ghost peaks
System performance
   acceptance criteria, 145, 148, 149
   acetone test, 146
   dwell volume. *See* Dwell volume
   flow rate check, 149
   gradient linearity, 146–147
   gradient rounding, 147
   gradient tests, 146–149
   measuring, 145–151
   peak area reproducibility, 151, 212–213
   pressure bleed-down, 150, 212
   retention reproducibility, 150–151, 212–213
   step-test, 147–148
   step-test failure, 206–208
   test failures, 205–213, 224
   typical parameters, 145
System suitability, 155, 160, 196, 223
   failure, 220
System tests. *See* System performance

t_0 peaks, 185–186
"Tag-along" effect, 301–302, 304, 317
Tailing peaks. *See* Peak tailing or peak shape
Tandem MS, 328–330
Temperature. *See also* Column temperature
   bias, 169–170
   effect on peptide separation, 253–256
   effect on protein recovery, 272
   effect on protein separation, 256–260
   effect on selectivity, 92–94
   equilibration, 169–170
   frictional heating, 170
   inadequate control, 155–156, 225
   isocratic, 28–29
   peak shape, 170
   problems, 224, 226
   programming, 3–4
Ternary-solvent gradients, 365–368
Tertiary structure of macromolecules, 236

Test failures, 205–213, 224. *See also* System performance
Tetrahydrofuran, 125
TFA. *See* Trifluoroacetic acid
Time constant, detector, 190, 225
Touching-peak separation, 283, 287–289, 306–315, 444
Transfer, method. *See* Method transfer
Trifluoroacetic acid (TFA)
   degradation, 201
   for peptide and protein separations, 252
   suggestions for use, 201
Troubleshooting. *See also* Specific problems and Chapter 5
   case studies, 213–222
   emergency instructions, 194–195
   rules of thumb, 204–205
   strategy, 195–197
   suggestions, 197–205
Tubing, 142
   minimizing length, 189
Two-dimensional separation, 128–130
   of peptides and proteins, 274
"Two-run" procedures, 119

Unretained peaks. *See* t_0 peaks
UPLC, 141, 144
USP tailing factor, 188, 418–419
UV absorbance of mobile phase, 82–83

Vacancy peaks. *See* Ghost peaks
Virus. *See also* Biomolecules
   chromatography, 448–449
   sample preparation, 447–448
   separations, 267–271
   structure, 445–447
Viscosity changes, 223
Viscous fingering, 383

Water
   cleanup, 201–203
   purity, 201–203
   scrubber column, 201–203

Xanax separation, 345

DATE DUE

GAYLORD   PRINTED IN U.S.A.

SCI QD 79 .C454 S579 2007

Snyder, Lloyd R.

High-performance gradient elution